Marine Chemistry

MARINE CHEMISTRY

The Structure of Water and the Chemistry of the Hydrosphere

R. A. HORNE

Woods Hole Oceanographic Institution

WILEY–INTERSCIENCE

A Division of John Wiley & Sons
New York · London · Sydney · Toronto

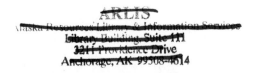

Library of Congress Catalog Card Number: 69-16120
ISBN 0471 40942 1
Printed in the United States of America

TO MY MOTHER AND FATHER

οἷς θέμις ὀφείλοντα ἀποτίνειν
τὰ πρῶτά τε καὶ μέγιστα ὀφειλήματα
χρεῶν πάντων πρεσβύτατα

PLATO, *Laws*, 717B

Preface

The oceans consist of water. Therefore marine chemistry forms the heart of oceanography. An understanding of the detailed nature of water and aqueous electrolytic solutions and of the physical chemistry of the processes occurring therein forms the heart of marine chemistry. Yet this is an area that has been conspicuously neglected in chemical oceanography. Part I of this book is intended to remedy this neglect. It deals with the physical chemistry of aqueous electrolytic solutions using seawater and related systems as examples whenever possible.

Part II presents materials traditionally described as chemical oceanography but hopefully in a more coherent and orderly manner than hitherto.

I feel that the chemistry of the marine interfaces—the relations and interaction of the ocean with its meteorological and geological environment—is becoming increasingly important. There is a large and fast-growing literature dealing with this subject, but it is largely scattered. Part III attempts to give a systematic presentation of this information, with particular attention to the basic principles involved.

Part IV is a selection of topics which could not be injected elsewhere without disrupting the development of ideas but which I feel are too important to omit.

I hope that the book will serve four functions:

1. A readable and useful presentation of its subject matter for the marine scientist.

2. A convenient reference handbook in the form of extensive tables of seawater properties in the Appendix.

3. An advance text for graduate students.

4. In its role as a text, with its emphasis on physical-chemical fundamentals, I hope that this book will help chemical oceanography to become more scientific, and less descriptive, to cease being, as has been all too often the case in the past, merely analytical chemistry, and to become instead preoccupied with what I feel is the genuine subject matter of chemical oceanography, namely, the elucidation of the physical-chemical processes, past and present, responsible for the chemical composition and changes in composition of the sea and its environs.

I wish to express my gratitude to Arthur D. Little, Inc., for providing the funds necessary for the preparation of this book and to the officers of the company responsible for the decision to allocate them. I am also deeply indebted to the Department of the Navy for its continued support over many years of the research program conducted in my laboratory, first by the then Bureau of Ships and in more recent years by the Office of Naval Research. This book would not be possible without the support and encouragement of the Navy personnel in these agencies, for it was through these studies that I first became concerned with the application of the principles of the physical chemistry of aqueous electrolytic solutions to marine chemistry.

I also wish to acknowledge the assistance of my colleagues whom I have plagued with questions, of Mr. Victor E. Cheah for helping me to prepare certain sections of the book, of the staffs of our libraries and art department, and of the typists who have struggled with my impossible script.

R. A. HORNE

Cambridge, Massachusetts

Contents

transitions. Cluster breakup by temperature. Effect of temperature on viscosity.

Seawater as an electrolytic solution. Synthesis of simplified seawater. Dissolution of NaCl. Heat of solution. Alteration of the colligative properties. The freezing point depression. Electrolytic dissociation. The hydration of ions. Hydration and charge density. Electrostriction. Hydration numbers from compressibilities. The viscosity and water structure makers and breakers. Model of the hydration atmosphere of ions.

Ideal, real, and dilute solutions. Is seawater ideal? Solvent properties and vapor pressure. Solute properties and conductivity. Comparison of NaCl solutions and seawater properties.

Specific volume and *PVT* properties. The Tait equation and its modification. Compressibility of seawater. The anomalous density maximum.

Calculation of thermodynamic quantities. Dynamic and geometric depth. Heat capacity. Adiabatic processes. Potential temperature.

Solute addition. Ionic activities and chemical potential. Activity coefficients. Temperature and pressure dependence. The Debye-Hückel theory. Ionic atmospheres. Ionic strength. Extension to more concentrated solutions. Osmotic coefficients.

Born-Haber cycle. Heats and entropies of solution. Dilution. Mixed electrolytic solutions.

Surface area and absorption. Water content and particle size. Consolidation and lithification. Surface properties and pore structure. Effect of organic coatings. Chemical composition of pelagic sediments.Comparison with land rocks. Rare earth content. $CaCO_3$ distribution and its depth dependence. The Ba/Al ratio in the East Pacific Rise. Atmospheric and artificial radionuclides. Distribution of natural series radioelements in marine sediments U, Th^{232}, Ra^{226}, Pa^{231}. Organic content of sediments. Biochemistry of marine sediments. Anaerobic conditions and their peculiar chemistry.

Coacervate droplets. Enter natural selection. Lipids and the partitioning and isolation of the biomaterial. Acceleration and evolution, biocatalysts. Energy sources, fermentation, photosynthesis, and respiration.

Marine Chemistry

Introduction

No mercy, no power but its own controls it.
Panting and snorting like a mad battle steed
that has lost its rider, the masterless ocean
overruns the globe.

Herman Melville
Moby Dick, LVIII

1 Definition of Marine Chemistry

Until the last half of the nineteenth century, men devoted to the study of
the world in which we find ourselves described their vocation as "natural
philosophy." There is much to recommend this proud old name, now unfortunately discarded, for it implies strongly the unity of human knowledge.
But, for the sake of convenience, for catalogs, faculties, curricula, and budgets,
natural philosophy has been fragmented into many and still proliferating
disciplines or sciences. The practitioner working in his narrow compartment
often forgets that the fabric of human knowledge is one and a whole, that no
science is independent of its sister disciplines, and that each science necessarily
incorporates within it and is in varying degree comprised of contributions from
the other sciences. This is especially true of oceanography which, at least in
its present state, is not so much an identifiable science as an aggregate of
sciences united by a common concern—the oceans of this planet and their
boundaries.

Chemistry has as its subject the properties of substances as such apart
from their form. This definition may sound strangely Aristotelian to modern
ears, but while like most useful things it is imperfect, it is in my experience a
very meaningful and practical one. It means, among other things, that the
chemist is occupied with the properties of atoms, molecules, and extended
structures of atoms, and in particular with their interactions with one another.
Marine chemistry or, if you prefer, chemical oceanography, thus becomes
simply the study of the properties and interactions of the substances present
in the marine environment.

1

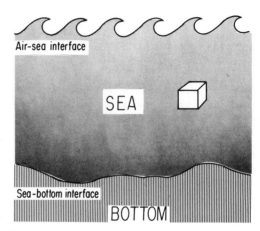

Figure I.1 The Sea as a System and Its Boundaries.

Marine chemistry is a very large and complex topic by itself, and we can be thankful that we shall have to touch upon such other formidable oceanographic topics as the geology of the ocean basins, water movement in waves, tides, and currents, and biopopulations only indirectly and insofar as they involve chemical processes.

In principle, marine chemistry could be entirely devoted to the description of the properties of a given representative element of seawater (Figure I.1). However, the oceans are not merely a chemical system. They are a phenomenon contained in a box. Accordingly we shall devote considerable attention, not only to the physical-chemical properties of seawater, but also to the interaction of the sea with its boundaries, that is, to the physical chemistry of the air-sea and sea-bottom interfaces.

2 The Ocean as a Chemical Environment

Today we are living amidst the excitement of man's first efforts to escape his native planet and explore its neighbors in space. This great adventure will expose him to the most extreme environmental conditions (Table I.1): the near-vacuum of space; the desiccation and temperature extremes of the lunar surface; cloud-shrouded Venus; the debilitated atmosphere of Mars; perhaps some day even the vast envelope of exceedingly cold ammonia, methane, hydrogen, and helium that surrounds Jupiter. Yet, strange as these might be,

Table I.1   Solar System Extreme Environments

Environment	Medium	Temperature, °C	Pressure, atm	Gravitation Field
Earth, sea	aq 0.5 M NaCl	0 to +25	1 to 1000	1
Earth, air	OO_2, N_2, CO_2	−607 to +57	0 to 1	1
Earth, center	Fe	+4000	3,700,000	0
Space	Near-vacuum	—	0	Variable
Moon	Near-vacuum	−153 to +134	0	0.165
Sun	H_2, He	+4700 to +5700	~1	—
Mercury	Near-vacuum	−253 to +340	0.003	0.37
Venus	CO_2, H_2O, N_2O_4	+427	0.1	0.89
Mars	CO_2, H_2O, N_2	−100 to +10	0.1	0.38
Jupiter	H_2, He, NH_3, CH_4	−138	—	27
Saturn	CH_4, NH_3	−153	—	11
Uranus	CH_4	−170	—	0.96
Neptune	CH_4	−170	—	15

I remain convinced that the most extraordinary environment in all of creation is the one down the road that chews at the wharves and throws its shoulder against the headlands—the oceans of Earth.

The oceans are some 1,370,000,000 cubic kilometers, or about 1,413,000,000,000,000,000,000 kilograms, of a moderately concentrated aqueous electrolytic solution. They cover about 361,000,000 square kilometers or 71% of the surface of this planet (see Table A.22 in the Appendix).

The substances contained in this enormous quantity of water may be conveniently categorized into two types (Figure I.2): dissolved substances including salts, organic compounds, and dissolved gases and substances present as a second phase such as gas bubbles and both inorganic and organic solids, the latter ranging in size from colloidal to particulate forms. Alternatively, it is sometimes useful to classify the dissolved substances into electrolytes and nonelectrolytes, that is, substances which result in charge-carrying species or ions when dissolved in water and thus increase the electrical conductivity of the system and those that do not. The amounts of both types of substances in an element of seawater can vary with geographic location, and with depth and time in a given location. Our element of seawater may well contain an important second phase that we have not mentioned—a fish.

The most obvious dissolved substances in seawater are the salts. The chemical composition of seawater is examined in detail in Part II, but suffice it here to say that a "typical" element of seawater 1 kg in weight contains about 19 g of chlorine as chloride ion, 11 g of sodium ion, 1.3 g of magnesium and 0.9 g of sulfur mostly as some form of sulfate ion. In other words,

IN SEAWATER

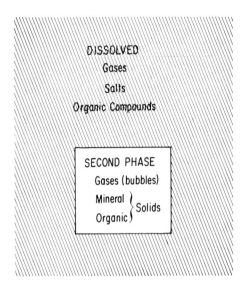

Figure I.2   Types of Substances Contained in Seawater.

roughly speaking, seawater is an aqueous 0.5 M NaCl solution,* 0.05 M in $MgSO_4$, and containing in addition a pinch or trace of just about everything imaginable.

The temperature on the sunlit side of Mercury is $+340°C$, hot enough to melt lead, while the dark side is so cold, $-253°C$, that air would freeze. On even our closest neighbor in space, the Moon, the temperature varies drastically from $+134°C$ at the noon of the lunar day to $-153°C$ during the night. In strong contrast the temperature of Earth's seas rarely varies by more than 25°C, either in a given location or from place to place. Clearly then, the oceans are an excellent thermostat. The consequences of this single, simple fact are enormous, for it is one of the necessary conditions for the genesis and continuation of life. Then, too, the ability of nearby seas to moderate climatic conditions has played an important role in the history of the development of human civilizations.

The thermal structure or temperature profile of the ocean can be very complex and change in space and time. So important is this subject, both to our understanding of the fundamentals of marine science and to the practical problem of the acoustic detection of enemy naval craft, that it forms a large

---

* Signs, symbols, definition, conventions, and conversion units used in this book are listed in the Appendix, Table A-1.

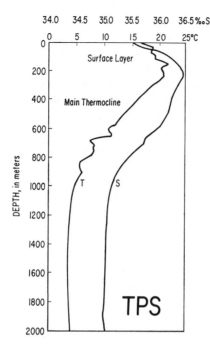

Figure I.3 Typical Temperature and Salinity Profiles.

and separate branch of oceanology by itself, but one with which we are not concerned in this book. Temperature profiles are classified into a dozen or more types which tend to be characteristic of particular ocean regions and seasons. A "typical" thermal profile is shown in Figure I.3.* The first 200 m are marked by a surface layer in which the temperature is quite variable, often exhibiting diurnal fluctuations, and strongly dependent on meteorological conditions. Beneath the surface layer the temperature drops off slowly with increasing depth. The more stable region, called the "main thermocline," may exhibit seasonal fluctuations. Beneath the main thermocline the temperature of the deep water tends to be very stable and usually remains between 0° and +4°C, the latter being the temperature of maximum density of pure water.

Perhaps the most unusual and striking feature of the sea as a chemical environment is the presence of hydrostatic pressures. Although these pressures are small compared to those presently available in the laboratory, they

---

* Figure I.3 also shows a salinity profile. Salinity is defined in detail in Chapter 4. For present purposes it can be taken to represent the total weight of salts in grams in one kilogram of seawater.

are certainly impressive by human standards, and I shall refer to them as "high." The pressures encountered in the sea range from 1 atm at the surface to roughly 1000 atm at the bottom of the greatest ocean deeps. The abscissa in Figure I.4 indicates the percentage of the area of the Earth's oceans, excluding adjacent seas, having a given depth and pressure range.

The oceans are a dynamic system, not too unlike a reactor or scrubber in a chemical processing plant. They are mixed by tidal action, by currents and eddies, upswelling and sinking. Every hour huge volumes of water are added by rivers, precipitation, and the melting of ice, and huge volumes are removed by evaporation and freezing. The oceans are continuously leaching the continents and washing the skies.

Upon first consideration the conditions characterizing the marine environment may seem mild. Every one of these conditions, however, conspires to create a system fantastically complex and, in consequence, fascinating.

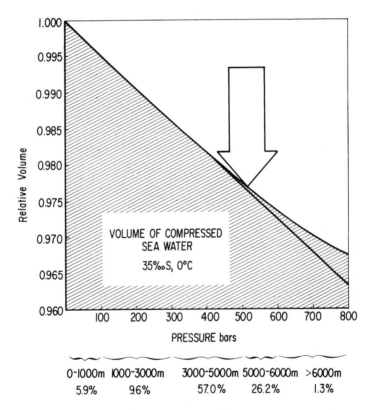

Figure I.4   Relative Volume of Sea Water as a Function of Pressure, and the Percentage of the Oceans Having a Given Depth and Pressure Range.

Water itself, as we shall see, is easily the most extraordinary and complex substance known to man, anomalous in all of its physical-chemical properties. NaCl, at least, is a well-behaved, "normal," 1:1 electrolyte, but its concentration in seawater more than negates this single simplification. We have satisfying theories of aqueous electrolytic solutions only for those cases in which the electrolyte is sufficiently dilute that the ions "see" only water molecules and not one another. As their concentration is increased, interionic interactions become significant and the situation becomes very much more complicated. A concentration of $0.5\ M$ lies in the region where ion-ion interactions are starting to become appreciable. $MgSO_4$, the constituent of seawater present in next greatest quantities after NaCl, like some other 2:2 electrolytes, is complex in its behavior in aqueous solution. This complexity is reflected in such disparate phenomena as the anomalous absorption of sound in seawater and the peculiar physiological effects of this salt.

Finally there are the low temperatures and high pressures which characterize the deep-sea environment. Not only is the structure of pure liquid water sufficiently complex to bewilder the efforts of the most modern experimental and theoretical techniques, but also this structure is altered by the presence of an electrolyte or a surface and by the application of temperature and pressure. In fact, as we shall see, there now appears to be a growing body of evidence indicating that the structure of water undergoes important and ill-understood transformations exactly in the temperature and pressure ranges of oceanographic interest.

In the light of the foregoing facts I can repeat, without fear of being found guilty of exaggeration, that the deep sea represents the most remarkable, the most mysterious, environment in the cosmos.

Part I
The Physical Chemistry
of Water and Aqueous
Electrolytic Solutions

I   The Structure of Liquid Water

1   Introduction

The oceans consist of water. Oddly enough, this central truism of oceanography in the past appears to have largely escaped the notice of oceanographers. Chemical oceanography for the most part has occupied itself with problems which are relatively peripheral and has conspicuously neglected the most fundamental and central problem—that of the structure and physical chemistry of liquid water and aqueous electrolytic solutions. My hope is that this book will correct this oversight and that, by examining in detail just what seawater is on a molecular level and how it behaves mechanistically in microscopic terms, a first step will be taken in the direction of raising chemical oceanography to an exact, analytic science.

2   The Water Molecule

Our knowledge of the configuration of the water molecule is based largely on the gaseous state. There is no reason, however, for believing that the structure of the molecule is greatly altered in the liquid.

The aspect presented by the electronic cloud of the water molecule is that of an abbreviated jack (Figure 1.1) contained in a distorted cube. The oxygen occupies the center and the two hydrogens, opposite corners of one face of this cube. The H—O—H angle is 104° 31′ (Mecke and Baumann, 1932; Darling and Dennison, 1940) rather than the 109.5° which would obtain if the cube were perfect. Two of the oxygen's eight electrons are near its nucleus, two are involved in the bonding of the hydrogens, and the two pairs of unshared electrons form arms directed toward opposite corners of the cube's face opposite the face containing the hydrogens. These arms of the electronic cloud are of particular interest to us, for they represent regions of negative electrification which can attract the positive partial charges of the hydrogen atoms of nearby water molecules and thus bind water molecules together. The OH distance in water vapor is 0.9568 Å and in ice somewhat greater, being 0.99 Å. The spectroscopy of water is discussed by Herzberg (1945), and Mulliken (1932, 1933) and Lennard-Jones and Pople (1950) have applied the molecular orbital method to the elucidation of its electronic structure. Rather than as its electronic cloud (Figure 1.1), water may also be pictured

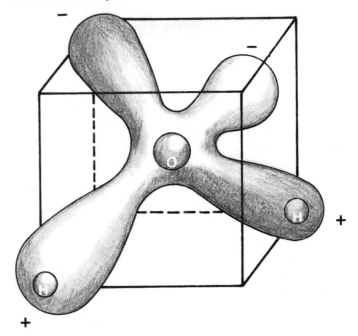

Figure 1.1   Electronic Cloud of the Water Molecule.

as a cluster of electronic orbitals (Figure 1.2). In the free oxygen atom the electronic configuration of the L shell is $(2s)^2\ (2p_z)^2\ (2p_y)\ (2p_x)$ with the charge density of the $(2s)^2$ pair of electrons distributed spherically about the inner shell electrons and with the charge densities of the $2(p_z)^2$ and $2p_y$ and $2p_x$ electrons distributed rotationally symmetrically about the mutually perpendicular x, y, and z axes. The binding of two hydrogen atoms by $2p_y$ and $2p_x$ widens the 90° angle as a result of electrostatic repulsion, and this perturbation gives rise to hybridization. The valence angle, that corresponding to the minimum potential energy of the molecule as the s contribution to the valence electron states, and the angle between the axes of maximum charge density both increase. The hybridization of the p and s states also changes the charge distribution of the two pairs of nonbinding oxygen electrons, causing them to project unsymmetrically from the oxygen nucleus away from the protons.

Although the bond angles and distances in the water molecule are accurately known, the question of the exact distribution of electric charge in the molecule remains somewhat unsettled. Bjerrum (1951), in order to account for ice properties, takes the positive charge of the oxygen nucleus to be wholly, and that of the hydrogen nucleii to be strongly, screened, whereas Verwey (1941),

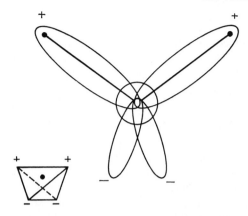

Figure 1.2   Electronic Orbitals of the Water Molecule.

for example, prefers a somewhat different charge distribution. In any event the important feature is that the charge distribution gives rise to a large dipole moment of 1.84×10^{-18} esu. This important parameter, and the bond angle and length, are summarized schematically in Figure 1.3. Just how important are these features? If the water molecules did not have their negative electronic cloud arms and dipole moments, they would not be able to interact with one another, the "oceans" of Earth would be gaseous, and life would be impossible!

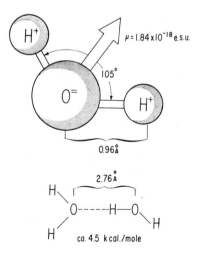

Figure 1.3   Schematic Summary of the Structural Features of the Water Molecule and the Hydrogen Bond.

3   Theories of Water Structure

Earlier I mentioned that liquid water is anomalous in all its physical-chemical properties. Table 1.1, taken from Sverdrup, Johnson, and Fleming's classic *The Oceans* (1942), lists some of these properties and their significance.*

Let us concern ourselves with one type of property—the temperatures at which the phase transition occurs. The melting points of the hydrides of the chemical elements in group VIA of the periodic table (see Table A.2 in the Appendix) increase with increasing molecular weight from H_2S to H_2Se to H_2Te (Figure 1.4). The exception to this regular progression is H_2O, which has the smallest molecular weight in this family of compounds, yet does not melt until a temperature nearly 60°C greater than the melting points of its sister compounds is reached. The same anomaly appears even more strikingly in the boiling point. By all reasonable comparison with similar compounds, water should be a gas at room temperature, yet in fact its boiling point is more than 160°C higher than it ought to be! Clearly something very strange is going on here. The water molecules in the liquid are not separating as easily as they should, something is preventing their escape into the vapor, they are sticking together. The force which binds the water molecules together, the hydrogen bond, is now fairly well understood and will be discussed presently. The geometrical configuration, the way in which the molecules stick together in the liquid state, is however quite a different matter. The structure of liquid water is not known, even though the existence of such a structure was recognized as early as 1884 (see Chadwell, 1927, for a review of early theories of water structure) and has been the subject of intensive scientific effort for more than thirty years.

Water vapor has no structure. The gas is mostly monomeric with an occasional dimer and, possibly, very rare trimers. Charged clusters such as $H_9O_4{}^+$ appear to be formed by electrical discharge in water vapor at low pressures (Anderson, Knight, and Winter, 1966). The solid state, ice, can be highly ordered and its structure has been adequately characterized (and will be described subsequently). The difficulty then centers around the liquid state, and here we are confronted with not one, not two, not a few, but many theories of water structure, theories which continue to proliferate at an alarming rate.† The confusion is compounded, not because these theories fail but because many of them work surprisingly well. The situation has been nicely summarized by Franks and Ives (1966) who write, "Increased efforts

* Further tables of water properties can be found in the Appendix.
† For recent reviews of theories of the structure of water see: R. A. Horne, *Surv. Progr. Chem.*, **4**, 1, 1968; J. L. Kavanau, *Water and Solute-Water Interactions*, Holden-Day, Inc., San Francisco, 1964; O. Y. Samoilov, *Structure of Aqueous Electrolyte Solutions and the Hydration of Ions*, Consultants Bureau, New York, 1965; E. Wicke, *Angew. Chem.*, **5**, 106, 122 (1966).

Table 1.1 Anomalous Physical Properties of Liquid Water
(From Sverdrup, Johnson, and Fleming, 1942, with permission of
Prentice-Hall, Inc.)

Property	Comparison with Other Substances	Importance in Physical-Biological Environment
Heat capacity	Highest of all solids and liquids except liquid NH_3	Prevents extreme ranges in temperature; heat transfer by water movements is very large; tends to maintain uniform body temperatures
Latent heat of fusion	Highest except NH_3	Thermostatic effect at freezing point due to absorption or release of latent heat
Latent heat of evaporation	Highest of all substances	Large latent heat of evaporation extremely important in heat and water transfer of atmosphere
Thermal expansion	Temperature of maximum density decreases with increasing salinity; for pure water it is at 4°C	Fresh water and dilute seawater have their maximum density at temperatures above the freezing point; this property plays an important part in controlling temperature distribution and vertical circulation in lakes
Surface tension	Highest of all liquids	Important in physiology of the cell; controls certain surface phenomena and drop formation and behavior
Dissolving power	In general dissolves more substances and in greater quantities than any other liquid	Obvious implications in both physical and biological phenomena
Dielectric constant	Pure water has the highest of all liquids	Of utmost importance in behavior of inorganic dissolved substances because of resulting high dissociation
Electrolytic dissociation	Very small	A neutral substance, yet contains both H^+ and OH^- ions
Transparency	Relatively great	Absorption of radiant energy is large in infrared and ultraviolet; in visible portion of energy spectrum there is relatively little selective absorption, hence is "colorless"; characteristic absorption important in physical and biological phenomena
Conduction of heat	Highest of all liquids	Although important on small scale, as in living cells, the molecular processes are far outweighed by eddy conduction

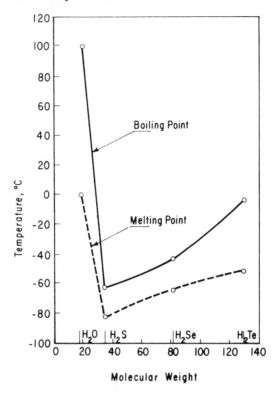

Figure 1.4 Transition Temperatures of the Group VIA Hydrides. From Horne (1968) with permission of Academic Press, Inc.

of recent years have provided an embarrassment of partially successful alternative models; these are . . . very likely, despite unresolved difficulties, to incorporate between them many ultimately acceptable features." The failure of modern techniques, techniques powerful enough to resolve the structure of such exceedingly complicated biomolecules as DNA and myoglobin, to disclose the structure of water gives us due warning that we are dealing with a system of horrendous complexity. At the present time the field of water structure is a confused and very controversial one. The reader therefore should be cautioned that much of what I shall say about the subject must be of a speculative and tentative nature.

Models for the structure of liquid water fall into two types, continuum theories and mixture theories, but both types take as their point of departure the structure of the solid. We now, accordingly, examine the structure of Ice-I_h —the common form of ice at 1 atm. The structure of Ice-I_h, as revealed by

X-ray and confirmed by neutron scattering studies (Dennison, 1921; Barnes, 1929; Wollan, Davidson, and Shull, 1949; Peterson and Levy, 1953, 1957; Chidambaram, 1961), is a hexagonal system (Figure 1.5) in which the O atoms are arranged in the same way as the Si atoms in tridymite. The interchangeability of these two lattices plays an important role in the hydration of certain silicate minerals (Macey, 1942). Each O atom in the structure is hydrogen-bonded to four other oxygen atoms arranged tetrahedrally at a distance of 2.76 Å from the central O (Figure 1.6). The hydrogen atoms lie on the line (or hydrogen bond) connecting O's but closer to one O atom than the other. The O atoms can be pictured as forming a layer network of puckered hexagonal rings, each layer being the mirror image of the adjacent layer. This is perhaps better shown in the more diagrammatic Figure 1.7 than in Figure 1.5. Fox and Martin (1940) have calculated the interoxygen distance, r, in ice on the basis of its density at 0°C, 0.9168 g/cm^3, using the relation that the tridymite-like structure gives $3\sqrt{3}/8r^3$ water molecules in 1 cm^3 of ice, and they have obtained a value of about 2.77 Å which is in good agreement with the distance 2.76 Å from X-ray measurements quoted above.

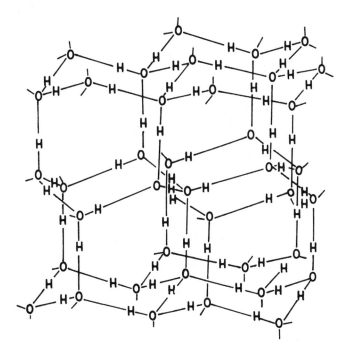

Figure 1.5   The Arrangement of Oxygen Centers in Ice-I$_h$. From Barnes (1929), with permission of The Royal Society.

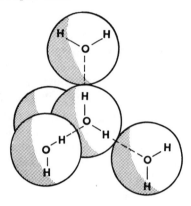

Figure 1.6   Tetrahedrally Arranged Neighboring Waters in the Ice-I_h Lattice.

The majority opinion appears to be that the hydrogens in the ice lattice are capable of a variety of motions* and these movements form the basis of current theories of the dielectric and conduction properties of ice. However, nuclear magnetic resonance measurements in single-crystal heavy water (D_2O) ice revealed only a small amplitude vibration of the hydrogens and no hindred rotation or hydrogen transfer (Waldstein, Rabideau, and Jackson, 1964). My own prejudice is that our understanding of transport processes in ice is far more imperfect than some authors would lead us to believe.

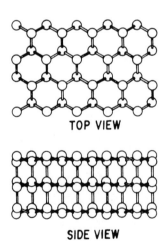

TOP VIEW

SIDE VIEW

Figure 1.7   Schematic Representation of the Ice-I_h Structure. From Davis and Litovitz (1965), with permission of American Institute of Physics.

* For a good semipopular discussion see L. K. Runnels, *Sci. Am.*, **215**, 118 (Dec., 1966).

Table 1.2   Hydrogen-bond Properties

Bond	Substance	Bond Energy, kcal/mole	Bond Length, Å
F—H—F	H_6F_6	6.7	2.26
O—H⋯O	H_2O (ice)	4.5	2.76
	H_2O_2	4.5	
	Alcohols, (ROH)	6.2	2.70
	$(HCOOH)_2$	7.1	2.67
	$(CH_3COOH)_2$	8.2	—
C—H⋯N	$(HCN)_2$	3.2	—
	$(HCN)_3$	4.4	—
N—H⋯N	NH_3	1.3	3.38
N—H⋯F	NH_4F	5	2.63
O—H⋯Cl	o—C_6H_5OHCl	3.9	—

The force that holds the water molecules one to another with surprising tenacity is the hydrogen bond (Figure 1.3). Hydrogen bonding also occurs in substances other than water; some typical values of bond strengths and lengths taken from Pauling (1948) are listed in Table 1.2. Notice that the stronger bonds tend to be shorter. Hydrogen-bond strengths lie intermediate between

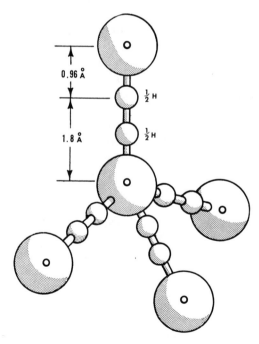

Figure 1.8   Position of the Hydrogen Atoms in the Ice Lattice.

weak van der Waals forces (about 0.6 kcal/mole) and typical ionic-bond energies (10's of kcal/mole). X-ray measurements are incapable of determining the position of the hydrogen atom in the ice structure; nevertheless Barnes (1929) assumed that they are equivalent between the two O's. Subsequently Pauling (1939) proposed that there are two preferred sites of the H between the two O's, about 0.96 Å from one O and 1.81 Å from the other, and that the H jumps back and forth from one O to the other so that each site may be considered as being occupied by $\frac{1}{2}$H (Figure 1.8). Neutron diffraction studies on powdered D_2O ice (Wollan, Davidson, and Shull, 1949) have established the validity of the Pauling model. The strength of the H bond in H_2O arises from coulombic interaction of the polar molecules and is strengthened by the enhancement of the molecules' charge separation by different resonance configurations such as

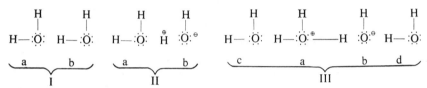

In this manner the formation and stabilization of one H bond facilitates H bond formation with neighboring waters. Thus, in a sense, the bond propagates itself, and the water structure tends to be cooperatively strengthened over a more extended region.

Let us now turn to the structure of liquid water. Inasmuch as we have examined the structure of Ice-I_h in detail, an obvious approach to the problem is to ask ourselves quite simply, "What happens when ice melts?" The answer, upon first examination, appears to be "Not much." Pauling (1948) argues that the small heat of fusion of ice (1.44 kcal/mole) shows that only about 15% of the hydrogen bonds present in ice are broken upon melting. The dielectric constant of materials is a property highly dependent on the spatial disposition of atoms and charge sites, yet water's anomalously high dielectric constant changes relatively little on melting/freezing, being about 74 (and strongly frequency-dependent) in ice and 88 in liquid water at 0°C (Dorsey, 1940). Similarly the X-ray radial distribution curve for liquid water (Figure 1.9) indicates that not only is the tetrahedral arrangement still very much in evidence, but also that the "radius" of a water molecule is only slightly altered upon melting (Morgan and Warren, 1938; Danford and Levy, 1962). Finally let us consider simply the density of liquid water. A cube 1 cm by 1 cm can contain $1/\sqrt{32}\, r^3$ closest-packed spheres of radius r and, the molecular weight being known, the densities of "normal" liquids like argon or methane can be readily calculated. In the case of liquid water, however, taking the distance between centers as 2.9 Å, we calculate a density of 1.7 g/cm³, or

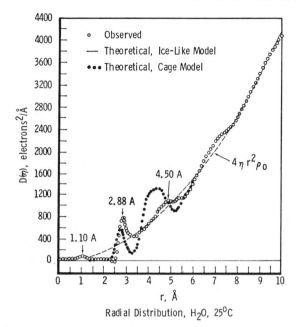

Figure 1.9 X-ray Radial Distribution Curve in Liquid Water. From Danford and Levy (1962), with permission of American Chemical Society.

nearly twice too large (Podolsky, 1960). Clearly, then, the water molecules, far from being closest-packed, are much more widely spaced, as they are in ice, with between 4 and 5 nearest neighbors (Danford and Levy, 1962) rather than the expected 12.

The picture of liquid water emerging from this discussion is one very similar to that for Ice-I_h. But in emphasizing the similarities we must not overlook the obvious differences between a liquid and a solid. The structure of Ice-I_h must be modified in some way in the liquid. Perhaps the increased thermal agitation results in bending and stretching of the bonds, thus rendering the structure more "fluid" (an alternative we discuss subsequently), or perhaps some of the water molecules "melt" away from the structure and enjoy a state of greater mobility. In the latter hypothesis, liquid water is represented as a "mixture" of two or more forms, one type of water being bound into some sort of ice-like lattice and the other type being "free" or not bound in the ice lattice. At the time of this writing (1967), the state of this "free" water is a subject of intense controversy. In the first place, what fraction of the water is "free"? Falk and Ford (1966) have pointed out with a fine sense of outrage that, depending on the experimental method used, values for the percentage of broken hydrogen bonds in liquid water at 0°C (a parameter closely related to the amount of "free" water) ranging from 3% to 72% have been reported in the literature. They take these discrepancies as a repudiation of any type of mixture model (see also Stevenson, 1965). But difficulties need not mean that a theory is wrong; they may simply indicate that life is not as simple as initially supposed, and the crucial question in this particular instance is what do we mean by "free" water? As to the non-ice-like water, opinion ranges the whole spectrum from free, gas-like water molecules on one extreme to a specific type of second structural lattice on the other. Pelah and Imry (1963) speak of the unbound water as a "gas." While this may be a useful conceptual formalism, I feel that it is entirely unreasonable to expect molecules to be as free to execute movements in a condensed phase as in the gaseous state, especially molecules such as water, which are notorious for the variety and strength of their intermolecular interactions. No water molecule in the liquid state is oblivious of its neighbors (Horne, 1968). That there should be no evidence for the existence of vapor-like water molecules in the liquid state (Wall and Hornig, 1965) is not surprising, and when we speak of "free" water the qualifications indicated by the quotation marks should be kept carefully in mind. An intermediate point of view pictures the "free" water as a sort of "normal" fluid in which the regions of structure swim about almost like particles—a model which Franks and Good (1966) have applied in an interesting way and with impressive success to the problem of accounting for the viscous flow properties of liquid water. Further restrictions upon the movements of the "free" water molecules are imposed by the theory of Samoilov (1965)—a model of water structure that has strongly influenced the very considerable Soviet experimental and theoretical effort on the physical chemistry of water and aqueous solutions. Figure 1.10 shows an attempt to represent diagrammatically some of the principal types of theory of water structure, including that of Samoilov. Samoilov (1965) points out that the Ice-I_h lattice is very spacious, having adequate room in the interstitial spaces to accommodate water molecules. When ice melts, the lattice becomes slightly distorted and these interstitial spaces become occupied by water molecules that have moved translationally out of their equilibrium positions in the lattice, thereby increasing the density. A somewhat similar model has been proposed by Forslind (1952) who estimates that at 0°C about 16% of the total number of water molecules in the liquid are nonassociated interstial molecules.

Another important type of mixture theory looks upon the non-ice-like water as in no sense "free" but rather forming a specific, second structural form. Of these theories one of special interest is that of Bernal and Fowler (1933) (see Figure 1.10), for its appearance in the first volume of *The Journal of Chemical Physics* marked the beginning of modern theories of the structure of water. According to these authors, who framed their hypothesis on the basis of the then recent X-ray information on liquid water, the melting process

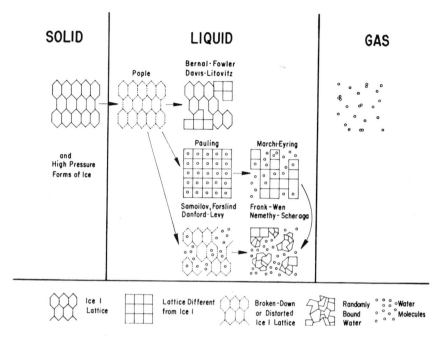

SOLID   **LIQUID**   **GAS**

Figure 1.10 Principal Theories of Water Structure. From R. A. Horne (1967), with permission of Academic Press, Inc.

is marked by the appearance of the second type of lattice—Water II. Water II is a quartz-like structure (Figure 1.11). Like the ice-tridymite Water I, its molecules are four-coordinated, but it has a greater density, 1.08 g/cm³ compared with 0.91 g/cm³ for Water I. But liquid water is not a mixture in the usual sense; the two structural "forms are not distinct, and they pass continuously into each other. "[The liquid is homogeneous, there is no] mixture of volumes with difference structures." More recently Wada (1961) has published a simplified mixture model, again based on an ice-like and a closely packed state of 0.9334 cm³/g specific volume. He estimates that, at 0°C, 0.4 mole fraction of the water is in the ice-like form.

Davis and Litovitz (1965), whose interest in water structure stems from a concern with the absorption of acoustic energy in that fluid—a topic to which we return subsequently in this book—have proposed still another variation on the two-state mixture model (Figure 1.10). Both forms are puckered, hexagonal ring structures related to Ice-I_h. In one state (Figure 1.12) the rings are open-packed with extensive H bonding between them, whereas in the second, more dense state this structure is compressed so that the resulting close-packed structure is nearly a body-centered cubic configuration. Our

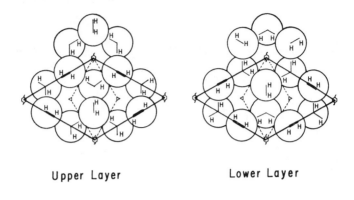

Upper Layer   Lower Layer

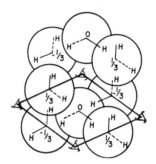

Figure 1.11 The Structure of Ice-Like Water I (above) and Quartz-Like Water II (below). From Bernal and Fowler (1933), with permission of American Institute of Physics.

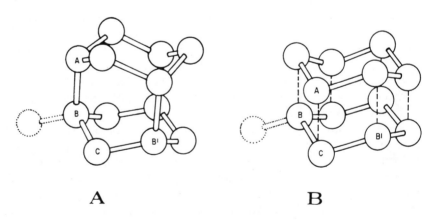

A   B

Figure 1.12 Water-Ring Structures. From Davis and Litovitz (1965), with permission of American Institute of Physics.

understanding of water structure is impaled on a sort of dilemma. There really is no very good, direct, unambivalent experimental evidence for the existence of particular lattice forms other than Ice-I_h in liquid water; yet, if we are not prepared to venture far from the Ice-I_h form, we are at a loss to explain a very obvious fact, namely, the fluidity of the liquid. The various theories of water structure can account for many of the properties of liquid water with alarming ease. There are a few properties, however, where they tend to fail. One of these properties is the fluidity of water. Thus an important argument in favor of the Davis-Litovitz model is its ability to yield a high volume to shear viscosity ratio such as observed in contrast to some of the competing theories (Table 1.3). In a comment upon the Davis-Litovitz model, Wall (1965) has pointed out that on purely geometrical considerations one might expect a closed "chair" ring even in normal liquids, and that the uniqueness of water may lie in the fact that hydrogen-bonding requirements result in a more open ring.

The water structure models that we have been discussing are all mixture theories; that is, they postulate at least two water forms, the one form some sort of lattice, the other either a second type of lattice or more or less "free" or unbound water. Another important and, in principle, quite different category of theory is enjoying more frequent mention—they are the uniformist or continuum theories mentioned earlier. They postulate that, when ice melts, the H bonds become more flexible and, although these bonds may be bent and distorted, they are not broken (Pople, 1951). In support of their viewpoint the proponents of the continuum theories cite evidence such as failure to observe the light-scattering phenomena expected to arise from any structural heterogeneities present in the liquid (Mysels, 1964). Infrared spectra studies indicate that "H-bonds in liquid water have a broad, smooth, single-peaked

Table 1.3   Ratio of Volume to Shear Viscosity in Liquid Water

Tempera-ture, °C	Observed (Pinkerton, 1947)	Calculated from Two-State Relaxation Parameters				
		Davis and Litovitz (1965)[a]	Nemethy and Scheraga (1962)	Grjotheim and Krogh-Moe (1954)	Wada (1961)	Eucken (1948, 1949)
0	2.33	3.97	0.98	0.53	0.44	0.80
10	2.18	3.75	0.86	—	0.39	—
20	2.10	3.48	0.73	0.40	0.32	0.34
30	2.10	3.20	0.59	—	—	—
40	2.01	2.92	0.46	0.23	—	0.11
60	2.01	2.64	—	—	—	—
80	2.04	2.37	0.26	0.14	—	0.03

[a] Based on a nearest-neighbor distance of 2.8 Å.

distribution of strengths which gradually shifts with temperature" (Falk and Ford, 1966). This finding, the investigators argue ". . . fully support(s) the 'continuum' models of the structure of liquid water and present(s) strong evidence against the 'mixture' models" (Falk and Ford, 1966).

If you now feel, as I hope you do, on the basis of the foregoing discussion that our knowledge of water's structure is badly confused, brace yourself. There appear to be even more serious difficulties! We have used the structure of Ice-I_h as our point of departure. We have said that, as ice melts, structurally speaking, relatively little happens. And, although the several models we have discussed have varied widely in detail for the greater part, they have all implied that there is a lot of Ice-I_h, or something very similar to it, left in liquid water. Although we have been able to represent melting as a simple matter, the reverse is not true. Water does not like to freeze. It supercools, solidifying, not just a few, but many, degrees below the freezing point. A familiar laboratory practice when working with chilled aqueous solutions is to scratch the side of the container with a glass rod in the hope of thereby creating small particles of suitable shape and size that can serve as nucleation centers. But, if water is permeated with clots or a contiguous network of ice-like substance, why doesn't this substance serve as nucleating centers or surfaces for the freezing process? We can fabricate rather lame *post facto* arguments—the aggregates are smaller than the minimum size required to initiate nucleation, their existence is too transitory, they are covered with a film of non ice-like structure—none of them really convincing. Then too, as we shall see, electrolytes are capable of enhancing the water structure in their immediate neighborhood. If they are indeed surrounded by a sort of "ice-berg," however, why are electrolytes excluded from the solid and concentrated in the still liquid phase, rather than being incorporated into the ice lattice when solutions are frozen at reasonable rates? Until we understand a great deal more about the freezing process than we do at the moment, the concept of Ice-I_h of ice-like structures existing in appreciable amounts in the liquid phase leaves me with grave misgivings, however facile an explanation it may provide for other phenomena.

Fortunately there are alternative theories of water structure which are not deeply committed to the postulation of specifically Ice-I_h-like forms in the liquid, and we now consider these models. Perhaps the most fascinating theory of the structure of liquid water to be proposed in recent years is that of Pauling (1961). Hundreds of substances produce anesthesia. They have quite disparate chemistry. Some, such as N_2, Ar, and Xe, are virtually inert; thus the possibility of anesthesia arising from the specific chemical reactions in the organism is rendered unlikely. But even the inert substances do form hydrates of the sort $Xe \cdot 5.75 \ H_2O$, $CHCl_3 \cdot 17 \ H_2O$, and $CHCl_3 \cdot 2 \ Xe \cdot 17 \ H_2O$, and Pauling (1961) and Miller (1961) attribute the anesthetizing efficacy of these substances to the formation of microcrystals of these hydrates in the brain fluids. These hydrates are definite, isolatable, and identifiable compounds.* In these clathrate or "cage-like" structures the guest atom or molecule is contained within sort of a Buckminster Fuller geodesic dome formed by a framework of H-bonded water atoms. Such structures, in addition to a molecular theory of general anesthesia, can also be evoked to provide a structural model for liquid water (Pauling, 1960). Liquid water is theorized to be a self-clathrate, a network of joined polyhedral cages formed of H-bonded waters and containing within their cavities entrapped, but unbound, water molecules (Figure 1.13).

Pauling's self-clathrate theory of water has been quantitatively developed by Frank and Quist (1961). The pentagonal dodecahedra of 20 water molecules enclose a vacancy with an unobstructed diameter of 5 Å. In addition there may be other polyhedra present, enclosing larger spaces of 6 Å or more. These interstitial spaces are sufficiently large to enable the guest water molecules to rotate without appreciable restriction; that is, the nonlattice water molecules are "free" (cf. above). Calculations based on this model, together with a few simple assumptions, "can give a satisfactory representation of the *P-V-T* properties of water over limited ranges of temperature and pressure" (Frank and Quist, 1961). In particular they can account for two of water's most characteristic features, the density maximum and the structural compressibility. As in the case of the other models mentioned earlier which postulate some kind of contiguous lattice or network extending throughout the substance, the fluidity of liquid water is problematical. Frank and Quist

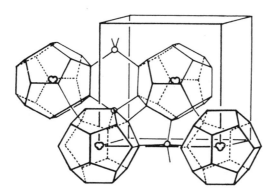

Figure 1.13 Self-Clathrate Model of Liquid Water. From Pauling (1961), with permission of AAAS, © 1961, and the author.

* The hydrate nature of "crystalline chlorine" was first demonstrated by no less a chemist than Sir Humphry Davy (1811) and analyzed some time later by the great Michael Faraday (1823).

(1961) avoid this difficulty by reiterating a concept introduced earlier by Professor Frank (Frank and Wen, 1957): the structural framework "flickers." It is constantly in a state of flux, old bonds vanishing, new bonds appearing. We shall return to this concept in a moment. Although solving one problem, the introduction of this idea creates a new one—the need to postulate a third "state" or collection of states to describe the condition of those waters that for an instant are survivors of the destruction of one lattice configuration but not yet incorporated into another. A further discussion of this "State III," which its creators regard as a "blemish" on the model, can be found in the review paper of Wicke (1966).

Marchi and Eyring (1967) have applied significant structure theory to water and published a water model which they describe as "qualitatively the same as those proposed earlier by Samoilov, Pauling, and Danford and Levy." When the solid melts, the Ice-I_h is replaced by a 20% more dense yet still tetrahedrally H-bonded form, perhaps similar to the Ice III, a high-pressure form of the solid. The vacancies in this network also increase and are filled with a second water species—rotating monomers, just as in the Pauling-Frank-Quist model.

Danford and Levy (1962) have undertaken a fresh X-ray diffraction study of liquid water. The radial distribution they obtained (Figure 1.9) is in agreement with earlier work of Morgan and Warren (1938) and Brady and Romanow (1960) but inconsistent with the Pauling pentagonal dodecahedral structure. They prefer a model similar to Samoilov's involving a distorted ice-like framework with interstitial water molecules. The failure to observe a radial distribution corresponding to the clathrate model is a bit discouraging but not entirely unambivalent, and we need not abandon this model in its light. Although present, under ordinary circumstances, in amounts too small to affect the X-ray radial distribution, the clathrate structure may still be responsible for certain anomalies that appear in water properties in the neighborhood of 30–40°C (Franks and Good, 1966; Frank and Quist, 1961; see also the recent review paper of Drost-Hansen, 1967). It also may be the predominant structural form in certain regions in the liquid such as those adjacent to interfaces and hydrophobic ions and segments of macromolecules (Drost-Hansen, 1965; Horne, Day, Young, and Yu, 1966). We shall return to this possibility in Chapters 3, 11, and 12.

Everyone today seems to have a favorite water model (and certainly there are enough alternatives to choose from). I am no exception. The model which I prefer and to which we largely adhere throughout the remainder of this book is that of Frank and Wen (1957). They picture liquid water as a mixture of "flickering clusters" of H-bonded water molecules swimming in more or less "free" water (Figure 1.14). The theory has two important features. In the first place, it does not postulate remnants of an Ice-I_h lattice; it makes no

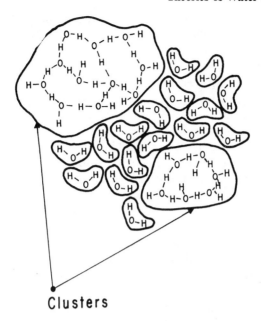

Clusters

Figure 1.14   Frank-Wen Flickering Cluster Model of Liquid Water. From Nemethy and Scheraga (1962), with permission of American Institute of Physics.

commitment as to the details of the organization of the water molecules in the clusters, thus avoiding entirely the questions raised by the supercooling, freezing, and fluidity phenomena. Frank and Wen (1957) in describing the clusters are careful to put the expression "ice-like-ness" in quotation marks, and Nemethy and Scheraga (1962), who are responsible for the quantitative development of the model, are even more insistent that ". . . the model is not dependent on the postulate of a particular semi-crystalline structure . . . . The tridymite-like arrangement of ordinary ice . . . is merely one of several possible, nearly equivalent structures. . . . It is to be noted that rather irregular arrangements of the molecules in the clusters are not excluded by the model. . . ." Nemethy and Scheraga (1962) are equally cautious in their description of the "free" water, which, as we have seen, is another bone of contention. Although they write of "monomeric 'unbounded water'" they are careful to qualify this by pointing out that, while the H bonds are broken in the noncluster water, "each molecule is participating in strong dipole-dipole and London interactions with neighboring molecules."

A second very important emphasis of the Frank-Wen-Nemethy-Scheraga model concerns the transitory or "flickering" nature of the clusters. We saw earlier how Frank and Quist (1961) had to introduce this concept into their analysis of the Pauling clathrate model in order to make allowance for the fluidity of liquid water. We also noted in our discussion of the H bond that the bonding tends to propagate itself throughout its immediate neighborhood. Thus cluster formation is a cooperative process and, as a consequence, small aggregates, the dimers, trimers, and other small polymers prominent in early theories of water structure (Chadwell, 1927) and recently reconsidered by Wicke (1966), are disfavored. Heraclitus would have appreciated clusters. They are constantly in a state of flux, continually forming and dissolving with the random thermal fluctuations in the fluid's microregions. Frank and Wen (1957) estimate that the half-life of a cluster is between 10^{-10} and 10^{-11} sec, a value which corresponds to the time required for relaxation processes in water and which, although very brief, is nevertheless 100–1000 times longer than the period of molecular vibrations and hence ". . . long enough . . . to constitute a meaningful existence of the cluster."

What shall we make of this largesse of models? Viewing them qualitatively and from an appropriate distance, I do not think they represent quite as much chaos as they may at first appear to do. When held up to close scrutiny many of the current controversies tend to take on the complexion of semantic rather than real difficulties. None of the theories is entirely satisfactory, yet none is devoid of truth. Each in its turn can serve to illustrate some half-truth of what is patently a physical-chemical system of horrendous complexity. Trying to cope with these difficulties can be frustrating, but the same difficulties can also make the physical chemistry of the deep ocean the most fascinating of studies. Aristotle somewhere observes that virtue is concerned with the difficult. Most genuine scientists would rather make a little progress on a big problem than solve a little one.

4   The Effects of Pressure and Temperature on the Structure of Pure Water

The effect of the application of hydrostatic pressure on the structure of pure liquid water appears to be more straightforward than that of increasing temperature, and so we treat it first. The Frank-Wen clusters are less dense than the "free" water in which they are immersed. The principle of Le-Chatelier is generally applicable; it states that a system at equilibrium, if disturbed, shifts in such a way as to minimize the effect of the disturbance. Applying this principle to water, one expects the application of pressure to favor the state of smaller specific volume and the more voluminous cluster regions to "dissolve" or "melt." And, in fact, as pressure is applied to seawater its volume decreases (Figure I.4 of the Introduction). In view of the presence of relatively rarefied regions we might expect water to be more compressible than other liquids. But we find it is not. Quite the contrary, it is less compressible than a great many common liquids. At 0°C and pressures less than 25 atm, its coefficient of compressibility is 53×10^{-6} compared to 79×10^{-6} and 96×10^{-6} unit vol/atm for methanol and ethanol, respectively. The comparison is rendered invalid by the fact that the branchy molecules of these alcohols exert intermolecular repulsive forces which distend the liquid and increase their compressibilities. These forces are also responsible for the decrease of these liquids in density with increasing molecular weight, 0.796 g/ml for CH_3OH but 0.7893 g/ml for C_2H_5OH. A truly "normal" liquid should not exhibit such intermolecular repulsion, and liquid mercury at 0°C does indeed have a very much smaller coefficient of compressibility than water, 4×10^{-6} unit vol/atm. Figure 1.15 compares the specific volume of pure water (from values tabulated in Dorsey, 1940) with the relative volume of liquid mercury (from values tabulated in *The International Critical Tables*). Not only is water more compressible, but also the pressure dependence of specific volume is markedly nonlinear. Comparison of Figure I.4 with Figure 1.15 reveals that the effect of electrolyte addition on the compressibility of water is small but, although small, it is important, and we shall say more about it later. But, if we expect to find any reflections in the pressure dependence of compressibility which might throw light upon water structure,

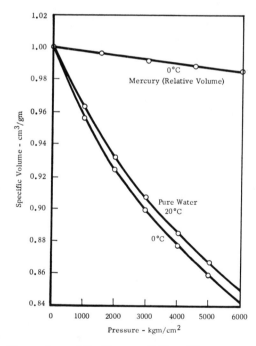

Figure 1.15   Comparison of the Compressibility of Pure Water and Mercury.

we are disappointed. There is a minimum in the isothermal compressibility of water near 50°C (Kell and Whalley, 1965) which may arise from a higher structural transition in water but, while possibly of great biological interest, it lies above the oceanographic temperature range.

The ease with which water molecules can move about should be strongly dependent on the extent to which they are bound, that is, on the amount of structure present in the liquid. We might expect, therefore, that the viscosity might be a useful measure of the degree of structuring in the liquid. And in

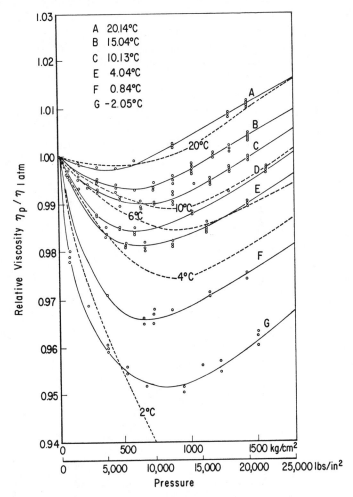

Figure 1.16 Relative Viscosities of Pure Water (dashed lines) and 35‰ Salinity Seawater (solid lines) under Pressure. From Horne and Johnson (1966), with permission of American Geophysical Union.

this expectation we are not disappointed. The pressure dependence of the viscosity of water is strikingly anomalous. While the viscosity of "normal" liquids increases as increasing pressure crowds the molecules together and hinders their movement, the viscosities of both pure water and $35\%_{00}$ S seawater (dotted and solid curves, respectively, in Figure 1.16) relative to their 1 atm values actually decrease, go through minima, and finally increase in the "normal" way (Cohen, 1892; Bett and Cappi, 1965; Horne and Johnson, 1965a,b). Increasing the temperature or adding electrolyte tends to reduce the depth of these minima (Figure 1.16). At pressures greater than about 1000 kg/cm^2 the relative viscosity, $\eta_P/\eta_{1\,atm}$, increases with increasing pressure as expected of a "normal" fluid. Unfortunately, highly specific molecular interactions strongly dependent on molecular structure play an important role in the viscous properties of fluids under pressure; hence we cannot make direct, unambiguous comparisons between the behavior of water and that of simple liquids. At 20°C and above 1500 kg/cm^2 the pressure coefficient of the relative viscosity, $\Delta(\eta_P/\eta_{1\,atm})/\Delta P$, is nearly equal to but opposite in sign from the pressure coefficient of the relative volume, $\Delta(V_P/V_{1\,atm})/\Delta P$. That is, the viscosity under pressure is behaving in just the same way as the reciprocal of the relative volume (Horne and Johnson, 1966a). This result could be fortuitous, but if it is not it indicates that, by the time pressures of about 1000–1500 kg/cm^2 are reached, the Frank-Wen clusters in liquid water have disappeared and water has become a simple, unassociated, "normal" liquid.

To the best of my knowledge, the only recent attempt to treat the high-pressure viscous properties of liquid water in a quantitative way in terms of a theory of water structure is an application of the theory of significant structures (Jhon, Grosh, Ree, and Eyring, 1966; Eyring and Jhon, 1966). The proponents of this theory use old experimental values of Bridgman for $\eta_P/\eta_{1\,atm}$ versus P at 0°C but, happily, these values are in fair agreement with more modern measurements (values at higher temperatures are not, Horne and Johnson, 1966a). Therefore their ability to match the theoretical and experimental curves is meaningful. They attribute the minima to the combined effect of decreasing the mole fraction of the clusters and squeezing out the vacancies. I should like to see the ideas of Franks and Good (1966) and the Frank-Wen-Nemethy-Scheraga treatment applied to the problem of the viscosity of compressed water.

To turn now to the effect of temperature on the structure of pure liquid water: the most obvious phenomenon which we must confront is the anomalous maximum in the density of water near 4°C (or minima in the specific and the molar volumes, Figure 1.17). The importance of this inflection has not escaped the attention of the theorists, and the proponents of the several water models have all taken pains to ensure that, by hook or by crook, their favorites encompass this anomaly. Hence the quantitative fit of theoretical

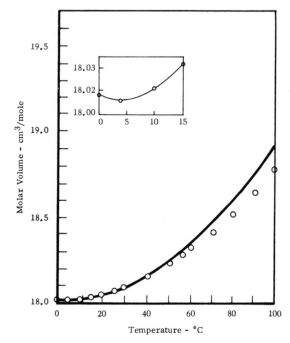

Figure 1.17   Temperature Dependence of the Molar Volume of Liquid Water Fitted by the Frank-Wen-Nemethy-Scheraga Model. From Nemethy and Scheraga (1964), with permission of American Institute of Physics.

to experimental P versus T curves is not very useful in discriminating among the several theories; they all fit or have been made to fit. Of the curve fitters, Jhon, Grosh, Ree, and Eyring (1966) and Nemethy and Scheraga (1964) have been particularly thorough. Heavy water, D_2O, is slightly more highly structured than H_2O, although this structure is more susceptible to thermal destruction (Thomas, Scheraga, and Schrier, 1965), and, as a consequence, the temperature of maximum density is shifted upward from 4° to about 11°C. These authors have shown that their theories can yield this shift. The temperature of maximum density of H_2O decreases with increasing pressure or with the addition of electrolytes.

In the classical water theory of Bernal and Fowler (1933), 4°C is interpreted as the temperature below which the less dense Water I (tridymite-like) predominates and above which the more dense Water II (quartz-like) is the principal form. Nemethy and Scheraga (1962) offer a more elegant explanation for the maximum at 4°C: it arises simply from the superposition of the density increase that results from cluster breakup on the density decrease due to normal thermal expansion. Recently I have developed some misgivings about this explanation; it may be too simple. There are symptoms of distressing complexity near 4°C. For example, the energetics of ion transport in sea-water (electrical conductivity) also show maxima near 4°C (Horne and Cour-ant, 1964), but the energetics of water transport (viscous flow) do not (Horne, Courant, Johnson, and Margosian, 1965) unless pressure is applied (Horne and Johnson, 1966a) and the slope of $\eta_P/\eta_{1\,atm}$ versus T changes abruptly near 4°C. These difficulties have tempted me to return to the notion of a definite structural transition near this temperature.

In addition to the maximum in density at 4°C, inflections in water proper-ties have also been reported at a number of other temperatures and attri-buted to higher structural transitions, but the reality of these phenomena is currently the subject of controversy and we shall not deal with them further.*

As a central part of their quantitative analysis of the Frank-Wen flickering cluster model, Nemethy and Scheraga (1962) have calculated the various structural parameters for water over the temperature range 0–100°C (Table 1.4). In addition to the parameters shown in Table 1.4, they also calculated the temperature dependence of the mole fraction of unbonded, singly bonded, doubly bonded, triply bonded, and tetrahedrally bonded water species. Table 1.4 exhibits several very important features. First notice that while the cluster size decreases the cluster concentration increases. Thus the effect of increased thermal energy is not only to "melt" or dissolve the clusters but also to fragment them. Second, in going from 0° to 100°C the fraction of unbroken H bonds decreases from 0.528 to only 0.325, remarkably little change

Table 1.4   Temperature Dependence of Structural Parameters in Liquid Water

Tempera-ture, °C	Av. No. H_2O's per Cluster	Cluster Concentration (mole fraction)	Mole Fraction of Unbound Water	Fraction Un-broken H Bonds
0	64.96	0.0084	0.2485	0.528
10	49.25	0.0102	0.2699	0.493
20	38.37	0.0124	0.2948	0.462
30	30.34	0.0147	0.3180	0.434
40	24.29	0.0172	0.3394	0.409
50	19.84	0.0198	0.3591	0.388
60	16.64	0.0224	0.3773	0.370
70	14.54	0.0243	0.3940	0.356
80	13.14	0.0257	0.4095	0.344
90	12.13	0.0264	0.4239	0.334
100	11.70	0.0268	0.4375	0.325

* For those interested in further reading on this topic, most of the relevant references can be found in Falk and Kell (1966) and Drost-Hansen (1967).

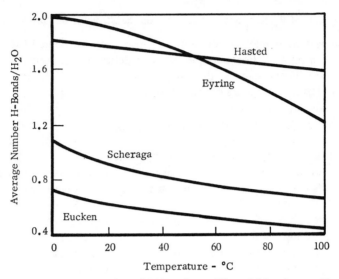

Figure 1.18 The Average Number of H Bonds per Water Molecule as a Function of Temperature according to Various Theories of the Structure of Liquid Water. From Wicke (1966), with permission of Verlag Chemie.

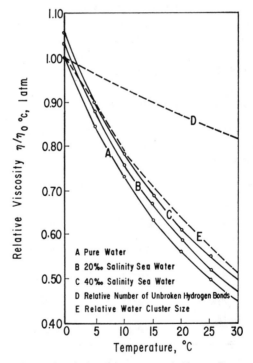

Figure 1.19 Comparison of Relative Transport and Cluster Properties. From Horne (1965), with permission of American Geophysical Union.

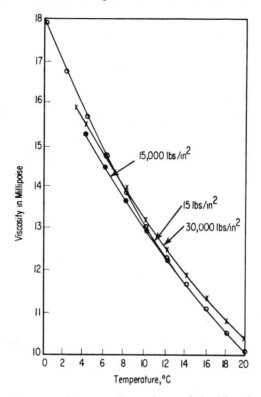

Figure 1.20 Temperature and Pressure Dependence of the Viscosity of Pure Water. From Horne and Johnson (1966a), with permission of American Chemical Society.

between the melting and boiling points. In other words, liquid water still contains an appreciable amount of structure, even at the boiling point.

The fraction broken H bonds and the monomer concentration in liquid water are, as I have mentioned, subjects of much current discussion. Thomas, Scheraga, and Schrier (1965) have recalculated the parameters listed in Table 1.4, using somewhat different initial assumptions, but with more or less the same results. Figure 1.18, taken from Wicke (1966), compares the average number of H bonds per H_2O estimated on the basis of various theories. The want of agreement, however, should not obscure the important qualitative fact that as the temperature increases the extent of H-bonded structure decreases. Despite the questions that have been raised, I still have a certain amount of faith in the values listed in Table 1.4. They make physical sense, and relative transport process parameters in liquid water, such as electrical conductivity and viscous flow, correlate very nicely with the relative size of the clusters (Figure 1.19) (Horne and Courant, 1964). Thus, although no one seems to know exactly what the parameters listed in Table 1.4 mean, there can be little doubt that, whatever they are, they, or parameters proportional to them, reflect real conditions obtaining in liquid water, conditions in turn capable of giving rise to observable phenomenological consequences.

Earlier we found that the viscosity gave us insight into the effect of pressure on liquid water. Finally we might note in passing that the effect of temperature on the viscosity (Figure 1.20) enforces the general conclusion we have already drawn: the decrease in viscosity with increasing temperature indicates that the injection of thermal energy into water tends to break up the regions of structure in that liquid.

5   The Effects of the Presence of Solutes on the Structure of Water

Hitherto in an attempt to examine first things first we have confined our attention to the structure of pure water and the effect of various environmental parameters on that structure. But, as we well know, the water of the oceans is far from pure. They are, roughly speaking, a moderately concentrated aqueous electrolytic solution. In our attempt to understand what seawater is and how it behaves, we must now come to grips with the question of how the addition of an electrolyte alters the structure of liquid water. And I should like to warn you in advance that the effects of temperature and pressure on water structure are simple compared with those produced by the presence of an electrolyte.

Because of the obvious importance of an understanding of aqueous electrolytic solutions to the chemical oceanographer our procedure in dealing with electrolytes will be as follows. In this section, I shall try to describe, for the greater part in a qualitative way, the effect of electrolyte addition on water structure in terms of pictorial imagery on the atomic-molecular or microscopic scale. In Chapter 2 I intend to return to the topic from a quantitative, thermodynamic macroscopic point of view and provide the basis for practical calculations of seawater property changes on processes involving changes in temperature, pressure, and salinity. Then in Chapter 3 further detail will be added to our description of the water structure near solutes.

Now let us make some *simplified* seawater. Let us take one-half a gram molecular weight or 29.22 g of NaCl and dissolve it in 1000 g of pure water to make a "sea water" sample 0.50 *m* (*Note:* molal, not molar) in salt. As we do this all kinds of peculiar things happen, and these phenomena provide clues to how the electrolyte substance is interacting with the water substance and altering the structure and properties of the latter. There is a heat (or enthalpy) change when an electrolyte is dissolved in water. The exact

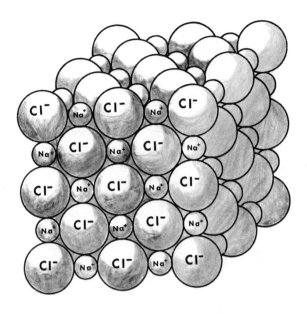

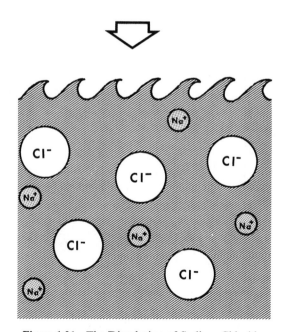

Figure 1.21 The Dissolution of Sodium Chloride.

magnitude of this change depends on the temperature and the concentration of the resulting solution; it is discussed further in the next chapter. Suffice it here to say that in the present instance roughly 0.6 kcal is absorbed. This heat change may be taken as evidence that the solution process involves the making and breaking of chemical bonds. The sodium and chloride ions in the crystal lattice of solid NaCl (Figure 1.21) are held together very strongly, and the observed absorption of energy indicates that the energy released by the formation of the new solute-solvent interaction bonds is inadequate to compensate for the energy required for the disruption of the ionic Na^+—Cl^- bonds in the crystal.

If we cool our "seawater" down, taking precautions to avoid supercooling, we shall find that, unlike pure water, it no longer freezes at $0°C$ and we must cool our solution to $-1.85°C$ before it solidifies. Similarly we find that the other so-called colligative properties—the boiling point, vapor pressure, and osmotic pressure—have also been altered. We are now dealing with a general solution phenomenon, not restricted to aqueous systems, and I think it might be worthwhile to digress a moment and derive the expression for the depression of the freezing points of liquid solvents by the addition of solutes.

At the freezing point the solid and liquid are in equilibrium and must therefore have the same vapor pressure (see Figure 1.22). The vapor pressure

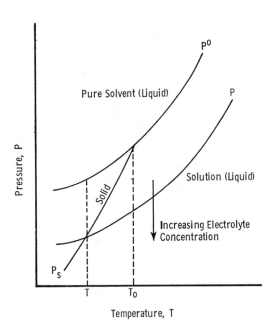

Figure 1.22   Freezing Point Depression.

curve for the solid, P_s in Figure 1.22, is steeper than the vapor pressure curve for the pure solvent, P^o in Figure 1.22, because the heat of sublimation of the solid is greater than the heat of vaporization of the liquid. The two curves, P_s and P^o, intersect at the freezing point of the pure liquid, T_o. The addition of solute lowers the vapor pressure of the pure solvent. For dilute solutions, in accordance with Raoult's law,

$$(1.1) \qquad P = N_1 P^o$$

where N_1 is the mole fraction of solvent. At the freezing point, equilibrium requires that

$$(1.2) \qquad P_s = P = N_1 P^o$$

Differentiating with respect to temperature, we obtain

$$(1.3) \qquad \frac{dP_s}{dT} = N_1 \frac{dP^o}{dT} + P^o \frac{dN_1}{dT}$$

Substituting on the basis of the Clausius-Clapeyron equation

$$(1.4) \qquad \frac{dP}{dT} = \frac{HP}{RT^2}$$

where H is the molar heat of the phase change, (1.3) becomes

$$(1.5) \qquad \frac{H_s P_s}{RT^2} = N_1 \frac{H_v P^o}{RT^2} + P^o \frac{dN_1}{dT}$$

where H_s and H_v are the molar heats of sublimation of the solid solvent and of the vaporization of the liquid solvent, respectively. Replacement of P_s by its equivalent $N_1 P^o$ (1.1), canceling, and rearranging give

$$(1.6) \qquad \frac{H_s - H_v}{RT^2} = \frac{d \ln N_1}{dT} = \frac{H_f}{RT^2}$$

where H_f, the heat of fusion, is equal to the difference between the heats of sublimation and vaporization. Expression (1.6) next can be integrated between the limits corresponding to the pure solvent, T_o and $N_1 = 1$, and the solution, T and N_1, to give

$$(1.7) \qquad \frac{H_f(T_o - T)}{RT_o T} = -\ln N_1$$

For a binary solution containing only one solute and solvent, $N_1 = 1 - N_2$, where N_2 is the solute mole fraction. Furthermore if the solution is dilute the higher terms in the MacLaurin series expansion can be neglected, so that

$$(1.8) \qquad -\ln N_1 = -\ln (1 - N_2) = N_2 + \tfrac{1}{2}N_2{}^2 + \tfrac{1}{3}N_2{}^3 + \cdots \approx N_2$$

and (1.7) becomes

(1.9)
$$N_2 = \frac{H_f(T_o - T)}{RT_oT_1}$$

As the solution is dilute, T_o and T will be nearly the same; hence T_oT becomes approximately T_o^2, and the number of moles of solute, n_2, is small compared to the number of moles of solvent, n_1, so that $N_2 = n_2/(n_1 + n_2)$ becomes approximately n_2/n_1. Finally the ratio n_2/n_1 can be replaced by the corresponding concentration expression, $m/1000\,M_1$, where M_1 is the molecular weight of the solvent and m the molality of the solution, to give the relationship between solute concentration and freezing point depression:

(1.10)
$$T_o - T = \frac{RT_o^2 n_2}{H_f n_1} = C_f m$$

where C_f is a constant, characteristic of the solvent, and has a value of 1.855 in the case of water. And, of course, for that solvent T_o is 0°C. A comparable expression can be derived for the boiling point elevation which, in the case of water, becomes

(1.11)
$$100 - T = -0.51\,m$$

The freezing points and other colligative properties of seawater are shown in Figure 1.23.

Although (1.10) and (1.11) are strictly applicable only for dilute solutions which obey Raoult's law and solvents whose vapors behave as ideal gases, they appear to work quite satisfactorily in aqueous solutions if our solute is, for example, sugar. If the solute is NaCl, however, something appears to be drastically amiss, for we observe very nearly *twice* the freezing point depression that we expect from (1.10). In other words, although we have added only 0.5 mole of NaCl there appear to be 1.0 mole of solute substance in solution. Each NaCl has dissociated into two species. If we measure the electrical conductivity of our "seawater" we are in for another surprise. If we have taken any care to use pure water the initial specific electrical conductivity will be less than 10^{-6} ohm^{-1} cm^{-1}, but our 0.5 m NaCl solution will have a specific conductance of 4.68×10^{-2} ohm^{-1} cm^{-1}. The addition of the salt has increased the electrical conductivity more than 10,000-fold, and clearly the dissociation of the electrolyte has produced many charge carriers. The solvent water itself dissociates:

(1.12)
$$H_2O \rightleftharpoons H^+ + OH^-$$

but very slightly, so that at 25°C the combined concentration of hydrogen and hydroxyl ions in pure water is only about 2×10^{-7} M. Thus we are led

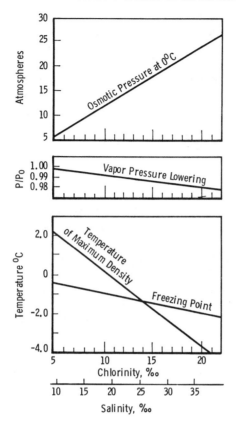

Figure 1.23 Some Colligative Properties of Seawater. From Sverdrup, Johnson, and Fleming (1942), with permission of Prentice-Hall, Inc.

to conclude that on addition to water the NaCl dissociates into two charged species or ions (Figure 1.21).

$$(1.13) \qquad Na^+Cl^- (s) \rightleftharpoons Na^+ (aq) + Cl^- (aq)$$

On the basis of data of Wicke and Eigen (1953), at 25°C in an 0.5 m aqueous solution the NaCl is about 93% dissociated (Horne and Courant, 1964). If the solution is diluted the dissociation becomes more nearly complete.

Today the Arrhenius theory of electrolytic dissociation is such a familiar and valuable concept that we forget just how remarkable it is! The coulombic attraction of small charged ions is very considerable. The crystal energy of Na^+Cl^- in the lattice, that is, the heat of formation of Na^+Cl^- (crystal) from the gaseous ions,

$$(1.14) \qquad Na^+ (g) + Cl^- (g) \rightarrow Na^+Cl^- (crystal)$$

is 181 kcal/mole (Helmholz and Mayer, 1934). Yet, in solution, not only are the Na^+ and Cl^- ions in the crystal torn apart as the salt dissolves, but also these ions seem content to exist side by side despite the very strong coulombic attraction one might expect between them. The dissolution process and the disruption of the tight NaCl lattice (Figure 1.21) must be accompanied by the formation of strong new bonds, and the resultant ions can coexist side by side in solution only if their coulombic fields are screened somehow and thereby attenuated. This is exactly what happens. The solute ions interact strongly with the solvent molecules ("solvation"), and in the case of aqueous solutions this "hydration" results in an envelope or atmosphere of water molecules surrounding the ions. Inasmuch as the forces of coulombic attraction diminish as the square of the distance, this envelope or atmosphere effectively screens the ionic charge. Many electrolytes hold their water of hydration so tenaciously that even when their solutions are evaporated the resulting solid crystalizes out with a definite number of water molecules of hydration, for example, $MgSO_4 \cdot 7\,H_2O$ (Epsom salt), $CuSO_4 \cdot 5\,H_2O$ (blue vitriol), $FeCl_2 \cdot 4\,H_2O$.

In chemical oceanography we are, needless to say, concerned with what chemical elements are present in seawater, but even more important is the *form* in which these elements occur. All of the dissolved and most of the insoluble constituents in seawater occur in the form either of hydrated species or hydrated ions in which some of the waters of hydration have been replaced by other ligands. The latter type of species are called complex ions. An understanding of ionic hydration is thus a necessary prerequisite for making any progress in understanding the chemistry of seawater. Unfortunately, as in most matters pertaining to water, despite very considerable effort over recent years, both experimental and theoretical, ionic hydration remains one of the most complex and perplexing topics in chemistry.

At this point I should like to digress a moment. I may seem to be perversely dwelling on difficulties. In science the written word and spoken word have two functions: to instruct and to challenge. In my experience, textbooks and lectures are very strong on the former and equally weak on the latter. I resent the book which gives the impression that everything is cut-and-dried, that everything is known, that one has only to feed the data into the appropriate equations and turn the crank. It seems to me that I have spent far too large a part of my adult career as a scientist discovering that the familiar textbooks with which I was trained are concatenations of half-truths and that the world is far more complex, and far less perfectly understood, than one would ever suspect from reading them. I should in particular like to prepare the younger readers of these pages for this painful discovery. But while the discovery may be personally disconcerting it should be only momentarily so. For science is a thing of the future, not of the past. Science is not knowledge, but the advancement of knowledge. And all the things we do not know are things we can know.

To return to the difficulties of hydration: a reasonable question would appear to be, "How many water molecules are associated with an Na^+ ion in aqueous solution?" Estimates of the "hydration number" of Na^+ in the literature (see Table A.6 in the Appendix) range from 2 to 70 water molecules! (Rutgers and Hendrikx, 1962). But, just as the controversy over the question of the concentration of monomeric water in pure liquid water does not mean that there is no monomeric water, these widely variant estimates of hydration numbers do by no means lead to the conclusion that the ionic hydration is a fiction. The hydration atmosphere of an ion in solution is not a simple discrete aggregate. It has a complex internal structure and its outer boundary is virtually impossible to establish. Within the hydration atmosphere there are in turn several types of water molecules correspondingly bound with various strengths. Some experimental methods measure only the most tightly bound water molecules, whereas other methods count the loosely bound waters as well. Little wonder, then, that various methods yield different hydration numbers. The Soviet school of solution physical chemists, of which Samoilov (1965) is a principal figure, has been so dismayed by the disparate "hydration numbers" that, rather than think in terms of a definite number of water molecules associated with an ion, they prefer to conceptualize hydration in terms of the residence time of a given water molecule near an ion. If an average water molecule loiters in a position near an ion for a longer time than it would loiter at some point a great distance from the ion, then the ion can be said to be positively hydrated: however, if the water molecule is more mobile near the ion than it would be at some great distance from it, the term "negative hydration" is applied. The residence time concept of hydration is by no means incompatible with the concept of "hydration number"; rather the two views nicely complement one another.

Another apparently straightforward question might be, "How much of its hydration atmosphere does Na^+ ion take with it when it moves?" Again we do not know, although Samoilov (1965) has published estimates of the quantity. As a matter of fact, inasmuch as all species in a liquid are in a state of constant agitation, we are not even sure that it is operationally meaningful to try to distinguish between the stationary and the kinetic hydration atmosphere of an ion.

Although when speaking of ionic hydration we can be quantitative only at our peril, there are some very useful qualitative generalizations which can be made and which seem to remain valid with relatively few exceptions. The bond joining some transition element ions to their waters of hydration can have appreciable covalent (or electron-sharing) character (Scrocco and Salvetti, 1954; Oshida and Horiguchi, 1956; Hofacker, 1957), but in the case of simple, monovalent cations such as the alkali metal cations the bond is almost entirely ionic in character and is amenable to relatively simple theoretical analysis in terms of electrostatic attraction. Thus for many ions (including those such as Na^+) important in seawater chemistry, the most important single factor in determining the extent and strength of hydration is the charge density. Generally speaking, the *higher the charge density, the more heavily hydrated is the ion*, and from this rule the following corollaries are obtained:

1. Cations tend to be more heavily hydrated than anions. The primary hydration numbers of the alkali metal cations are in the neighborhood of 4, whereas those for the halide anions seem to be somewhere between 1 and 4.

2. The greater the charge, the more heavily hydrated is the ion. The primary hydration number of doubly charged Mg^{2+} appears to be somewhere between 6 and 12 whereas that of Li^+, even though this ion has very nearly the same crystal radius, is near 4.

3. In a given charge type, the smaller the crystal radius of the ion, the more heavy is the hydration. Glueckauf (1964) has reported hydration numbers of 3.2, 2.1, and 1.1 for Li^+, Na^+, and K^+, respectively, and 2.2, 1.4, and 1.1 for F^-, Cl^-, and I^-. respectively. (Note the nonintegral values.)

As I have mentioned, different experimental methods tend to give different hydration numbers, yet, significantly, the numbers yielded by a given experimental method are usually self-consistent in the sense that they exemplify the preceding hydration "rules." For example, Rutgers and Hendrikx (1962) used a method based on ion transport through a cellophane membrane and, although their values seem uniformly high when compared to those obtained from other types of measurements, they nicely illustrate the expected trends: 22, 13, 7, and 6 for Li^+, Na^+, K^+, and Cs^+; 7, 5, and 5 for F^-, Cl^-, and Br^-; 12 for doubly charged SO_4^{2-} compared to 5 for singly charged Cl^-; and 29 and 36 for doubly charged Sr^{2+} and Mg^{2+} compared to 13 for singly charged Na^+.

Because of the importance of ionic hydration in seawater chemistry we shall return to this subject again and again in subsequent pages as our treatment of topics such as thermodynamics and transport processes enables us to add more and more detail to our picture of ionic hydration.

The formation of our 0.5 m NaCl "seawater" solution is accompanied by still another very peculiar and certainly unexpected phenomenon—*electrostriction*. At 25°C the density of a solid NaCl is 2.165 g/cm^3 and that of water is 0.997 g/cm^3. If the volumes of the two substances, salt and water, were simply additive, the resulting 1000 g solution should have a volume of 13.50 + 973.70 or 987.20 cm^3. The actual volume, calculated from *International Critical Tables* densities or from the much more recent apparent molal volume tables of Vaslow (1966), is between 982 and 983 cm^3.* The system has shrunk, and the volume of solution is *less* than the combined volumes of solute and solvent. The coulombic fields of the ions interact with the partial charges of the water dipoles with sufficient strength to draw the water molecules in close to them, thus compressing the solvent in their immediate neighborhood (Figure 1.24). The specific volume (cm^3/mole) of the water is greater near an ion than at some great distance from it. Electrostriction plays two important roles in marine chemistry: (a) in its influence on the hydration of ions and polar species; and (b) through hydration, it is the property which is largely respon-

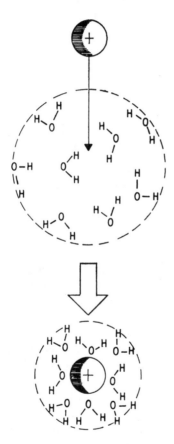

Figure 1.24 Electrostriction. From Horne (1967), with permission of Academic Press, Inc.

* Duedall and Weyl (1967) have recently published the results of dilatometer measurements of the partial equivalent volumes of the major salts in seawater.

sible for the effects of hydrostatic pressure on the dissociation of weak electrolytes in seawater.

If we assume that the electrostricted water molecules are crowded so tightly together that they are incompressible, then the compressibility of the solution under pressure can be ascribed to the remaining nonelectrostricted solvent, and we have in hand a method of calculating hydration numbers from compressibilities. It can be shown that the hydration number of say NaCl, n_{NaCl}, is given by

$$(1.15) \qquad n_{NaCl} = \frac{(\text{moles of } H_2O)}{(\text{moles of NaCl})} (1 - \beta_s/\beta_{H_2O}^\circ)$$

where β_s and $\beta_{H_2O}^\circ$ are the compressibilities of the solution and of pure water, respectively (Barnartt, 1953).

Ultrasonic velocity measurements represent a convenient technique of determining the compressibilities. Alternatively, Padova, (1961, 1963, 1964) has proposed that the average theoretical electrostriction per mole of water is 2.1 ml, and this enables him to calculate hydration numbers from the total observed electrostriction ΔV, using the simple relation

$$(1.16) \qquad n = \frac{\Delta V}{2.1}$$

This method yields 3, 3, 1, 1, and 1 for Li^+, Na^+, K^+, Rb^+, and Cs^+; 3, 1, 1, and 1 for F^-, Cl^-, Br^-, and I^-; 14 for Mg^{2+}; 26 for La^{3+}; and 1 for NO_3^- but 4 for SO_4^{2-}, which again nicely illustrates the hydration trends described above.

Earlier we discovered that a property which can give us a useful hint regarding the structural state of affairs in liquid water is the viscosity. If, finally, we measure the viscosity of our "seawater" sample, we find that it is somewhat greater than that of pure water at the same temperature. To pursue the same interpretation as before: this decrease in fluidity implies that the addition of the electrolyte has somehow enforced the structure of liquid water. More detailed studies of the electrolyte concentration dependence of the viscosity of aqueous solutions (Figure 1.25) show that some electrolytes, such as NaCl, increase the viscosity, whereas others, such as KCl, actually make the water less viscous. The former electrolytes are called "structure makers" and the latter "structure breakers." Notice in Fig. 1.25 that 2:2 electrolytes such as $MgSO_4$ tend to be very strong structure makers. Notice too in Figure 1.25 that the presence of an electrolyte is appreciably affecting water structure as reflected in the viscosity, even at low electrolyte concentrations where the ions are widely separated. The hydration phenomena discussed previously, including electrostriction, provide evidence for strong, short-range solute-solvent interactions. Now in the viscosity effects we find evidence for more

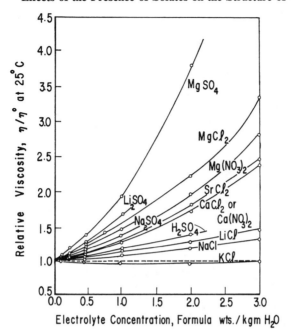

Figure 1.25   Effect of Electrolytes on the Viscosity of Water. From Horne (1965), with permission of American Geophysical Union.

long-range interactions. Other types of measurements on aqueous solutions as a function of electrolyte concentration, such as proton nuclear magnetic resonance and infrared studies (Wicke, 1966; Yamatera, Fitzpatrick, and Gordon, 1964; Fabricand, Goldberg, Leifer, and Ungar, 1963) yield results which tend to parallel the viscosity dependence.

The theories of water structure which postulate a distorted ice or ice-like lattice in the liquid attribute the structure-making or structure-breaking character of ions to the ease or difficulty, respectively, with which they fit into the empty spaces or with which they can occupy a water site in the matrix. These views are, I feel, insufficiently flexible to embrace all the experimental facts and, just as we preferred the less explicit, more versatile Frank-Wen flickering cluster model of liquid water, we shall now treat the water-structure-perturbing influence of solutes in terms of the ideas that have been presented by the same authors (Frank and Wen, 1957; Frank and Evans, 1945). An ion in solution is represented as being surrounded by two zones (Figure 1.26). An inner layer (A), which I feel we can equate to what is often called the "primary" hydration shell, is composed of dense, electrostricted, and immobilized water molecules strongly bound by the coulombic field of the ion. At some great distance from the ion (C) the water structure is "normal,"

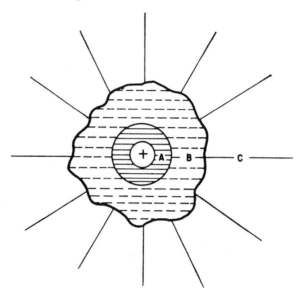

Figure 1.26 Frank-Evans-Wen Two-Zone Model for the Hydration Atmosphere ion in an Aqueous Solution. From Frank and Wen (1957), with permission of Faraday Society.

although the molecules may be ever so slightly polarized by the ubiquitous charge field. Of special interest is the intervening zone (*B*). In this volume the coulombic field of the ion is still strong enough to disrupt the "normal" structure of liquid water; yet its strength is inadequate to reorient the water molecules in some new configuration (such as in zone *A*). Consequently zone *B* is a region of comparative randomness, of disrupted water organization, of broken structure. Whether or not an ion is a structure maker or breaker depends on the relative importance of these two zones. Frank and Evans (1945) incline to the view that the structure-enhanced *A* zone is present and intact in all ions, while the particular characteristic of different types of ion arise from the variability of the structure-broken *B* zone. Gurney (1953), however, appears to be of the opinion that, in the case of the structure-breaking ions, the *B* region can encroach on and diminish *A*.

In Chapter 3 further detail will be added to our picture of the ionic hydration atmosphere.

6   Summary

The structure of liquid water is a subject of intense study and controversy. The polar nature of the water molecule and its ability to form strong intermolecular hydrogen bonds result in the cooperative association into multimolecular aggregates. Liquid water is a mixture of these aggregates or "clusters" and more or less "free" or monomeric water. As the temperature is increased, the clusters "melt" and, inasmuch as their specific volume is larger than that of the unassociated water, the application of hydrostatic pressure is effective in destroying the structured regions in liquid water. The presence of an electrolyte profoundly alters the structure of water. The local perturbation of water structure is called the hydration atmosphere of an ion and is itself complex, being composed of an inner, structure-enhanced region and an outer region of broken water structure. Quantitatively the hydration of an ion may be described in terms of the hydration number of the ion or in terms of water residence times. The experimental determination of the former presents certain difficulties; nevertheless, generally speaking, the larger the charge density on a cation, the heavier is the hydration and, usually, cations are more hydrated than corresponding anions.

REFERENCES

Anderson, A. R., B. Knight, and J. A. Winter, *Trans. Faraday Soc.*, **62**, 359 (1966).

Barnartt, S., *Quart. Rev. (London)*, **7**, 84 (1953).

Barnes, W. H., *Proc. Roy. Soc. (London)*, **A125**, 670 (1929).

Bernal, J. D., and R. H. Fowler, *J. Chem. Phys.*, **1**, 515 (1933).

Bett, K. E., and J. B. Cappi, *Nature*, **207**, 620 (1965).

Bjerrum, N., *Dan. Nat. Fys. Medd.*, **27**, No. 1 (1951).

Brady, G. W., and W. J. Romanow, *J. Chem. Phys.*, **32**, 306 (1960).

Chadwell, H. M., *Chem. Rev.*, **4**, 375 (1927).

Chidambaram, R., *Acta Cryst.*, **14**, 467 (1961).

Cohen, R., *Ann. Phys.*, **45**, 666 (1892).

Danford, M. D., and H. A. Levy, *J. Am. Chem. Soc.*, **84**, 3965 (1962).

Darling, B. T., and D. M. Dennison, *Phys. Rev.*, **57**, 128 (1940).

Davis, C. M., Jr., and T. A. Litovitz, *J. Chem. Phys.*, **42**, 2563 (1965).

Dennison, D. M., *Phys. Rev.*, **17**, 20 (1921).

Dorsey, N. E., *Properties of Ordinary Water-Substance*, Reinhold, New York, 1940.

Drost-Hansen, W., in W. Stumm, ed., *Equilibrium Concepts in Natural Water Systems*, *Advan. Chem. Ser.* No. 67 (1967).

Drost-Hansen, W., *Ind. Eng. Chem.*, **57**, No. 4, 18 (1965).

Duedall, I. W., and P. K. Weyl, *Limnol. Oceanog.*, **12**, 52 (1967).

Eucken, A., *Z. Elektrochem.*, **52**, 255 (1948); **53**, 102 (1949).

Eyring, H., and M. S. Jhon, *Chemistry*, **39**, No. 9, 8 (1966).

Fabricand, B. P., S. S. Goldberg, R. Leifer, and S. G. Ungar, *Mol. Phys.*, **7**, 425 (1963–64).

Falk, M., and T. A. Ford, *Can. J. Chem.*, **44**, 1699 (1966).

Falk, M., and G. S. Kell, *Science*, **154**, 1013 (1966).

Forslind, E., *Acta Polytech.*, **115**, 9 (1952).

Fox, J. J., and A. E. Martin, *Proc. Roy. Soc. (London)*, **A174**, 234 (1940).

Frank, H. S., and M. W. Evans, *J. Chem. Phys.*, **13**, 507 (1945).

Frank, H. S., and A. S. Quist, *J. Chem. Phys.*, **34**, 604 (1961).

Frank, H. S., and W. Y. Wen, *Disc. Faraday Soc.*, **24**, 133 (1957).

Franks, F., and W. Good, *Nature*, **210**, 85 (1966).

Franks, F., and D. J. G. Ives, *Quart. Rev.*, **20**, 1 (1966).

Glueckauf, E., *Trans. Faraday Soc.*, **60**, 1637 (1964).

Grjotheim, K., and K. Krogh-Moe, *Acta Chem. Scand.*, **8**, 1193 (1954).

Gurney, R. W., *Ionic Process in Solution*, McGraw-Hill Book Co., New York, 1953.

Helmholz, L., and J. E. Mayer, *J. Chem. Phys.*, **2**, 245 (1934).

Herzberg, G., *Molecular Spectra and Molecular Structure*, D. Van Nostrand Co., Princeton, N.J., 1945.

Hofacker, L., *Z. Elektrochem.*, **61**, 1048 (1957).

Horne, R. A., *Surv. Progr. Chem.*, **4**, 1, 1968.

Horne, R. A., *Water Resources Res.*, **1**, 263 (1965).

Horne, R. A., and R. A. Courant, *J. Phys. Chem.*, **68**, 1258 (1964).

Horne, R. A., and R. A. Courant, *J. Geophys. Res.*, **69**, 1152 (1964).

Horne, R. A., and R. A. Courant, *J. Geophys. Res.*, **69**, 1971 (1964).

Horne, R. A., R. A. Courant, D. S. Johnson, and F. F. Margosian, *J. Phys. Chem.*, **69**, 3988 (1965).

Horne, R. A., A. F. Day, R. P. Young, and N. T. Yu, "Interfacial Water Structure," Arthur D. Little, Inc., Tech. Rept. No. 23, Sept. 30, 1966. Office of Naval Research Contract Nonr-4424(00).

Horne, R. A., and D. S. Johnson, *J. Phys. Chem.*, **70**, 2182 (1966a).

Horne, R. A., and D. S. Johnson, *J. Geophys. Res.*, **71**, 5275 (1966b).

Jhon, M. S., J. Grosh, T. Ree, and M. Eyring, *J. Chem. Phys.*, **44**, 1465 (1966).

Kell, G. S., and E. Whalley, *Phil. Trans. Roy. Soc. London*, **A258**, 965 (1965).

Lennard-Jones, J., and J. A. Pople, *Proc. Roy. Soc. (London)*, **A202**, 166, 323 (1950).

Macey, H. H., *Trans. Brit. Ceram. Soc.*, **41**, 73 (1942).

Marchi, R. P., and H. Eyring, *J. Phys. Chem.*, **60**, 22, (1964).

Mecke, R., and W. Baumann, *Physik. Z.*, **33**, 833 (1932).

Miller, S. L., *Proc. Natl. Acad. Sci. U.S.*, **47**, 1515 (1961).

Morgan, J., and B. E. Warren, *J. Chem. Phys.*, **6**, 666 (1938).

Mulliken, R. S., *Phys. Rev.*, **41**, 756 (1932); **43**, 279 (1933).

Mysels, K. J., *J. Am. Chem. Soc.*, **86**, 3503 (1964).

Nemethy, G., and H. A. Scheraga, *J. Chem. Phys.*, **36**, 3382 (1962).

Nemethy, G., and H. A. Scheraga, *J. Chem. Phys.*, **41**, 680 (1964).

Oshida, I., and O. Horiguchi, *J. Phys. Soc. Japan*, **11**, 330 (1956).

Padova, J., *Bull. Res. Council Israel*, **A10**, 63 (1961).

Padova, J., *J. Chem. Phys.*, **39**, 1552 (1963).

Padova, J., *J. Chem. Phys.*, **40**, 691 (1964).

Pauling, L., *J. Am. Chem. Soc.*, **57**, 2680 (1935).

Pauling, L., *Science*, **134**, 15 (1961).

Pauling, L., *The Nature of the Chemical Bond*, 2nd ed., Cornell University Press, Ithaca, N.Y., 1948.

Pauling, L., *The Nature of the Chemical Bond*, 3rd ed., Cornell University Press, Ithaca, N.Y., 1960.

Pelah, I., and J. Imry, *Strong Hydrogen Bonds and the Structure of Liquid Water*, Israel Atomic Energy Commission Rept. No. IA-875 (Oct: 1963).

Peterson, S. W., and H. A. Levy, *Phys. Rev.*, **92**, 1082 (1953).

Peterson, S. W., and H. A. Levy, *Acta Cryst.*, **10**, 70 (1957).

Pinkerton, J. M. M., *Nature*, **160**, 128 (1947).

Podolsky, R. J., *Circulation*, **21**, 818 (1960).

Pople, J. A., *Proc. Roy. Soc. (London)*, **A205**, 163 (1951).

Rutgers, A. J., and Y. Hendrikx, *Trans. Faraday Soc.*, **58**, 2184 (1962).

Samoilov, O. Y., *Structure of Aqueous Electrolyte Solutions and the Hydration of Ions*, Consultants Bureau, New York, 1965.

Scrocco, E., and O. Salvetti, *Gazz. Chim. Ital.*, **84**, 1093 (1954).

Stevenson, D. P., *J. Phys. Chem.*, **69**, 2145 (1965).

Sverdrup, H. U., M. W. Johnson, and R. H. Fleming, *The Oceans*, Prentice-Hall, Englewood Cliffs, N.J., 1942.

Thomas, M. R., H. A. Scheraga, and E. E. Schrier, *J. Phys. Chem.*, **69**, 3722 (1965).

Vaslow, F., *J. Phys. Chem.*, **70**, 2286 (1966).

Verwey, E. J. W., *Rec. Trav. Chim.*, **60**, 837 (1941).

Wada, G., *Bull. Chem. Soc. Japan*, **34**, 955 (1961).

Waldstein, P., S. W. Rabideau, and J. A. Jackson, *J. Chem. Phys.*, **41**, 3407 (1964).

Wall, T. T., *J. Chem. Phys.*, **43**, 4187 (1965).

Wall, T. T., and D. F. Hornig, *J. Chem. Phys.*, **43**, 2079 (1965).

Wicke, E., *Angew. Chem.*, **5**, 106, 122 (1966).

Wicke, E., and M. Eigen, *Z. Elektrochem.*, **57**, 319 (1953).

Wollan, E. O., W. L. Davidson, and C. G. Shull, *Phys. Rev.*, **75**, 1348 (1949).

Yamatera, H., B. Fitzpatrick, and G. Gordon, *J. Mol. Spectr.*, **14**, 268 (1964).

2   The Thermodynamics of Seawater

1   Ideal and Real Solutions

An ideal solution may be defined phenomenologically as one which obeys Raoult's law (1.1).* Very dilute solutions tend to approach ideality. In these very dilute solutions the concentration of the solute is too low to appreciably affect the properties of the solvent, since the molecules of the latter represent an overwhelming majority. By the same token, a solute species so infrequently encounters a brother that solute-solute interactions are unimportant. As the concentration of electrolyte increases, however, the solvent begins to appreciate its presence and solute-solute interactions become increasingly evident.

In our attempt to understand marine chemistry two important questions confronting us are the following:

"Can seawater be considered a dilute solution?"
"Can seawater be considered a single electrolyte system?"

Why are these two questions so important? Our experimental and theoretical knowledge of ideal and near-ideal (dilute) solutions is quite adequate for a great many purposes. We have many fewer experimental data for concentrated solutions, despite their greater practical significance, and our understanding of them is presently in a state of confusion. Contemporary theories of more concentrated solutions are for the most part based on absurdly oversimplified models. These theories have yielded equations which enjoy some reputation of success in curve fitting, but these equations often consist of the expressions for dilute solutions with various subtractive and additive terms entrained, and in this sense they are in danger of being degraded to little more than empirical expressions. Then, too, the applications of these theories of more concentrated solutions, especially the more sophisticated ones, to practical situations tend to be so complicated as to render them useless.

So much for concentrated solutions. The situation with respect to aqueous mixed electrolytic solutions is even more discouraging. The first hesitant steps have been taken toward understanding mixed solutions of the most simple and well-behaved 1:1 electrolytes, and the difficulties that have been encountered are more than enough to warn us that an asymmetrical 1:1, 2:2

---

* In its most exact form, however, Raoult's law is expressed in terms of gas fugacities rather than partial pressures.

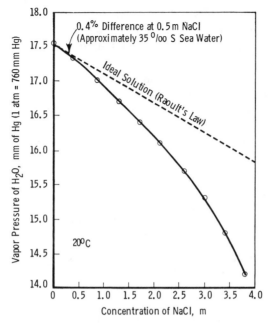

Figure 2.1   Water Vapor Pressure above NaCl Solutions at 20°C.

(NaCl and $MgSO_4$) electrolyte system such as seawater is, I feel it safe to say, hopelessly complicated.

Clearly, then, if we can get by with representing seawater as a dilute, single 1:1 electrolyte system, there is a great deal we can say about it, but if we cannot

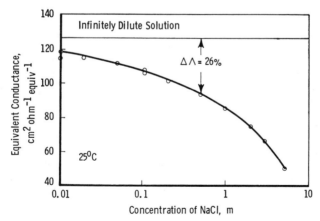

Figure 2.2   The Equivalent Conductance of Aqueous NaCl Solutions at 25°C as a Function of Electrolyte Concentration.

we are in very serious straits and any exact and detailed understanding of the nature of seawater must be postponed to the unforeseeable future. Figure 2.1 shows the water vapor pressure over salt solutions at 20°C as a function of NaCl concentration. Seawater of 35‰ salinity is roughly $0.5\ M$ (or $0.5\ m$) in NaCl, and the deviation from ideality at this concentration is only 0.4%— an encouraging result. But the vapor pressure is a solvent property, and we might expect solute properties to be more sensitive to the increased solute-solute interactions with increasing electrolyte concentration. The mobility of ions under the influence of an applied electric field should be particularly sensitive to both ion-water and ion-ion interactions. At 25°C the equivalent electrical conductivity of aqueous NaCl solutions by the time a concentration of $0.5\ M$ has been reached has dropped 26% lower than its value at infinite dilution (Figure 2.2). In terms of Raoult's law, seawater or an $0.50\ M$ NaCl solution would appear to be fairly ideal, whereas in terms of electrical conductivity both would appear to be relatively concentrated solutions. Table 2.1 compares transport phenomena in pure and 35‰ S seawater, and Table 2.2 compares a number of properties of 35‰ S seawater with $0.50\ M$ aqueous NaCl solution. If the total weight of salts in seawater were all NaCl, the concentration of NaCl could be about $0.6\ M$, and this concentration is also included in the table. With the exception of the electrical conductivity and the freezing point depression, the errors entailed in approximating seawater by an $0.5\ M$ NaCl solution are only a few percent. Thus Table 2.2 appears to lead

Table 2.1   Comparison of Transport Phenomena in Pure Water and of Seawater at 1 atm
(From Montgomery, 1957, with permission of McGraw-Hill Book Co.)

Name, Symbol, Units	Pure Water		Seawater Salinity 35/mille	
	0°C	20°C	0°C	20°C
Dynamic viscosity, η, g cm^{-1} sec^{-1} = poise	0.01787	0.01022	0.01877	0.01075
Thermal conductivity, k, watt cm^{-1} °C^{-1}	0.00566	0.00599	0.00563	0.00596
Kinematic viscosity, $\nu = \eta/\rho$, cm^2 sec^{-1}	0.01787	0.01004	0.01826	0.01049
Thermal diffusivity, $\kappa = k/c_p\rho$ cm^2 sec^{-1}	0.00134	0.00143	0.00139	0.00149
Diffusivity, D, cm^2 sec^{-1}				
NaCl	0.0000074	0.0000141	0.0000068	0.0000129
N$_2$	0.0000106	0.0000169	—	—
O$_2$		0.000021	—	—
Prandtl number, $N_P = \nu/\kappa$	13.3	7.0	13.1	7.0

Table 2.2   Comparison of Pure Water and Seawater Properties

Property	Seawater, 35‰ S	Aqueous NaCl Solution, 0.50 M	Aqueous NaCl Solution, 0.6 M	Pure Water
Density, g/cm³, 25°C	1.02412	1.01752	1.02172	1.0029
Equivalent conductivity, 25°C, cm² ohm⁻¹ equiv⁻¹	—	93.62	91.58	—
Specific conductivity, 25°C, ohm⁻¹ cm⁻¹	0.0532	0.0468	0.0458	—
Viscosity, 25°C, millipoise	9.02	9.32	9.41	8.90
Vapor pressure, mm Hg at 20°C	17.4	17.27	17.18	17.34
Isothermal compressibility, 0°C, unit vol/atm	46.4×10^{-6}	46.6×10^{-6}	45.9×10^{-6}	50.3×10^{-6}
Temperature of maximum density, °C	−3.52	—	—	+3.98
Freezing point, °C	−1.91	−1.72	−2.04	0.00
Surface tension, 25°C, dyne/cm	72.74	72.79	72.95	71.97
Velocity of sound, 0°C, m/sec	1450	—	—	1407
Specific heat, 17.5°C, joule/g°C	3.898	4.019	3.998	4.182

up to the conclusion that, for many practical purposes, representing seawater by an aqueous NaCl might be a permissible approximation. And in the last analysis we really have no other workable alternative. Our procedure then will be to treat seawater and an 0.5 M aqueous NaCl solution as synonymous, but always taking care to indicate the complexities and the analytic approaches dictated by the inescapable fact that seawater is not a dilute, but a moderately concentrated electrolytic, solution and that it contains not a single simple 1:1 electrolyte, but rather a horrendously complex soup of just about everything imaginable.

2   The Equation of State for Pure Water and Seawater

Of the thermodynamics of seawater, Craig (1960) has justly written, "Few subjects are more fundamental to the study of the sea and more neglected in application. ..." Those who are theoretically inclined view any suggestion that thermodynamics is impotent to deal with real problems as a personal affront, whereas those who deal daily and directly with the seas are content to work along as best they can with tables and empirical expressions. Fofonoff's (1959, 1962) treatment of the subject, which has been criticized on two specific points by Craig (1960), exemplifies these difficulties. It begins with an array of elegant appearing partial differential equations, but ends with the familiar empirical expressions. But at least there are four very good reasons why thermodynamics is so conspicuously ineffective in assisting the marine chemist: the system is not ideal; it is a multicomponent one; and it is rarely at equilibrium. The fourth reason is one to which Fofonoff (1962) calls attention and one which is surprising: the fundamental thermodynamic properties of seawater have not been thoroughly and carefully measured!

The purpose of this chapter is not to display the thermodynamic formalism which *in principle* forms a description of seawater. Rather it is a more modest and useful one, namely, to repeat the simple thermodynamic relationships and to set forth the means by which the thermodynamic properties can be conveniently estimated.

In the standard textbooks on oceanography such as *The Oceans* by Sverdrup, Johnson, and Fleming (1942), procedures are outlined for calculating the PVT properties of seawater, in particular the density as a function of temperature, pressure, and salinity, with the aid of compilations such as the hydrographical tables of Knudsen (1901) and a variety of empirical expressions. In 1958 Eckart gave a critical review of seawater specific volume data, and the problem has been more recently examined by Crease (1962) and Newton and Kennedy (1965). The former author concludes that the old specific volume data of Ekman (1908) are not seriously in error over most of the PTS range of oceanographic interest and that the errors are largest at higher temperatures and lower salinities. He also finds the variation in specific volume of seawater under pressure with salinity to be insignificant in the oceanographic range. Many years ago Tait (1888) discovered that the PVT relationships for pure water can be represented by the simple expression

$$(2.1) \qquad \beta_{o,P} V_{o,P} = -\left(\frac{\delta V_{o,P}}{\delta P}\right)_T = \frac{0.4343C}{B + P}$$

where $\beta_{o,P}$ is the compressibility and $V_{o,P}$ the specific volume of water at pressure P (in bars) and C and B are positive constants. Gibson (1934, 1935) subsequently extended the Tait equation to solutions. Although empirical in its origin, (2.1) is now seen to relate closely to the structure of liquids and can be derived from the theory of molecular association to form clusters and from consideration of the unoccupied volumes or "holes" in the liquid (Ginell, 1961). In a recent reexamination of the equation of state of water and seawater, Li (1967), after an examination of both compressibility and sound velocity data, has concluded that the PVT relationships of pure water are well represented by the Tait equation in its integrated form,

$$(2.2) \qquad V_{o,P} = V_{o,1\ atm} - C \log \frac{B + P}{B + 1}$$

and of seawater by the Tait-Gibson equation,

$$(2.3) \qquad V_P = V_{1\ atm} - (1 - S \times 10^{-3})C \log \frac{B^* + P}{B^* + 1}$$

where S is the salinity and

$$(2.4) \qquad C = 0.315 V_{o,1\ atm}$$

$$(2.5) \qquad B = 2668.0 + 19.867T - 0.311T^2 + 1.778 \times 10^{-3}T^3$$

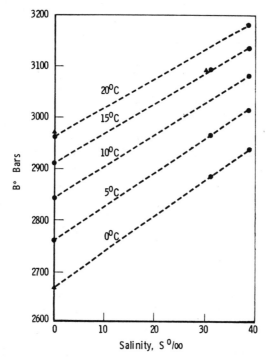

Figure 2.3   B* for Seawater as a function of Salinity. From Li (1967), based on data of V. W. Ekman, *Conseil Perm. Intern. Explor. Mer, Copenhagen*, **431** (1908) with permission of American Geophysical Union and the author.

where T is the temperature in centigrade degrees for the ranges $0 \le T \le 45°C$ and $1 \le P \le 1000$ bars, and

$$(2.6) \quad B^* = (2670.8 + 6.89656S) + (19.39 - 0.0703178S)T - 0.223T^2$$

for the ranges $0 \le T \le 20°C$, $1 \le P \le 1000$ bars, and $30 \le S \le 40‰$. Although perhaps more cumbersome in application than the familiar purely empirical expressions, the physical significance attachable to the Tait and related equations make their use preferred. Values of B and B^* are given in Table 2.3 and Figure 2.3. In the figure, values of B^* from sound velocity measurements are indicated by dashed lines.

Table 2.3 B of Pure Water in Units of Bars, $C/V_{o,1\ atm} = 0.3150$
[From Li, 1967 (based on data of E. N. Amagat, *Ann. Chim. Phys.*, **29**, 68,505 (1893), with permission of American Geophysical Union and the author]

P, atm	Temperature, °C					
	0	5	10	15	20	30
100	2656.2	2761.2	2820.2	2877.7	2911.9	2965.0
200	2672.1	2769.8	2840.9	2906.8	2949.3	3015.6
300	2670.0	2781.3	2839.1	2906.9	2960.0	3027.5
400	2669.1	2776.6	2837.7	2904.1	2958.2	3032.5
500	2666.3	2763.3	2836.3	2901.2	2956.3	3029.5
600	2668.5	2760.4	2835.9	2900.2	2955.4	3033.7
700	2668.2	2757.9	2837.7	2902.2	2954.5	3031.5
800	2667.7	2758.0	2837.2	2902.0	2951.4	3032.7
900	2669.2	2761.2	2836.3	2901.8	2954.3	3027.5
1000	2667.6	2762.2	2833.9	2905.1	2956.5	3028.3
Average	2668.1	2762.8	2836.4	2902.4	2955.2	3030.8
$\pm \sigma$	1.0	6.1	1.4	1.5	2.1	2.3
	40	50	60	70	80	90
100	3035.9	3035.1	2996.1	2947.1	2890.6	2879.6
200	3065.2	3078.5	3040.5	2982.0	2921.8	2893.9
300	3082.5	3088.6	3064.8	2998.6	2933.4	2890.0
400	3085.2	3099.1	3074.7	3021.8	2945.1	2884.2
500	3084.4	3095.2	3083.0	3033.3	2955.7	2885.3
600	3082.5	3092.6	3084.7	3038.2	2960.6	2892.4
700	3080.6	3095.0	3080.6	3043.3	2967.3	2900.0
800	3077.1	3093.8	3082.8	3044.1	2971.9	2908.8
900	3078.7	3093.5	3083.2	3044.0	2981.9	2914.6
1000	3081.1	3094.2	3080.0	3042.1	2988.2	2923.7
Average	3081.4	3094.8	3081.3	3038.1	2967.2	2901.3
$\pm \sigma$	2.6	1.8	2.5	7.5	13.8	14.2

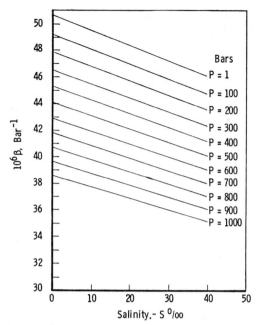

Figure 2.4 The Compressibility of Seawater at 0°C. From Wilson and Bradley (1966).

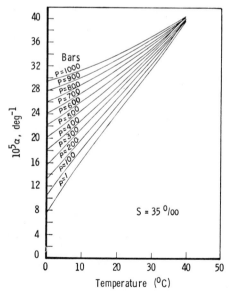

Figure 2.5 Coefficient of Thermal Expansion of Seawater as a Function of Temperature and Pressure. From Wilson and Bradley (1966).

Tables A.7, A.8, and A.9 in the Appendix present data of Wilson and Bradley (1966) which, I believe, represent the most careful and recent measurements of the specific volume of seawater over the ranges $0 \leq T \leq 40°$, $1 \leq P \leq 1000$ atm, and $10 \leq S \leq 40‰$. It might be worthwhile to pause a moment and look at this information in graphical form in order to get a qualitative sense of how properties such as the compressibility and thermal expansion of seawater vary with temperature, pressure, and salinity. The compressibility, $\beta = (1/V) \times (\Delta V/\Delta P)$, decreases with increasing pressure (Figure 2.4; also evident in the upward curvature in Figure I.4 in the Introduction) and increasing salinity. The coefficient of thermal expansion, $\alpha = (1/V)(\Delta V/\Delta T)$, increases with increasing temperature (Figures 2.5 and 2.6), pressure (Figure 2.5), and salinity (Figure 2.6).

The isothermal compressibility of pure water exhibits an anomalous minimum near 50°C (Kell and Whalley, 1965) which persists to high pressures (see Dorsey, 1940) and which may arise from a higher structural transition in liquid water. Equally fascinating is the familiar anomalous density maximum

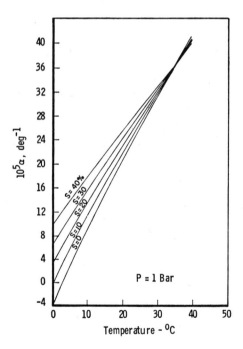

Figure 2.6 Coefficient of Thermal Expansion of Seawater as a Function of Temperature and Salinity. From Wilson and Bradley (1966).

near 4°C for pure water. Again this inflection presumably arises from some kind of structural transition in liquid water, and it represents the cardinal fact of life that any theory of the structure of liquid water (Chapter 1) must take into account. As the pressure is increased, the temperature of maximum density, T_{max}, decreases in a regular way, but more rapidly than does the depression of the freezing point. Similarly the addition of electrolyte also depresses T_{max} to an extent proportional to the concentration but, again, more pronounced than the freezing point depression. This T_{max} for seawater decreases with increasing salinity (Table 2.4 and Figure 2.7). The maximum density is given by $(1 + 10^{-4} \Delta_m)$, and the lowering of the maximum by $DT_m \equiv (3.947 - T_m)$.

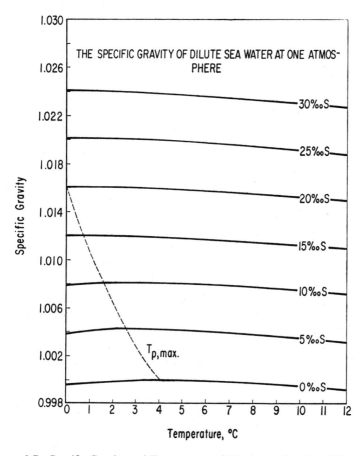

Figure 2.7 Specific Gravity and Temperature of Maximum Density of Seawater.

Table 2.4   Temperature of Maximum Density of Seawater[a]
(From Dorsey, 1940), with permission of Reinhold Book Corp.,
a subsidiary of Chapman-Reinhold, Inc.)

S	Δ_m	T_m	$D_m S$	S	Δ_m	T_m	D_m/X
0	0	3.947		20	160.7	−0.310	2.13
1	8.5	3.743	2.04	21	168.7	−0.529	2.13
2	16.9	3.546	2.00	22	176.7	−0.744	2.13
3	25.1	3.347	2.00	23	184.8	−0.964	2.14
4	33.3	3.133	2.04	24	192.9	−1.180	2.14
5	41.5	2.926	2.04	25	201.0	−1.398	2.14
6	49.6	2.713	2.06	26	209.1	−1.613	2.14
7	57.7	2.501	2.07	27	217.2	−1.831	2.14
8	65.8	2.292	2.07	28	225.3	−2.048	2.14
9	73.8	2.075	2.07	29	233.4	−2.262	2.14
10	81.8	1.860	2.09	30	241.5	−2.473	2.14
11	89.7	1.645	2.09	31	249.7	−2.687	2.14
12	97.6	1.426	2.10	32	257.8	−2.900	2.14
13	105.6	1.210	2.11	33	265.9	−3.109	2.14
14	113.5	0.994	2.11	34	274.0	−3.318	2.14
15	121.3	0.772	2.12	35	282.2	−3.524	2.14
16	129.2	0.562	2.12	36	290.4	−3.733	2.13
17	136.9	0.342	2.12	37	298.6	−3.936	2.13
18	144.8	0.124	2.12	38	306.8	−4.138	2.13
19	152.7	−0.090	2.13	39	315.0	−4.340	2.13
20	160.7	−0.310	2.13	40	323.2	−4.541	2.12
21	168.7	−0.529	2.13	41	331.4	−4.738	2.12

[a] limit of $S = 1$ g salt/kg seawater; of $\rho_m = 1$ g/cm^3; of $Dt_{m/S} = 0.1°C/g/kg$.

3   Thermodynamics of PVT Changes in Seawater

Fabuss, Korosi, and Hug (1966) have measured the densities of binary and ternary solutions of NaCl, Na$_2$SO$_4$, and MgSO$_4$ and of seawater and seawater concentrates at 1 atm over the temperature range 25–175°C, and their tables of data are reproduced in the Appendix (Table A.24).

By way of illustration, let us return now to a numerical application of the PVT relationships for water and the calculation of the changes in thermo-dynamic properties on changing the conditions of a seawater sample. Let us suppose that we collect a 1-liter sample at a depth of 3000 m and that the temperature is 5°C and the salinity of the water 35‰. The PVT properties and other thermodynamic properties depend *only* on the initial and final states of the sample, but for purposes of illustration we bring the sample to its surface condition (25°C and 1 atm) by two alternative paths:

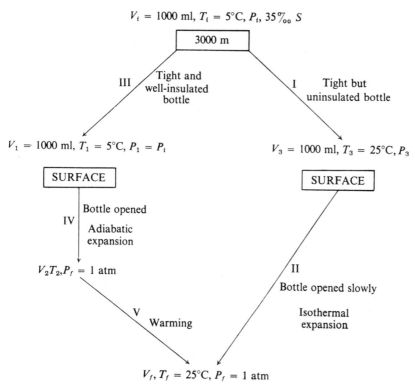

$V_i = 1000$ ml, $T_i = 5°C$, P_i, $35‰$ S

3000 m

III Tight and well-insulated bottle

I Tight but uninsulated bottle

$V_1 = 1000$ ml, $T_1 = 5°C$, $P_1 = P_i$

$V_3 = 1000$ ml, $T_3 = 25°C$, P_3

SURFACE

SURFACE

IV Bottle opened

Adiabatic expansion

$V_2 T_2, P_f = 1$ atm

II

Bottle opened slowly

Isothermal expansion

V Warming

$V_f, T_f = 25°C$, $P_f = 1$ atm

Salinity, it should be noted, is defined on a weight, rather than a volume, basis (as is, for example, the molal concentration, m) and hence is independent of temperature and pressure and remains constant. The first problem then is to calculate the initial pressure, P_i, corresponding to a depth of 3000 m in the seas. At a depth D the pressure on a unit area is given by

$$(2.7) \qquad P_i = g \int_{x=0}^{x=h} \rho \, dx$$

where the density ρ, because of the compressibility of seawater, is itself a function of depth, x. The situation is further complicated by the fact that the vertical water column is virtually never of either constant temperature or salinity (Figure I.3). Nevertheless a variety of tables and procedures are available for calculating density as a function of depth, salinity, and temperature (for example, Bjerknes and Sandström, 1910). Examination of the depth dependence of pressure thus evaluated (Figure 2.8) yields an interesting result: over the range 30–40‰ S and 0–20°C the effects of S and T are barely appreciable, for nearly all physical-chemical purposes the simple rule of

Geometric Depth, m

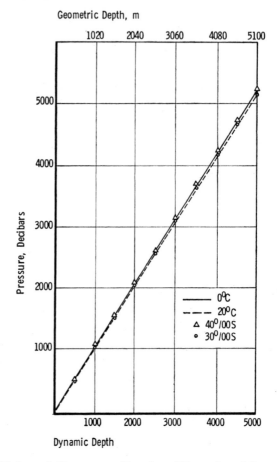

Figure 2.8 Hydrostatic Pressure as a Function of Dynamic and Geometric Depth in the Seas.

thumb that the pressure in decibars is numerically equal to the geometric depth in meters appears to be entirely adequate. Notice that Figure 2.8 gives the dynamic as well as the geometric depth. The dynamic depth is a useful concept in treating the gravity, mass, and pressure fields in the sea. The dynamic depth is given by $gh/10$, where g is the acceleration of gravity and h is the geometric depth. Numerically 1 dynamic meter corresponds to about 1.02 geometric meters, but the units of the dynamic depth are those of work. The value of P_i read from Figure 2.8 is 306 kg/cm².

Starting with step I, which is a nonadiabatic constant volume warming,

$$(2.8) \qquad V_i, T_i, P_i \rightarrow V_3 = V_i, T_3 = 25°C, P_3$$

P_3 could be evaluated by using the formulas and tables for compressibility and thermal expansion to calculate the pressure that must be applied in order to reduce the volume of the seawater sample to $V_3 = 1000$ ml at 25°C. But a much simpler procedure is to look up the specific volume of 35‰ S seawater at 306 kg/cm² and 5°C in Table A.7 and then the pressure corresponding to the same specific volume (0.9605 ml/g) at 25°C. The value for P_3 thus obtained, interpolating table values when necessary, is 423 kg/cm². Notice that, because of the smaller compressibility of seawater, the value is appreciably higher than the corresponding value for pure water: $P_3 = 396$ kg/cm².

Next we should like to calculate the changes in the thermodynamic quantities—ΔH, the change in heat content or enthalpy; ΔF, the change in free energy, and ΔS, the entropy change—accompanying step I (2.8). The changes in these quantities are related by the familiar equation

(2.9) $$\Delta F = \Delta H - T\Delta S$$

For step I

(2.10) $$\Delta H_I = m \int_{T_i}^{T_3} C_V \, dT + \int_{P_i}^{P_3} V \, dP$$

where m is the mass of the sample, but no adequate data are available for C_V, the heat capacity at constant volume

(2.11) $$C_V = \left(\frac{\delta E}{\delta T}\right)_V$$

where E is the energy content. Fortunately, however, values of the heat capacity of seawater at constant pressure are available:

(2.12) $$C_p = \left(\frac{\delta H}{\delta T}\right)_p$$

and C_V and C_p are related by the expression

(2.13) $$C_V = C_p - \frac{\alpha^2 VT}{\beta}$$

where α is the coefficient of thermal expansion and β is, as before, the compressibility. Using values of α from Table A.9 and values of β from Table A.8, we find that the difference between C_V and C_p is small. At 5°C and 306 kg/cm², $C_p - C_v$ is 0.0001 cal/g °C and, at 25°C and 423 kg/cm², 0.001 cal/g °C for seawater. The corresponding values for pure water are 0.00003 and 0.001 cal/g °C, respectively. The specific heat of seawater has been measured at 1 atm by Cox and Smith (1959), and more recently Bromley and colleagues (1967) have reported the heat capacity of seawater at 1 atm over a wide salinity and temperature range. The two sets of data are in agreement, and we

Table 2.5   Heat Capacity of Seawater at 1 atm (cal/g °C)

Temperature °C	S, ‰				
	0	10	20	30	40
0	1.005	0.989	0.974	0.959	0.946
10	1.002	0.987	0.973	0.960	0.947
20	1.000	0.986	0.973	0.961	0.949
30	0.999	0.986	0.974	0.962	0.950
40	0.998	0.986	0.974	0.963	0.951

shall use the latter (Table 2.5). Bromley and co-workers (1967) summarize their results in an empirical equation,

$$(2.14) \quad C_p = 1.0049 - 0.0016210S + (3.5261 \times 10^{-6}S^2)$$
$$- [(3.2506 - 0.14795S + 7.7765 \times 10^{-4}S^2) \times 10^{-4}T]$$
$$+ [(3.8013 - 0.12084S + 6.121 \times 10^{-4}S^2) \times 10^{-6}T^2]$$

where T is the temperature in centigrade degrees and where S, in both (2.14) and Table 2.5, has been converted from weight percent as given originally by the authors to salinity.

An important feature illustrated by Table 2.5 and first pointed out by Cox and Smith (1959) is that the sign of the temperature dependence of C_p reverses in going from pure water to seawater. These authors also point out that the relationship between the specific heat of pure water and structural changes is very inadequately understood and may provide an explanation for a number of presently rather mysterious observations; for example, the minimum in C_p versus T for pure water near 40°C and the tendency of C_p values to scatter below 5°C. There appear to be no reliable data for the pressure dependence of C_p either for pure water or for seawater. Sverdrup and co-workers (1942) give a table of differences between C_p at some pressure and C_p at 1 atm for seawater based on a computation of Ekman (1914) but, inasmuch as Ekman assumed that the temperature coefficient of C_p in seawater is the same as for pure water (see above), this table is suspect. Fofonoff (1959) gives the pressure dependence of C_p as

$$(2.15) \qquad \left(\frac{\delta C_p}{\delta P}\right)_{T,S} = -T\left(\frac{\delta^2 V_s}{\delta T^2}\right)_{P,S}$$

where V_s is the specific volume and the value of $(\delta^2 V/\delta T^2)$ at 0°C and 35‰ is $+13.2 \times 10^{-6}$, so that $(\delta C_p/\delta P) \approx 3 \times 10^{-3}$, a rather small correction, fortunately, which we shall neglect. I hope that this illustrates the need, even in this day and age when so much scientific spending is lavished on sometimes silly things, of fundamental thermodynamic data on pure water and aqueous electrolytic solutions including seawater, especially under high hydrostatic pressure. Not only does an immediate need for such data exist, but I think that such data will almost certainly yield an additional dividend in terms of increased understanding of the structure of liquid water and the effects of temperature, pressure, and electrolytes on that structure.

In our example, on substituting the values of $C_v \approx C_p$ from (2.14) or Table 2.5 into (2.10), a value of 22.95 kcal is obtained for ΔH for step I. The entropy change ΔS for step I is given by

$$(2.16) \qquad \Delta S_I = m \int_{T_i}^{T_3} C_v \, d(\ln T) - \int_{P_i}^{P_3} \left(\frac{\delta V}{\delta T}\right)_P dP$$

and its value in the present instance is 67.38 cal/°K.

Step II is an isothermal expansion, and ΔH is given by

$$(2.17) \qquad \Delta H_{II} = -m \int_{P_3}^{P_4} \mu C_p \, dP$$

where m is the mass (1037 g in our example from specific volumes given in Table A.7) and μ is the Joule-Thomson coefficient

$$(2.18) \qquad \mu = \left(\frac{\delta T}{\delta P}\right)_H$$

obtained from information given in Table A.30, and the value of H_{II} is +955 cal. ΔS for step II is given by

$$(2.19) \qquad \Delta S_{II} = - \int_{P_3}^{P_f} \left(\frac{\delta V}{\delta T}\right)_P dP$$

where $(\delta V/\delta T)_p$ is $V\alpha$ and has a value of +3.15 cal/°K, thus giving a ΔF_{II}, by (2.9), of +15 cal.

Let us now consider the second path. For step III, ΔH, ΔS, and ΔF are all zero. For step IV, which is an adiabatic expansion, ΔH and ΔS are given by

$$(2.20) \qquad \Delta H_{IV} = -m \int_{P_i}^{P_f} \mu C_p \, dP + m \int_{T_i}^{T_2} C_p \, dT = -86 \text{ cal}$$

and

$$(2.21) \qquad \Delta S_{IV} = -m \int_{T_i}^{T_2} C_v \, d \ln T - \int_{P_i}^{P_f} \left(\frac{\delta V}{\delta T}\right)_P dP = 1.12 \text{ cal/°K}$$

Notice that the first term on the right side of (2.21) is nearly zero since the temperature change (from Table A.30) is only about 0.3°C. The last step of this path is simply a warming up of the sample,

$$(2.22) \qquad \Delta H_v = m \int_{T_2}^{T_f} C_p \, dT = 19,850 \text{ cal}$$

Over the range of interest the temperature dependence of C_p is linear to within less than 0.3%; consequently, rather than C_p expressed as a function of T, simply the average value of C_p has been used in (2.22). The corresponding value of ΔS is

$$(2.23) \qquad \Delta S_v = m \int_{T_2}^{T_f} C_p d \ln T = 68.5 \text{ cal/}°\text{K}$$

Although we have neglected the small changes in the samplers' dimensions with changing temperatures and pressure, we have chosen our two paths because we take them to approximate more realistically actual conditions. They do not, however, represent the most convenient steps for calculating the change in thermodynamic properties. For the overall process we could have chosen any path thermodynamically convenient since, to repeat, the properties depend only on the initial and final states, for example, constant temperature decompression to 1 atm

$$(2.24) \qquad \Delta H_{VI} = -m \int_{P_i}^{P_f} \mu C_p \, dP = 211.0 \text{ cal}$$

$$(2.25) \qquad \Delta S_{VI} = - \int_{P_i}^{P_f} \left(\frac{\delta V}{\delta T} \right)_P dP = +1.12 \text{ cal/}°\text{K}$$

with a ΔF_{VI} of -101 cal, followed by warming at constant pressure (step V above).

The agreement in the total ΔS's for all three paths is good, 70.5, 69.6, and 69.6 cal/°K, but the agreement in ΔH is poor, 23,905, 19,764, and 20,061 cal, the discrepancies arising largely in the choice of values for the Joule-Thomson coefficient over different ranges.

It is also interesting to compare the above values for 35‰ S seawater with values calculated for distilled water from data given in Dorsey (1940): $\Delta H_I = 21,284$ cal, $\Delta S_I = 71.17$ cal/°K, $\Delta H_{II} = 940$ cal, $\Delta S_{II} = 17$ cal, $\Delta H_T = 22,224$ cal, and $\Delta S_T = 73.8$ cal/°K. The differences are quite appreciable.

In an adiabatic process no heat enters or leaves the system. Such processes are not confined to our example of removing a seawater sample from the ocean depths. Adiabatic processes play a very important role in physical oceanography; as a water mass sinks, in the absence of good circulation or thermal conduction, the compression results in an increase in temperature, while, on the other hand, the temperature of a upwelling water mass can decrease as a consequence of adiabatic decompression. The adiabatic temperature change for each centimeter of vertical displacement is given by

$$(2.26) \qquad \Delta T = 10^5 \frac{T\alpha}{JC_p} g\rho$$

where α is the coefficient of thermal expansion, ρ the density, g the acceleration of gravity, T the temperature in degrees Kelvin, and J the mechanical equivalent of heat. If a seawater sample is transported from some depth to the surface, the temperature which it would attain as a consequence of this adiabatic decompression is called the *potential temperature.* Table A.30 contains values of the adiabatic cooling from which potential temperatures can be computed.

4   Activities, Activity Coefficients, and the Debye-Hückel Theory

In the foregoing section we were concerned with the changes in the thermodynamic properties of seawater with temperature and pressure, and very little was said concerning the effect of the third major marine variable—salinity. The avoidance was deliberate, for it turns out that for more rigorous purposes the concentration, expressed as salinity for example, of an aqueous electrolytic solution, especially one as concentrated as seawater, is not very useful and the introduction of two additional concepts is necessary. The first of these, the partial molal quantity of a species, is a straightforward thermodynamic formalism, but the second, the activity of a species, is far from being straightforward.

As a solute is added to water, the physical-chemical properties of the water are altered. But the alteration is *not* directly proportional to the amount of solute added. This becomes particularly conspicuous at large concentrations of solute. In general the solute becomes increasingly less effective at producing the changes in solution properties. This phenomenon occurs for strong as well as weak electrolytes and cannot therefore be attributed to the partial dissociation of the solute. In order to describe aqueous electrolytic solutions such as seawater, what we really want is not the concentration but rather the *effective* concentration of a species, and this is just the need which the concept of activity is intended to fill.

As the concentration of electrolyte, such as NaCl, in an aqueous solution is increased, even in quite dilute solutions departures from a linear dependence on concentration soon became evident (Figure 2.1 and 2.2). Electrolyte species interact with solvent species (Figure 2.1) and with one another (Figure 2.2). The chemical efficacy, so to speak, of an ion in solution depends on electrolyte concentration in some complex way. For this reason the concentration of an ion itself does not represent a very precise way of describing the effective presence or influence of the ion, and it is necessary to introduce a new concept—the *activity* of an ion. The activity α_i of the solute ion i is defined by

$$(2.27) \qquad \bar{F}_1 \equiv \left(\frac{\delta F}{\delta n_1}\right)_{n_s, n_j, T, \rho} \equiv \bar{F}_i^\circ + RT \ln \alpha_i$$

where $\bar{F}_i$ is the partial molal Gibbs free energy or chemical potential, and the subscripts s and j refer to the solvent and all other ions.

Activities can be determined by any of a number of experimental methods including electromotive force, freezing point, and osmotic and vapor pressure measurements. In order to relate the activity of a species in solution to a readily establishable parameter such as the species' concentration a second concept, the *activity coefficient*, is defined. The *rational activity coefficient* γ_R, so called because it is dimensionless and always equal to unity in an ideal solution, is defined by

$$(2.28) \qquad \alpha_1 \equiv \gamma_{R,1} N_1$$

where N_1 is the mole fraction of the solute i. But the concentrations of solutions are more often expressed in molality (or molarity) rather than in mole fractions; hence a more useful definition is

$$(2.29) \qquad \alpha_i \equiv \gamma_i m_i$$

where γ_i is the *practical activity coefficient* and m_i the molal concentration of species i. An activity coefficient can also be defined in terms of the molar concentration, but it is not so readily relatable to other thermodynamic quantities.

Definition (2.27) represents an imaginary procedure, namely, changing the concentration of a given ion in solution in defiance of the principle of preservation of electrical neutrality. Consequently, from the standpoint of rigor, despite the continuing controversy in the literature, Guggenheim (1949) is correct in insisting that activities and activity coefficients of individual ions in solution are operationally meaningless. Only the mean activity of an electrolyte (including *both* cation and anion) can be measured experimentally; nevertheless we shall adopt the widespread practice of using individual ionic activities and their coefficients as if they were real. Consider the dissociation of a generalized electrolyte,

$$(2.30) \qquad C_n A_m \rightleftharpoons n C^{z_C} + m A^{z_A}$$

where z_C and z_A are the charges on the cation C and anion A, respectively. The *mean activity*, $\alpha_\pm$, is given by

$$(2.31) \qquad \alpha_\pm^{n+m} \equiv \alpha_C{}^n \alpha_A{}^m$$

and the *mean activity coefficient* by

$$(2.32) \qquad \gamma_\pm^{n+m} \equiv \gamma_C{}^n \gamma_A{}^m$$

and the *mean ionic molality* by

$$(2.33) \qquad m_\pm^{n+m} \equiv m_C{}^n m_A{}^m$$

The activity is defined with respect to a standard state $[(\bar{F}_i^\circ$ in (2.27)]. In the case of aqueous electrolytic solutions the standard state of the solute is chosen such that

$$(2.34) \qquad \lim_{m \to 0} \gamma_\pm = 1$$

for all temperatures and pressures. In its standard state $\alpha_\pm$ is equal to unity, but $\alpha_\pm$ may accidentally be numerically equal to unity, so a value of unity for $\alpha_\pm$ does not necessarily mean that the solute is in its standard state.

Table A.15 in the Appendix summarizes values that have been reported for the activities and activity coefficients of some seawater constituents. Figure 2.9, based on the extensive tables of activity coefficients given by Robinson and Stokes (1959), shows the concentration dependence of the mean ionic activity coefficients of a few electrolytes. The temperature dependence of the activity is given by

$$(2.35) \qquad \left(\frac{\delta \ln \alpha_i}{\delta T}\right)_{m,P} = -\frac{\bar{H}_i - \bar{H}_i^\circ}{RT^2}$$

where $\bar{H}_i$ is the partial molal heat content. For an aqueous NaCl solution the temperature dependence of γ is small: its value is 0.638 at 0°C, goes through a maximum of about 0.655 near 50°C, and at 100°C has fallen off to 0.622 (Robinson and Harned, 1941). The pressure dependence is given by

$$(2.36) \qquad \left(\frac{\delta \ln \alpha_i}{\delta P}\right)_{T,m} = \frac{\bar{V}_i}{RT}$$

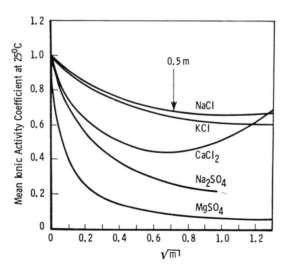

Figure 2.9 Concentration Dependence of Some Mean Ionic Activity Coefficients.

where $\bar{V}_i$ is the partial molal free volume of constituent i. Tables of partial molal volumes for ions and compressibilities can be found in Owen and Brinkley's (1941) paper on the effect of pressure on ionic equilibria in pure water and seawater, and in the more recent compilations of Zen (1957), Mukerjee (1961), and Duedall and Weyl (1967).

The solvent as well as the solute has an activity coefficient. The activity of pure water by choice of standard state is unity and is not drastically lowered by the presence of an electrolyte. Thus at 25°C the activity coefficient of water falls from 0.99665 to 0.98355 in going from 0.1 to 0.5 m NaCl.

For practical purposes the activity coefficient can be considered simply as a "fudge factor," that is, a number which when multiplied by the concentration of a species yields the species' activity, which in turn can be substituted in the appropriate expressions for the thermodynamic equilibria constant (Chapter 6). Nevertheless it would be satisfying to be able to calculate activity coefficients from theory. Such a theory exists—the Debye-Hückel (1923) theory—and has been extensively tested and found to work very well indeed for *dilute* solutions and is, as a consequence, universally accepted and used. Although seawater is not a dilute solution with respect to NaCl, the Debye-Hückel theory will serve as a starting point for any theoretical analysis of more concentrated solutions and is certainly important enough to warrant its main features being outlined here.

The Debye-Hückel theory postulates that, if an ion had no charge, its activity coefficient would be unity and its solutions would obey Henry's law,

$$(2.37) \qquad \alpha_i = CN_i$$

where C is a constant and N_i is the mole fraction. In other words, the deviation from ideality exhibited by real solutions is attributed to the mutual interaction of the electric charges carried by the ions. Each ion in solution is surrounded by an ionic atmosphere of opposite sign. The charge density at any point in this atmosphere is greatest near the ion and falls off with increasing distance from the ion, but the total charge of the atmosphere is exactly equal to but opposite in sign from that of the ion. Using a Boltzmann distribution and Poisson's more general form of Coulomb's inverse square law, Debye and Hückel showed that the electrical potential ψ_i at an ion due to its surrounding atmosphere is given by

$$(2.38) \qquad \psi_i = -\frac{z_i \bar{e}}{\epsilon} \frac{\kappa}{1 + \kappa \mathring{a}}$$

κ being given by

$$(2.39) \qquad \kappa^2 = \frac{4\pi \bar{e}^2 N}{1000 \epsilon kT} \sum c_j z_j^2$$

where N is here Avogadro's number (see Table A-1), $\bar{e}$ is the unit electronic charge, ϵ is the dielectric constant of the medium, assumed to be continuous, $\mathring{a}$ is the mean distance of closest approach of the ions in solution, and c_j and z_j are the molar concentrations and charges, respectively, of all ions in the solution. In dilute solutions the effect of the ionic atmosphere is the equivalent of its charge all being placed at a distance $1/\kappa$ from the central ion: $1/\kappa$ is thus a measure of the apparent radius of the ionic atmosphere. By consideration of the electrical work necessary to charge all the ions in solution at a given concentration, it can be shown that the activity coefficient is given by

$$(2.40) \qquad \ln \gamma_i = -\frac{N\bar{e}^2 \kappa z_i^2}{2\epsilon RT}$$

The ionic strength I can be defined in terms of either molality or molarity (and for dilute solutions the two concentration terms tend to become numerically equal) and in the latter case is

$$(2.41) \qquad I \equiv \tfrac{1}{2} \sum c_j z_j^2$$

Combining (2.39) and (2.40), converting to logarithms to the base 10, and substituting in the values of the universal constants yields

$$(2.42) \qquad \log \gamma_i = -\frac{1.824 \times 10^6}{(\epsilon T)^{3/2}} z_i^2 \sqrt{I} = -A z_i^2 \sqrt{I}$$

where A at a given temperature is a constant characteristic of the solvent, and for a single, strong, 1:1 electrolyte in water at 25°C this becomes simply

$$(2.43) \qquad \log \gamma_{\pm} = -0.509 \sqrt{I}$$

Because it is applicable only to solutions more dilute than $I = 0.01$ and is strictly obeyed only as infinite dilution is approached, equations of the form (2.42) or (2.43) are called the Debye-Hückel limiting law. In the derivation of (2.42) the term $\kappa\mathring{a}$ in the denominator of (2.38) was deleted since it is negligible compared to unity in very dilute solutions. If this term is retained, expressions are obtained which work up to about 0.1 m:

$$(2.44) \qquad \log \gamma_{\pm} = -\frac{A z_C z_A \sqrt{I}}{1 + \mathring{a}\kappa}$$

or

$$(2.45) \qquad \log \gamma_{\pm} = -\frac{A z_C z_A \sqrt{I}}{1 + \mathring{a}B\sqrt{I}}$$

where B is composed of universal constants, ϵ, and T. The value of B for water is 0.325×10^8 and 0.329×10^8 at 0° and 25°C, respectively. The distance of closest approach parameter, $\mathring{a}$, which may be taken to be a measure of the effective radius of the ions in solution, is 4.4×10^{-8} cm for NaCl and 3.4×10^{-8} cm for $MgSO_4$ (Harned and Owen, 1943). For still more concentrated solution a "salting-out" term can be added to (2.45) to yield the Hückel equation,

$$(2.46) \qquad \log \gamma = -\frac{A z_C z_A \sqrt{I}}{1 + \mathring{a} B \sqrt{I}} + C \sqrt{I}$$

which works for many electrolytes up to concentrations as high as 3 M, but the constant C must be determined experimentally.

In conclusion let me return to the topic of determining activity coefficients. The principal methods are:

Freezing point measurement.
Boiling point measurement.
Isopiestic comparison of vapor pressure.
Osmotic coefficient measurement.
Electromotive force measurement.
Solubility measurement.

Table A.15 summarizes the result of activity coefficient measurements on seawater. Freezing point data for seawater are given in Tables 13.3 and A.31, and boiling point data in Table A.32. Tables A.31 and A.33 give vapor pressure lowering data, Table 11.10 the vapor pressure of 35‰ S seawater as a function of temperature, and Tables A.31, A.33, and A.34 give osmotic pressures. Rush and Johnson (1966) have found that the osmotic coefficient of synthetic seawater is closely approximated by the expression for an NaCl solution at the same ionic strength if a correction is made for the divalent ions present in seawater. At 0°C the osmotic pressure of seawater, ϕ_{sw}, in atmospheres, is given by

$$(2.47) \qquad \phi_{sw,0°C} = -12.08 T_f$$

where T_f is the freezing point in centigrade degrees, and ϕ_{sw} at temperature T (in centigrade degrees) is given by

$$(2.48) \qquad \phi_{sw,T°C} = \phi_{sw,0°C} \left(\frac{273 + T}{273}\right)$$

where $\phi_{sw,0°C}$ can be calculated from (2.47) or read from Table A.31 (Cox, 1965). The osmotic pressure of seawater and other saline solutions such as body fluids are of fundamental concern to the biologist, since it is one of the principal factors in controlling water and solute transport through bio-membranes.

5 Heats of Solution, Dilution, and Mixing

In the preceding chapter we examined in some detail what happens *structurally* when NaCl is dissolved in water to make a "seawater" sample. Now let us turn our attention to what happens thermodynamically or *energetically* on electrolyte dissolution.

The energy of the formation of a substance can often be estimated by use of the so-called Born-Haber cycle, for example in the case of NaCl,

$$
\begin{array}{ccc}
\text{NaCl (s)} & \xleftarrow{\quad U_0 \quad} & \text{Na}^+\text{ (g)} + \text{Cl}^-\text{ (g)} \\
\Big\uparrow{\scriptstyle -Q} & & \Big\uparrow{\scriptstyle +I} \quad \Big\uparrow{\scriptstyle -E} \\
\text{Na (s)} + \tfrac{1}{2}\text{Cl}_2\text{ (g)} & \xrightarrow{\; +S+\frac{1}{2}D \;} & \text{Na (g)} + \text{Cl (g)}
\end{array}
$$

Standard State

where U_0 is the crystal lattice energy of NaCl, I the ionization potential of Na, E the electron affinity of Cl, S the sublimation energy of Na, D the dissociation energy of Cl_2, and Q the heat of formation of NaCl (s), and for NaCl these quantities have the values 180.4, 118.0, 90.7, 26.0, 57.8, and 90.2 kcal/mole, respectively. This cycle can be extended to include the dissolution of the substance in a suitable solvent:

$$
\begin{array}{ccc}
\text{NaCl (s)} & \xrightarrow{\; U_0 \;} & \text{Na}^+\text{ (g)} + \text{Cl}^-\text{ (g)} \\
{\scriptstyle \pm L}\searrow & & \swarrow{\scriptstyle -H_s} \\
& \text{Na}^+\text{ (solv)} + \text{Cl}^-\text{ (solv)} &
\end{array}
$$

where H_s is the solvation energy of the gaseous ions and L the observed heat of solution at infinite solution. Thermodynamic quantities for some ions that are seawater constituents are summarized in Table 2.6. The energy required to transfer an ion from the environment with a dielectric constant of ϵ_1 to water of dielectric constant ϵ_2 is given by the Born equation,

$$
(2.49) \qquad H_s = \frac{z^2}{2r}\left(\frac{1}{\epsilon_1} - \frac{1}{\epsilon_2}\right)
$$

where z is the charge and r the radius of the ion. This expression makes immediately obvious the great importance of the very high dielectric constant of water but, apart from this qualitative insight, its usefulness is very much limited by its complete failure to take water structural effects, in particular hydration, into consideration. Glueckauf (1964), for example, has calculated the heats and energies of ions in aqueous solution and for the alkali and alkaline earth cations; Al^{3+}, Sc^{3+}, and La^{3+}; and the halide anions; and he is

Table 2.6   Standard Free Energies, Heats of Formation, and Entropies of Ions That are Seawater Constituents (25°C)

Ion	$\Delta F°$ kcal/mole	$\Delta H°$ kcal/mole	$\Delta S°$ cal/deg mole
Na^+	− 62.59	− 57.5	14.0
K^+	− 67.43	− 60.3	24.2
Mg^{2+}	− 107.8	− 110.2	− 31.6
Ca^{2+}	− 132.7	− 129.5	− 11.4
Sr^{2+}	− 133.2	− 130.0	− 7.3
Cl^-	− 31.33	− 39.9	13.50
Br^-	− 24.58	− 28.7	19.7
SO_4^{2-}	− 176.1	− 216.3	4.4
CO_3^{2-}	− 126.4	− 160.5	− 13.0
OH^-	− 37.59	− 54.8	− 2.49
H^+	0	0	0
			(standard state)

able to obtain the observed decrease in ΔH and ΔS with increasing ionic radius, but in the case of ΔS the agreement becomes poorer for the polyvalent ions as the water-structuring effects become increasingly important. Equation (2.49) has been mended and patched repeatedly, but "at the present time, continuum theories of ionic hydration based on the Born equation seem to have reached an asymptotic limit of usefulness or applicability" (Conway, 1966).

Nor can the heat of solution of a simple electrolyte such as NaCl in water be determined by direct measurement, for the heat change on the addition of a Na^+ and a Cl^- ion to water depends not only on the temperature (and presumably pressure), but also on how many Na^+ or Cl^- (and any other ions) are already there. The experimental difficulty, however, is more readily circumvented than the theoretical ones. Although we cannot add an infinite amount of water to make a solution of given concentration infinitely dilute, we can dilute by incremental stages and, inasmuch as the heat content changes are additive, extrapolate a plot of $\Delta H_{c_1 \to c_2}$ versus $\sqrt{C}$ to $C = 0$ in order to determine the heat of dilution to infinite dilution, $\Delta H_{c_1 \to 0}$. If H is the heat content of the solution and H_s and H_e are the molar heat contents of pure solvent and solute, respectively, then the change of heat content on mixing to form the solution is

$$(2.50) \qquad \Delta H = H - (n_s H_s + n_e H_e)$$

where η_s and η_e are the moles of solvent and solute. The quantity $\Delta H/\eta_e$ is the *total* or *integral heat of solution*. Differentiation of (2.50) with respect to η_e

at constant T, P, and η_s gives the *partial* or *differential heat of solution* of the solute,

$$(2.51) \qquad \left[\frac{\delta(\Delta H)}{\delta n_e}\right]_{T,P,\eta_s} = \left(\frac{\delta H}{\delta n_e}\right)_{T,P,\eta_s} - H_e = \bar{H}_e - H_e$$

where $\bar{H}_e$ by definition is the *partial molal heat content* of the solute. The heats of dilution of aqueous NaCl solutions were determined by Gulbransen and Robinson (1934), and the experimental values they obtained for the intermediate heats of dilution in going from c_1 to c_2 are given in Table 2.7. Values of the integral heat of dilution, $\Delta H_{1 \to 0}$, and of relative partial molal heat contents of NaCl and H_2O derived from the data in Table 2.7 are tabulated in Table 2.8.

A crude estimate based on the data in these tables indicates that the heat changes involved on dilution in the sea as the result of the mixing of more saline with less saline seawater must be very small. For 1 kg of seawater at 10°C a change in salinity from 35‰ to 34‰ corresponds to a heat change of less than one-tenth of a calorie. Because of the small magnitude of the effect, in this instance it is probably permissible to approximate seawater properties with those of an aqueous NaCl solution of comparable concentration. In fact there is little else that can be done, for there are relatively few data on the heats of solution of 2:2 electrolytes such as $MgSO_4$, and the problem is complicated by the inclusion in the heat of solution of the heat of dissociation

Table 2.7   Heats of Dilution of Aqueous NaCl Solutions (in cal/mole) (From Gulbransen and Robinson, with permission of American Chemical Society)

c_1	c_2	25°	20°	15°	10°
0.816 m	0.02025 m	+39.6	+73.7		
0.816	0.0425	+54.0	+86.8		
0.404	0.01025	−30.6	−6.45	−25.5	+60.8
0.404	0.02025	−18.1	+6.2	−34.4	+68.3
0.2	0.00512	−57.1	−41.8	−20.7	−1.9
0.2	0.0101	−48.8	−32.7	−14.2	+4.9
0.1	0.00257	−65.1	−50.3	−40.3	−26.3
0.1	0.00507	−55.9	−43.5	−32.7	−22.1
0.05	0.001285	−56.2	−49.2	−39.1	−28.9
0.05	0.002535	−51.1	−43.9	−34.7	−24.6
0.025	0.000642	−46.4	−45.9	−38.0	30.6
0.025	0.00128	−41.1	−39.9	−33.1	−28.1
0.0125	0.000322	−34.6	−32.2	−32.5	−24.7
0.0125	0.000634	−30.8	−30.4	−29.0	23.3
0.00625	0.000161	−27.6	−25.3	−23.7	
0.00625	0.000318	−24.6	−20.2	−20.2	

Table 2.8 Integral Heats of Dilution and Relative Partial Molal Heat Capacities of Aqueous NaCl Solutions (From Gulbransen and Robinson, 1934, with permission of American Chemical Society)

M	Integral Heat of Dilution, cal/mole, NaCl				Partial Molal Heat Content, cal/mole, NaCl				Partial Molal Heat Content, cal/mole, H_2O			
	25°	20°	15°	10°	25°	20°	15°	10°	25°	20°	15°	10°
0.0001	-4.2	-3.7	-3.4	-2.3	6.3	5.3	4.7	3.0	-0.0000038	-0.0000036	-0.0000024	-0.0000013
0.0005	-9.4	-8.2	-7.7	-4.3	14.8	13.1	12.0	8.3	-0.000049	-0.000044	-0.000039	-0.000036
0.001	-13.2	-11.6	-10.8	-7.5	20.1	17.9	16.5	11.5	-0.00012	-0.00011	-0.00010	-0.000072
0.003	-29.6	-26.0	-24.3	-16.9	42.2	38.1	35.8	25.6	-0.0011	-0.0011	-0.0010	-0.00078
0.01	-39.3	-36.2	-33.3	-23.9	55.8	51.0	46.0	34.8	-0.0030	-0.0027	-0.0029	-0.0020
0.05	-71.3	-63.5	-54.2	-40.6	93.7	76.3	65.0	43.2	-0.020	-0.012	-0.0098	-0.0028
0.1	-86.5	-70.0	-58.3	-38.6	99.6	76.6	57.7	30.2	-0.024	0.051	0.00011	0.015
0.2	-87.6	-68.3	-46.3	-18.3	85.0	54.2	9.4	-20.0	0.0094	0.44	0.13	0.14
0.404	-69.4	-42.3	-6.6	27.2	30.0	-19.4	-89.9	-130.0	0.28	2.11	0.70	0.65
0.816	-11.7	+23.5	—	—	-123.3	-169.7	—	—	1.19	—	—	—

of these weak electrolytes. A little work has been done on the mixing of electrolytic solutions but only for relatively simple systems, not for such a complex asymmetrical system as Na^+-Mg^{2+}-Cl^--SO_4^{2-}. Wood and Smith (1965) have shown that the cross-square rule, which states that for mixing at constant ionic strength the sum of the heats represented by the sides equals the sum of the heats represented by the diagonals for the 1:1 electrolytes

MX, MY, NX, and NY, holds for the Cl^-, Br^-, and NO_3^- of Li^+, Na^+, and K^+ at 25°C over the 0.1–0.5 m concentration range. They describe this result as being in good agreement with Friedman's (1960) theoretical treatment of mixed electrolyte solutions. This work has also been extended to mixtures of the alkaline earth halides (Wood and Anderson, 1966). As a consequence of our government's interest in the desalination of seawater, however, efforts to determine the fundamental thermodynamic properties of seawater and related solutions are now in progress, and such badly needed information can be expected to be forthcoming in the not too distant future. No adequate studies have ever been made on the effect of pressure on these properties; in view of the strong water-structure altering capabilities of pressure, the results of such an investigation could be most interesting. At present the theory of the thermodynamics of aqueous solution is restricted to very dilute solutions ($\sqrt{c} < 0.1$), and in the region of great dilution properties such as the heat of solution appear to obey the familiar $\sqrt{c}$ limiting dependence (Harned and Owen, 1958). Hopefully, with the influx of new information, more sophisticated theories will evolve that will take into consideration solute–solute and solute-solvent interactions and the structure of water. In the case of the dissolution of gases in water (see below), inasmuch as the solute is uncharged, there are no dielectric effects to mask the structural contribution. The thermodynamics of these processes have yielded valuable insight into changes in the local water structure.

6   Dissolution of Gases in Aqueous Solution

We have seen that, when a charged species is added to water, the energetics of the process often can be accounted for in terms of transfer of the charge from a medium of one dielectric constant to another and that, at least in dilute solutions, water structural effects may be hardly evident. On the other hand, if an uncharged species is added to water, we might expect the structural changes to dominate the thermo-dynamics of the process. But we must remember that the hydrophobic hydration around such a species in all probability represents a local structuring of water quite different from the coulombic hydration around a charged species. As our uncharged species we shall focus our attention on the noble gases, thereby avoiding the added complication of any chemical reactions, such as in the case of the dissolution of CO_2. Our remarks with regard to the noble gases will also be relevant to the more immediate problem of the dissolution of inert atmospheric constituents such as O_2 and N_2 in seawater. The thermodynamics of the dissolution of the noble gases in pure water and aqueous electrolytic solution is quite anomalous; in particular it is characterized by pronounced negative enthalpy and entropy changes. This is in contrast to the dissolution of these gases in "normal" liquids, where the subsequent loosening of solvent intermolecular forces results in an increase of the partial molal entropy. These phenomena have been interpreted in terms of the effect of these species on the structure of liquid water in their immediate vicinity.

Table 2.9   Heats and Entropies of Vaporization of Gases from Their Aqueous Solution
(From Frank and Evans, 1945, with permission of American Institute of Physics)

Solute	Temperature	ΔH	$\Delta S°$	Solute	Temperature	ΔH	$\Delta S°$
H_2	25°	1280	26.0	C_2H_2	25°	3360	25.6
	40°	426	23.7	He	25°	840	26.5
	80°	−170	21.8		50°	710	26.1
N_2	25°	2140	29.8		80°	550	25.5
	40°	1990	29.3	Ne	25°	1880	28.8
	80°	180	23.9		50°	1280	26.6
CO	25°	3910	29.8		80°	580	24.7
	40°	1970	28.5	A	25°	2730	30.2
	80°	10	22.6		50°	1840	27.2
O_2	25°	2990	31.3		80°	710	23.5
	40°	2070	30.1	Kr	25°	3550	32.3
NO	25°	2680	29.4		50°	2350	27.8
	40°	2270	28.1		80°	740	23.2
CO_2	25°	4730	30.6	Xe	25°	4490	33.6
	40°	3870	28.2		50°	2680	28.0
COS	25°	5800	35.1		80°	720	22.1
N_2O	25°	4840	31.6	Rn	25°	5050	34.3
CH_4	25°	3180	31.8		50°	3100	28.2
C_2H_6	25°	4430	35.4		80°	720	21.1
C_2H_4	25°	3790	31.3				

In order to account for the observed negative enthalpy of solution of the noble gases in water, Eley (1939) considered a two-step dissolution process: the creation of a "hole" or cavity in the solvent; then the introduction of a gas molecule into the cavity. But, although this idea could account for the sign of ΔH_s, Lange and Watzel (1938) had earlier pointed out that the value of $|\Delta H_s|$ was so large that the formation of definite chemical bonds between solvent and solute appeared to be indicated.

Table 2.9 lists heats and entropies, not of dissolution but of vaporization of gases from their aqueous solution as calculated by Frank and Evans (1945). When the molal entropies of vaporization, ΔS_v, for gases from water are compared with the same thermodynamic quantity for other "normal" solvents (Figure 2.10), the values are not only very much higher but also in

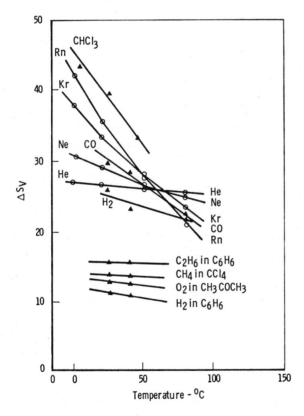

Figure 2.10   Comparison of the Molal Entropies of Vaporization of Gases from Water with other Solvents at 25°C. From Frank and Evans (1945), with permission of American Institute of Physics.

general they exhibit a much stronger temperature dependence. Frank and Evans (1945) concluded that "when a rare gas atom or a non-polar molecule dissolves in water . . . it modifies the water structure in the direction of greater 'crystallinity'—the water, so to speak, builds a microscopic iceberg around it," and they were careful to add that they did not wish to imply that the structure of the iceberg is specifically that of Ice-I_h. More recently Ben-Naim (1965) has modified the Frank-Evans model somewhat: he has proposed that new water structure does not build up around the solute molecules, but rather that the structure is already there and the introduction of the solute causes it to shift locally.

As for the nature of this local structure, Claussen and Polglase (1952) have compared it to certain crystalline hydrates, while Pauling (1961) prefers to describe it as some sort of clathrate (see Chapter 1, especially Figure 1.13). In any event, I believe that it represents another example of the hydrophobic hydration discussed in Chapter 1. The thermodynamics of the tetraalkyl-ammonium ions in water bears many resemblances to that of inert gases. As the size of R increases in R_4N^+, the charge is in effect diluted so that this series of species covers the whole spectrum from an ionic to an essentially nonpolar solute, from coulombic to hydrophobic hydration (Figure 3.6).

The most detailed study of the effect of the presence of electrolytes on the dissolution of an inert* gas in water is a recent investigation of argon solubility (Ben-Naim and Egel-Thal, 1965). As the electrolyte concentration increases, the standard free energy $\Delta F°$ for the transfer of the gas from pure water to the solution increases yet drops off more steeply with increasing temperature. In general the presence of the electrolyte interferes with the gas dissolution processes and the organization of the solvent about the gas solute molecules, thereby making the nonpolar solutes less soluble—"salting out" (Long and McDevitt, 1952). Green and Carritt (1967) have examined the "salting out" of O_2 in seawater. They give the excess partial molar thermodynamic quantities for the solution of O_2 in seawater listed in Table 2.10 and report that the ratio of the solubility of O_2 in seawater to pure water, β, is best represented by an exponential relation (Figure 2.11). The "salting out" decreases with increasing temperature. The authors conclude that the solute gas produced "a more highly structured order of the water molecules in its vicinity . . . [and that] the addition of sea salt results in disorganization of this induced structure." The coulombic hydration envelope of the electrolyte is sufficiently successful in competing for water molecules to disrupt the hydrophobic hydration envelope of the nonpolar solute. The latter tends to mask

---

* In view of the interaction of these gases with water discussed above, it should be clear that the term "inert" has lost much of its stricter meaning.

Table 2.10   Excess Partial Molar Thermodynamic Quantities[a] for the
Solution of Oxygen in Seawater
(From Green and Carritt, 1967, with permission of AAA.S and the
authors)

Temperature, °C	Cal (mole O_2)$^{-1}$ (unit Cl)$^{-1}$		Cal deg^{-1} (mole O_2)$^{-1}$ (unit Cl)$^{-1}$	
	$\Delta \widetilde{F}_{O_2}^{XS}$	$\Delta \widetilde{H}_{O_2}^{XS}$	$\Delta \widetilde{S}_{O_2}^{XS}$	$\Delta \widetilde{C}_{P O_2}^{XS}$
0	6.92	24.3	0.064	(−0.7)
10	6.43	17.3	0.038	(−0.7)
20	6.14	10.0	0.013	(−0.7)
30	6.12	2.5	−0.012	(−0.8)

[a] These excess values represent the change of the total thermodynamic quantity
resulting from the transfer of one mole of oxygen from a hypothetical solution
in which there is one mole of oxygen per litre to a hypothetical molar solution
of seawater of unit chlorinity. Unit Cl: unit chlorinity, sea salt.

the hydrophobic character of the dissolved gas molecules and, as it is diminished, this hydrophobic character becomes more evident to the solvent and the
gas is accordingly excluded with greater vigor from the liquid phase; that is,
its solubility decreases and it is "salted out."

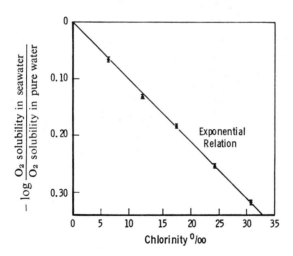

Figure 2.11   Variation of the Oxygen Solubility Ratio with Chlorinity. From Green and
Carritt (1967), with permission of AAAS and the authors.

7 Summary

With respect to processes depending largely on the solvent it is often possible to approximate the properties of seawater by an 0.5 M NaCl solution or even, more rarely, by pure water, but with respect to processes depending largely on the solute this is hardly ever the case. Empirical expressions are available for calculating the PVT and thermodynamic properties of seawater, but exact theoretical expression and in many instances the fundamental data are lacking. The properties of a solution are not directly dependent on the concentration of electrolyte, hence the concept of the activity of the solute is introduced, and for dilute solutions this quantity can be estimated from the Debye-Hückel theory. For dilute solutions the heat of solution can be estimated from the energy involved in the transfer of an ion from one dielectric constant medium to another (the Born equation), but not for more concentrated, mixed electrolyte solutions. Fortunately the heats of dilution for processes likely to occur in the seas are very small. In the case of the dissolution of inert gases, such as O_2, in seawater, in the absence of a charge on the solute, structural effects dominate the thermodynamics of the process.

REFERENCES

Ben-Naim, A., *J. Phys. Chem.*, **69**, 3240 (1965).
Ben-Naim, A., and M. Egel-Thal, *J. Phys. Chem.*, **69**, 3250 (1965).
Bjerknes, V., and J. W. Sandström, "Dynamic Meteorology and Hydrography," *Carnegie Inst. Washington, Publ.*, No. 88, Part I (1910).
Bromley, L. A., V. A. Desaussure, J. C., Clipp, and J. S. Wright, *J. Chem. Eng. Data*, **12**, 202 (1967).
Claussen, W. F., and M. F. Polglase, *J. Am. Chem. Soc.*, **74**, 4817 (1952).
Conway, B. G., *Ann. Rev. Phys. Chem.*, **17**, 481 (1966).
Craig, H., *Proc. Natl. Acad. Sci. (U.S.)*, **46**, 1221 (1960).
Crease, J., *Deep-Sea Res.*, **9**, 209 (1962).
Cox, R. A., in J. P. Riley and G. Skirrow, eds., *Chemical Oceanography*, Academic Press, London, 1965, Vol. I, Chap. 3.
Cox, R. A., and N. D. Smith, *Proc. Roy. Soc. (London)*, **252**, 21 (1959).
Debye, P., and E. Hückel, *Physik. Z.*, **24**, 185 (1923).
Dorsey, N. E., *Properties of Ordinary Water-Substance*, Reinhold Book Corp., New York, 1940.
Duedall, I. W., and P. K. Weyl, *Limnol. Oceanog.*, **12**, 52 (1967).
Eckart, C., *Am. J. Sci.*, **251**, 225 (1958).
Ekman, V. W., *Ann. Hydrog. Berlin*, **42**, No. 6, 340 (1914).
Ekman, V. W., *Publ. Circ. Conseil Exploration Mer.*, No. 43, 1 (1908).
Eley, D. D., *Trans. Faraday Soc.*, **35**, 1281, 1421 (1939).
Fabuss, B. M., A. Korosi, and A. K. M. S. Hug, *J. Chem. Eng. Data*, **11**, 325 (1966).
Fofonoff, N. P., "Interpretation of Oceanographic Measurements—Thermodynamics," in *Physical and Chemical Properties of Sea Water*, Natl. Acad. Sci.—Natl. Res. Council Publ., No. 600, Washington, D.C., 1959.

Fofonoff, N. P., "Physical Properties of Sea-Water," in M. H. Hill, ed., *The Sea*, Interscience New York, 1962, Vol. I.

Frank, H. S., and M. W. Evans, *J. Chem. Phys.*, **13**, 507 (1945).

Friedman, H. L., *J. Chem. Phys.*, **22**, 1134 (1960).

Gibson, R. E., *J. Am. Chem. Soc.*, **56**, 4 (1934); **57**, 284 (1935).

Ginell, R., *J. Chem. Phys.*, **34**, 1249 (1961).

Glueckauf, E., *Trans. Faraday Soc.*, **60**, 572 (1964).

Green, E. J., and D. E. Carritt, *Science*, **157**, 191 (1967).

Guggenheim, E. A., *Thermodynamics*, North-Holland Pub. Co., Amsterdam, 1949.

Gulbransen, E. A., and A. L. Robinson, *J. Am. Chem. Soc.*, **56**, 2637 (1934).

Harned, H. S., and B. B. Owen, *The Physical Chemistry of Electrolytic Solutions*, 3rd ed., Reinhold, New York, 1958.

Kell, G. S., and E. Whalley, *Phil. Trans. Roy. Soc. (London)*, **A258**, 565 (1965).

Knudsen, M., *Hydrographical Tables*, G. E. C. Gad, Copenhagen, 1901.

Lange, E., and R. Watzel, *Z. Physik. Chem.*, **A182**, 1 (1938).

Li, Y.-H., *J. Geophys. Res.*, **72**, 2665 (1967).

Long, F. A., and W. F. McDevitt, *Chem. Rev.*, **51**, 119 (1952).

Montgomery, R. B., in D. E. Gray, ed., *American Institute of Physics Handbook*, McGraw-Hill (Book Co.), New York, 1957.

Mukerjee, P., *J. Phys. Chem.*, **65**, 740 (1961).

Newton, M. S., and G. C. Kennedy, *J. Marine Res. (Sears Found. Marine Res.)*, **23**, 88 (1965).

Owen, B. B., and S. R. Brinkley, Jr., *Chem. Rev.*, **29**, 461 (1941); see also B. B. Owen, *Natl. Bur. Std. Cir.*, No. 524, 193 (1953).

Pauling, L., *Science*, **134**, 15 (1961).

Robinson, R. A., and H. S. Harned, *Chem. Rev.*, **28**, 419 (1941).

Robinson, R. A., and R. H. Stokes, *Electrolyte Solutions*, 2nd ed., Butterworths Sci. Pub., London, 1959.

Rush, R. M., and J. S. Johnson, *J. Chem. Eng. Data*, **11**, 590 (1966).

Sverdrup, H. U., M. W. Johnson, and R. H. Fleming, *The Oceans*, Prentice-Hall, Englewood Cliffs, N.J., 1942.

Tait, P. G., *Rept. Sci. Results Voyage H.M.S. Challenger*, **2**, Pt. IV, 1 (1888).

Wilson, W., and D. Bradley, "Specific Volume, Thermal Expansion, and Isothermal Compressibility of Sea Water," U.S.N. Ord. Lab. Rept. No. NOLTR-66-103 (June 2, 1966) (Unclass.), AD-635-120.

Wood, R. H., and H. L. Anderson, *J. Phys. Chem.*, **70**, 992 (1966).

Wood, R. H., and R. W. Smith, *J. Phys. Chem.*, **69**, 2974 (1965).

Zen, E. A., *Geochim. Cosmochim Acta.*, **12**, 103 (1957).

3   Transport Processes in Solution

1   Introduction

A water molecule or a solute ion in the sea may move in space in two different ways. It may move in the company of its neighbors in a macroscopic water mass as the result of ocean currents, eddies, tides, and convection, or it may move microscopically and individually relative to its neighbors by diffusion. The Gulf Stream current off Chesapeake Bay at times transports water at speeds as great as 300 cm/sec, but the self-diffusion coefficient of pure water at 25°C is only 2.57×10^{-5} cm^2/sec (Wang, 1965). Clearly, then, mass transfer by the former type of process plays a far more important role in mixing in the oceans than does the latter. However, this is by no means to say that diffusional processes play an insignificant role in chemical oceanography. On the contrary. Chemical species encounter each other and react in the region defined by the boundaries of a moving water mass, but within the mass as well species collide and interact with their neighbors; thus the rate of some chemical reactions can still be diffusion-controlled in the oceans. Furthermore many important chemical reactions occur in "quiet zones" in the immediate neighborhood of the surface (see Chapter 11) or in the interstitial water in marine sediments (see Chapter 12) where diffusional processes are rate-determining. Finally, investigations of the various movements of solvent and solute species have proved to be most valuable in increasing our understanding of the structure of electrolytic solutions such as seawater.

2   Dielectric Relaxation

The water molecule as a whole can exercise two types of movement, translational and rotational. The molecule is distinguished, as we have seen, by a strong dipole moment; consequently, if an electric field is applied, the water molecules tend to rotate and align themselves with the imposed field (Figures 3.1A and 3.1B). The relaxation time τ for the rotational process can be determined from ac measurements of the frequency dependence of the complex dielectric constant, and the results (Table 3.1) are very revealing (Collie, Hasted, and Ritson, 1948). At a given temperature, only one relaxa-

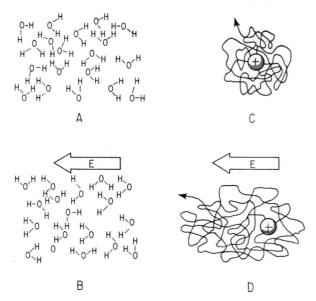

Figure 3.1   The Effect of an Imposed Electric Field on Solvent and Solute Movements.

tion time is observed—an indication that only one orientable molecular species is involved. The results obey exactly the theoretical relation

$$(3.1) \qquad \tau = \frac{4\pi\eta r^3}{kT}$$

where η is the viscosity and r the radius of the orientable species. The value of r calculated from this expression is not only nearly constant, independent of temperature, but also is in good agreement with the 1.38 Å radius for the water molecule from X-ray measurements, again indicating that the sole orientable species in liquid water is the monomer and not any polymeric forms,

Table 3.1   Dielectric Relaxation Time of Water

Temperature, °C	Relaxation Time, τ, sec	Radius, Å	τ_{D_2O}/τ_{H_2O}	η_{D_2O}/η_{H_2O}
0	17.7×10^{-12}	1.44	—	—
10	12.6	1.44	1.30	1.29
20	9.5	1.45	1.27	1.25
30	7.4	1.45	1.24	1.22
40	5.9	1.46	1.21	1.19
50	4.8	1.46	—	—
60	4.0	1.46	—	—

$(H_2O)_n$. A repetition of the relaxation measurements in heavy water showed that the ratios of relaxation times, τ_{D_2O}/τ_{H_2O} was very nearly the same as the ratios of viscosities η_{D_2O}/η_{H_2O} (Table 3.1). On the basis of this finding Collie, Hasted, and Ritson (1948) suggest that dielectric relaxation and viscous flow in water have the same mechanism.

The dielectric relaxation results strongly support the presence of an appreciable quantity of "free," rotable, monomeric molecules in pure liquid water. In the presence of an electrolyte the coulombic field of the ion should considerably perturb the ability of the water molecules in their immediate, and perhaps even their more distant, neighborhoods to exercise such rotational movements. As electrolyte is added to water, both the dielectric constant and the dielectric relaxation time decrease in an approximately linear fashion.

$$(3.2) \qquad \epsilon = \epsilon_0 + \frac{c}{2} \sum n_i \delta_i$$

where ϵ_0 is the static dielectric constant of pure water (79.30 at 25°C), c is the molar concentration of the electrolyte, and $\eta_i \delta_i$ is the product of the number of a given ion formed (i.e., 1, 2, 3, etc.) on ionization, and δ_i a constant characteristic of the ion. The equation continues to be valid even in solutions far more concentrated (2 M for NaCl) than seawater. The effect is approximately additive and, by assigning the values $\delta_{Na^+} = -8$ and $\delta_{Cl^-} = -3$, it is possible to arrive at the values of other individual ions listed in Table 3.2. Hasted, Ritson, and Collie (1948) have estimated cationic hydration numbers on the basis of the dielectric constant depressions, obtaining values of 4 and 14 for Na^+ and Mg^{2+}, respectively. But, inasmuch as the water molecules in the innermost hydration sheath are, they claim, only *partially* prevented from participating in the dielectric process, these hydration numbers represent only minimum values.

The relaxation time, expressed in terms of the relaxation wavelength, also exhibits an electrolyte dependence analogous to (3.2) and correlates nicely

Table 3.2   Ionic Dielectric Constant Depression Coefficient

Ion	δ_i
Li^+	-11
Na^+	-8
K^+	-8
Mg^{2+}	-24
Cl^-	-3
I^-	-7
SO_4^{2-}	-7

with the effects of the electrolytes on the viscosity (Hasted, Ritson, and Collie, 1948). That is, the dielectric properties of aqueous electrolytic solutions can be accounted for, at least qualitatively, in terms of the water structure-making or -breaking properties of the ions.

Up to a distance of 2 Å from a point charge there is complete dielectric saturation, then the dielectric constant increases rapidly so that, by the time a distance of only about 4 Å is reached, it has attained its ordinary bulk solution value. Only the first hydration layer lies in this region of appreciable dielectric saturation; hence in the case of monovalent ions the observed lowering of the bulk dielectric constant must arise for the most part from the primary hydration shell. Schellman (1957) has estimated that even at a distance as close as 2 Å the dielectric constant is lowered by only about 17%; thus we would appear to be quite safe in using the bulk dielectric constant in the overwhelming majority of our ion-ion interaction calculations.

Nuclear magnetic resonance (NMR) in recent years has proved to be another powerful technique for investigating the rotational movements of water molecules in solution, "more reliable with regard to structure forming and breaking in electrolyte solutions than ... the dielectric method" (Wicke, 1966) but leading to the same qualitative conclusions, namely, that anions and K^+ and Cs^+ tend to increase the fluidity of water and facilitate rotation whereas Na^+ and especially ions like Mg^{2+} and Ca^{2+} hinder the rotation of water molecules.

3   Diffusion

The water molecules and dissolved solute species in seawater are in a state of ceaseless translational motion. In a microscopically homogenous element of seawater this fretful agitation takes the form of a "random walk" (Figure 3.1C) but, if a thermal or concentration gradient is present (the earth's magnetic and gravitational fields are insignificant) or if the species is an ion, an electric field is imposed, and *on the whole* the walk ceases to be completely random and becomes prejudiced in some favored direction (Figure 3.1D).

The diffusion coefficient D mentioned in the opening paragraph of this chapter is defined by Fick's first law of diffusion

$$(3.3) \qquad\qquad J = -D\frac{\delta c}{\delta x}$$

where the flux of matter, J, is the amount of a particular substance crossing a unit area perpendicular to the direction of flow in unit time expressed in moles/cm^2/sec^{-1}, and $\delta c/\delta x$ is the concentration gradient in the direction of flow in moles/cm^3/cm. Equation (3.3) is readily extended to three dimensions.

Sverdrup, Johnson, and Fleming (1942) point out that the value of D for water is nearly constant within the temperature and salinity ranges of oceanographic interest.

Using isotopic tracer techniques, Wang (1951, 1965) has measured the self-diffusion of water and obtained values of 1.43, 1.68, 1.97, and 2.57 × 10^{-5} cm²/sec for D at 5°, 10°, 15°, and 25°C, respectively. He interprets his findings to support a quasicrystalline structure for liquid water, although the diffusion itself appears to involve only the movement of individual water molecules despite the tetrahedrally H-bonded structure. Diffusion and the other transport processes presently of concern to us can be treated as rate processes, and energies of activation, E_a, can be calculated by the Arrhenius equation

$$\text{Rate} = Ce^{-E_a/RT} \tag{3.4}$$

where C is a constant, from slopes of plots of log D versus $1/T$ (Figure 3.2) (Glasstone, Laidler, and Eyring, 1941), where T is the temperature in degrees Kelvin. E_a for the self-diffusion of water evaluated in this manner averages 4.5 kcal/mole over the 10–50°C range, but below 5°C it increases rapidly as the temperature decreases, reaching a value of 6.4 kcal/mole at 0°C

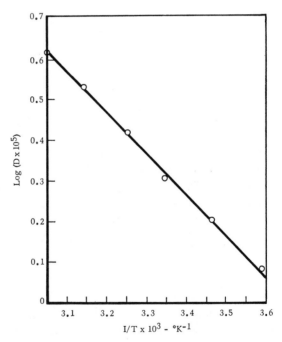

Figure 3.2 Temperature Dependence of the Self-Diffusion Coefficient of Water. From data of Wang (1951).

(Glasstone, Laidler, and Eyring, 1941). If dielectric relaxation and self-diffusion in water involve the same mechanism, then

$$(3.5) \qquad\qquad D = \frac{d^2}{\tau}$$

where τ is the relaxation time and d is the average distance between two successive equilibrium positions of a diffusing molecule. Using the Collie, Hasted, and Ritson (1948) values of τ and his own values for D, Wang (1965) has obtained a mean water jump distance of 3.7 Å and has further found that the distance is independent of temperature over the range 5–25°C, just as one would expect from the small specific volume change.

The effect of electrolytes on the self-diffusion of water (Devell, 1962; McCall and Douglass, 1965) is just what one might expect—those electrolytes which make water less fluid lower D, those which make it more fluid raise D—and the concentration dependence of the diffusion coefficient relative to that of pure water, D/D_o, parallels that of the reciprocal of the relative viscosity, η_o/η (Figure 3.3). In Figure 3.3 notice the effectiveness of $MgSO_4$ in reducing the self-diffusion of water.

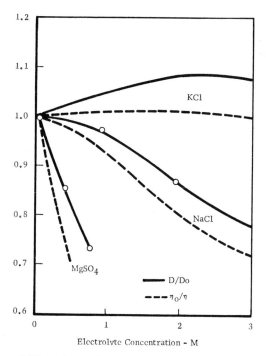

Figure 3.3 Effect of Electrolyte Addition on the Relative Self-Diffusion of Water at 25°C. From Devell (1962).

If we now turn our attention to the diffusion of electrolytes in water (Table 3.3), on first inspection we notice some familiar features. Those ions which enhance the water structure in their neighborhood move more slowly and, again, $MgSO_4$ is distinguished by its particularly low diffusion coefficient. Cs^+ tends to make the nearby water more fluid and, while it is not an important constituent of seawater, CsCl has been included in Table 3.3. It has, as we might have expected, a higher D than any of the other electrolytes listed. But, on closer examination of Table 3.3, some differences become apparent. For example, D versus concentration goes through a minimum for NaCl, KCl, and CsCl, which has no fluidity analog inasmuch as the relative viscosity versus concentration curve exhibits a maximum for NaCl, but minima for KCl and CsCl (Figure 1.25). The theory of ion diffusion in aqueous electrolytic solutions, especially mixed, concentrated solutions, is both very complex and imperfect. The changes in both ionic and self-diffusion coefficients in seawater over the oceanographic temperature and concentration ranges are slight, and I feel that we can achieve a clearer picture of the microscopic phenomena involved in transport processes in seawater in our subsequent discussion of viscosity and electrical conductivity. Consequently we shall not pursue our examination of diffusion beyond this point. In conclusion, the attempt of Richardson, Bergsteinsson, et al. (1964) to measure the pressure dependence of D_{NaCl} in artificial "seawater" consisting of aqueous 3.5 wt of NaCl should be mentioned. The experiment proved to be a difficult one and, although there *appears* to be nearly a 7% increase in going from 1 to 1000 bars, the experimental uncertainty was so great that the investigators were forced to conclude that the effect is negligible while at the same time being careful not to eliminate the possibility of pressure effects in a complex multicomponent system such as real seawater.

Some recent values of gas diffusion coefficients in pure water are listed in Table A.27.

Table 3.3   Diffusion Coefficients of Some Seawater Electrolytes at 25°C

Concentration, M	NaCl	KCl	CsCl	$CaCl_2$	$SrCl_2$	Na_2SO_4	$MgSO_4$
				$cm^2/sec \times 10^5$			
0.005	1.560	1.934	1.978	1.179	1.219	1.123	0.710
0.01	1.545	1.917	1.958	—	—	—	—
0.05	1.507	1.864	—	—	—	—	—
0.10	1.483	1.844	1.971	—	—	—	—
0.50	1.474	1.850	1.860	—	—	—	—
1.00	1.404	1.892	1.902	—	—	—	—
1.50	1.473	1.943	—	—	—	—	—

4   Viscosity

Viscosity is a familiar concept which has the property of being clear and simple on first glance but increasingly obscure and complex the more one tries to think about it. In dealing with seawater we are interested in the internal fluid friction or the forces of drag which its molecular and ionic constituents exert on one another. Yet the classical techniques of determining viscosity all determine shear viscosity with respect to some foreign surface. Everybody uses the latter viscosities—they appear to work—and we will not depart from this usage, but I continue to nurture some misgivings.

Already we have found several occasions to comment that the viscosity is a parameter especially sensitive to fluid structure, and one which appears to give us an insight into the nature of liquids in a relatively straightforward way. The viscosities of pure water and seawater (Figure 1.19 and 3.4, Tables 3.4 and A.10) decrease with increasing temperature in a manner closely paralleling the average size of the Frank-Wen clusters (Figure 1.19). The Arrhenius

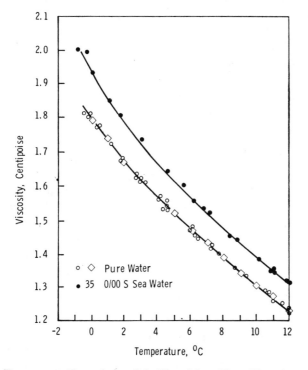

Figure 3.4   Temperature Dependence of the Viscosities of Pure Water (open circles) and 35‰ S Seawater (closed circles). From Horne, Courant, Johnson, and Margosian (1965), with permission of American Chemical Society.

Table 3.4   Relative Viscosity of Seawater at 1 atm[a]
(From Dorsey, 1940, with permission of Reinhold Pub. Corp.)

Temperature, °C	5‰ S	10‰ S	20‰ S	30‰ S	40‰ S
0	1.009	1.017	1.032	1.056	1.054
5	0.855	0.863	0.877	0.891	0.905
10	0.738	0.745	0.785	0.772	0.785
15	0.643	0.649	0.662	0.675	0.688
20	0.568	0.574	0.586	0.599	0.611
25	0.504	0.510	0.521	0.533	0.545
30	0.454	0.460	0.470	0.481	0.491

[a] η/η_0, where η_0 is the viscosity of pure water at 0°C (1.787 centipoise).

activation energies of viscous flow of pure water and 35‰ S seawater are shown in Figure 3.5 (Miller, 1963; Horne, Courant, Johnson, and Margosian, 1965; Franks and Good, 1966). While the viscosity versus temperature curves for pure water and 35‰ S seawater are nearly parallel, as the temperature decreases below about 4°C, E_a of viscous flow of seawater begins to increase much more steeply than the comparable quantity for pure water. Evidently the combined water structure-enhancing effects of the low temperature and the presence of structure-making electrolytes makes the slow rate-determining step of the viscous flow process much more difficult. As Miller (1963) has pointed out, a startling feature of Figure 3.5 is the absence of any kind of discontinuity or inflection near 4°C, a structurally important temperature. Nor, for that matter, is there an inflection at 0°C or 100°C! He suggests that "the molecular motion required for flow occurs only in the unbounded state." In a recent interesting paper, Franks and Good (1966) concur that ". . . in a discussion of viscous flow, the nature and concentration of non-hydrogen bonded water must be considered a major factor." And they argue that the mean lifetime of the Frank-Wen clusters (10^{-11} sec) is sufficiently long that liquid water can be treated as a suspension of incompressible spheres in a dipolar fluid of viscosity η_0. Employing an extension of the Einstein (1906) equation

$$(3.6) \qquad \eta = \eta_0(1 + 2.5V_c + 7.17V_c^2 + 16.2V_c^3)$$

developed for concentrated suspensions (Saito, 1950; Vand, 1948), where V_c is the volume fraction of the clusters, they obtain linear Arrhenius plots of η_0 versus $1/T$ (but with discontinuities near 40°C) corresponding to activation energies approximating those for liquid NH_3 and SO_2, and values of η_0 which are not only of the same order of magnitude as those for dipolar nonassociated liquids, but also which, when substituted in the Debye equation (3.1) yield a

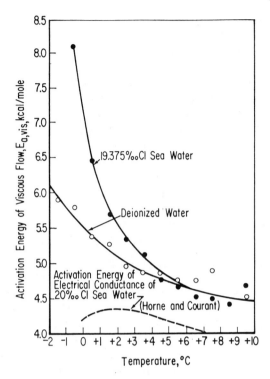

Figure 3.5   Arrhenius Activation Energy of Viscous Flow of Pure Water and Seawater at 1 a.m. From Horne, Courant, Johnson, and Margosian (1965), with permission of American Chemical Society.

relaxation time of 1×10^{-12} sec at 40°C (compare Table 3.1). They conclude that the rate-determining process of viscous flow in liquid water is the decay of the H-bonded clusters.

The effect of hydrostatic pressure on the viscosity of pure water was examined in Chapter 1. The relative viscosity, $\eta_P/\eta_{1\mathrm{atm}}$, decreases as the less dense Frank-Wen clusters are destroyed; then, after going through a minimum at about 1000 kg/cm² , it increases in the "normal" way (Figure 1.16) (Horne and Johnson, 1966a). The anomalous minima tend to be smeared out as the temperature increases, and they are less pronounced for 35‰ S seawater (Figure 1.16). The high-pressure viscosity measurements on seawater (Horne and Johnson, 1966b) would appear to indicate that the presence of electrolytes has less influence on these minima than suggested by earlier work on aqueous NaCl solutions (Cohen, 1892), a finding substantiated by fresh measurements on 1.0 and 2.0 M NaCl solutions (Horne and Johnson, 1967). It remains unclear, however, why the presence of a structure-making electrolyte such as

NaCl should tend to make water more "normal," that is, more like an unstructured liquid. Other mysteries also remain: increasing the pressure decreases E_a of viscous flow as expected (Horne and Johnson, 1966a; Horne and Johnson, 1967), but so does the addition of NaCl even though, again, this electrolyte is supposed to be a structure maker. Stranger still, as the pressure is increased, a maximum appears in E_a versus T near 4–6°C (Horne and Johnson, 1966a). Then, too, isobaric plots of η_P/η_{1atm} versus T not only show inflections near 4°C but also form distinctly different families of curves above and below 1000 kg/cm². A temperature of 4°C and a pressure of 1000 kg/cm², both lying in the range of oceanographic interest, evidently correspond to some important form of structural transitions in liquid water, although the details of such transitions and their full significance in oceanographic and life processes remain to be grasped.

The dependence of the viscosities of solutions on electrolyte concentration (Figure 1.25) can be represented by the Jones-Dole (1929) equation

$$(3.7) \qquad \eta = \eta_0(1 + A\sqrt{c} + Bc)$$

where η_0 is the viscosity of the pure solvent, c the concentration of solute, and A and B are constants characteristic of the solute. The viscosity B coefficient, which has been treated in some detail by Gurney (1953), is of particular interest to us here, for it provides a quantitative measure of the water structure altering properties of ions. Miyake and Koizumi (1948) have given viscosity coefficients for seawater. Ions which are structure makers and whose solutions tend to be less fluid than the pure solvent have positive B coefficients, whereas the structure breakers which tend to make water more fluid have negative coefficients. Table 3.5 lists viscosity coefficients of a few ions in aqueous solutions at 25°C and should be studied in conjunction with Figure 1.25. That figure indicates that 2:2 electrolytes such as $MgSO_4$ are very powerful structure makers with large positive B coefficients. The individual ionic viscosity B coefficients are additive; thus the B coefficient for NaCl is $+0.086 - 0.007$ or $+0.079$. The viscosity B coefficients become more positive with increasing temperature (Gurney, 1953). Their pressure dependence is not known owing to the paucity of reliable viscosity data on aqueous

Table 3.5   Individual Ionic Viscosity B
Coefficients in Water at 25°C

Li^+	$+0.147$	Cl^-	-0.007
Na^+	$+0.086$	Br^-	-0.032
K^+	-0.007	I^-	-0.080
Rb^+	-0.029		
Cs^+	-0.045		

solutions at high pressure, but it is certainly a subject of sufficient interest to warrant investigation.

The viscosity B coefficient is a very interesting and useful parameter theoretically. It correlates with many solution properties including ionic entropies (see Gurney, 1953, Chapter 10). It is a descriptive parameter of an ion's hydration atmosphere as useful as the hydration number or water residence time. Inasmuch as it is a measure of hydration, it depends on the charge density of the ion, decreasing with increasing crystal ionic radii (see the alkali metal cations in Figure 3.6).

Figure 3.6 introduces us to a new topic. There are many types of solute in seawater. Hitherto we have restricted our attention to the strongly polarizing ions formed on the dissolution of electrolyte substances. But there are also weakly polar and nonpolar solutes in seawater—dissolved gases, organic molecules, etc. A series of solute cations of the form NR_4^+ enables a gradual transition to be made from strongly to weakly polar substances. The ammonium ion, NH_4^+, the smallest member of this series, is a perfectly well-behaved, strongly polarizing ion, and it lies nicely on the curve of B versus ionic radius r for the alkali metal cations (Figure 3.6). But the next member of the series, tetramethylammonium ion, $N(CH_3)_4^+$, clearly falls off the curve. As we move on to the next member of the series, the B coefficient becomes very large and we are confronted by a new hydration phenomenon. For ordinary cations such as NH_4^+ the hydration is coulombic, that is, the water molecules are bound tightly to the ion by electrostatic forces. In the case of the tetraalkylammonium ions the charge is screened and its density greatly decreased. At the same time the ion is being enshrouded by an envelope of strongly hydrophobic hydrocarbon groups. The anomalous behavior of the tetraalkylammonium ions shown in Figure 3.6 could arise even if they were completely unhydrated, simply if they are large enough that the Einstein equation, the first two terms of (3.6), is applicable. Or we may have in hand evidence of a completely new, noncoulombic type of hydration. Activity coefficient (Wen, Saito, and Lee, 1966) viscosity B coefficient (Evans and Kay, 1966; Kay, Vituccio et al., 1966), heats and entropies of solvation (Lindenbaum, 1966), deuteron nuclear quadrupole resonance (Wicke, 1966; Hertz and Zeidler, 1964), ultrasonic (Allam and Lee, 1966) and osmotic coefficient (Lindenbaum and Boyd, 1964) measurements—all indicate that these hydrophobic ions strongly perturb the nearby water structure. Yet it is equally clear that, in the words of Conway, Verrall, and Desnoyers (1966), the NR_4^+ ions "are not hydrated in the normal sense involving significant electrostriction. . . ." Polyhedral clathrate hydrates of these compounds have been isolated (McMullan, Bonamico, and Jeffrey, 1963). It is not unreasonable, then, to theorize that even in solution nonpolar and/or hydrophobic solutes are surrounded by a cage-like or clathrate hydration atmosphere (Wen, Saito,

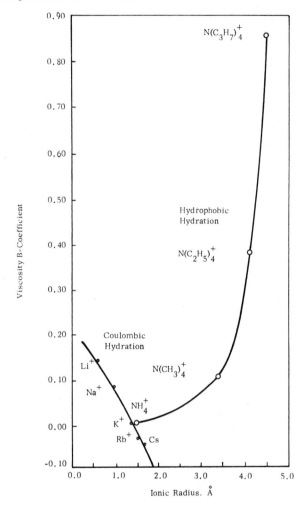

Figure 3.6   Coulombic and Hydrophobic Hydration: The Dependence of Ionic Viscosity *B* Coefficients on Ionic Radii. From Horne and Young (1968).

and Lee, 1966). The presence of dissolved gases also increases the "degree of crystallinity" of the surrounding water (Ben-Naim, 1965), and this structure again may well be clathrate-like rather than ice-like as Frank and Evans (1945) had earlier suggested.

In the structured regions in liquid water the molecules are joined together by hydrogen bonds. In the primary hydration atmosphere of "normal" polarizing solutes the water molecules are bound by electrostatic forces. In the case of nonpolar solutes we are dealing with a third type of water bonding —the hydrophobic bond. Water molecules are incurable bigots, and they try to exclude dissimilar foreigners from their midst. They consolidate and try to reduce the disturbance introduced by the intruder solutes by segregating them and, if possible, minimize the disruption of the water structure by minimizing their volume. In 1962 Nemethy and Scheraga published a series of papers exploring the nature of the hydrophobic bond, especially as applied to biopolymers.

Figure 3.7 summarizes estimates of ion and water molecule rotational and residence times for the salt $(C_2H_5)_4N^+Br^-$ at 25°C. The values beside the straight arrows are average times spent by the species at a given site between two jumps.

To return now to "normal" polarizing electrolytes: liquid water is only about 1000/18.02 or 55.5 M in water, yet there are many water molecules in the hydration envelope of ions. Consequently, as electrolyte is added, the availability of water molecules decreases and the competition for them becomes more keen. As the concentration of electrolyte increases, a point is reached at which the various hydration zones of the ions begin to overlap. The outermost zone (Figure 1.26) is a region of broken water structure and enhanced fluidity, whereas the inner total zone is one of reinforced water structure and decreased fluidity. The onset of overlap of these zones should

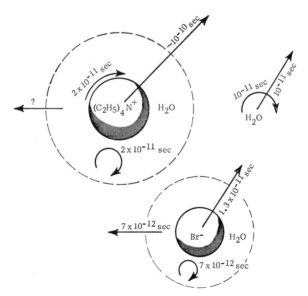

Figure 3.7 Rotation and Residence Times of Ions and Water Molecules in $(C_2H_5)_4$ $N^+Br^- \cdot nH_2O$. From Wicke (1966), with permission of Verlag Chemie.

give rise to drastic alterations in viscous properties of the solution. The concentration dependence of the viscosity itself (Figure 1.25) gives little hint of any such changes, but Good (1964) has shown that the concentration dependence of the Arrhenius activation energy of viscous flow for aqueous NaCl solution exhibits a minimum (Figure 3.8). At low concentrations E_a decreases with increasing electrolyte concentration as the overlap of the structure-broken zones becomes more extensive; then, as more electrolyte is added and the structure-enhanced zones begin to overlap, movement becomes more difficult and E_a begins to increase. By interpreting the minima in Fig. 3.8 as representing the onset of overlap of the structure-enhanced zones and neglecting the anions, the *total* number of water molecules in the structure-enhanced hydration atmosphere of Na^+, n, can be estimated from the simple relation

$$(3.8) \qquad\qquad n = \frac{1000 - 58.54c_m}{18c_m}$$

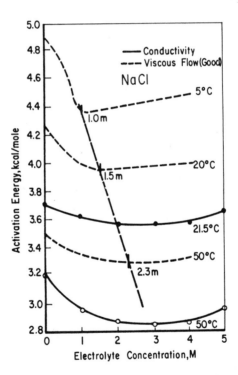

Figure 3.8   Activation Energy of Viscous Flow as a Function of Electrolyte Concentration for Aqueous NaCl Solutions. From Horne and Birkett (1967), with permission of Pergamon Press, Ltd.

Table 3.6   Number of Water Molecules in the
Total Structure-Enhanced Zone of the Hydration
Atmosphere of Na$^+$ Ion

Temperature, °C	c_m	n	n_c
5	1.0	52	57
20	1.5	34	38
50	2.3	21	20

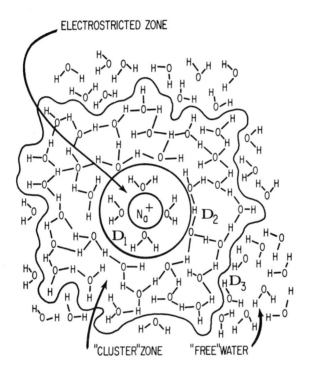

At 1atm   (Density)$_1$ > (Density)$_3$ > (Density)$_2$

Figure 3.9   Two-Dimensional Representation of the Total Structure-Enhanced Hydration Atmosphere of the Sodium Ion. From Horne (1968), with permission of Academic Press, Inc.

where c_m is the molal concentration at which E_a is a minimum (Horne and Birkett, 1967). The values thus obtained (Table 3.6) are in remarkably good agreement with the average number of water molecules, n_c, in the Frank-Wen clusters in bulk water at the same temperature as estimated by Nemethy and Scheraga (1962). This observation leads to the conclusion that hydrated Na^+, viewed from afar, is simply an ion-counting cluster, a conclusion supported by several other lines of evidence (Horne and Birkett, 1967).

The results summarized in Table 3.6 add more detail to the picture of the hydration atmosphere of ions in solution. The values of n contained therein represent the *total* number of water molecules in the structure-enhanced region surrounding Na^+. The Frank-Wen two-zone hydration model (Figure 1.26) has now become a three-zone model (Figure 3.9) with the inner structure-enforced region being in turn divided into two subzones. The inner zone consists of dense, tightly bound, electrostricted water molecules, and it is surrounded by what is in effect a Frank-Wen cluster. It may be identified with the so-called primary hydration sheath, and the consensus appears to be that in the case of Na^+ it contains about 4 water molecules (see Table A.6) while the total structure-enhanced region is not too dissimilar from Gurney's (1953) concept of an ion's "co-sphere"; that is, the volume of solvent surrounding an ion which by virtue of the ion's presence is modified or different from "normal" solvent at some great distance from the ion.

In the foregoing discussion we have been concerned only with viscosity as it applies to laminar flow. In many respects, in the oceans turbulent flow is of far greater practical importance. However, turbulent motion falls outside our definition of the subject of chemical oceanography (Chapter 1) as the properties of seawater apart from its form, and thus lies outside the scope of this book.

5   Electrical Conductivity

The viscosity is a property arising primarily from the movement of solvent molecules and provides accordingly, as we have seen, useful insight into solvent structure. The electrical conductivity of a solution, on the other hand, is primarily a solute property and provides a great deal of information about the nature and movement of electrolyte species.

Electrical conductivity is of special interest to the oceanographer. After temperature, it is very probably the most frequently measured physical-chemical property of seawater. The hows and whys of this measurement are examined in some detail in Chapter 4.

Scandalously enough, the mechanism of electrical conduction in aqueous electrolytic solutions is not known. Two reasons, one good and one bad, are largely responsible. First, the system, as we have seen many times before, is an extremely complex one. The second reason is a curious one. There is a faulty model of the conductive process which has worked so well that it has to a certain extent obviated the necessity of developing a more realistic model.

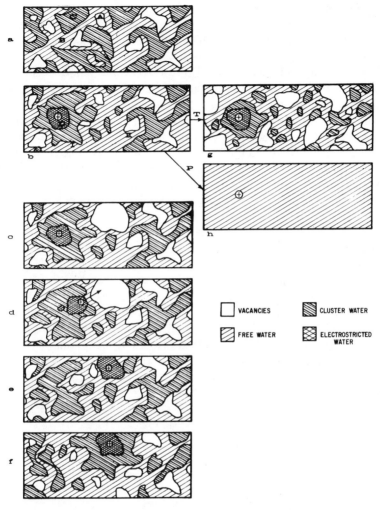

Figure 3.10 The Movement of an Ion through an Aqueous Solution.

Aristotle somewhere mentions that, for a particle to move in a condensed phase, it must either shove its neighbors aside or find a nearby space into which it can slip. Glasstone, Laidler, and Eyring (1941) have theorized that the rate-determining step for transport processes in liquids is the formation of a "hole" or vacancy into which a particle can move. The normal movement of an ion in solution, such as, for example, Na^+ in seawater, is a ceaseless, fretful, more or less random walk (Figure 3.1C). The superposition of an electric field prejudices this walk in a direction dictated by the field (Figure

3.1D) so that to a distant observer on the average an ion appears to be moving persistently in a single direction. Over a wide range the measured electrical conductivity appears to be independent of field strength, leading to the conclusion that to the distant observer the ion's motion is almost entirely terminal velocity motion and that the fraction of accelerated motion is relatively negligible. When viewed more closely, however, the ion is actually jerking along in a series of starts and stops. Figure 3.10 attempts to describe pictorially the sequence of events that occur as an ion moves. The various types of water are not represented in the proper relative quantitative proportion and, in this cinematic representation and for the sake of simplicity, the protean nature of the clusters is not shown. In Figure 3.10a we see the "normal" state of affairs in pure water. The presence of an ion Figure (3.10b) disrupts this system to form the structure-enhanced, structure-broken hydration atmosphere of the ion. As a consequence of microthermal fluctuations a vacancy appears near the ion—the slow or rate-determining step (Figure 3.10c). In Figures 3.10d and 3.10e the ion moves into this vacancy. How much of its hydration atmosphere the ion carries with it when it jumps is an important question which has not received the attention it deserves. Samoilov (1965), however, has approached this problem and obtained the values for the ratio of ionic jumps with and without the hydration shells, j_h/j, listed in Table 3.7. This table also lists the ratio of the residence times, τ_h/τ, of a water molecule adjacent to and at some great distance from an ion.

On one hand, transport of the total hydration atmosphere seems unlikely, yet on the other hand it seems equally unlikely that an ion could shed its strongly bound primary hydration sheath.

The specific electrical conductivity κ is obtained from a direct experimental measurement of the resistivity R of the sample:

$$(3.9) \qquad \kappa = \frac{C}{R} \text{ ohm}^{-1} \text{ cm}^{-1}$$

Table 3.7   Hydration Shell Jumps and Water Residence Times at 21.5°C

	Li$^+$	Na$^+$	K$^+$	Cs$^+$	Cl$^-$	Br$^-$	I$^-$	Mg^{2+}	Ca^{2+}
j_h/j	0.23	0.10	0.04	0.03	0.04	0.04	0.04	5.65	0.14
τ_h/τ	3.48	1.46	0.65	0.57	0.63	0.61	0.58	86.31	2.16
	Structure makers				Structure breakers			Strong structure makers	

where the cell constant C is a characteristic of the geometry of the apparatus. The equivalent conductivity Λ is given by

$$(3.10) \qquad \Lambda = \frac{1000\kappa}{c} \text{ ohm}^{-1} \text{ equiv}^{-1} \text{ cm}^{-2}$$

where c is the molar concentration.

At 1 atm, in the temperature range of oceanographic interest the specific conductivity of an aqueous electrolyte solution increases almost linearly with temperature and somewhat less linearly with electrolyte concentration (Figure 3.11). For more detailed data see Table A.11 in the Appendix (Thomas, Thompson, and Utterback, 1939; Malmberg, 1965; Reeburgh, 1965). Increasing the temperature not only increases the size and/or concentration of vacancies into which an ion can move (Figure 3.10b and 3.10g), but also it increases the available energy of the species necessary to make the jump.

Closer inspection of κ versus T curves based on the data of Thomas, Thompson, and Utterback (1934) reveals an upward curvature as the temperature decreases below 10°C. This curvature gives rise to a maximum in the Arrhenius activation energy versus temperature curve (Figure. 3.12) (Horne and Courant, 1964). E_a of electrical conduction in ice is about 10–12 kcal/mole (Bradley, 1957), yet Figure 3.12 shows that above 4°C, as the temperature decreases, E_a versus T appears to be aimed at a value of only 4.8 kcal/mole or less at 0°C and, below about 4°C, E_a actually decreases with decreasing temperature. The value of E_a at its maximum decreases, and the temperature

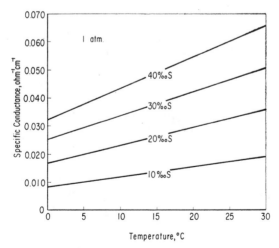

Figure 3.11 The Effect of Temperature on the Specific Electrical Conductivity of Seawater of Various Salinities at 1 atm. From Horne and Courant (1964), with permission of American Geophysical Union.

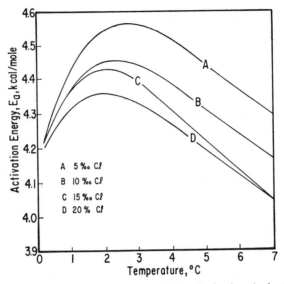

Figure 3.12 The Temperature Dependence of the Arrhenius Activation Energy of Electrical Conduction in Seawater at 1 atm. From Horne and Johnson (1966), with permission of American Institute of Physics.

at which this occurs shifts downward as the salinity of the seawater increases (Figure 3.12). The phenomenon of the maximum E_a then seems to be associated with the temperature of maximum density and presumably takes its rise in the same structural changes responsible for the latter phenomenon. Maxima in E_a of electrical conduction versus temperature have also been reported for KCl, LiCl, and CsCl solutions (Horne and Courant, 1964; Horne and Johnson, 1965a) and for concentrated HCl solutions but not for dilute HCl solutions (Horne and Courant, 1965) or for $MgSO_4$ solutions (Horne and Johnson, 1966).

At this point I should like to digress and say a few words about $MgSO_4$, the electrolyte present in seawater in the next greatest amounts after NaCl, and an electrolyte which differs very markedly from NaCl in its behavior. NaCl is a strong electrolyte, and in seawater it is very nearly completely dissociated. $MgSO_4$, on the other hand, is a weak electrolyte. If, for example, we were to make up an 0.5 m $MgSO_4$ solution just as we made an 0.5 m NaCl solution in Chapter 1 by dissolving half a gram molecular weight in 1000 g of seawater, we would find that the solution freezes before a temperature of $-1.85°C$ is reached. That is, $MgSO_4$ mole for mole is less effective at depressing the freezing point than is NaCl. The obvious conclusion to be drawn is that $MgSO_4$ is less completely dissociated and does not therefore yield as many species when it goes into solution. In fact, from the freezing point depression $MgSO_4$ produces

(3.11) $$MgSO_4 \rightleftharpoons Mg^{2+} + SO_4^{=}$$

the percent dissociation can be estimated (see the discussion of chemical equilibrium in Chapter 6). In addition to its character as a weak electrolyte, in our discussion of viscosity and diffusion we have noticed repeatedly that $MgSO_4$ has a powerful influence on the local water structure, behaving as one of the strongest coulombically hydrated structure makers. Figure 3.13 shows that E_a of electrical conductivity for $MgSO_4$ is, as might be expected, greater than for KCl or seawater, its concentration dependence is large, opposite that for seawater (cf. Figure 3.12), it has no maximum, and, in general, its temperature dependence bears much more resemblance to that for the viscous flow of

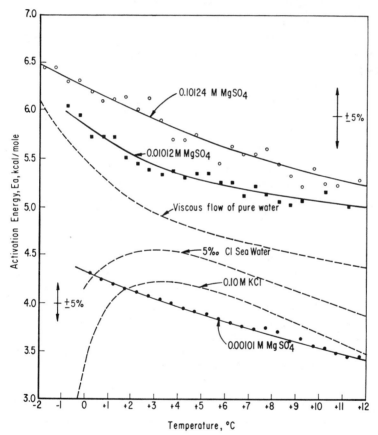

Figure 3.13 The Temperature Dependence of the Arrhenius Activation Energy of Electrical Conduction of Aqueous $MgSO_4$ Solution at 1 atm. From Horne and Courant (1964), with permission of American Institute of Physics.

water than to that for the electrical conduction of "normal" strong, 1:1 electrolytes. Samoilov (1965) estimates (Table 3.7) that the likelihood of Mg^{2+} jumping with its hydration shell is nearly 60 times greater than for Na^+. Elsewhere (Horne and Johnson, 1966b) I have proposed that the species in aqueous $MgSO_4$ solutions are so strongly hydrated that they are too bulky to fit into the available distribution of hole sizes and, as a result, must lumber through the solution by some different mechanism—perhaps one similar to that for viscous flow.

Figure 3.10 represented an attempt to examine microscopically just what happens when an ion in solution moves. Let us now step back a bit and re-examine the process from a greater distance, so to speak, so that the details are obliterated and we see only a very gross image of what is happening. We now see what is in effect a charged sphere moving through a viscous continuum. Of course, water is anything but continuous; rather, it is exceedingly complex and highly structured. Yet, fantastically enough, so crude a model works very much better than we might have dared hope. The system is so complex, it consists of so many individuals, that the very complexity and multiplicity tend to "smear out" and give rise to phenomenological simplicity. The velocity v of a spherical particle of radius r under the influence of a force F is given by Stokes' law,

$$(3.12) \qquad v = \frac{F}{6\pi\eta r}$$

where η is the viscosity of the continuous medium. In an electrical conductivity measurement, F is constant and the conductance or mobility is directly proportional to v. Hence, if we assume that the r remains constant, we can write

$$(3.13) \qquad \Lambda\eta = \text{constant} \neq f(T)$$

Relation (3.13) is known as Walden's rule, and it was first discovered simply as an empirical observation. Walden's rule is a very useful relationship, for it says that the conductivity simply varies as the reciprocal of the viscosity. Although strictly applicable only to infinitely dilute solutions of large ions, it continues to work well even for solutions as concentrated as 35‰ S seawater (Table 3.8) (Horne and Courant, 1964).

Table 3.8   Constancy of the Walden Product

	% Change in Going from 0° to 20°C		
Solution	Equiv. Cond., Λ	Viscosity, η	Walden Product, $\Lambda\eta$
Inf. dil. NaCl	68	44	6.0
0.5 M NaCl	62	42	3.7
35‰ S seawater	61	49	5.1

In going from 0° to 20°C, the conductivity of seawater increases by 61%, the viscosity decreases by 49%, but their product only changes by 5.1%. But Table 3.6 says that the number of water molecules in the *total* hydration atmosphere of Na$^+$ decreases with increasing temperature, and on this basis we might expect the Walden product to increase with increasing temperature. The constancy of the Walden product, however, implies that the assumption of temperature-invariant radius is a valid one. In fact the Walden product, if anything, tends to decrease slightly with increasing temperature (Robinson and Stokes, 1959). One way out of this impasse is to conclude that an ion takes only part of its total hydration atmosphere with it when it moves, presumably the tightly bound primary hydration sheath, thereby exhibiting a hydrodynamic radius independent of temperature. This approach leads to a further difficulty: it appears possible to estimate total hydration numbers on the basis of ionic transport processes (Horne and Birkett, 1967). And I am at a loss for an explanation of that.

In the range of oceanographic interest the effect of pressure on the electrical conductivity of seawater amounts to no more than 12% (Hamon, 1958; Groh, 1963; Horne and Frysinger, 1963. Bradshaw and Schleicher, 1965) (Table A.12) but, although the effect is comparatively modest in magnitude, its theoretical significance is considerable and throws much light on the problem of the physical chemistry of the great ocean depths. As the hydrostatic pressure increases, the conductivity of a strong 1:1 electrolyte relative to its 1 atm value, κ_P/κ_{1atm}, first increases, goes through a maximum at about 2000 kgm/cm^2, and then decreases (Horne, 1968b). Inasmuch as the viscosity of water under pressure exhibits a minimum (Figure 1.16), a logical approach might be to apply Walden's rule (3.13) under the condition of variable pressure. The results are disappointing: in going from 1 to 665 bars the viscosity decreases by 2.9%, the conductivity increases by 6.79%, but the Walden product increases by 3.7% (Horne and Courant, 1964). In other words, the Walden product is not significantly more constant than its constituent terms. The product of the conductivity and the *square* of the viscosity, however, increases by only 0.6%. We might on the basis of this finding reformulate Walden's rule as (Horne, 1963)

(3.14) $$\Lambda \eta^{n+1} = \text{constant} \neq f(P)$$

but this is purely an empirical relation devoid of theoretical meaning.

In order to fit the observed pressure dependence of the electrical conductivity of 35‰ S seawater (closed circles in Figure 3.14). a number of corrections must be applied. On the assumption that all the electrolytes are strong and completely dissociated, we can make a crude estimate of the relative contribution to the total conductivity of seawater of its major ionic constituents: Cl$^-$, 64%; Na$^+$, 29%; Mg^{2+}, 3%; SO$_4^{2-}$, 2%; and each of the

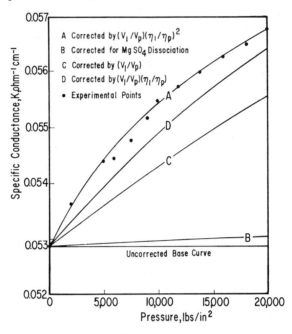

Figure 3.14   Analysis of the Electrical Conduction of 35‰ S Seawater under Pressure at 25°C. From Horne and Courant (1964).

remaining constituents less than 1% (Horne and Courant, 1964; Park, 1964). The contribution of $MgSO_4$ is not negligible and must be taken into consideration if we are to account for the conductive properties of seawater under pressure. $MgSO_4$ is, as we have seen, a weak electrolyte. In the case of equilibria such as (3.11) involving ionic species in aqueous solutions, that side of the equilibrium representing the maximum charge is usually the side of smaller volume owing to the electrostriction of the solvent. And, by the principle of LeChatelier, the application of hydrostatic pressure should shift the equilibrium in the direction of the low volume side. Hence the application of hydrostatic pressure increases the dissociation of weak electrolytes such as $MgSO_4$. Fisher (1957) has measured the pressure dependence of equilibrium (3.11), and with his results the increase in conductivity attributable to the increased dissociation can be calculated (curve B in Figure 3.14). Water is compressible. As pressure is applied, the volume of seawater decreases (Figure I.4 in the Introduction), the concentration of the ions it contains is thereby increased, and this in turn increases the conductivity. If we superimpose a correction for the compressibility, we obtain curve C in Figure 3.14. The superposition of a third correction for the pressure dependence of the viscosity yields curve D in Figure 3.14, which still falls below the observed values. The electrical conductivity of seawater increases more rapidly with pressure than it should. The experimental points can be nicely fitted by (3.14) if $n = 1$ (curve A in Figure 3.14). But such a higher viscosity dependence makes no sense in terms of (3.12). Let us take a more careful look at this equation. It embraces an important assumption, namely, that the radius of the hydrated ion, r, is independent of pressure. A number of years ago, Zisman (1932) measured the conductivities of a number of electrolytes under pressure and found that he was unable to interpret his findings in terms of the Debye-Hückel theory. He concluded that ". . . there must be a large change in the diameters of the ions due to the pressure if the theory is to agree with the data." If the application of hydrostatic pressure is indeed capable of stripping away the hydration atmospheres of ions in solution, thereby reducing their effective radii, then the effect should be more pronounced in the case of the more heavily hydrated ions. This is exactly what happens: Li^+ is the most heavily hydrated alkali cation, Cs^+ the least, and the order of $\Delta(\Lambda_P/\Lambda_{1\ atm})/\Delta P$, the slope of the pressure dependence, is $LiCl > NaCl$ and $KCl > RbCl > CsCl$ (Horne, 1963; Horne, 1968b).

To turn back now to the picture of the hydration atmosphere of Na^+ (Figure 3.9): the specific volume of the cluster zone is greater than that of the "free" water. As the pressure is increased, just as the Frank-Wen clusters in the bulk water disappear, this cluster zone is destroyed. The high specific volume, electrostricted water is highly incompressible. As the pressure continues to increase, a point is reached at which the specific volume of the "free" bulk water is less than that of the electrostricted water and, when this happens, the electrostricted zone falls victim to the effect of pressure. By the time a pressure of 5000 kg/cm² is reached, the cluster zone has long since disappeared and the number of water molecules in the primary hydration sheath of Na^+ is reduced from 4 to 1 and that of Cl^- from 1 to none. Some authors have argued that because of electrostriction the application of pressure should stabilize rather than destroy the hydration atmosphere of an ion. For electrostriction to be observed it is necessary only that the volume decrease on formation of the primary hydration and that outermost structure-broken zones exceed the volume increase on formation of the ion's cluster zone. The effect of high pressure on an aqueous NaCl solution is summarized in Figure 3.15, and the differences between the effects of temperature and pressure were compared in Figures 3.9g and 3.9h. To summarize:

100°C moderately breaks up the Frank-Wen clusters in bulk water and the cluster zone in the hydration atmospheres of ions but has virtually no effect on that part of an ion's solvation atmosphere which travels with it (presumably the primary hydration sheath).

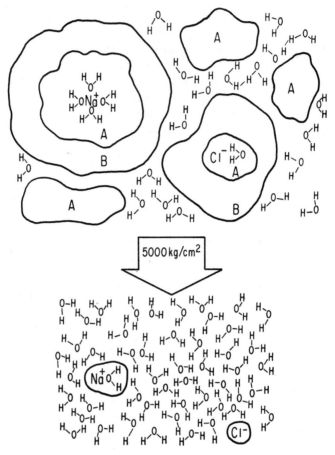

Figure 3.15   The Effect of Pressure on an Aqueous NaCl Solution. *A*, water structure enhanced region; *B*, water structure broken region.

2000 kg/cm² completely destroys the Frank-Wen clusters in bulk liquid water and the cluster zone in the hydration atmosphere of ions and is capable of considerably reducing even the primary hydration zone.

The Arrhenius activation energy of electrical conductivity of seawater decreases with increasing pressure from a value of about 4.5 at 1 atm to a little more than 4.0 kcal/mole at 1000 kg/cm² (Horne and Johnson, 1964). At a still higher pressure of 2000 kg/cm² the E_a for a strong 1:1 electrolyte (for example, KCl) goes through a minimum and then starts to increase with increasing pressure (Horne, Myers, and Frysinger, 1963). Notice that this minimum occurs at a higher pressure than the minimum in η versus P (Figure 1.16). Evidently the pressure-induced destruction of the structured regions in liquid water facilitate hole formation but, as the pressure is further increased, the solvent molecules are crowded more closely together, hole formation becomes increasingly difficult, and E_a increases.

Generally speaking, in the ocean depths the low temperatures seem to have a more profound effect on transport processes than the high pressures because of the appearance of T as an exponential in the Arrhenius equation (3.4). Thus, phenomena ranging from battery performance, to marine corrosion, to organism growth tend to be slower in the depths. Also, in addition to the effect of temperature on rate processes, its effect on chemical equilibria is usually greater than the pressure effects. However, I hesitate to say that pressure effects are relatively unimportant. The crucial step in many of the most fundamental biological processes involves the transport of ions through membranes, a process strongly dependent on the radii of the hydrated species.

Even for strong electrolytes such as NaCl, as the concentration increases the equivalent conductivity falls off markedly (Figure 2.2). The ions move more and more slowly than they should; evidently their mobility is being retarded by forces arising from ion-ion interactions. The forces may be so strong as to constitute chemical bonds and the resultant formation of weak electrolytes and complex ions (see Chapter 6 on chemical equilibria), or they may be sufficiently weak that the resulting species is best described as an ion pair. There is a tendency among solution physical chemists, especially those occupied with nonaqueous solvent systems, to attribute any discrepancy between theory and measured conductivity to complex or ion-pair formation. However, to the best of my knowledge, no one has yet had the temerity to attribute the falloff in the conductivity of an aqueous NaCl solution (Figure 2.2) to ion-pair formation, and we must turn elsewhere for an explanation. Figure 2.2 and Table 2.1 both point to the same inescapable conclusion: seawater is *not* a dilute solution as far as solute transport processes are concerned, and we cannot sneak by attempting to treat it as such. Still I am not convinced that a detailed understanding of the theory of conductivity in more concentrated solutions is essential to the marine chemist. To those who wish to pursue the subject I recommend the well-known works of Robinson and Stokes (1959), Harned and Owen (1958), and Fuoss and Accascina (1959), and the references cited therein. But I do feel that a qualitative appreciation of the physical processes postulated by these theories for the reduction in ionic mobility is most desirable, and the following, somewhat superficial, discussion is directed to this purpose.

The retardation of ionic movement in solutions is largely attributed to two effects: the electrophoretic effect and the relaxation effect.

As an ion, the anion in Figure 3.16 for example, moves through a viscous medium, it tends to drag the nearby solution along with it. As a consequence, neighboring ions find themselves moving along, not in a stationary medium,

ELECTROPHORETIC EFFECT (η)

RELAXATION EFFECT (I/D)

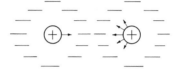

Figure 3.16   The Electrophoretic and Relaxation Effects.

but either with or against the stream produced by the first ion. The theoretical treatment of this phenomenon of electrophoresis has been developed by Onsager and Fuoss (1932).

For a given ion in solution the field exerted by all the other ions in solution (in Figure 3.16 only the closest neighbors of opposite sign are shown) is, on the average, symmetrically distributed and exerts no net force. If the given ion is now displaced to an off-center position, a finite time is required by the neighboring ions to make compensating movements, and the ion experiences a short-lived restoring force—the relaxation effect. Robinson and Stokes (1959) describe the analysis of this effect as "mathematically the most difficult part of electrolyte theory." Significant contributions to the theory have been made by Debye and Hückel (1923), Onsager (1927), Falkenhagen, Leist *et al.* (1952, 1953), Pitts (1953), and Fuoss and Onsager (1957). The resulting equation takes the form of the subtraction of two concentration-dependent terms (actually ionic strength, I) from the limiting equivalent conductivity Λ°,

$$(3.15) \quad \Lambda = \Lambda^\circ - \left[\underbrace{\frac{2.801 \times 10^6 |z_1 z_2| q \Lambda^\circ}{(\epsilon T)^{3/2}(1 + \sqrt{q}}}_{\text{Relaxation term}} + \underbrace{\frac{41.25(|z_1| + |z_2|)}{n(\epsilon T)^{1/2}}}_{\text{Electrophoretic term}} \right] \sqrt{I}$$

where z_1 and z_2 are the charges on the ions 1 and 2, ϵ is the dielectric constant, η the viscosity, and q is defined by

$$(3.16) \quad q = \frac{|z_1 z_2|}{|z_1| + |z_2|} \cdot \frac{\lambda_1^\circ + \lambda_2^\circ}{|z_2|\lambda_1^\circ + |z_1|\lambda_2^\circ}$$

and has a value of $\frac{1}{2}$ for symmetrical electrolytes such as NaCl, where $|z_1| = |z_2|$. The limiting conductivity is obtained by extrapolation of plots of Λ versus $\sqrt{c}$ to infinite dilution.

Inasmuch as it incorporates a first-order electrophoretic correction based on the Onsager limiting law, (3.15), which, incidentally, has the simple form

$$(3.17) \qquad \Lambda = \Lambda^\circ - C\sqrt{c}$$

is valid only for dilute solutions. Above 0.001 M various terms such as c, $c^{3/2}$, and c log c, must be added. One such equation proposed by Shedlovsky (1932, 1934a, 1934b) takes the form

$$(3.18) \qquad \Lambda = \Lambda^\circ - (C\Lambda^\circ + C')\sqrt{c} + C''c(1 - C\sqrt{c})$$

where the C's are constants, and appears to work satisfactorily up to about 0.1 M. Wishaw and Stokes (1954) have introduced the relative viscosity and published an equation

$$(3.19) \qquad \Lambda \frac{\eta}{\eta_0} = \left(\Lambda^\circ - \frac{C\sqrt{c^1}}{1 + \kappa a} \right)\varphi$$

where κ is proportional to $\sqrt{I}$, a is an arbitrary parameter, and φ is Falkenhagen's relaxation expression. The equation appears to work very well for strong 1:1 electrolytes up to concentrations of 4 M and greater, but unfortunately the equation has so little theoretical justification that it is hardly better than an empirical expression.

Seawater is a particularly complex, mixed electrolytic solution because of its asymmetry. The principal component, NaCl, is a strong 1:1 electrolyte, and the second component, which may be taken to be $MgSO_4$, is a weak 2:2 electrolyte. Kohlrausch's law (1893, 1898) of the additivity of conductivities

$$(3.20) \qquad \Lambda^\circ = \sum \lambda_i^\circ$$

postulates the complete independence of the individual ionic conductances and is valid only in the limiting case. Solutions of finite concentration exhibit an appreciable departure from the law. For dilute solutions the equivalent conduction λ_i of an ion i in the presence of an arbitrary mixture of ions is given by

$$(3.21) \qquad \lambda_i = \lambda_i^\circ \left[\frac{1.981 \times 10^6}{(\epsilon_0 T)^{3/2}} \lambda_i^\circ z_i \sum_0^\infty C_n r_i^{(n)} + \frac{29.16 z_i}{\eta_0 (\epsilon T)^{1/2}} \right] \sqrt{I}$$

where the terms not defined previously have the same meaning as in Harned and Owen (1958). The similarity of (3.15) and (3.21) should be noted. The "mixture effect" or departure from Kohlrausch's principle becomes most marked for mixtures of ions of like sign but very different mobilities such as

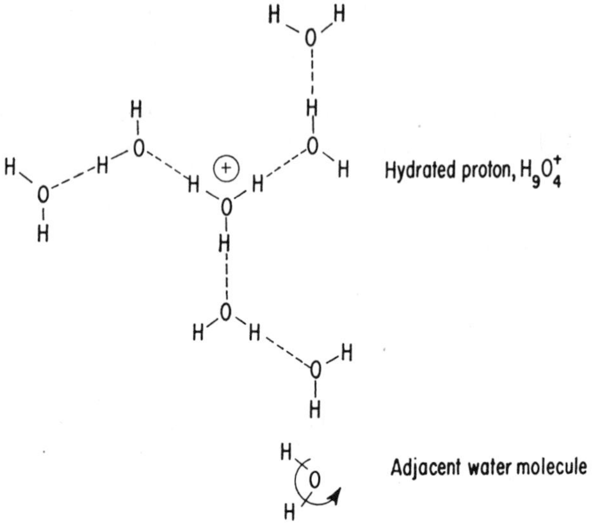
Hydrated proton, H9O4+
Adjacent water molecule

Figure 3.17 The Hydrated Proton.

HCl-KCl (Longsworth, 1930). Bremner (1944) has measured the conductivity of NaCl, KCl, and $MgSO_4$ solutions and their mixtures over the range 0–25°C and gives an empirical expression for the temperature coefficients. The conductivity of the systems KCl-HCl and $SrCl_2$-HCl has been measured under pressure; the ratio of observed to calculated, [from (3.20)] conductivity can

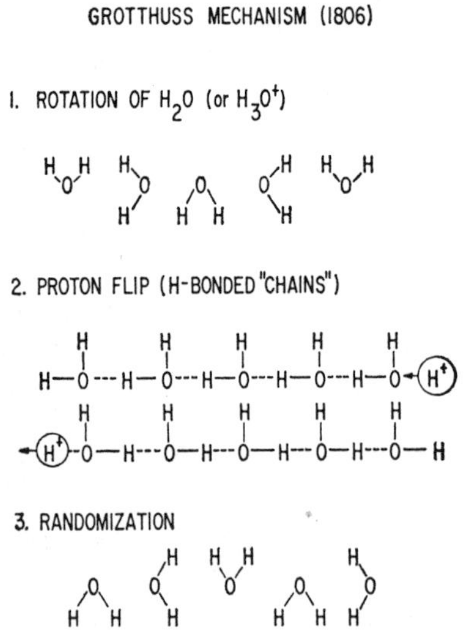
GROTTHUSS MECHANISM (1806)
1. ROTATION OF H2O (or H3O+)
2. PROTON FLIP (H-BONDED "CHAINS")
3. RANDOMIZATION

Figure 3.18 The Grotthuss Mechanism.

be as low as 0.957 and appears to exhibit some odd behavior near a pressure of 1300 kg/cm² (Horne and Courant, 1964). Data on the system artificial seawater with added $MgCl_2$ (Faletti and Gackstetter, 1967) at 1 atm and $NaCl$-$MgCl_2$ under pressure (Horne, Bannon, Sullivan, and Frysinger, 1963) have been published, but no attempt has been made to analyze them.

The concentration of hydrogen ions or protons, H^+, in seawater is very small, and their contribution to the total conductivity is entirely negligible. Protonic conduction, however, may play a crucial role in biological systems and in the origin and evolution of life in the seas, so it will not be wasteful if we briefly say something about it here. At 25°C the limiting equivalent conductivities of most ions fall in the range 20–80 ohm^{-1} equiv^{-1} cm^{-2}, yet $\lambda^o_{H^+}$ is 350 ohm^{-1} equiv^{-1} cm^{-2}. The proton has a much greater mobility than any other ion even though it is strongly hydrated, being best represented as $H_9O_4^+$ (Figure 3.17). At ordinary temperatures and pressures and for dilute solutions the "normal" conductive contribution to the total protonic conduction is small, the majority of the conduction being by the anomalous Grotthuss mechanism (Conway Bockris, and Linton, 1956). This very rapid *apparent* proton movement consists of three steps (Figure 3.18). The first step is a rotational movement to form an H-bonded chain, the second is the passage of the proton, possibly by quantum mechanical tunneling through the process

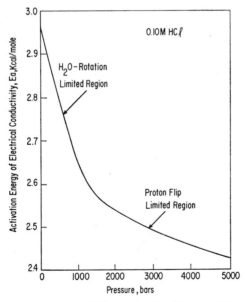

Figure 3.19 Pressure Dependence of the Activation Energy of Protonic Conduction. From Horne, Courant, and Johnson (1966), with permission of Pergamon Press, Ltd.

potential barrier (see Figure 6.2) along this chain, and the third is a rotational randomization of the water molecules. At 1 atm the slow step is believed to be a rotational one. Under pressure, as rotation is facilitated by the destruction of the structured regions, E_a of electrical conductivity decreases (Figure 3.19) until at somewhere between 1000 and 2000 kg/cm² rotation has been so facilitated that it ceases to be the slow step, the actual proton flip becomes rate-determining, and the slope of E_a versus P changes sharply (Figure 3.19). As might be expected, the protonic conductivity by the Grotthuss mechanism is restricted to the "free", rotatable, monomeric water (Horne and Courant, 1965; Horne, 1965). The possible biological significance of the Grotthuss mechanism is expanded by the likelihood that this process is capable of transmitting chemical redox energy as well as electrical energy (Horne and Axelrod, 1964).

6   Comparison of Transport Mechanisms

The similarity of the activation energies of dielectric relaxation, self-diffusion, ion mobilities, and viscous flow in water has prompted earlier authors to suggest that all the processes involve the same mechanism with the

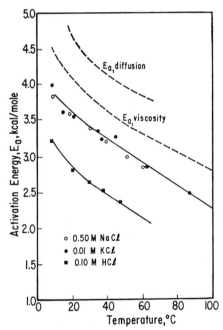

Figure 3.20   Arrhenius Activation Energies of Electrical Conductivity of Some Aqueous Solutions. From Horne (1965), with permission of American Physical Union.

same rate-determining step (Glasstone, Laidler, and Eyring, 1941; Wang, Robinson, and Edelman, 1953), although they have differed about the nature of the rate-determining step, some preferring hole formation and others the breaking of H bonds. These conclusions, however, were for the most part based on E_a's calculated from measurements at 15°, 25°, and 35°C. The low-temperature region, especially near 4°C, despite its obvious oceanographic importance, was neglected. Above 10°C the E_a's, even when not equal, do nicely parallel one another in their temperature dependence (Figure 3.20). But in the crucial region from 0° to 10°C the temperature, pressure, and electrolyte concentration dependencies, as we have seen, exhibit along with

Table 3.9   Summary of Activation Energies and Mechanisms of Processes in Ice and Aqueous Solution
(Further explanation and the appropriate literature citations can be found in Horne and Axelrod, 1964, and Horne, Courant, and Johnson, 1966.)

Process	Possible Mechanism	E_a kcal/mole
Rupture of H bonds in liquid water		1.3–2.6
Proton transfer over a water bridge		2.4
Electrical conductivity of 0.10 M HCl solutions at 2000 kg/cm²	Proton tunneling	2.45
Proton transfer reactions in acid solution		2.6
Electrical conductivity of 0.10 M HCl solution at 1 atm	Grotthuss mechanism (rotation of H_2O or H_3O^+)	2.97
Proton jumps in ice		3.1
Electrical conductivity of 0.10 M KCl solution at 1 atm	Hole formation	3.6
Classical H^+ transfer		3.85
Dielectric relaxation	Rotation of H_2O	3–5
O—H—O bond energy in ice		4.5
Viscous flow	Tumbling (?)	4.5
Self-diffusion	Tumbling (?)	4.6
Zero-point O—H bond stretching		5.3
Electrical conductivity of $MgSO_4$	Tumbling (?)	4.0–5.5
H bond energy		6.8
Rupture of H bonds in water (heat of sublimation of ice)		9
Fe(II)-Fe(III) electron exchange	Grotthuss mechanism	8.5–9.9
Electrical conductivity of pure water		9.8
Electrical conductivity of ice (calc.)		9.95
Rotation of water molecules		10
Electrical conductivity of ice (obs.)		12.3
Proton transfer in pure ice		24.2

certain similarities equally important differences suggesting a multiplicity of mechanisms (Horne, Courant, and Johnson, 1966). Liquid water is sufficiently complex that there may be more than one alternative mechanism for a given process, all with closely spaced activation energies so that the slow step may shift under the influence of relatively mild external forces (such as the change of mechanism of protonic conduction, Figure 3.19). Then, too, a number of processes such as H-bond rupture, hole formation, and the rotation of water molecules all basically disrupt the same forces of solvent-solvent interaction and might therefore be expected to have similar energetics. Table 3.9 summarizes the Arrhenius activation energies of various processes and also tabulates guesses regarding their possible mechanisms.

7 Summary

Transport processes in aqueous solutions are intimately related to the structure of liquid water. Dielectric relaxation studies indicate that there is only one rotatable species in water—presumably the monomer. Both self-diffusion and ionic diffusion in water depend strongly on the viscosity and correlate directly with the effects of different electrolytes on water structure. The viscosity is particularly useful in providing valuable insights into water structure. The pressure dependence of viscous flow, for example, indicates that hydrostatic pressure is very effective in destroying the higher specific volume Frank-Wen clusters.

The variation of viscosity with electrolyte concentration gives us a quantitative measure of the effects of different ions on water structure. Structure makers such as Na^+ and Mg^{2+} make water more viscous, while structure breakers such as Cs^+ and Cl^- make water more fluid. The viscous properties of solutions of the tetraalkylammonium salts represent evidence for a second type of hydration, hydrophobic hydration, while the concentration dependence of the energetics of viscous flow enable us to estimate the number of water molecules in the *total* coulombic hydration atmosphere of ions, thereby yielding a more detailed picture of the local water structure in their hydration atmosphere. For example, the primary hydration shell of Na^+ contains four electrostricted waters, but its total hydration atmosphere contains very nearly the same number of water molecules as an average Frank-Wen cluster in pure water at the same temperature.

Its electrical conductivity is an important and frequently measured seawater property. The rate-determining step of ionic conduction in aqueous solutions is believed to be the formation of a "hole" or vacancy in the solvent. Walden's rule, which states that the conductivity varies as the reciprocal of the viscosity, is a good approximation even for solutions as concentrated and complex as seawater. The pressure dependence of conductivity shows that that parameter, unlike temperature, is capable of destroying the hydration atmospheres of ions. The classical theory of electrical conduction is based on the model of a charged sphere moving through a viscous continuum, and the retardation of ionic mobility with increasing electrolyte concentration is explained in terms of two drag phenomena, the relaxation and the electrophoretic effects. Despite the unrealistic model, this theory works well for dilute solutions and can be extended to more concentrated solutions by the addition of empirical terms.

The several transport processes—dielectric relaxation, diffusion, viscous flow, "normal" conduction, and protonic conduction—involve several different but sometimes very closely related mechanisms.

REFERENCES

Allam, D. S., and W. H. Lee, *J. Chem. Soc.*, **A426** (1966).

Ben-Naim, A., *J. Phys. Chem.*, **69**, 3245 (1965).

Bradley, R. S., *Trans. Faraday Soc.*, **58**, 687 (1957).

Bradshaw, A., and K. E. Schleicher, *Deep-Sea Res.*, **12**, 151 (1965).

Bremner, R. W., *J. Am. Chem. Soc.*, **66**, 444 (1944).

Conway, B. E., J. O'M. Bockris, and H. Linton, *J. Chem. Phys.*, **24**, 834 (1956).

Cohen, R., *Ann. Physik.*, **45**, 666 (1892).

Collie, C. M., J. B. Hasted, and D. M. Riston, *Proc. Phys. Soc. (London)*, **60**, 145 (1948).

Conway, B. E., R. E. Verrall, and J. E. Desnoyers, *Trans. Faraday Soc.*, **62**, 2738 (1966).

Debye, P., and E. Hückel, *Physik. Z.*, **24**, 305 (1923).

Devell, L., *Acta Chem. Scand.,* **16**, 2177 (1962).

Dorsey, N. E., *Properties of Ordinary Water-Substance*, Reinhold Pub. Corp., New York, (1940).

Einstein, A., *Ann. Physik.*, **19**, 289 (1906).

Evans, D. F., and R. L. Kay, *J. Phys. Chem.* **70**, 366 (1966).

Faletti, D. W., and M. A. Gackstetter, *J. Electrochem. Soc.*, **114**, 299 (1967).

Falkenhagen, H., M. Leist, and G. Kelbg, *Ann. Phys. (Leipzig)* [6], **11**, 51 (1952); H. Falkenhagen and G. Kelbg, *Z. Electrochem.*, **58**, 43 (1953).

Fisher, F. H., *J. Phys. Chem.*, **66**, 1607 (1972).

Frank, H. S., and M. W. Evans, *J. Chem. Phys.*, **13**, 507 (1945).

Franks, F., and W. Good, *Nature*, **210**, 85 (1966).

Fuoss, R. M., and F. Accascina, *Electrolytic Conductance*, Interscience, New York, 1959.

Fuoss, R. M., and L. Onsager, *J. Phys. Chem.*, **61**, 668 (1957).

Glasstone, S., K. L. Laidler, and H. Eyring, *The Theory of Rate Processes*, McGraw-Hill Book Co., New York, 1941, Chaps. 9, 10.

Good, W., *Electrochem. Acta*, **9**, 203 (1964).

Groh, G., *Z. Angew. Phys.*, **15**, 181 (1963).

Gurney, R. W., *Ionic Processes in Solution*, McGraw-Hill Book Co., New York, 1953, Chap. 9.

Hamon, B. V., *J. Marine Res. (Sears Found. Marine Res.)*, **16**, 83 (1958).

Harned, H. S., and B. B. Owen, *The Physical Chemistry of Electrolytic Solutions*, 3rd ed. Reinhold, New York, 1958.

Hasted, J. B., R. M. Ritson, and C. N. Collie, *J. Chem. Phys.*, **16**, 1 (1948).

Hertz, H. G., and M. D. Zeidler, *Ber. Bunsenges. Physik. Chem.*, **68**, 821 (1964).

Horne, R. A., *Advan. High Pressure Res.*, **2**, in press (1968b).

Horne, R. A., *J. Electrochem. Soc.*, **112**, 857 (1965).

Horne, R. A., *Nature*, **200**, 418 (1963).

Horne, R. A., *Surv. Progr. Chem.*, **4**, 1 (1968).

Horne, R. A., *Water Resources Res.*, **1**, 263 (1965).

Horne, R. A., E. H. Axelrod, *J. Chem. Phys.*, **40**, 1518 (1964).

Horne, R. A., W. J. Bannon, E. Sullivan, and G. R. Frysinger, *J. Electrochem. Soc.*, **110**, 1282 (1963).

Horne, R. A., and J. D. Birkett, *Electrochim. Acta*, **12**, 1153 (1967).

Horne, R. A., and R. A. Courant, *J. Phys. Chem.*, **68**, 1258 (1964).

Horne, R. A., and R. A. Courant, *J. Geophys. Res.*, **69**, 1152 (1964).

Horne, R. A., and R. A. Courant, *J. Geophys. Res.*, **69**, 1971 (1964).

Horne, R. A., and R. A. Courant, *J. Phys. Chem.*, **69**, 2221 (1965).

Horne, R. A., and R. A. Courant, *J. Chem. Soc.*, 3548 (1964).

Horne, R. A., R. A. Courant, and D. S. Johnson, *Electrochim. Acta*, **11**, 987 (1966).

Horne, R. A., R. A. Courant, D. S. Johnson, and F. F. Margosian, *J. Phys. Chem.*, **69**, 3988 (1965).

Horne, R. A., and G. R. Frysinger, *J. Geophys. Res.*, **68**, 1967 (1963).

Horne, R. A., and D. S. Johnson, *J. Chem. Phys.*, **45**, 21 (1966a).

Horne, R. A., and D. S. Johnson, *J. Chem. Phys.*, **44**, 2946 (1966b).

Horne, R. A., and D. S. Johnson, *J. Geophys. Res.*, **71**, 5275 (1966c).

Horne, R. A., and D. S. Johnson, *J. Phys. Chem.*, **70**, 2182 (1966d).

Horne, R. A., and D. S. Johnson, *J. Phys. Chem.*, **71**, 1147 (1967).

Horne, R. A., B. R. Myers, and G. R. Frysinger, *J. Chem. Phys.*, **39**, 2666 (1963).

Horne, R. A., and R. P. Young, *J. Phys. Chem.*, **72**, 1763 (1968).

Jones, G., and M. Dole, *J. Am. Chem. Soc.*, **51**, 2950 (1929).

Kay, R. L., T. Vituccio, C. Zawoyski, and D. F. Evans, *J. Phys. Chem.*, **70**, 2336 (1966).

Kohlrausch, F., *Ann. Physik.*, **50**, 385 (1893); **66**, 785 (1898).

Lindenbaum, S., *J. Phys. Chem.*, **70**, 814 (1966).

Lindenbaum, S., and G. E. Boyd, *J. Phys. Chem.*, **68**, 911 (1964).

Longsworth, L. G., *J. Am. Chem. Soc.*, **52**, 1897 (1930).

McCall, D. W., and D. C. Douglass, *J. Phys. Chem.*, **69**, 2001 (1965).

McMullan, R. K., M. Bonamico, and G. A. Jeffrey, *J. Chem. Phys.*, **39**, 3295 (1963).

Malmberg, C. G., *J. Res. Natl. Bur. Stat. USA*, **A69**, 39 (1965).

Miller, A. A., *J. Chem. Phys.*, **38**, 1568 (1963).

Miyake, Y., and M. Koizumi, *J. Marine Res. (Sears Found. Marine Res.)*, **27**, 63 (1968).

Nemethy, G., and H. A. Scheraga, *J. Chem. Phys.*, **36**, 3382, 3401 (1962); *J. Phys. Chem.*, **66**, 1773 (1962).

Onsager, L., *Physik. Z.*, **28**, 277 (1927).

Onsager, L., and R. M. Fuoss, *J. Phys. Chem.*, **26**, 2689 (1932).

Park, K., *Deep-Sea Res.*, **11**, 729 (1964).

Pitts, E., *Proc. Roy. Soc. (London)*, **A217**, 43 (1953).

Reeburgh, W. S., *J. Marine Res. (Sears Marine Found. Res.)*, **23**, 187 (1965).

Richardson, J. L., P. Bergsteinsson, R. J. Getz, D. L. Peters, and R. W. Sprague, "Sea Water Mass Diffusion Coefficient Studies," Philco Aeronutronic Div., Publ. No.

U-3021, W.O. 2053 (Dec. 1964), Office of Naval Res. Contract No. Nonr-4061(00) (Unclass.).

Robinson, R. A., and R. H. Stokes, *Electrolyte Solutions*, Butterworths Sci. Publ. London, 1959.

Saito, N., *J. Phys. Soc. Japan*, **5**, 4 (1950).

Samoilov, O. Y., *Structure of Aqueous Electrolyte Solutions and the Hydration of Ions*, Consultants Bureau, New York, 1965.

Schellman, J. A., *J. Chem. Phys.*, **26**, 1225 (1957).

Shedlovsky, T., *J. Am. Chem. Soc.*, **54**, 1411 (1932); T. Shedlovsky, A. S. Brown, and D. A. MacInnes, *Trans. Electrochem. Soc.*, **66**, 165 (1934); T. Shedlovsky and A. S. Brown, *J. Am. Chem. Soc.*, **56**, 1066 (1934b).

Sverdrup, H. U., M. W. Johnson, and R. H. Fleming, *The Oceans*, Prentice-Hall, Englewood Cliffs, N.J., 1942.

Thomas, B. D., T. G. Thompson, and C. L. Utterback, *J. Conseil, Conseil. Perm. Intern. Exploration Mer.*, **9**, 28 (1934).

Vand, V., *J. Phys. Colloid Chem.*, **52**, 277, 300, 314 (1948).

Wang, J. H., *J. Am. Chem. Soc.*, **73**, 510 (1951).

Wang, J. H., *J. Phys. Chem.*, **69**, 4412 (1965).

Wang, J. H., C. V. Robinson, and I. S. Edelman, *J. Am. Chem. Soc.*, **75**, 466 (1953).

Wen, W. Y., S. Saito, and C. Lee, *J. Phys. Chem.*, **70**, 1244 (1966).

Wicke, E., *Angew. Chem.*, **5**, 106 (1966).

Wishaw, B. F., and R. H. Stokes, *J. Am. Chem. Soc.*, **76**, 2065 (1954).

Zisman, W. A., *Phys. Rev.*, **39**, 151 (1932).

Part II
The Chemical Composition
of Seawater

4 Instruments and Techniques

1 *In situ* Measurements

A physical-chemical experiment with seawater may be approached in one of three ways:

1. *In situ* measurement
2. Shipboard measurement   ↑ increasing   ↑ increasing
3. Land laboratory measurement   pertinence   difficulty

With suitable instrumentation one can attempt a measurement in the actual marine environment, that is, in the ocean's depths, on its surface, or in the air above. Alternatively one can collect a sample, transport it to a shipboard laboratory, and proceed expeditiously to perform the desired experiment. Or, finally, one can collect a sample, bring it on shipboard, store it in some suitable container under suitable conditions, bring it back to a well-equipped land laboratory, and execute the desired measurements when convenient at some future date. All this may seem obvious and trivial, but it is not. On the contrary, the three choices are not necessarily equivalent, and the selection of the best approach is a matter of the greatest importance, for the success—the accuracy, productivity, and usefulness—of the experiment is dependent on the selection of the most appropriate approach.

Let us suppose that we collect a seawater sample from some great depth. As this sample is brought to the surface, its temperature and pressure change (Chapter 2). These changes displace rapid equilibria (see Chapter 6) so that by the time our sample has reached the surface its chemical composition has already changed. For example, the ratios of the concentrations $(MgSO_4)/(Mg^{2+})$ and $(HCO_3^-)/(CO_3^{2-})$ have changed. Not only can the exact nature of the species alter but, if there is loss of dissolved gases on warming and depressurization, the total amount of a given chemical element can change. Now let us put our sample into storage. Even if we do not get contamination from and loss through the container, the chemical nature of the seawater continues to change. Suspended material may coagulate and settle out; slow equilibria may continue to react old species and form new ones. Finally we must not forget that seawater is a living soup. The metabolic processes of organisms remove some substances and replace them with others and, when

129

the organisms die, their decomposition produces other new chemicals. Storing seawater is like trying to store a rat: when we uncork the bottle, we must be prepared for some unpleasant chemical surprises.

Why not, then, perform all measurements *in situ*? The answers are that experiments *in situ* are often so difficult as to be unreliable, and in the seas we usually have an imperfect characterization of, and total lack of control over, relevant environmental conditions. Thus to try to measure, for example, the electrical conductivity of seawater under pressure *in situ* is as absurd as trying to measure the seasonal fluctuations of a North Atlantic salinity profile without going to sea. Careful consideration and mature judgment are necessary to decide which chemical experiments are best performed at sea and which in the laboratory.

An *in situ* chemical experiment may consist of an isolated measurement in a given place at a given time and, in the past, marine chemistry has been dominated by such measurements. At present the trend is to make a series of measurements or to monitor continuously at a fixed position over an extended period of time, or to make measurement along a line, vertical or parallel to the sea's surface, at a given moment or over an extended period. The procedure for obtaining the most detailed information, however, would be to make measurements at a great number of fixed points describing a *volume* of the ocean and over an extended period of time. The development of arrays of fixed buoys and platforms is making measurements of this kind feasible, and in the future we can expect such systems to bear much of the burden of work now being done in an unsatisfactory manner by oceanographic vessels.

The most common types of *in situ* measurement are temperature, electrical conductivity, pressure (or depth), and materials testing. Marine corrosion is discussed in Part IV, Section 1. Temperature and pressure are auxiliary measurements. The former is measured traditionally with a thermometer, but bimetallic strips, thermocouples, quartz crystals, and especially thermistors are enjoying increasing popularity because of the ease with which it is possible to telemeter their information to the surface (consult Chapter 10 in Svedrup, Johnson, and Fleming, 1942, for a description of the classical oceanographic thermometer). Direct pressure measurement can be accomplished with a Bourdon coil, a Sylphon bellows, or a manganin element. Of particular interest to us here is the electrical conductivity, inasmuch as it is a measure of a very important chemical property of seawater, namely, its salinity or total salt concentration.

In the laboratory the electrical conductivity of aqueous solutions is obtained by measuring the resistance between two platinized platinum electrodes immersed in the solution (Figure 4.1, left). Descriptions of the apparatus and techniques can be found in any standard electrochemical textbook such as Kortüm and Bockris (1951). The specific conductivity is calculated directly

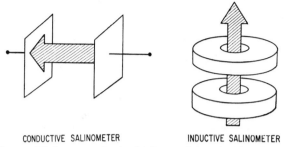

CONDUCTIVE SALINOMETER   INDUCTIVE SALINOMETER

Figure 4.1   Types of Salinometer Probe for Determining the Electrical Conductivity of Seawater.

from the measured resistance and the cell constant by (3.9). The cell constant is the ratio of the length to cross-sectional area of the electrical path, and its value is usually established by calibrating the conductivity cell with a standard solution of well-known conductance such as 0.10 M KCl. As the conductance is highly temperature-dependent, the cell must be carefully thermostated. When determining the conductance of seawater in the laboratory, it is now common practice, rather than attempting an absolute determination, to compare the sample with seawater standard of known salinity, thereby obviating the necessity of such stringent temperature control. Park (1964) has considered the use of Standard Seawater (see below) as a salinometer reference.

Paquette (1959) and Carritt (1963) have prepared brief reviews summarizing the rather considerable effort, including the development of a temperature-compensating system, to devise an *in situ* salinometer patterned after the laboratory apparatus for the direct measurement of conductivity. This effort has not been entirely successful and, with such equipment, "even with frequent standardization and recalibration, which is difficult to do routinely at sea, the precision appears to be limited to 0.05‰ to 0.10‰ in chlorinity (Carritt, 1963)." This is not surprising. The slightest deformation of the electrodes by mechanical shock or other abuse can appreciably change the cell constant. Then, too, in such a dirty soup as the sea it is only a matter of time before the electrode surface becomes poisoned, corroded, or fouled by organisms.

The obvious solution to these difficulties lies in the development of an apparatus for measuring conductivity without sea-exposed electrodes. Such a device is the ingenious induction salinometer (Pritchard, 1959; Brown and Hamon, 1961; Brown, 1964). A current is induced in the seawater by excitation of a wound, iron-core toroid (Figure 4.1, right) and is detected by a second toroid immediately adjacent to and coaxial with the first. Circuits can be added to compensate for both temperature and pressure, thus enabling the

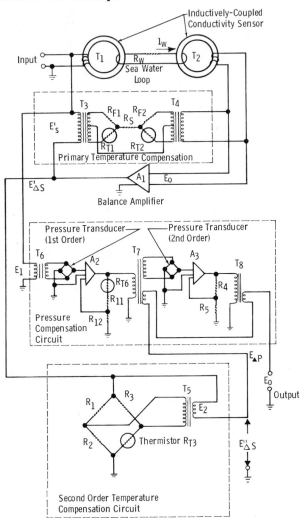

Figure 4.2   Induction Salinometer Circuit. From Brown (1964), with permission of Instrument Society of America.

sensor to be used in the deep oceans (Brown, 1964). The circuit diagram is shown in Figure 4.2, and Figure 4.3 is a photograph of a sensor unit. Carritt (1963) concludes that "... salinometers of this type are extremely stable over long periods" and their sensitivity and precision are most impressive.

Park et al. (1964, 1965) have examined the effects of dissolved CO_2 and $CaCO_3$ and photosynthesis and respiration on the electrical conductivity of seawater.

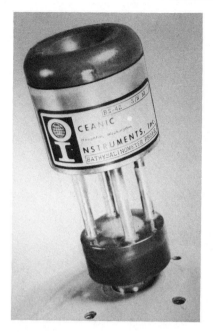

Figure 4.3 Inductive Salinometer Sensor Unit. Photograph courtesy of R. M. Van Haagen.

Salinometers measure what is in effect the total salt content of seawater. Clearly we should also like to be able to measure *in situ* the concentrations of individual seawater components and, in recent years, instruments for this purpose have begun to appear. Animal life in lakes, rivers, and the sea is dependent on dissolved oxygen. Waters devoid of oxygen are often devoid of life, and there are in the oceans "dead" seas characterized by lack of oxygen (see Chapters 7 and 9). The Carritt-Kanwisher (1959) dissolved-oxygen probe (Figure 4.4) can be used to continuously monitor dissolved oxygen in seawater. The half-cell reaction on the platinum indicator electrode is

(4.1) $$2 H_2O + O_2 + 4 e^- \rightarrow 4 OH^-$$

and its potential is given by

(4.2) $$E \approx E^\circ - \frac{RT}{n\mathscr{F}} \log \frac{(OH^-)^4}{P_{O_2}}$$

The potential is established with respect to a silver-silver oxide standard electrode, and (4.2) shows that the value obtained depends on the temperature (Carritt and Kanwisher, 1959; Rayment, 1962) and the partial pressure of dissolved oxygen, P_{O_2}. The oxygen diffuses from the seawater to the cell electrolyte through an O_2-permeable membrane and the current is proportional to the rate of diffusion.

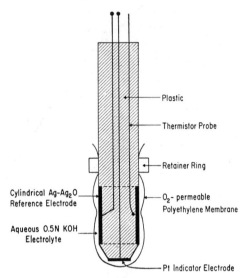

Plastic

Thermistor Probe

Retainer Ring

Cylindrical Ag-Ag$_2$O
Reference Electrode

O$_2$- permeable
Polyethylene Membrane

Aqueous 0.5N KOH
Electrolyte

Pt Indicator Electrode

Figure 4.4   Dissolved Oxygen Electrode. From Carritt and Kanwisher (1959).

Repeatedly we have mentioned that the hydrogen ion concentration in the seas, although small, is very important. It is, for example, a controlling factor in the CO_2-HCO_3^--CO_3^{2-} cycle in the seas (see Chapter 7). The conventional hydrogen electrode cannot be used for seawater since the stream of hydrogen sweeps gaseous CO_2 from the sample, thereby perturbing its hydrogen ion concentration. (Skirrow, 1965). Disteche and co-workers (1959, 1960, 1962, 1965) have developed a glass electrode system capable of withstanding high hydrostatic pressures for *in situ* measurements of the hydrogen ion concentration in the ocean depths from the French bathyscape. At this point it might be worthwhile to note that in *situ* measurements can be made, not only with devices lowered from ships or mounted on buoys, but also from submerged manned stations such as used in the U.S. Navy *Sealab* experiments or from deep submersibles.

More recently electrolyte systems have been developed for measuring Mg^{2+} (Thompson, 1966), Ca^{2+} (Thompson and Ross, 1966), and Cl^- (Koske, 1963) in seawater, and the adaption of these devices to *in situ* measurements may prove to be feasible. The principle of operation of, for example, the Ca^{2+} electrode is similar to that of the conventional glass pH electrode and is based on the electrical potential developed across a thin layer of ionically conductive materials. In the pH electrode this material is a thin layer of glass; in the Ca^{2+} electrode it is a water-immiscible liquid ion exchanger, specifically an organophosphoric acid with a very high specificity for Ca^{2+}. The potential generated is measured against a Ag-AgCl reference electrode. The manufacturer of the electrode probes used by Thompson, Orion Research Inc., of Cambridge, has utilized the same principle of a liquid membrane ion exchanger to develop anion-specific electrodes for Cl^-, NO_3^-, and ClO_4^-, and a second type of system using a solid state ionic conductor in place of the ion exchanger for I^-, Br^-, Cl^-, F^- and S^{2-}. Carritt (1965) has suggested that stripping voltammetry (Ariel and Eisner, 1963)—a technique applicable to the determination of a number of trace metal ions in seawater—might be adapted for *in situ* work. In stripping voltammetry, the cations are first plated out of the solution sample onto a suitable electrode by an appropriate applied potential. Then the potential is slowly lowered so that each of the plated-out metals is reoxidized and put back into solution, and the areas under the successive maxima in the current versus voltage curve accompanying the redissolution give the concentration of the elements in the original sample.

2   Sample Collection and Storage

If an *in situ* measurement is impossible or inexpedient, a sample of seawater must be collected, brought to the surface, and possibly stored until physical-chemical tests can be made. Traditionally the sample is collected in a bottle (this is not by any means as simple as it might sound), and for this purpose a number of bottles bearing the names of their inventors have been developed—Knudsen, Kemmerer, Foerst, VanDorn, Frautschy, Kitahara, Gleason, etc. Perhaps the most familiar one is the Nansen bottle (Figure 4.5). This collection bottle consists of a brass tube of about 1200 ml volume with valves on each end operated synchronously by a connecting rod. The bottle is lowered with both valves open so the seawater can pass through. At the desired depth the release mechanism is triggered by a weight or "messenger" dropped down the suspension cable. On release the bottle falls over, turning through 180° and shutting the valves. The turning of the bottle releases a second messenger which then drops to activate the next bottle on the cable. Commonly a reversing thermometer is attached to the bottle to record the temperature when the sample is taken. If two thermometers, one protected and the other unprotected, are used, a comparison of their readings enables the depth at which the sample was taken to be estimated.

In addition to their volume, other important characteristics of collection bottles are (a) their flushing constant, (b) their tightness, and (c) their chemical inertness. The larger the mouth compared to the cross-sectional area and length of the bottle, the faster it fills. As the sample is brought to the surface, the contents should not leak out and, even more important, the sample should not be contaminated by shallower seawater seeping in. Finally, every precaution must be taken to ensure that the bottle is chemically inert with

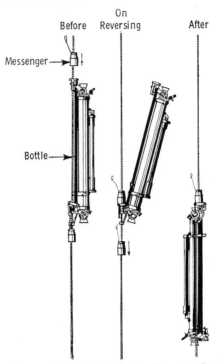

Figure 4.5   The Nansen Bottle for Collecting Seawater Samples at Depth.

respect to seawater. A noble metal electroplate on the inside of the bottle tends to become scratched and otherwise damaged after repeated use and cleaning, thus exposing the brass base to corrosion. Samples collected at great depths may stand in the bottle an hour or more—enough time to reduce appreciably the dissolved oxygen content and increase the trace metal concentrations by corrosion. The use of epoxy resins, Teflon, or nylon liners may correct these difficulties, and in studies of the trace metal constituents of seawater a metal collecting bottle should be avoided entirely in favor of plastic or glass collectors.

Sorption and desorption processes on suspended solid material and the presence of microorganisms can alter the concentrations of seawater constituents; therefore expeditious filtration of the sample is necessary. Appropriate filtering procedures have been reviewed briefly by Riley (1965).

Seawater samples can also be brought up from the ocean depths by pumping, by either vacuum, centrifugal, turbine, or jet pumps. For example, Carritt and Kanwisher (1959) obtained vertical dissolved oxygen profiles in Chesapeake Bay by pumping water up and through a dissolved oxygen cell, and Sandels (1956) described a system of continuously monitoring the temperature and electrical conductivity (salinity) of estuary water pumped through the analyzing equipment.

The minor chemical constituents of seawater present a peculiar collection problem. Because their concentrations are often so low, a very large sample, difficult to handle and process, is necessary. A number of different techniques have been described for concentrating the seawater-sample. Carritt (1965) has proposed an ingenious solution to this inconvenience. The trace elements are collected and concentrated by absorption on a suitable ion exchange resin column. In this way the trace constituents of 2000 liters of seawater can be stored in less than 100 ml. Then, when it is convenient, they can be washed out of the resin column with eluants and analyzed. In this way an enormous sea-water sample can be carried about, literally in one's vest pocket. Minor constituents can also be concentrated by evaporation, coprecipitation and cocrystallization, and solvent extraction.

The chemistry of the boundaries of the sea, the air above it, and the ocean floor beneath it is every bit as interesting as that of the seawater itself, and the samples collected from these zones present their special problems. I shall deal with them only in passing here, since they fall more properly in the domains of geology and meteorology. Samples of the ocean floor can be gathered with a variety of dredges, scoops, and grabs (Hopkins, 1964). Corers are of particular interest because they retrieve a less disturbed sample that enables the chemical composition of the bottom material to be studied as a function of its depth. Present-day practice is to refrigerate the cores, even while still in the plastic liner of the corer barrel, in order to minimize biological activity and subsequent chemical changes. The coring process probably disturbs the degree of compaction in the sample, but there is no reason to believe that the composition of the interstitial water is seriously altered thereby.

The chemical constituents in the air above seawater can be collected on precipitation in the meteorologist's familiar rain collector. May (1945) has described a cascade impactor for sampling coarse aerosols, and Duce, Winchester, and van Nahl (1965) used such an airplane-mounted device in their study of the halogen content of the Pacific atmosphere. The cascade impactor (Figure 4.6 left) consists of a series of vanes of different width which serve to separate the different aerosol particle sizes. The air is drawn through the impactor, then a filter, and finally collected in K_2CO_3 solution (Figure 4.6 right).

Seawater, I repeat, is a living soup, and the troublesome nature of storing the stuff has already been hinted at. The oceanographic literature is strewn with rather pathetic technical notes describing how hours of work were wasted before it was discovered that the seawater samples were being contaminated or otherwise altered on collection or storage. If one is concerned only with the salinity and the major constituents, airtight storage (to prevent

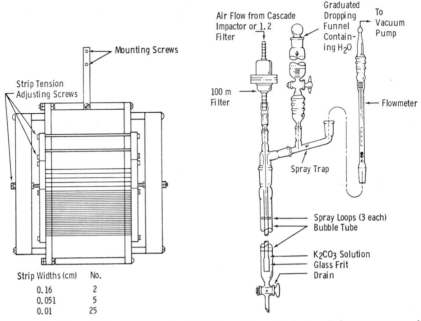

Figure 4.6   Aerosol Sample Collector. Left, aircraft-mounted cascade impactor aerosol collector; Right, gas collection train. From Duce, Winchester, and van Nahl (1965), with permission of American Geophysical Union and the authors.

evaporation) in hard glass or high-density polyethylene bottles is satisfactory. Appreciable alkalis can be leached from soft glass, and water vapor can diffuse through low-density polyethylene. However, if one is concerned with dissolved oxygen, trace elements, or constituents of biological importance, more elaborate precautions must be taken (Riley, 1965).

Further material on *Apparatus and Methods of Oceanography*, especially the more classical procedures, can be found in Barnes' (1959) book of that name.

3   Analytical Chemistry of Seawater

Riley (1965) has prepared an excellent review of the analytical chemistry of seawater, thereby sparing us of the need of examining this topic in detail here. It is a highly specialized area and it represents merely a tool of marine chemistry, albeit a very crucial one. I say this because I feel that in the past there has been some tendency to confuse chemical oceanography with fancy analytical chemistry. A few words are appropriate here, however, on the methods of analysis for the major constituents (for more detailed reviews see Riley, 1965, Culkin, 1965, and Barnes, 1959). The National Oceanographic

Data Center is preparing a bibliography of techniques for determining the chemical constituents of seawater.

Sodium and Potassium. Unfortunately, Na and K are among the most difficult elements to determine quantitatively. Na can be precipitated with zinc uranyl acetate, and potassium can be determined gravimetrically by precipitation of the chloroplatinate from water-ethanol solutions.

Calcium and Magnesium. Ca can be determined gravimetrically by precipitation as the oxalate, and Mg as the ammonium phosphate. Both cations can also be determined volumetrically by utilizing complex formation with aminopolycarboxylic acids (complexones) and suitable indicators.

Sulfate. Classically SO_4^{2-} is determined gravimetrically by precipitation of the barium salt.

Halogen. Chloride and other halides may be determined gravimetrically by the standard Volhard method

$$(4.3) \qquad\qquad Ag^+ + Cl^- \rightarrow \underline{AgCl}$$

Generally, because volumetric analyses are more convenient than gravimetrical ones, this basic reaction (4.3) has been adapted to titration methods. For this end a variety of indicators have been developed, including chromate ion, which forms a red Ag_2CrO_4, and adsorption indicators. The end point of the titration can also be determined potentiometrically.

Oxygen. The Winkler method, often modified, is used to determine dissolved O_2. The reaction of the O_2 with fresh $Mn(OH)_2$ forms a reagent which reacts stoichiometrically with iodide ion to liberate iodine. The iodine is then titrated with standard thiosulfate, starch being used as an indicator.

4   Simulation of the Sea in the Laboratory

Traditionally the oceanographer has brought his laboratory to the sea. As long as oceanography remained a qualitative and descriptive science, this procedure, although sometimes inconvenient, was for the most part adequate. In recent years, however, as the science has struggled to mature into a more quantitative and analytic discipline concerned with the "whys" as well as the "hows" of phenomena, it has become increasingly clear that adequate experimental controls for exact experiments are rarely achievable at sea. Science is a highly refined type of human experience, and an experiment is a controlled experience. Commonly the goal of an experiment, implicit or explicit, is to vary a single parameter, holding all other parameters fixed, in order to isolate the effect of that particular parameter on the phenomenon under investigation. This is virtually impossible to accomplish under natural

Table 4.1 Recipes for Artificial Seawater (19.00‰ chlorinity)

From McClendon, Gault, and Mulholland (1917)

NaCl	26.726 g	H_3BO_3	0.088 g
$MgCl_2$	2.260	Na_2SiO_3	0.0024
$MgSO_4$	3.248	$Na_2Si_4O_9$	0.0015
$CaCl_2$	1.153	H_3PO_4	0.0002
KCl	0.721	Al_2Cl_6	0.013
$NaHCO_3$	0.198	NH_3	0.002
NaBr	0.058	$LiNO_3$	0.0013

Add water to make a total weight of 1000 g.

From Subow (1941)

NaCl	26.518 g	$CaCl_2$	0.725 g
$MgCl_2$	2.447	$NaHCO_3$	0.202
$MgSO_4$	3.305	NaBr	0.083

Add water to make a total weight of 1000 g.

From Lyman and Fleming (1940)

NaCl	23.476 g	$NaHCO_3$	0.192 g
$MgCl_2$	4.981	KBr	0.096
Na_2SO_4	3.917	H_3BO_3	0.026
$CaCl_2$	1.102	$SrCl_2$	0.024
KCl	0.664	NaF	0.003

Add water to make a total weight of 1000 g.

From Kalle (1945)

NaCl	28.014 g	$CaCO_3$	0.1221 g
$MgCl_2$	3.812	KBr	0.1013
$MgSO_4$	1.752	$SrSO_4$	0.0282
$CaSO_4$	1.283	H_3BO_3	0.0277
K_2SO_4	0.8163		

Add water to make a total weight of 1000 g.

From Kester, Duedall, Conners, and Pytkowicz (1967)

NaCl	23.926 g	53.27 ml of a 1.0 M $MgCl_2$ solution
Na_2SO_4	4.008	10.33 ml of a 1.0 M $CaCl_2$ solution
KCl	0.677	0.90 ml of a 0.1 M SrO_2 solution
$NaHCO_3$	0.196	
KBr	0.098	
H_3BO_3	0.026	
NaF	0.003	

(After equilibration and aeration the pH should be 7.9–8.3).

conditions at sea—there are far too many uncontrollable and even unknown variables. Little wonder, then, that more and more the younger generation of marine scientists are bringing the sea to their laboratories and performing seawater research under stringently controlled laboratory conditions.

There is an old adage that chemists perform sloppy experiments on very pure substances and that physicists, with the greatest elegance and precision, investigate the properties of dirt. All too frequently oceanographers, I regret to say, have seemed to occupy an extreme position in this scale considerably beyond that of the physicists. Out of charity I will not cite the work, but one series of measurements comes to mind in which a measured property is reported to *six* significant figures, but the seawater sample was scooped up in a bucket, presumably from the surface, at least filtered, but characterized only by its salinity. No one will ever have in hand this same material. Even if some remains in storage, it will have changed irreversibly by now. In a word,

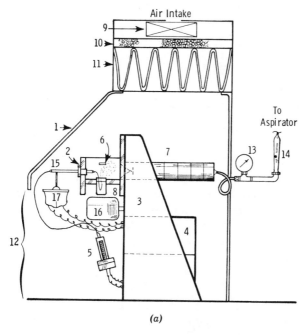

(a)

Figure 4.7A "Micro-Ocean" for Examining Ionic Fractionating in the Air-Sea Interface. Overall Schematic Diagram of the Apparatus: 1, dust-free enclosure; 2, working volume; 3, instrument rack; 4, electrical gear; 5, gas bubble generator control; 6, jet drops collector; 7, aerosol impactor; 8, "micro-ocean"; 9, blower; 10, activated charcoal bed; 11, "absolute" filter; 12, access opening; 13, vacuum gauge; 14, flowmeter; 15, oscillating horizontal capillary; 16, magnetic stirrer; 17, loudspeaker driving unit for vibrating capillary. From MacIntyre (1965), with permission of the author.

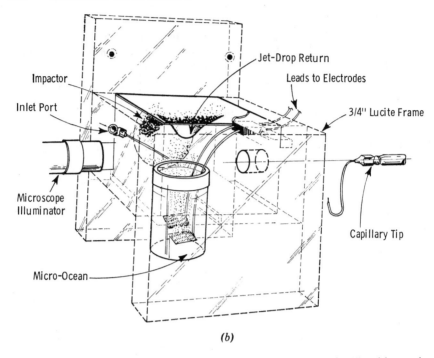

Impactor

Inlet Port

Microscope
Illuminator

Micro-Ocean

Jet-Drop Return

Leads to Electrodes

3/4'' Lucite Frame

Capillary Tip

(b)

Figure 4.7B   Detail View of the "Micro-Ocean." From MacIntyre (1965), with permission of the author.

it was a totally nonreproducible, inadequately characterized sample. Do the very careful measurements have any meaning? Who knows? But the fault is as much that of seawater as of the scientist. Because of its complexity, because of its instability, both inorganic and biological, I strongly suspect that in the final analysis a meaningful physical-chemical measurement cannot be made on a real seawater sample, nor do I foresee any entirely satisfactory resolution of this overriding difficulty.

At the present time the only two possible avenues of escape from this sad state of affairs would seem to be the universal use of a standard seawater (so that at least we are all talking about the same dirty stuff) or the use of artificial seawater. The Standard Sea-water Service of the International Association of Physical Oceanography provides oceanographers with Standard Seawater samples with the chlorinity specified to five significant figures, and in 1962 a UNESCO committee advised that the conductivity as well as the chlorinity be certified. The availability of such Standard Seawater has at least provided a common ground for the oceanographers of the world which has enabled them to resolve many of the previous squabbles over calibration and such matters. As I write these lines, there is a five-year-old sample of Standard Seawater lying in front of me on my desk. A considerable amount of solid material has coagulated out and, although I have great faith in the value of the chlorinity specified on the label, I still do not feel that I know what is in the stuff.

The second alternative is to use completely synthetic seawater, devoid of the suspended, organic, and biological material which so complicates real seawater, and made up of various inorganic chemicals, added in the correct amount, to simulate the concentrations of the major constituents. For this purpose a simple NaCl solution is not always adequate (Table 2.1) and a number of recipes have been proposed (Table 4.1).

The greatest variety of equipment for simulating marine conditions in the laboratory is available, and new apparatus and techniques appear weekly as an appreciation of the advantage of laboratory experiments grows. Here I should like to confine our attention to two examples designed to simulate extremes of the marine environment—the sea-air interaction zone and the great ocean depths. The first example is an experiment which at first thought might be considered to be ill-adapted for anything other than *in situ* studies: the formation of the salt-containing aerosol in the air above the sea from breaking white-cap bubbles. The second one is a experiment which would be horrendously difficult to perform *in situ* with proper controls but which is

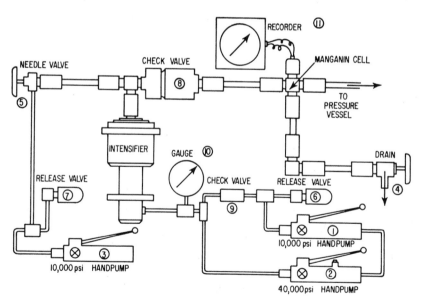

Figure 4.8a Apparatus for Simulating the Deep-Ocean Environment. The manual high-pressure-producing system.

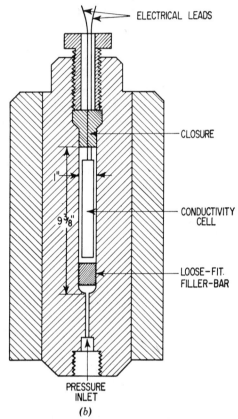

Figure 4.8b Apparatus for Simulating the Deep-Ocean Environment. The high-pressure vessel used for conductivity measurements.

easily accomplished in the laboratory: molecular and ionic transport processes in seawater under high hydrostatic pressure.

The ratios of the concentrations of ionic constituents in the atmosphere above the sea differ appreciably from those in the seawater itself (this phenomenon is discussed in detail in Chapter 11). In order to study the relative enrichment of the aerosol above seawater in phosphate, MacIntyre (1965) constructed an apparatus which he has aptly called a "micro-ocean" (Figure 4.7). The experiment consists in tagging the seawater with $Na_3P^{32}O_4$ and $Na^{22}Cl$, blowing bubbles through it, catching the resulting spray, classified by size, in an aerosol impactor, and β- and γ-counting the samples to obtain PO_4^{3-}/Na^+ ratios. The whole apparatus was enclosed (Figure 4.7a) in order to minimize contamination. The bubbles were produced by either of two methods: a vibrating horizontal capillary, or direct electrolysis with platinum gauze electrodes (Figure 4.7b). The large jet drops were caught on a curved surface above the "micro-ocean" and so returned. The smaller aerosol droplets were sucked into the impactor with mild vacuum from an aspirator.

MacIntyre (1965) also described his apparatus as a "bubble microtome." That is, it provides a technique of sampling the very thin surface layer of a solution such as seawater. In this capacity the device might turn out to be an extremely useful tool in future studies of solution interface physical chemistry.

For our second example I shall describe apparatus in my own laboratory which is capable of simulating conditions of temperature and pressure anywhere in the oceans. In addition to being designed for oceanographic studies, this equipment was designed to be used in fundamental research on transport processes in aqueous solution under high pressure and, as a consequence, working volume (about 100 cm³) has been sacrificed for a pressure capacity (10,000 kg/cm³) considerably in excess of that of oceanographic interest. The manually operated pressure-producing, -controlling, and -measuring system

Figure 4.9a Apparatus for Simulating the Deep-Ocean Environment (manufactured by Harwood Eng. Inc., Walpole, Mass.). Front View. Hand pumps in the foreground, manganin cell recorder on top of the high-pressure cabinet, and precision conductivity bridge to the left.

Figure 4.9b Rear View of the Apparatus. Intensifier and main check valve on the left, manganin cell in center T connection, and thermostatic bath containing the high-pressure vessel in the right foreground. The bath regulating and cooling units are not shown.

is diagramed in Figure 4.8*a*, and the tapered shrink construction pressure vessel in Figure 4.8*b*. The general aspect of the equipment is shown in Figures 4.9*a* and 4.9*b*. The electrical conductivity and viscosity of seawater have been investigated with the equipment (Horne and Frysinger, 1963; Horne and Johnson, 1966). Other recent investigations in which the high pressures of the deep sea were simulated in the laboratory include the electrical conductivity measurements of Bradshaw and Schleicher (1965), the seawater diffusion studies of Richardson and co-workers (1965), the emf measurements of Disteche and colleagues (1959, 1960, 1962, 1965, 1967), and the $CaCO_3$ equilibrium studies of Pytkowicz and coworkers (1963, 1964, 1965).

5   Definitions—Salinity, Chlorinity, and pH

A single parameter for describing seawater chemically, which could be used in conjunction with temperature and pressure for calculating other properties such as the density, would be most useful. The concepts of salinity and chlorinity, often mentioned previously, represent attempts to establish such a chemical parameter, and we now define these terms in detail. *Chlorinity (Cl)* was originally defined as *the total amount of chlorine, bromine, and iodine in grams contained in one kilogram of seawater, assuming that the bromine and the iodine have been replaced by chlorine.* This definition, however, is contingent on the accepted values of the atomic weights of the elements involved, and in 1940 Jacobsen and Knudson proposed the following definition, independent of atomic weights: "*The number giving the chlorinity in per mille of a seawater sample is by definition identical with the number giving the mass with unit gram of atomic weight silver just necessary to precipitate the halogens in* 0.3285234 *kilograms of seawater sample.*" Both definitions are obviously based on the classical techniques of chemical quantitative analysis used to determine the halide content (4.3).

The *salinity (S)* is defined as *the total amount of solid material, in grams, contained in one kilogram of seawater when all the carbonate has been converted to oxide, the bromide and iodine replaced by chlorine, and all organic matter completely oxidized.* As we mentioned earlier, the salinity is, roughly speaking, a measure of the total salt content of the seawater.

Both chlorinity *(Cl)* and salinity *(S)* are expressed in terms of grams per kilogram of seawater (parts per thousand, or per mille, the notation being ‰), and they are related by the empirical expression

(4.4) $S = 0.03 + 1.805\ Cl$

Less frequently the term *chlorosity* is employed. It is a concentration term based on the volume rather than the weight of seawater, and it consists of the product of the chlorinity and the density of the sample at 20°C.

Although the total salt content of seawater can be variable, the ratios of the major constituents remain *nearly* constant (see Chapter 5 for a more detailed discussion). Insofar as the ratios are constant, the concept of salinity is a very useful one and represents an adequate chemical description of seawater for the overwhelming majority of practical and theoretical purposes. Insofar as the ratios are not constant, the concept of salinity, let alone being useless, can lead to a virtual nightmare of confusion. One can immediately see, for example, that (4.4) is not valid for dilute solutions of seawater since $s \nrightarrow 0$ as $Cl \rightarrow 0$. Over recent years with the refinement of analytic techniques a specter has come into being and has grown: *nearly* constant may no longer be constant enough. With a fine show of understatement, Cox (1962, 1963a, 1963b) has referred to the resulting agony as the "salinity problem." Physical oceanographers are interested principally in the salinity as a means of calculating the density of seawater—a property whose accurate direct measurement *in situ* or even in the laboratory is relatively difficult. Yet Cox and co-workers (1962) have found a significantly better correlation between density and salinity calculated from conductivities (Figure 4.10A) than between density and salinity calculated from chlorinities (Figure 4.10B). At the conclusion of their paper they opine, "We think . . . that the time is coming when oceanographers may have to abandon the somewhat artificial concept of 'salinity' and define seawater by its conductivity or density." Earlier we mentioned that the 1962 UNESCO special commission also recommended that the electrical conductivity as well as the chlorinity of Standard Seawater be certified. The electrical conductivity is a property which can be measured with very great accuracy both in the laboratory and *in situ*. Carritt (1963) compares the uncertainty in chlorinity from the widely used Knudsen titration to that for laboratory salinometers, $\pm 0.01\%_{00}$ Cl compared to ± 0.0005 to $0.002\%_{00}$ Cl. I am entirely in sympathy with the tenor of these remarks. The concept of chlorinity has outlived its usefulness and should be allowed to die a natural death before it becomes an impediment. I feel, however, that the concept of salinity, as a measure of the total salt content of seawater, is a handy one if not taken too seriously, and should be retained until something better is forthcoming—an unlikely event.

Water dissociates to form hydrogen (protons) and hydroxyl ions:

$$(4.5) \qquad H_2O \rightleftharpoons H^+ + OH^-$$

At 25°C the ion product is 10^{-14}; that is, the concentrations (or more accurately the activities) of H^+ and OH^- are both 10^{-7} M. The pH is defined by the equation

$$(4.6) \qquad pH = -\log(H^+)$$

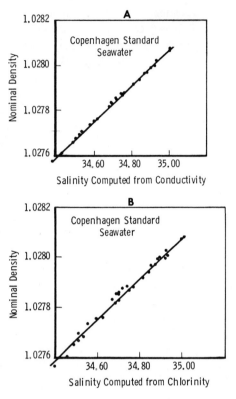

Figure 4.10 The Relationships between the Density of Seawater and the Chlorinity and Electrical Conductivity. From Cox, Culkin, Greenhalgh, and Riley (1962), with permission of Macmillan, Ltd., and the authors.

where (H^+) is the hydrogen ion concentration.* In all solutions at 25°C, pH + pOH = 14, and in neutral solutions pH = pOH = 7. If pH > 7 the solution is alkaline; if pH < 7 the solution is acidic. Seawater is slightly alkaline with a pH between 7.5 and 8.4.

6   Summary

The physical-chemical properties of a specimen of seawater can be measured *in situ*, on a ship, or in a land-based laboratory. Suitable precautions must be taken during collection, preparation, and storage of the sample in order to eliminate chemical changes. A great number of instruments and techniques are available for chemical analyses and other measurements on seawater.

* One sometimes finds pOH, pK, etc., defined similarly.

Also, judgment must be exercised to determine which types of physical-chemical experiments are best performed at sea and which under carefully controlled laboratory conditions.

The chemical content of seawater is commonly specified in terms of the chlorinity or salinity. The latter is approximately a measure of the total electrolyte concentration in seawater and is now usually determined by measuring the electrical conductivity of the sample.

REFERENCES

Ariel, M., and U. Eisner, *J. Electroanal. Chem.*, **5**, 362 (1963).
Barnes, H., *Apparatus and Methods of Oceanography*, George Allen & Unwin, Ltd., London, 1959, Part I, Chemical.
Bradshaw, A., and K. E. Schleicher, *Deep-Sea Res.*, **12**, 151 (1965).
Brown, N. L., "An *in situ* Salinometer for Use in the Deep Oceans," *Instr. Soc. Am. Reprint*, No. 812-3-64 (1964).
Brown, N. L., and B. V. Hamon, *Deep-Sea Res.*, **8**, 65 (1961).
Carritt, D. E., "Chemical Instrumentation," in M. N. Hill, ed., *The Sea*, Interscience, New York, 1963, Vol. II, Chap. 5.
Carritt, D. E., "Marine Geochemistry: Some Guesses and Gadgets," in D. R. Schink and J. T. Corless, eds., *Marine Geochemistry*, Narragansett Marine Lab., Univ. Rhode Island, Occ. Publ., No. 3 (1965).
Carritt, D. E., and J. W. Kanwisher, *Anal. Chem.*, **31**, 5 (1959).
Cox, R. A., *Nature*, **197**, 785 (1963a).
Cox, R. A., *Progr. Oceanog.*, **1**, 243 (1963b).
Cox, R. A., F. Culkin, R. Greenhalgh, and J. P. Riley, *Nature*, **193**, 518 (1962).
Culkin, F., "The Major Constituents of Sea Water," in J. P. Riley and G. Skirrow, eds., *Chemical Oceanography*, Academic Press, London, 1965, Vol. 1, Chap. 4.
Disteche, A., *J. Electrochem. Soc.*, **109**, 1084 (1962).
Disteche, A., *Rev. Sci. Instr.*, **30**, 474 (1959).
Disteche, A., and S. Disteche, *J. Electrochem. Soc.*, **112**, 350 (1965).
Disteche, A., and S. Disteche, *J. Electrochem. Soc.*, **114**, 330 (1967).
Disteche, A., and M. Dubuisson, *Bull. Inst. Oceanog.*, No. 1174 (25 March, 1960).
Duce, R. A., J. W. Winchester, and T. W. van Nahl, *J. Geophys. Res.*, **70**, 1775 (1965).
Hopkins, T. L., *Progr. Oceanog.* **2**, 213 (1964).
Horne, R. A., and G. R. Frysinger, *J. Geophys. Res.*, **68**, 1967 (1963).
Horne, R. A., and D. S. Johnson, *J. Geophys. Res.*, **71**, 5275 (1966).
Jacobsen, J. P., and M. Knudsen, *Assoc. d'Oceanog. Phys.*, U.G.G.I. Publ. Sci., No. 7 (1940).
Kalle, K., in *Probleme der kosmischem Physik.*, Leipzig, 1945, Vol. 23.
Kester, D. A., I. W. Duedall, D. N. Conners, and R. M. Pytkowicz, *Limnol. Oceanog.*, **12**, 176 (1967).
Kortüm, G., and J. O'M. Bockris, *Textbook of Electrochemistry*, Elsevier, Amsterdam, 1951, Vol. II.
Koske, P. H., *Kiel. Meereforsch.*, **19**, 182 (1963).
Lyman, J., and R. H. Fleming, *J. Marine Res.*, **3**, 134 (1940).
McClendon, J. F., C. C. Gault, and S. Mulholland *Carnegie Inst. Washington, Publ.*, No. 251 (1917).

MacIntyre, F., "Ion Fractionation in Drops from Breaking Bubbles," Ph.D Thesis, M.I.T., Cambridge, Mass., 1965.

May, K. R., *J. Sci. Instr.*, **22**, 187 (1945).

Paquette, R. G., "Conf. Phys. Chem. Properties of Sea Water," Easton, Maryland, 1958," *Natl. Acad. Sci.-Natl. Res. Council Publ.*, No. 600, 128 (1959).

Park, K., *Deep-Sea Res.*, **11**, 85 (1964).

Park, K., *Science*, **146**, 56 (1964).

Park, K., and H. C. Curl, Jr., *Nature*, **205**, 274 (1965).

Park, K., P. K. Weyland, and A. Bradshaw, *Nature*, **201**, 1283 (1964).

Pritchard, D. W., "Conf. Phys. Chem. Properties of Sea Water, Easton, Maryland, 1958," *Natl. Acad. Sci.-Natl. Res. Council Publ.*, No. 600, 146 (1959).

Pytkowicz, R. M., *Deep-Sea Res.*, **10**, 633 (1963).

Pytkowicz, R. M., *Limnol. Oceanog.*, **10**, 220 (1965).

Pytkowicz, R. M., and D. N. Conners, *Science*, **144**, 840 (1964).

Rayment, R. B., *Anal. Chem.* **34**, 1089 (1962).

Richardson, J. L., P. Bergsteinsson, R. J. Getz, D. L. Peters, and R. W. Sprague, "Sea Water Mass Diffusion Coefficient Studies," Philco Corp. Publ., No. U-3021 (26 Feb., 1965) Office of Naval Research Contract Nonr-4061(00).

Riley, J. P., "Analytic Chemistry of Sea Water," in J. P. Riley and G. Skirrow, eds., *Chemical Oceanography*, Academic Press, London, 1965, Vol. 2, Chap. 21.

Sandels, E. G., *J. Sci. Instr.*, **33**, 424 (1956).

Skirrow, G., "Dissolved Gases–Carbon Dioxide," in J. P. Riley and G. Skirrow, eds., *Chemical Oceanography*, Academic Press, London, 1965, Vol. I, Chap. 7.

Subow, N. N., *Oceanographic Tables*, U.S.S.R. Oceanog. Inst., Moscow, 1941.

Sverdrup, H. U., M. W. Johnson, and R. H. Fleming, *The Oceans*, Prentice-Hall, Englewood Cliffs, N.J., 1942.

Thompson, M. E., *Science*, **153**, 866 (1966).

Thompson, M. E., and J. W. Ross, Jr., *Science*, **154**, 1643 (1966).

5 Elemental Composition of the Seas and Salinity Variations

1 Marcet's Principle

In 1819 Marcet reported the results of analyses of Arctic and Atlantic oceans and Mediterranean, Black, Baltic, China, and White seas waters in a paper presented to the Royal Society. On the basis of his findings he concluded that "all the specimens of seawater contain the same ingredients all over the world, these bearing very nearly the same proportions to each other, so that they differ only as to the total amount of their saline contents." Nearly fifty years later, in his book, *The Physical Geography of the Sea*, Maury (1855) states that "... as a general rule ... the constituents of sea water are as constant in their proportions as are the components of the atmosphere." In the course of its famous oceanographic cruise (1873–1876) the *H.M.S. Challenger* collected some 77 seawater samples from various depths from the world's major oceans. These specimens were carefully analyzed by Dittmar (1884) and, although he noted some minor discrepancies, the results for the most part confirmed Marcet's original generalization.

Marcet's principle may be stated as follows: although the total salt content of seawater, or the salinity, can be variable, the ratios of the major constituents (see Table 5.1) remain *nearly* constant. This generalization is, as we have said before, a very useful one, and it is responsible for whatever utility the concept of salinity can claim. However, the qualifying term "nearly" must be given emphasis (see Carritt and Carpenter, 1959). Two other important

Table 5.1   Major Seawater Constituents

Constituent	Concentration, gm/kg of 35°/$_{oo}$ seawater	Constituent	Concentration, gm/kg of 35°/$_{oo}$ seawater
Cl^-	19.353	K^+	0.387
Na^+	10.76	HCO_3^-	0.142
SO_4^{2-}	2.712	Br^-	0.067
Mg^{2+}	1.294	Sr^{2+}	0.008
Ca^{2+}	0.413		

Table 5.2 Major Constituent Concentration-to-Chlorinity Ratios for Various Oceans and Seas

Ocean or Sea	$\dfrac{Na}{‰\,Cl}$	$\dfrac{Mg}{‰\,Cl}$	$\dfrac{K}{‰\,Cl}$	$\dfrac{Ca}{‰\,Cl}$	$\dfrac{Sr}{‰\,Cl}$	$\dfrac{SO_4}{‰\,Cl}$	$\dfrac{Br}{‰\,Cl}$
N. Atlantic	—	—	0.02026	—	—	—	0.00337–0.00341
Atlantic	0.5544–0.5567	0.0667	0.01953–0.0263	0.02122–0.02126	0.000420	0.1393	0.00325–0.0038
N. Pacific	0.5553	0.06632–0.06695	0.02096	0.02154	—	0.1396–0.1397	0.00348
W. Pacific	0.5497–0.5561	0.06627–0.0676	0.02125	0.02058–0.02128	0.000413–0.000420	0.1399	0.0033
Indian	—	—	—	0.02099	0.000445	0.1399	0.0038
Mediterranean	0.5310–0.5528	0.06785	0.02008	—	—	0.1396	0.0034–0.0038
Baltic	0.5536	0.06693	—	0.02156	—	0.1414	0.00316–0.00344
Black	0.55184	—	0.0210	—	—	—	—
Irish	0.5573	—	—	—	—	0.1397	0.0033
Puget Sound	0.5495–0.5562	—	0.0191	—	—	—	—
Siberian	0.5484	—	0.0211	—	—	—	—
Antarctic	—	—	—	0.02120	0.000467	—	0.00347
Tokyo Bay	—	0.0676	—	0.02130	—	0.1394	—
Barents	—	0.06742	—	0.02085	—	—	—
Arctic	—	—	—	—	0.000424	0.1395	—
Red	—	—	—	—	—	—	0.0043
Japan	—	—	—	—	—	—	0.00327–0.00347
Bering	—	—	—	—	—	—	0.00341
Adriatic	—	—	—	—	—	—	0.00341

points must also be borne in mind. Many more surface samples than deep-water samples have been analyzed; thus our knowledge of the chemical composition of seawater may be thereby prejudiced. Some constituents, especially those involved in life cycles in the sea, such as nitrate, phosphate, dissolved oxygen, commonly have highly variable ratios.

Table 5.2 summarizes the ratios of major constituent concentration (in grams per kilogram) to chlorinity for a number of oceans and seas. Further details and the relevant references can be found in Culkin (1961).

Marcet's principle suggests that the waters of the world's oceans are well mixed. And so indeed they are but, fortunately, they are not so vigorously stirred that individual water masses are indistinguishable. Sufficient chemical, thermal, radiological, and biological characteristics persist so that certain waters can be identified and their travels from their sources and their meanderings throughout the oceans of the world can be followed.

2   The Elemental Composition of Seawater

Table 5.3, taken from Goldberg (1963), but with values for the rare earth elements from his subsequent writing (Goldberg, 1965) added, summarizes the abundances of the elements in seawater, their residence times, and the principal species present. (For comparison with the distributions in the universe and the earth's crust, see Table A.17.) This table says just about all that can be said concerning the elemental composition of the sea. The particular forms in which some of the elements appear in seawater are discussed in Chapter 6 on chemical equilibria. The abundance of any particular element in seawater, especially elements of geological or biological significance, often varies both in time and space (horizontally and vertically) in a complex way. Such variations are properly the subject of detailed monographs. While some of the more important examples are treated in Chapter 6, any attempt to consider all of them here would deflect us from our present purpose. Therefore the remainder of this chapter, rather than being devoted to such specialized knowledge, addresses itself to the more important features of the salinity structure of the world's oceans.

Table 5.3   The Elemental Composition of Seawater

Element	Abundance, mg/l	Principal Species	Residence Time, years
H	108,000	H_2O	—
He	0.000005	He (g)	—
Li	0.17	Li^+	2.0×10^7
Be	0.0000006	—	1.5×10^2
B	4.6	$B(OH)_3$; $B(OH)_4{}^-$	—

Table 5.3 (*continued*)

Element	Abundance, mg/l	Principal species	Residence time, years
C	28	HCO_3^-; H_2CO_3; CO_3^{2-}; organic compounds	—
N	0.5	NO_3^-; NO_2^-; NH_4^+; N_2 (g) organic compounds	—
O	857,000	H_2O; O_2 (g); SO_4^{2-} and other anions	—
F	1.3	F^-	—
Ne	0.0001	Ne (g)	—
Na	10,500	Na^+	2.6×10^8
Mg	1350	Mg^{2+}; $MgSO_4$	4.5×10^7
Al	0.01	—	1.0×10^2
Si	3	$Si(OH)_4$; $Si(OH)_3O^-$	8.0×10^3
P	0.07	HPO_4^{2-}; $H_2PO_4^-$; PO_4^{3-}; H_3PO_4	—
S	885	SO_4^{2-}	—
Cl	19,000	Cl^-	—
A	0.6	A (g)	—
K	380	K^+	1.1×10^7
Ca	400	Ca^{2+}; $CaSO_4$	8.0×10^6
Sc	0.00004	—	5.6×10^3
Ti	0.001	—	1.6×10^2
V	0.002	$VO_2(OH)_3^{2-}$	1.0×10^4
Cr	0.00005	—	3.5×10^2
Mn	0.002	Mn^{2+}; $MnSO_4$	1.4×10^3
Fe	0.01	$Fe(OH)_3$	1.4×10^2
Co	0.0005	Co^{2+}; $CoSO_4$	1.8×10^4
Ni	0.002	Ni^{2+}; $NiSO_4$	1.8×10^4
Cu	0.003	Cu^{2+}; $CuSO_4$	5.0×10^4
Zn	0.01	Zn^2; $ZnSO_4$	1.8×10^5
Ga	0.00003	—	1.4×10^3
Ge	0.00007	$Ge(OH)_4$; $Ge(OH)_3O^-$	7.0×10^3
As	0.003	$HAsO_4^{2-}$; $H_2AsO_4^-$; H_3AsO_4; H_3AsO_3	—
Se	0.004	SeO_4^{2-}	—
Br	65	Br^-	—
Kr	0.0003	Kr (g)	—
Rb	0.12	Rb^+	2.7×10^5
Sr	8	Sr^{2+}; $SrSO_4$	1.9×10^7
Y	0.0003	—	7.5×10^3
Zr	—	—	—
Nb	0.00001	—	3.0×10^2
Mo	0.01	MoO_4^{2-}	5.0×10^5
Tc			
Ru			
Rh	—	—	—
Pd	—	—	—
Ag	0.00004	$AgCl_2^-$; $AgCl_3^{2-}$	2.1×10^6

Table 5.3 (*continued*)

Element	Abundance, mg/l	Principal species	Residence time, years
Cd	0.00011	Cd^{2+}; $CdSO_4$; $CdCl_n^{2-n}$; $Cd(OH)_n^{2-n}$	5.0×10^5
In	<0.02	—	—
Sn	0.0008	—	1.0×10^5
Sb	0.0005	—	3.5×10^5
Te	—	—	—
I	0.06	IO_3^-; I^-	—
Xe	0.0001	Xe (g)	—
Cs	0.0005	Cs^+	4.0×10^4
Ba	0.03	Ba^{2+}; $BaSO_4$	8.4×10^4
La	1.2×10^{-5}	—	4.4×10^2
Ce	5.2×10^{-6}	—	8.0×10^1
Pr	2.6×10^{-6}	—	3.2×10^2
Nd	9.2×10^{-6}	—	2.7×10^2
Pm	—	—	—
Sm	1.7×10^{-6}	—	1.8×10^2
Eu	4.6×10^{-7}	—	3.0×10^2
Gd	2.4×10^{-6}	—	2.6×10^2
Tb	—	—	—
Dy	2.9×10^{-6}	—	4.6×10^2
Ho	8.8×10^{-7}	—	5.3×10^2
Er	2.4×10^{-6}	—	6.9×10^2
Tm	5.2×10^{-7}	—	1.8×10^3
Yb	2.0×10^{-6}	—	5.3×10^2
Lu	4.8×10^{-7}	—	4.5×10^2
Hf	—	—	—
Ta	—	—	—
W	0.0001	WO_4^{2-}	1.0×10^3
Re	—	—	—
Os	—	—	—
Ir	—	—	—
Pt	—	—	—
Au	0.000004	$AuCl_2^-$	5.6×10^5
Hg	0.00003	$HgCl_3^-$; $HgCl_4^{2-}$	4.2×10^4
Tl	<0.00001	Tl^+	—
Pb	0.00003	Pb^{2+}; $PbSO_4$; $PbCl_n^{2-n}$; $Pb(OH)_n^{2-n}$	2.0×10^3
Bi	0.00002	—	4.5×10^5
Po	—	—	—
At	—	—	—
Rn	0.6×10^{-15}	Rn (g)	—
Fr	—	—	—
Ra	1.0×10^{-10}	Ra^{2+}; $RaSO_4$	—
Ac	—	—	—
Th	0.00005	—	3.5×10^2
Pa	2.0×10^{-9}	—	—
U	0.003	$UO_2(CO_3)_3^{4-}$	5.0×10^5

3 The Salinity Structure of the Seas

The salinity structure of the seas in its broadest aspects bears certain resemblances to the temperature structure. At the surface both exhibit seasonal and diurnal fluctuations, both are modified by currents and mixing, both tend to be highly variable near the surface and become more constant with increasing depth. On first inspection, salinity variations may appear to be less than temperature fluctuations but, when we remember that the temperature should be expressed in an absolute scale (in degrees Kelvin), we find that the percentage changes of both variables in the oceans are comparable.

For the study of subsurface water masses, plots of the temperature versus salinity, the so-called *TS* curves, have proved helpful in identifying water masses (Figure 5.1). A brief discussion of such diagrams can be found in Sverdrup, Johnson, and Fleming (1942).

The vertical salinity structure of the sea at a particular time and location can be represented by a *salinity profile*, which is analogous to a *temperature profile* (Figure I.3 of the Introduction). The salinity profile, like the temperature profile, tends to fluctuate in the surface layer; it changes in a marked and regular way in a region sometimes called the "halocline"; and then, finally, changes very slowly in the deep waters (Figure 5.2). The salinity profile shown in Figure 5.2 is not a "typical" one. Given places at given seasons tend to have

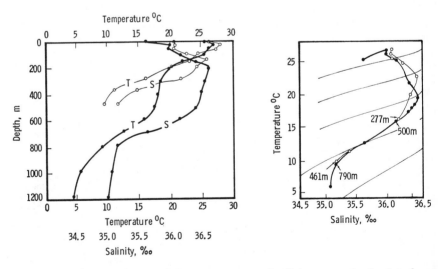

Figure 5.1 The *TS* Diagram. Left, temperature and salinity versus depth plots from data on the Gulf Stream off Onslow Bay. Right, the same information plotted as a *TS* diagram. From Sverdrup, Johnson, and Fleming (1942), with permission of Prentice-Hall, Inc.

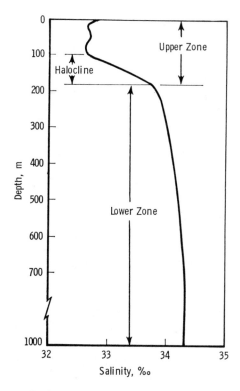

Figure 5.2 Salinity Profile in the Subarctic Pacific Ocean, February 1957. From Tully and Barber (1961), with permission of AAAS and the authors.

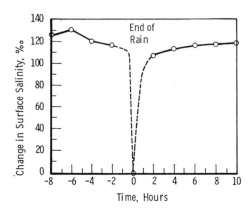

Figure 5.3 Reduction of Surface Salinity due to Precipitation. From Defant (1961), with permission of Pergamon Press, Ltd.

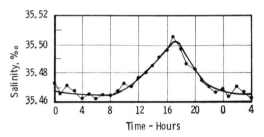

Figure 5.4 Diurnal Salinity Variations.

characteristic salinity profiles, and this particular profile is for the subarctic Pacific Ocean in winter. In this region, precipitation exceeds evaporation, giving rise to reduced salinity at the surface. The lower boundary of the halocline represents the limit of the downward transfer of fresh water (Tully and Barber, 1961). The immediate effect of rain on surface salinity is shown in Figure 5.3, where the change in salinity is given as deviations ($1/100\%_0$) of the value of salinity relative to its value at the end of the rain (time = 0, $1/100\%_0$ = 0). After the rain ceases, turbulent mixing brings the salinity back to very nearly its original value within 2–3 hours (Defant, 1961). The phenomenon is amenable to quantitative analysis. Similarly an influx of fresh water from the melting of icebergs can reduce the salinity.

Apart from occasional perturbations such as precipitation or melting ice, the salinity profile in a given location also may exhibit regular diurnal and seasonal fluctuations. Figure 5.4 shows the variation in the salinity of the surface waters of an Atlantic station over a 24-hour period. After sunrise the salinity begins to rise, more rapidly as evaporation increases, going through a maximum in the late afternoon and falling off rapidly after darkness (Defant, 1961). Figure 5.5 shows the seasonal variation at a location in the North Atlantic. Seasonal variations in the salinity of the surface waters

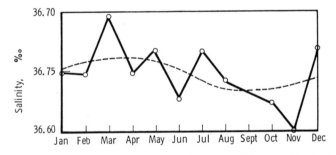

Figure 5.5 Seasonal Salinity Variations. From Sverdrup, Johnson, and Fleming (1942), with permission of Prentice-Hall, Inc.

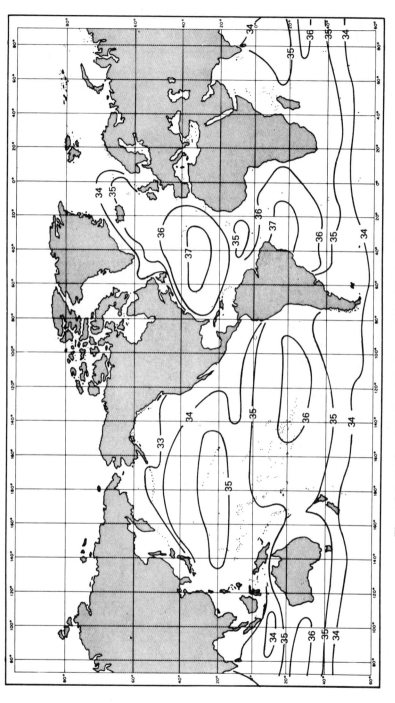

Figure 5.6 Surface Salinity of the Oceans in the Northern Summer.

159

depend on the evaporation-precipitation balance. In this particular region the salinity variations closely parallel the annual variations in evaporation, but much more complicated interrelations are frequently obtained.

The most spectacular features of the global variations of surface salinity are a pronounced minimum near the equator and two maxima at latitudes of 20°N and 20°S (Figures 5.6 and 5.7). In the tropical regions high temperatures and strong trade winds are responsible for high evaporation rates and the high surface salinities. Toward the extreme north and south the salinity falls off as rainfall exceeds evaporation. The salinity minimum in the equatorial zone is due to greater rainfall and diminished wind speed.

In most oceans salinity decreases with depth, that is, the vertical salinity profile resembles Figure I.3 rather than Figure 5.2. Figure 5.8 is a cross section of the salinity structure in the Atlantic Ocean at about 23°S. The salinity in the open sea varies over a relatively narrow range. Large temperature and salinity gradients are sometimes encountered in the zones marking the boundaries of the great ocean currents. Fresh water runoff from the continents tends to reduce the salinity near land masses. The salinity structure of the subarctic Pacific Ocean (Figure 5.2) has been described as estuarine (Tully and Barber, 1961) with its light fresh water flowing away from land over more dense and highly saline waters. Just the reverse occurs outside the Straits of Gibraltar, where heavy, highly saline water from the Mediterranean Sea flows out through the straits and down the side of the land mass *under* the less saline Atlantic water (Figure 5.9).

Sometimes special conditions create extraordinary salinities. One of the most spectacular of such anomalies has come to light recently with the discovery of very hot (45°C), very saline (43‰ S and greater) waters at the bottom of the Red Sea (Miller, 1964; Swallow and Crease, 1965). Chemical

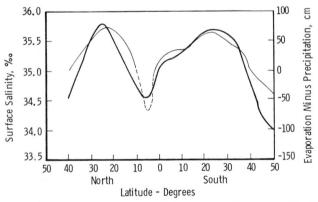

Figure 5.7 Average Surface Salinity for All Oceans as a Function of Latitude. From Sverdrup, Johnson, and Fleming (1942), with permission of Prentice-Hall, Inc.

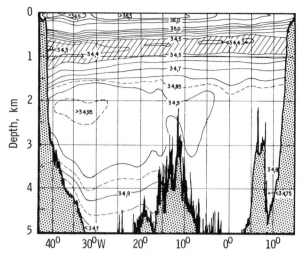

Figure 5.8 Cross Section of the Salinity Structure in the Atlantic Ocean at about 23°S. From Defant (1961), with permission of Pergamon Press, Ltd.

analyses (Brewer, Riley, and Culkin, 1965) of these waters are equally surprising—the ratios of the constituents to the chlorinities: 0.5980 for Na, 0.0139 for K, 0.0303 for Ca, 0.0052 for Mg, 0.0048 for $SO_4{}^{2-}$, and 0.00079 for Br^-—are quite different from the corresponding ratios for seawater (compare Table 5.2), Na^+ and Ca^{2+} enrichment being most obvious; so different, in fact, that the concepts of salinity loses its meaning in connection with these waters. Their composition is, however, very similar to that found in oil fields and deep-well brines (Neumann and Chave, 1965; Craig 1966) (see Table A.23 in the Appendix for the composition of a natural brine solution) which has led to the hypothesis that these are connate waters. Isotopic measurements with D and O^{18} have suggested that these highly saline waters are the consequence of leaching processes in the sedimentary rock in a shallow sill about 1000 km deep near the Red Sea basin where they are found.

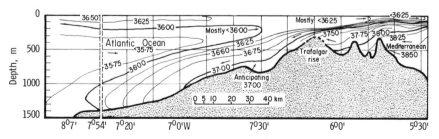

Figure 5.9 Salinity Profile through the Straits of Gibraltar in Spring-Summer. From Defant (1961), with permission of Pergamon Press, Ltd.

4   Summary

Although the total electrolyte content of seawater is variable, the relative amounts of the major ionic constituents tend to be the same in all the oceans of the world; that is, the oceans, like the atmosphere, are well mixed. All the naturally occurring chemical elements are found in seawater— from hydrogen, the most abundant (1.08×10^5 mg/l), to radon (6×10^{-6} mg/l).

The total salt content or salinity tends to be most highly variable near the surface, the boundary of ocean currents, or coastal areas. The surface salinity is dependent on the balance of evaporation and precipitation and, as a consequence, shows systematic diurnal and seasonal fluctuation as well as characteristic features dependent on geographic location.

REFERENCES

Brewer, P. G., J. P. Riley, and F. Culkin, *Nature*, **206**, 1345 (1965); *Deep-Sea Res.*, **12**, 497 (1965).

Carritt, D. E., and J. H. Carpenter, "Conf. on Phys. Chem. Properties of Sea Water," *Easton, Maryland*, 1958, *Natl. Acad. Sci.-Natl. Res. Council, Publ.*, No. 600 (1959).

Craig, H., *Science*, **154**, 1544 (1966).

Culkin, R., "The Major Constituents of Sea Water," in J. P. Riley and G. Skirrow, eds., *Chemical Oceanography*, Academic Press, London, 1961, Chap. 4.

Defant, A., *Physical Oceanography*, Pergamon, New York, 1961, Vol. 1, Chap. 4.

Dittmar, W., in J. Murray, ed., *The Voyage of the H.M.S. Challenger*, H.M. Stationery Office, London, 1884, Vol. I.

Goldberg, E., "Minor Elements in Sea Water," in J. P. Riley and G. Skirrow, eds., *Chemical Oceanography*, Academic Press, London, 1965, Chap. 5.

Goldberg, E., "The Oceans as a Chemical System," in M. N. Hill, ed., *The Sea*, Interscience, New York, 1963, Vol. 2, Chap. 1.

Marcet, A., *Phil. Trans. Roy. Soc. (London)*, **109**, 161 (1819).

Maury, M. F., *The Physical Geography of the Sea*, Sampson, Low, Son & Co. London, 1855.

Miller, A. R., *Nature*, **203**, 590 (1964).

Neumann, A. C., and K. E. Chave, *Nature*, **206**, 1346 (1965).

Sverdrup, H. U., *Oceanography for Meteorologists*, G. Allen & Unwin, Ltd., London, 1945.

Sverdrup, H. U., M. W. Johnson, and R. H. Fleming, *The Oceans*, Prentice-Hall, Englewood Cliffs, N.J., 1942, Chap. 4.

Swallow, J. C., and J. Crease, *Nature*, **205**, 165 (1965).

Tully, J. P., and F. G. Barber, "An Estuarine Model of the Sub-Arctic Pacific Ocean," in M. Sears, ed., *Oceanography*, Am. Assoc. Advan. Sci., Publ. No. 67, Washington, D.C., 1961.

6 Chemical Equilibria in the Seas

1 The Types of Species in Seawater

More important even than the elemental composition of seawater are the specific chemical species present. Garrels and Thompson (1962) have prepared a table (Table 6.1) giving the species in which some of the major constituents are found at 25°C and 1 atm in seawater of 19‰ Cl and pH 8.1. Although Na, K, and even Ca are present as simple hydrated cations, in the case of Mg and nonhalide anions more complex species account for appreciable fractions of their total quantities. In concentrated brine solutions an appreciable fraction of the Na and K may even be complexed (Table A.23). Inspection of the third column in Table 5.3 leads to the same conclusion, namely, elements present in seawater in the form of simple ions are a minority and complex species are of great importance. The species in seawater fall into several types and, inasmuch as a certain sloppiness of nomenclature has found its way into the literature, I think some words clarifying the difference in nature of these several types of species are appropriate here.

Table 6.1 Species Distribution of some Major Seawater Constituents (From Garrels and Thompson, 1962, with permission of *American Journal of Science*)

Ion	Molality	Free Ion, %	With Sulfate, %	With Biocarbonate, %	With Carbonate, %
Ca^{2+}	0.0104	91	8	1	0.2
Mg^{2+}	0.0540	87	11	1	0.3
Na^+	0.4752	99	1.2	0.01	—
K^+	0.0100	99	1	—	—

			With Ca, %	With Mg, %	With Na, %	With K, %
SO_4^{2-}	0.0284	54	3	21.5	21	0.5
HCO_3^-	0.00238	69	4	19	8	—
CO_2^{2-}	0.000269	9	7	67	17	—

If the constituent ions of a strong electrolyte such as NaCl are added to water, apart from the trauma represented by the hydration process, relatively little happens (Figure 6.1). The ions swim through the medium more or less independently of one another's presence. On the opposite extreme, if calcium and carbonate ions are added to water, they immediately combine and a definite solid compound precipitates out:

$$(6.1) \qquad Ca^{2+} + CO_3^{2-} \rightarrow CaCO_3 \text{ (s)}, \quad K = 1/(Ca^{2+})(CO_3^{2-}) = 1/K_{sp}$$

where K_{sp}, by the way, is called the solubility product of $CaCO_3$. If the ions of a weak electrolyte are added to water, only a certain fraction of them combine with one another:

$$(6.2) \qquad Mg^{2+} + SO_4^{2-} \rightleftharpoons MgSO_4, \quad K = (MgSO_4)/(Mg^{2+})(SO_4^{2-})$$

It should be clear that, throughout this book, when we speak of seawater as consisting of dissolved NaCl and $MgSO_4$, we do so only for the sake of convenience with no intent to imply that these are the species actually present. In

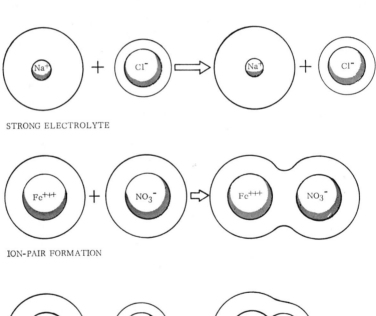

STRONG ELECTROLYTE

ION-PAIR FORMATION

COMPLEX ION FORMATION (Fe$(H_2O)_6^{+++}$ + Cl$^-$ $\rightleftharpoons$ Fe$(H_2O)_5$ Cl^{++} + H_2O)

Figure 6.1 Ion-Pair and Complex-Ion Formation.

a sense, $MgSO_4$ is a "complex ion" even though it does not have a charge, but a cation better than Mg^{2+} for describing complex ion formation is Fe^{3+}. Ferric ion has a coordination number of 6, that is, it is capable of binding six ligands. These ligands may be charged or neutral atoms or groups of atoms; for example, the ligand may be water to give $Fe(H_2O)_6^{3+}$. Inasmuch as the Fe-H_2O bond has a lot of covalent character, it is perhaps an over-simplification to consider these H_2O's as being simply waters of hydration. Ligands can be replaced in whole or in part by other ligands. Thus, if we add chloride ion to a solution of ferric ion, we may get successive formation of a series of complex ions.

(6.3)

$$\xrightarrow{Cl^-} Fe(H_2O)_2Cl_4{}^- + H_2O$$

$$\xrightarrow{Cl^-} Fe(H_2O)Cl_5{}^{2-} + H_2O$$

$$\xrightarrow{Cl^-} FeCl_6{}^{3-} + H_2O$$

Mixed complex ions such as $Fe(H_2O)_4ClBr^+$ also exist. The anionic complexes of iron(III) are relatively weak and are present only in concentrated aqueous HCl solutions. Other complexes are very strong and can exist even in dilute solutions; examples are $Fe(CN)_6{}^{3-}$ and $AuCl_2{}^-$. The latter species may be the form in which gold is present in seawater.

A complex ion is a definite, identifiable species. It may even have a characteristic color: Fe^{3+} is colorless but the chloride complexes are yellow; Cu^{2+} is pale blue but $Cu(NH_3)_4^{2+}$ is an intense blue. A more casual form of combination is possible and is frequently exemplified in marine chemistry, and that type of species is the ion pair. The partners in an ion pair are electrostatically attracted to one another and tend to loiter near one another—a liaison but not a marriage. The dividing line between a true complex ion and a mere ion pair is not sharply defined. One criterion for distinguishing between the two types of species is the following: in an ion pair the hydration atmospheres of the partners remain more or less intact although somewhat joined (Figure 6.1); in a complex ion, on the other hand, the separate hydration sheaths of the reactants merge to form a joint hydration sheath (Figure 6.1). In the former case the change in volume is slight, in the latter case appreciable. Consequently ion-pair formation should be independent of pressure, but complex ion formation should be pressure-dependent. Thus the formation of $FeNO_3^{2+}$ ion pair shows little pressure dependence, whereas the formation of the $FeCl^{2+}$ complex ion is pressure-dependent (Horne, Myers, and Frysinger, 1964). Equilibrium (6.2) is, as we shall see, also strongly pressure-dependent, hence we are justified in describing $MgSO_4$ as a complex "ion" rather than as an ion pair.

In seawater the more important ligands are H_2O, OH^-, and Cl^-, and to a lesser extent forms of sulfate and carbonate (Martin, 1967). Partially protonated polyvalent anions such as HCO^{3-}, HPO_4^-, and HPO_4^{2-} are not ordinarily considered to be complex ions.

2   Principles of Chemical Equilibria

Consider the generalized chemical reaction

$$(6.4) \qquad bB + cC + \cdots \underset{R_r}{\overset{R_f}{\rightleftharpoons}} xX + yY + \cdots$$
$$\text{Reactants} \qquad \text{Products}$$

The rate of the forward reaction, R_f, is proportional to the concentrations of the reactants, or

$$(6.5) \qquad R_f = k_f(B)^b(C)^c, \quad \text{etc.}$$

where k_f is the specific rate constant for the forward reaction and where the parentheses indicate concentrations of the species symbolized within them. Similarly the rate of the reverse reaction is given by

$$(6.6) \qquad R_r = k_r(X)^x(Y)^y, \quad \text{etc.}$$

Steady state equilibrium is reached when the rates of the forward and reverse reactions are equal:

(6.7)
$$R_f = R_r$$

from which it follows that

(6.8)
$$\frac{(X)^x(Y)^y, \text{ etc.}}{(B)^b(C)^c, \text{ etc.}} = \frac{k_f}{k_r} = K$$

where K is the *equilibrium constant*.

The foregoing paragraph does not constitute a derivation of the equilibrium constant; rather it is a qualitative description which gives some insight into the meaning of the concept. Next the rigorous derivation of the equilibrium constant will be outlined. The partial molar free energy of the species i, $(\partial F/\partial n_i)_{T,P,n_j}$, is sometimes called the *chemical potential* μ of the species. Let reaction (6.4) reach equilibrium at constant temperature and total pressure. Now displace the system to the right by an infinitesimal amount. The accompanying free energy change is

(6.9)
$$dF_{T,P} = \left(\sum \mu_i \, dn_i \right)_{\text{products}} - \left(\sum \mu_i \, dn_i \right)_{\text{reactants}}$$

In a closed system at equilibrium the free energy change $dF_{T,P}$ is equal to zero. Also the changes in the number of moles, dn_i, is proportional to the corresponding stoichiometric coefficients in (6.4). Thus (6.9) becomes

(6.10)
$$(x\mu_X + y\mu_Y + \ldots) - (b\mu_B + c\mu_C + \ldots) = 0$$

The left side of this equation is equivalent to the free energy change; thus we can write

(6.11)
$$\Delta F = 0$$

and these two expressions, (6.10) and (6.11), represent the fundamental condition for chemical equilibrium.

The chemical potential of the constituent species i is given by

(6.12)
$$\mu_i = \mu_i^\circ + RT \ln a_i$$

where μ_i° is the chemical potential of that constituent species in a chosen standard state of unit activity and a_i is the activity of species i in the system under consideration. Substituting expressions of the form (6.12) in (6.10) gives

(6.13)
$$x(\mu_X^\circ + RT \ln a_X) + y(\mu_Y^\circ + RT \ln a_Y) + \ldots$$
$$- b(\mu_B^\circ + RT \ln a_B) - c(\mu_C^\circ + RT \ln a_C) - \ldots = 0$$

and

(6.14)
$$RT \ln \frac{a_X^x \times a_Y^y \times \ldots}{a_B^b \times a_C^c \times \ldots} = (b\mu_B^\circ + c\mu_C^\circ + \ldots) -$$
$$- (x\mu_X^\circ + y\mu_Y^\circ + \ldots) = -\Delta F^\circ$$

At constant temperatures the right side of equation (6.14) and RT are constant; hence

(6.15)
$$\frac{a_X^x \times a_Y^y \times \ldots}{a_B^b \times a_C^c \times \ldots} \equiv \text{constant} \equiv K^\circ$$

where K° is the *thermodynamic equilibrium constant*. K° is expressed in terms of activities; in practice it is much easier to deal with concentrations. The practical equilibrium constant K is related to the thermodynamic equilibria constant:

(6.16)
$$K^\circ = \frac{a_X^x \times a_Y^y \times \ldots}{a_B^b \times a_C^c \times \ldots} = \frac{\gamma_X^x \times \gamma_Y^y \times \ldots}{\gamma_B^b \times \gamma_C^c \times \ldots} \cdot \frac{(X)^x(Y)^y \ldots}{(B)^b(C)^c \ldots}$$

$$= \frac{\gamma_X^x \times \gamma_Y^y \times \ldots}{\gamma_B^b \times \gamma_C^c \times \ldots} \cdot K$$

where the γ's are activity coefficients, from which it follows that

(6.17)
$$\lim_{c \to 0} K = K^\circ$$

where c is the concentration of the solution. K° is a true constant at a given temperature and total pressure, but K is not, being concentration-dependent. K is constant enough for many purposes, however. In calculating K the concentrations of gases are expressed as their partial pressures (or activities as fugacities for K°) and the concentration of water in aqueous solution and of solids is taken to be unity.

The temperature dependence of the equilibrium is given by the Van't Hoff equation

(6.18)
$$\frac{d \ln K^\circ}{dT} = \frac{\Delta H^\circ}{RT^2}$$

or approximately

(6.19)
$$\frac{d \ln K}{dT} \approx \frac{\Delta H}{RT^2}$$

where ΔH is the heat of the reaction. The pressure dependence of K will be considered in connection with particular equilibria. Inasmuch as

(6.20)
$$-\Delta F = nE\mathscr{F} = RT \ln K$$

where n is the number of Faradays, $\mathscr{F}$ is Faraday's number, and E is the electromotive force or cell voltage, it is possible to relate electrical work and the equilibria constants:

(6.21)
$$E = E^\circ - \frac{RT}{n\mathscr{F}} \ln Q$$

where Q is the quotient of the product of resultant activities and the product of reactant activities and is usually synonymous with K° and often can be approximated by K.

3 Chemical Kinetics

The sign of ΔF (or ΔF°) can be used as a criterion for determining whether a reaction should go or not. If ΔF is negative, then $K > 1$; that is, the reaction should proceed as written. If, however, ΔF is positive, the equilibrium lies in the sense opposite to that in which it is written. The sign of ΔF tells us whether or not the reaction *should* go thermodynamically. It does not tell us whether or not it *will* go. To choose a well-known example, the reaction

(6.22)
$$H_2 \text{ (g)} + \tfrac{1}{2}O_2 \text{ (g)} \to H_2O \text{ (g)}$$

at 25°C has a free energy of -54.507 kcal/mole, yet a mixture of H_2 and O_2 gases can stand around indefinitely. The *rate* of the reaction is infinitely slow. However, if an electric spark is passed through the mixture, it explodes violently. The rate of a chemical reaction is given by (3.4) and, if the Arrhenius activation energy, E_a, is very high, the rate can be slow even though ΔF is negative and large. E_a thus corresponds to an energy barrier (Figure 6.2) which must be overcome (for a much more detailed discussion consult Glasstone, Laidler, and Eyring, 1941).

Most simple ionic equilibria in aqueous solution tend to be very rapid, their rates often being controlled by diffusional and mixing processes. We are probably safe in assuming that most equilibria of marine chemistry are rapid,

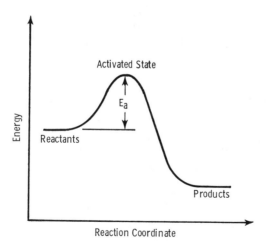

Figure 6.2 The Activation Energy.

and equilibrium models of seawater such as that proposed by Sillen (1967) are based on this important assumption. There do appear to be some slower processes, however; the precipitation of $CaCO_3$ and the deposition of manganese nodules are examples. Chemical investigation of a system passes through three distinct phases:

Identification and quantitative analysis of the chemical species.
Elucidation of the chemical equilibria.
Examination of the kinetics and mechanism of the chemical processes.

In the past, chemical oceanography has been concerned largely with the quantitative analysis of the elemental constituents in seawater. Contemporary marine chemists have recognized the importance of determining the species in which these elements occur and of the equilibria relating them. Virtually nothing is known about the kinetics of chemical reactions in the sea, but investigation directed to this end can be expected to play an important role in experimental marine chemistry in the future.

4   Chemical Species and Equilibria in Seawater

The remainder of this chapter is devoted to an element-by-element description of the more important constituents of seawater, the species in which they are present, and the equilibria involving them, together with more detailed information relevant to their concentration in seawater. The discussion of gaseous substances and their equilibria is deferred to a later chapter. The study of chemical equilibria is greatly facilitated by the compilations of stability complexes by Bjerrum, Schwarzenbach, and L. G. Sillen (1958) and Sillen and Martell (1964). The stability constants in Table A.13 in the Appendix are abstracted from this source.

Hydrogen Ion. Two-thirds of the atoms in the oceans are hydrogens. Hydrogen, obviously, is easily the most abundant element in seawater. The concentration of the hydrogen ion, or proton, H^+, in seawater is very small, between 10^{-7} and 10^{-8} M; yet, despite its small concentration, it is not an exaggeration to say that H^+ is one of the most significant ionic constituents of seawater for two reasons. It is involved in many equilibria (including most oxidation-reduction reactions) involving anions of weak acids, such as SO_4^{2-}, CO_3^{2-}, PO_4^{3-}, S^{2-}, SiO_4^{4-}, and BO_3^{3-}:

$$H^+ + A^{n-} \rightleftharpoons HA^{(n-1)-}$$

(6.23)

$$H^+ + HA^{(n-1)-} \rightleftharpoons H_2A^{(n-2)-}, \quad \text{etc.}$$

and the hydrolysis of metal cations:

$$M^{n+} + H_2O \rightleftharpoons MOH^{(n-1)+} + H^+$$

(6.24)

$$MOH^{(n-1)+} + H_2O \rightleftharpoons M(OH)_2^{(n-2)+} + H^+, \quad \text{etc.}$$

especially cations with oxidation numbers greater than 2 (Sillen, 1961). And H^+ plays an important role in many of the life processes; living things are very dependent on, and sensitive to, the pH.

In view of this sensitivity of the life processes to the H^+ concentration, a necessary condition for the genesis and evolution of life in the seas is that that medium cannot be subject to wild variations in pH. Seawater is a buffered system; that is, it is a system containing salts of weak (slightly dissociated) acids. The anions of such salts combine with the protons, thereby buffering the system so that the resulting acidity varies relatively little as H^+ or OH^- ions are added. Most body fluids and serums, including blood and milk, are also buffered, and any fluids injected into a living organism must either be neutral or buffered.

Like other cations, the "free" protons in water are hydrated, $H(H_2O)_n^+$, and, although usually written H^+ or H_3O^+ (hydronium ion), the best available modern evidence points strongly to $H_9O_4^+$ as the species present in greatest amounts (Figure 3.15).

The equilibrium constant for

(6.25) $$H_2O \rightleftharpoons H^+ + OH^-, \quad K_w = (H^+)(OH^-)$$

decreases with decreasing temperature (Figure 6.3) so that the H^+ concentration decreases at low temperatures. Earlier suggestions by Owen and Brinkley

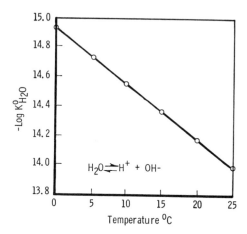

Figure 6.3   Temperature Dependence of the Dissociation Constant for Water. Based on data of Harned and Hamer (1933).

(1941), Owen (1953), Bodanszky and Kauzmann (1962), and Disteche (1962) that the dissociation of water increases with increasing pressure have been experimentally confirmed by Hamann (1963), who found that, at 25°C and 1000 atm, the dissociation constant is 2.14 times greater than at 1 atm. Hamann and his co-workers (1959, 1966) have further found that, at very high pressures (> 100,000 atm), pure water is highly conductive, and they theorize that the dissociation constant is increased to 10^{12} times its value at 1 atm.

The oceanographic literature contains many references to pH measurements both in samples and *in situ* (where its value is dependent on temperature and pressure), the majority pertaining to the $CO_2^- $-$HCO_3^- $-$CO_3^{2-}$ equilibrium in the sea—a topic to which we shall return. Because of its involvement in so many equilibria, within the limits set by the buffering action of seawater, the H^+ concentration often shows many interesting local variations but, before we turn to such variations, we must first address ourselves to the large problem of what, generally speaking, controls the pH of the oceans. While the importance of the buffering action of the $CO_2^- $-$HCO_3^- $-$CO_3^{2-}$ and also the H_3BO_3-H_4SiO_4 systems cannot be denied, Sillen (1961) points out that, if the oceans were a result of simply the titration of geological acids with geological alkali, then the resulting solution is remarkably near neutrality, within 0.5% of the "end point," and the pH of the ocean should be precarious. All of which suggests some further buffering factor. Garrels

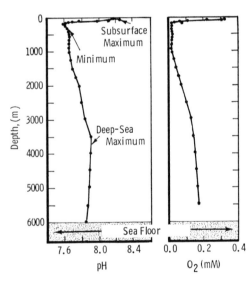

Figure 6.4 Deep-Sea pH Profile. From Park (1966), with permission of AAAS and the author.

(1965) and Mackenzie and Garrels (1965) have emphasized the importance of silica minerals in the control of the hydrogen ion concentration, and Sillen (1961, 1967) advocates a similar point of view. Park (1966) has published a pH profile taken in deep northeastern Pacific waters which exhibit several distinct inflections (Figure 6.4). He ascribes the observed pH maximum of 8.2–8.3 in the first 100 m to air-sea CO_2 exchange and biological activity. The pH minimum of 7.5–7.7 at 200–1200 m corresponds closely to the observed minimum in the oxygen profile (Figure 6.4) and so is attributed to biochemical processes. The causes of the second maximum of about pH 7.9 near 400 m is less clear. The author expresses the opinion that the effect of hydrostatic pressure on the dissociation constants of H_3CO_3 are primarily responsible, but allows that there may be a dependence on the composition of the bottom sediments. He sees the deep maximum as marking the boundary between two zones: in the upper regions the influence of biochemical processes dominate; in the lower, physiochemical and possible geochemical processes.

Unfortunately the availability of modern instrumentation has made the measurements of "pH" so facile that values of dubious meaningfulness often

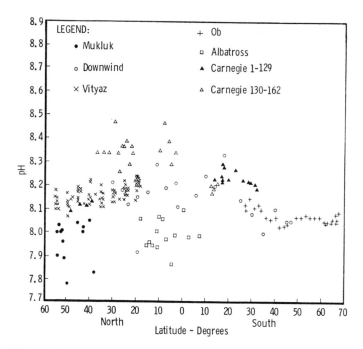

Figure 6.5   pH of Surface Waters on the Western Pacific. From Postma (1964), with permission of Nederlands Institut voor Onderzoek der Zee.

find their way into the literature, and as a consequence there is considerable scatter (Figure 6.5). But closer examination of Figure 6.5 also reveals variations which are almost certainly real. This is not surprising, for the pH depends on many factors, in particular, as we have seen, on the amount of dissolved CO_2, which in turn depends on photosynthetic activity and thence on the amount of solar radiation. It is not surprising, then, to find systematic seasonal and diurnal variations in pH (Figures 6.6 and 6.7).

Figure 6.4 is characteristic inasmuch as high pH tends to occur near the surface where the water is in equilibrium with the CO_2 of the atmosphere. Even higher pH's sometimes occur in the shallow water of tidal pools, bays, and estuaries (Figure 6.6). On the other hand, pH's as low as 7.0 (neutral) and even less (acid) are encountered in diluted seawater and in basins where H_2S is produced.

Oceanographers sometimes use the term *alkalinity* to describe the number of milliequivalents of H^+ necessary to titrate the anions of weak acids in one liter of seawater. Unfortunately this usage is confusing because chemists have long used the same term to describe the hydroxyl ion concentration.

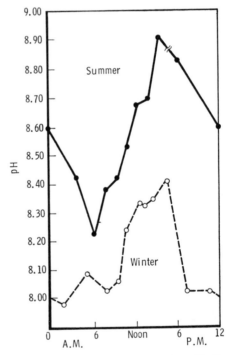

Figure 6.6 Diurnal and Seasonal Variations of pH in a Shallow Texas Bay. From Park, Hood, and Odum (1958), with permission of Plenum Press.

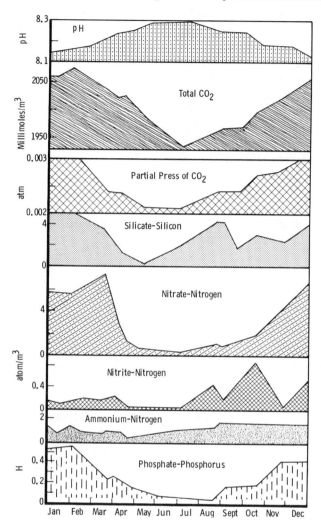

Figure 6.7 Seasonal Average pH and Concentrations of Various Constituents in the English Channel. From Dietrich (1963) (after Cooper), with permission of John Wiley & Sons.

Sodium, Potassium, and the Alkali Metal Cations. The salts, even the hydroxides and carbonates of Na^+, K^+, and the other alkali metal cations, Li^+, Rb^+, and Cs^+, are, with very few exceptions, strong electrolytes and completely dissociated in aqueous solution at all temperatures and pressures in the oceanographic range. It is possible, however, to exceed their solubility products in the concentrated brines formed on evaporation and/or freezing.

As Table 6.1 shows, in seawater at least 99% of the Na and of the K is in the form of the "free" ions, while the consensus summarized in Table A.6 of the Appendix would appear to be that these two elements are best represented as $Na(H_2O)_4^+$ and $K(H_2O)_4^+$. A number of years ago Adams and Hall (1931) proposed that the anomalous increase in the electrical conductivity of aqueous NaCl solution with increasing pressure was due to increased dissociation but, inasmuch as the electrolyte is almost completely dissociated at 1 atm, this explanation is untenable and now, in the light of pressure-induced ionic dehydration (see Chapter 3), redundant.

Platford (1965a, 1965b) has measured the activity coefficient of NaCl in seawater as a function of concentration and temperature (see Table A.15) and found it to be slightly higher in synthetic seawater than a plain NaCl solution of the same ionic strength. Nevertheless he concludes that the γ_{NaCl} in seawater is adequately approximated by γ_{NaCl} of an NaCl solution of the same ionic strength. Electrode determinations of $\gamma_{Na_2SO_4}$ in seawater lead to the conclusion that 1.5% of the Na^+ and 27% of the SO_4^{2-} are tied up as $NaSO_4^-$ (Platford and Defoe, 1965), which is in agreement with the earlier findings of Garrels and Thompson (1962) as listed in Table 6.1.

The oxidation-reduction potentials of these elements are so high (see Table A.14) that they do not undergo redox reactions under naturally occurring conditions. They are not metabolized by organisms, and their concentration in the body fluids of living creatures does not differ widely from that in seawater. All of this says that their concentrations in seawater are not subject to special perturbations except by evaporation and rainfall, and that they vary only as the salinity varies. Table 5.2 shows that the ratio of their concentrations to the salinity stays constant throughout the oceans of the world. The only exceptions, such as the hot saline waters of the Red Sea discussed earlier, represent the conspiracy of very special circumstances. The same constancy of the relative amount of the less abundant alkali metal cations is expected and has been confirmed for lithium (Riley and Tongudai, 1964). Folsom, Feldman, and Rains (1964) have examined the cesium content of seawater and, unlike Smales and Salmon (1955), they find a 14% average increase in the cesium content at 500–1800 m compared to the surface waters which they attribute to downward transport by particulate material.

The average lithium content of 19.374‰ Cl seawater is 183μg/l, and for the major oceans and seas the concentration of lithium is proportional to the chlorinity (Riley and Tongudai, 1964). The rubidium content of seawater down to 3900 m is uniform and averages about 120 μg/l (Smith, Pillai, et al., 1965).

Magnesium. Unlike Na^+ or K^+, for Mg^{2+} the possibility of the formation of a number of complex ions and ion pairs with anions contained in seawater must be considered. For purposes of this discussion, unless otherwise stated, we take the seawater to have a salinity of 35‰ and a pH of 8, and the temperature to be 25°C and the pressure 1 atm. Bjerrum, Schwarzenbach, and Sillén (1958), in their useful compilation of stability constants, mention that a weak complex of magnesium and chloride ions is expected on the basis of activity coefficient data

(6.26) $Mg^{2+} + Cl^- \rightleftharpoons MgCl^+,$ $K = (MgCl^+)/(Mg^{2+})(Cl^-)$

but that no experimental evidence for such a species has been forthcoming from conductivity (Righellato and Davies, 1930) and water vapor pressure (Stokes, 1945) measurements. As the concentration of $MgCl_2$ added to aqueous 0.47 M NaCl solution increases, the electrical conductivity increases more steeply as the pressure is increased (Horne and Frysinger, 1963) (Figure 6.8).

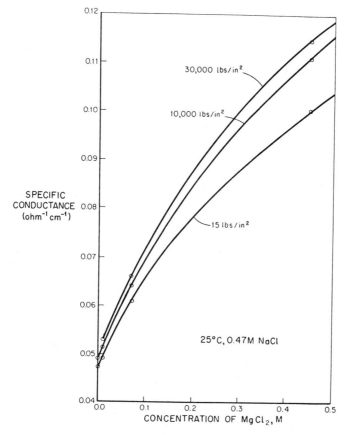

Figure 6.8 Concentration-Pressure Dependencies of the Electrical Conductivity of $MgCl_2$-NaCl Solutions. From Horne and Frysinger (1963).

This increase can be attributed to the increased dissociation of $MgCl^+$ under pressure although, of course, it is by no means conclusive or even good evidence for the existence of such a species.

Mg^{2+} can hydrolyze in seawater:

$$(6.27) \quad Mg^{2+} + H_2O \rightleftharpoons MgOH^+ + H^+, \quad K_1 = (MgOH^+)(H^+)/(Mg^{2+})$$

which, alternatively, can be treated as a stepwise complex ion formation:

$$(6.28) \quad Mg^{2+} + OH^- \rightleftharpoons MgOH^+, \quad K_2 = (Mg(OH)_2)/(MgOH^+)(OH^-)$$

$$(6.29) \quad MgOH^+ + OH^- \rightleftharpoons Mg(OH)_2, \quad K_3 = (Mg(OH)_2)/(MgOH^+)(OH^-)$$

Notice that equilibria (6.25), (6.27), and (6.28) are related by

$$(6.30) \qquad K_2 = \frac{K_1}{K_w}$$

The ratio $(MgOH^+)/(Mg^{2+})$ is given by $K_2(OH^-)$ or $K_2 K_w/(H^+)$ and has a value of about $(10^2)(10^{-14})/(10^{-8})$ or 10^{-4}. In other words, the concentration of $MgOH^+$, and certainly that of $Mg(OH)_2$, are negligible (10,000 times lower) compared with the concentration of Mg^{2+}.

Two further possibilities are equilibrium (6.2) and

$$(6.31) \quad Mg^{2+} + HSO_4^- \rightleftharpoons MgHSO_4^+, \quad K = (MgHSO_4^+)/(Mg^{2+})(HSO_4^-)$$

The constant for

$$(6.32) \quad HSO_4^- \rightleftharpoons H^+ + SO_4^{2-}, \quad K = (H^+)(SO_4^{2-})/(HSO_4^-)$$

is 0.095 (Eichler and Rabideau, 1955). Hence the $(SO_4^{2-})/(HSO_4^-)$ ratio is $(0.095)/(10^{-8})$ or about 10^7 and we need, therefore, to concern ourselves only with equilibrium (6.2). The ratio $(MgSO_4)/(Mg^{2+})$ is given by $K(SO_4^{2-})$ and has a value of *roughly* $(2.6)(0.027)$ or 0.07, indicating that an appreciable amount of the magnesium is in the form of $MgSO_4$. The foregoing estimates of species ratios are very crude, ignoring as they do activity coefficients. They are adequate, however, for identifying species deserving of more detailed attention. Garrels and Thompson (1962) give the activity coefficients of Mg^{2+} and $MgSO_4$ as 0.36 and 1.13, respectively (see Table A.15), and they calculate (Table 6.1) that 11% of the magnesium is in the form $MgSO_4$. They describe Mg^{2+} as "the major cation [in seawater] in terms of effectiveness in complexing anions." Platford (1965) has determined the activity coefficient of Mg^{2+} in seawater by a method involving precipitation of $Mg(OH)_2$ and has obtained values ranging from 0.17 in 5‰ S seawater to 0.0063 in 35‰ S seawater. His results indicate that only about 3% of the magnesium is in the form $MgSO_4$. Pytkowicz, Duedall, and Connors (1966), on the other hand, point out that the true state of affairs is unclear, and as much as 39–53% of the magnesium in seawater may be in the form $MgSO_4$. Such a conclusion, they add, is compatible with Fisher's (1962) determination of the formation constant of $MgSO_4$. Subsequently Thompson (1966) published the results of electrode measurements of magnesium ion in Standard Seawater, indicating that 90% of the element is in the form of free "ion," Mg^{2+}, a finding in agreement with the model of Garrels and Thompson (1962). What can be concluded from these contradictory findings? Not very much, I am afraid, about the state of magnesium in seawater. But they do contain a most important lesson: familiar physical-chemical concepts and techniques which work well on dilute, single electrolyte solutions can bring us to grief when we attempt to apply them to so complex a soup as seawater.

ΔH for the formation of $MgSO_4$ is 5.7 kcal/mole (Jones and Monk, 1952). The effect of pressure on equilibrium (6.2) has been examined by Fisher (1965) (Table 6.2). As a result of electrostriction the side of the equilibrium with the greater number of charged species represents the state of smaller volume and is therefore the state favored by the application of hydrostatic pressure. Table 6.2 clearly shows that, as in the case of other weak electrolytes, when the pressure is increased, dissociation increases. The difference in the partial molal volumes of products and reactants for equilibria (6.2), $\Delta \bar{V}°$, is $+7.3$ cm^3/mole.

Although the most important one, SO_4^{2-} is not the only anion in seawater capable of combining with Mg^{2+}. Garrels and Thompson (1962) (Table 6.1) estimate that 1% and 0.3% of the magnesium in seawater are in the forms $MgHCO_3^+$ and $MgCO_3$, respectively. Pytkowicz, Duedall, and Connors (1966) suggest that as much as 14% of the magnesium may be in the form of $MgHCO_3^+$, $MgCO_3$, and [despite our conclusion above in regard to equilibrium (6.28)] $MgOH^+$.

Magnesium, like sodium, is not a nutrient or product; hence, apart from pH effects, it is not perturbed by biological activity and its ion-to-salinity ratio in the seas tends to be both steady and uniform (Table 5.2). Because it is

Table 6.2   Molal Dissociation Constant for $MgSO_4$ in Aqueous Solution at 25°C (From Fisher, 1962, with permission of American Chemical Society)

$MgSO_4$ Concentration, 1 atm, M	K_{MgSO_4} at				
	1 atm	500 atm	1000 atm	1500 atm	2000 atm
0.0005	0.067	0.059	0.056	0.043	0.037
0.0010	0.107	0.096	0.089	0.075	0.068
0.0020	0.158	0.144	0.133	0.115	0.105
0.0100	0.314	0.290	0.271	0.248	0.230
0.0200	0.386	0.360	0.340	0.315	0.297

a very important cation geologically, the possibility of its perturbation by suspended matter or bottom sediments remains. Berner (1966) has demonstrated mole-by-mole cation exchange between aqueous $MgCl_2$ solutions and a fine-grained $CaCO_3$ sediment.

We return to the $MgSO_4$ equilibrium again in our discussion of the physical chemistry of underwater sound transmission in Part IV.

Calcium and Strontium. As for Mg^{2+}, complex and ion-pair formation is much more important for divalent ions such as Ca^{2+} and Sr^{2+} than for Na^+ or K^+. Equilibrium (6.1) is of particular importance in marine chemistry, together with

$$(6.33) \quad Ca^{2+} + HCO_3^- \rightleftharpoons CaHCO_3^+, \quad K = (CaHCO_3^+)/(Ca^{2+})(HCO_3^-)$$

and it is discussed in Chapter 7. Berner (1965) has determined γ_{Ca}^{2+} in seawater at 25°C and 1 atm (see Table A.15). The values he obtains (0.203 for standard artificial seawater of 19‰ Cl and 0.223 for Atlantic waters, taken near Woods Hole, of 17‰ Cl) are commensurate with the Garrels and Thompson (1962) chemical model of seawater, which gives the percentages of Ca^{2+} as $CaHCO_3^+$ and $CaCO_3$ as 1% and 0.2%, respectively (Table 6.1). Estimates of the $(CaOH^+)/(Ca^{2+})$ ratio from the reaction

$$(6.34) \quad Ca^{2+} + OH^- \rightleftharpoons CaOH^+, \quad K = (CaOH^+)/(Ca^{2+})(OH^-)$$

indicate that $CaOH^+$ formation is negligible at seawater pH's. Bianucci (1963) has calculated solubility and activity coefficients of $CaSO_4$ in seawater, and indeed the formation of this substance

$$(6.35) \quad Ca^{2+} + SO_4^{2-} \rightleftharpoons CaSO_4, \quad K = (CaSO_4)/(Ca^{2+})(SO_4^{2-})$$

is appreciable, representing as much as 8% of the total calcium in seawater (Table 6.1). The formation constant for $SrSO_4$ is much smaller than for $CaSO_4$, and for this reason I wish to take exception to Table 5.3, which lists this species as being present in appreciable amounts.

The $Ca^{2+}/‰$ Cl ratio (Table 5.2) is quite constant in the oceans of the world, and Chow and Thompson (1955) have made the same claim for the $Sr^{2+}/‰$ Cl ratio. Mackenzie (1964), however, has reported a distinct maximum in the strontium concentration between 500 and 800 m in the Sargasso Sea (Figure 6.9). He advances two possible explanations: the release of Sr by the solution of the calcareous skeletons of organisms at these depths and/or the aggregation of Sr-containing organic matter on the collapse of rising bubbles. This is also about the same depth range where cesium enrichment occurs (Folsom, Feldman, and Rains, 1964) and the minimum in the dissolved O_2 concentration in the same region (Figure 6.9) may be significant. For further discussion of Na, K, Mg, Ca, and Sr, see Culkin and Cox (1966).

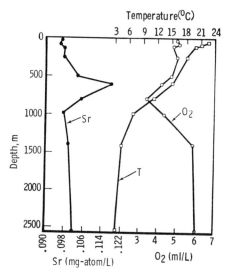

Figure 6.9 Vertical Distribution of Strontium, Oxygen, and Temperatures in the Sargasso Sea. From Mackenzie (1964), with permission of AAAS.

Iron. All the cationic seawater constituents that we have examined thus far have been of elements which exist in a single, highly stable oxidation or valence state, and we have had to concern ourselves only with simple ionic equilibria which do not involve oxidation-reduction reactions and consequent changes in valence state. Many of the minor constituents of seawater can exist in two or more valence states. Iron is one of the more important minor constituents of seawater, and we shall use it as an example. This element commonly exists in two valence states, ferrous, Fe^{2+}, and ferric, Fe^{3+}, but higher valence states are known. Before we can examine its equilibria we must determine which of these forms is the dominant one under the oxidative conditions obtaining in seawater (Cooper, 1948).

Under certain circumstances, such as in the presence of H_2S or in some anaerobic sediments, a reducing environment can exist in the sea but, under normal circumstances, owing for the most part to the dissolved oxygen, seawater represents a mildly oxidative environment. Analogous to the representation of the proton activity by pH, the oxidative condition or electron activity in seawater can be conveniently expressed as the pE, defined as

(6.36) $$pE \equiv -\log(e^-)$$

In practice the value of pE is obtained by dividing the oxidation potential of the solution (on the hydrogen scale) by $(RT/\mathscr{F})\ln 10$ (which is equal to 0.05915 v at 25°C and 0.05419 v at 0°C). In seawater, pE is determined by the half-cell reaction

(6.37) $O_2 (g) + 4 H^+ + 4 e^- \rightleftharpoons 2 H_2O (l)$, $E° = +1.229$

At 25°C Sillen (1961) takes the value of pE to be 12.5, and this value is not very sensitive to the usual small variations in pH or oxygen concentration in seawater.

The standard potentials, $E°$, for the valence states of iron are

(6.38) $Fe \rightarrow Fe^{2+} + 2 e^-$, $E° = +0.440$ v

(6.39) $Fe \rightarrow Fe^{3+} + 3 e^-$, $E° = +0.036$ v

(6.40) $Fe^{2+} \rightarrow Fe^{3+} + e^-$, $E° = -0.771$ v

in acid, and

(6.41) $Fe + 2 OH^- \rightarrow Fe(OH)_2 + 2 e^-$, $E° = +0.877$ v

(6.42) $Fe(OH)_2 + OH^- \rightarrow Fe(OH)_3 + e^-$, $E° = +0.56$ v

in basic solution (Latimer, 1952). Taking the acidic solution potential for the sake of simplicity (alkaline solution potentials lead to the same conclusion) and combining potentials (6.37) and (6.40), the potential for the oxidation of Fe(II) to Fe(III) by dissolved oxygen is +0.458 v. Therefore, by (6.20), $\Delta F°$ is negative and the reaction should go as written, namely,

(6.43) $4 Fe^{2+} + O_2 \text{ (dissolved)} + 4 H^+ \rightarrow 4 Fe^{3+} + 2 H_2O$

Although we have neglected factors such as pH, ionic strength, and complex ion formation, it is safe to say that the dominant form of iron in seawater is the +3 valence state. Sillen (1961) has estimated the concentration of Fe^{2+} in seawater to be 10^{-20} M. Redox equilibria can affect the pH of natural waters (Boström, 1967) but, in the sea, carbonate and silicate equilibria are far more important.

Having established the oxidation state of the element in seawater, we now can concern ourselves with the several possible ionic equilibria:

(6.44) $Fe^{3+} + Cl^- \rightleftharpoons FeCl^{2+}$ $K_1 = (FeCl^{2+})/(Fe^{3+})(Cl^-)$

(6.45) $FeCl^{2+} + Cl^- \rightleftharpoons FeCl_2^+$, etc., $K_2 = (FeCl_2^+)/(FeCl^{2+})(Cl^-)$

(6.46) $Fe^{3+} + OH^- \rightleftharpoons FeOH^{2+}$ $K_3 = (FeOH^{2+})/(Fe^{3+})(OH^-)$

(6.47) $FeOH^{2+} + OH^- \rightleftharpoons Fe(OH)_2^+$ $K_4 = (Fe(OH)_2^+)/(FeOH^{2+})(OH^-)$

(6.48) $Fe(OH)_2^+ + OH^- \rightleftharpoons Fe(OH)_3$ $K_5 = 1/Fe(OH_2^+)(OH^-)$

(6.49) $Fe^{3+} + HSO_4^- \rightleftharpoons FeHSO_4^{2+}$ $K_6 = (FeHSO_4^{2+})/(Fe^{3+})(SO_4^{2-})$

Sillén (1961) points out that in seawater the hydroxide complexes are important for all cationic species of oxidation states greater than 2. This may be surprising, for the Cl^- concentration is very much greater than the OH^- concentration (0.54 compared to 10^{-6} M) in seawater. Still using iron as an example, K_3 above is so much greater than K_1 that the ratio $(FeOH^{2+})/(FeCl^{2+})$ is still about 10^5 despite the very large $(Cl^-)/(OH^-)$ ratio. This ratio is estimated in the following ways: consider the equilibria

(6.50) $Fe^{3+} + H_2O \rightleftharpoons FeOH^{2+} + H^+$,

$$K_7 = (FeOH^{2+})(H^+)/(Fe^{3+})$$

(6.51) $FeCl^{2+} + H_2O \rightleftharpoons FeOH^{2+} + H^+ + Cl^-$,

$$K_8 = (FeOH^{2+})(H^+)(Cl^-)/(FeCl^{2+})$$

Multiplying both numerator and denominator of the expression for K_8 by (Fe^{3+}), we see that

(6.52) $$K_8 = \frac{K_7}{K_1}$$

For a rough approximation, ignoring activity coefficients, K_8 is estimated to be $10^{-3.6}$ from values of K_7° and K_1°, given by Bjerrum and co-workers (1958),

(6.53) $$(FeOH^{2+})/(FeCl^{2+}) = 10^{-3.6}/(Cl^-)(H^+)$$

A similar line of argument leads to the conclusion that $FeHSO_4^{2+}$ (or bicarbonate or phosphate complexes), like $FeCl^{2+}$, is also negligible compared to the hydroxide complexes resulting from the hydrolysis of Fe^{3+} (6.52) at pH 8. Although F^- complexes strongly with Fe^{3+}, Cooper (1948) has concluded that such complexes are unimportant in the sea. The question of the dominant Fe(III) species now resolves itself to which of the hydrolysis products is the more prevalent, $FeOH^{2+}$, $Fe(OH)_2^+$, or $Fe(OH)_3$. The solubility product of $Fe(OH)_3$

(6.54) $Fe(OH)_3 \rightleftharpoons Fe^{3+} + 3 OH^-$,

$$K_{sp} = (Fe^{3+})(OH^-)^3/(Fe(OH)_3) = (Fe^{3+})(OH^-)^3$$

is very small (10^{-37}), so small that the $(Fe(OH)_3)/(FeOH^{2+})$ ratio is large, and we can conclude, in agreement with Table 5.3, that the iron in seawater is in the form of $Fe(OH)_3$, probably suspended as a colloid. But, although most of the Fe(III) is in this form, about 10^{-6} M iron appears to be in true solution on the basis of analyses of carefully filtered samples (Lewis and Goldberg, 1954). The concentration of Fe^{2+} (see above) is much too small to account for this dissolved iron. Sillen (1961) suggests several forms for this dissolved iron including $FeOH^{2+}$, $Fe(OH)_2^+$, $Fe_2(OH)_2^{4+}$, and organic complexes. He also points out that too little is known about the phosphate complexing of Fe(III) to exclude it as a contributor. Earlier Cooper (1948) pointed out that Fe(III) phosphate and organic complexes should play a significant role in marine biochemistry.

The formation constant of $FeCl^{2+}$ is pressure-dependent (Horne, Myers, and Frysinger, 1964) and so, presumably, are the hydrolysis reactions; yet it seems unlikely that the pressures encountered in the oceans are great enough to affect seriously the conclusions that have been drawn.

Chloride. Hydrochloric acid is a very strong electrolyte and completely dissociated in aqueous solution. A number of the minor constituents of sea-water form chloride complex ions, and Mg^{2+} may, as we have seen, form a weak complex, but they represent the removal of only very small amounts of chloride, leaving the overwhelming majority in seawater as free Cl^-. The same is also true of the other halide anions (except possibly F^-, since HF is a partially dissociated acid and MgF^+ may be important (Sillen, 1961) (cf. Table 5.3). Nor is the concentration of Cl^- perturbed by the life processes. Thus the comparison of other ion concentrations to the chlorinity as a standard probably represents a meaningful procedure.

Sulfate. The first dissociation of sulfuric acid

(6.55) $H_2SO_4 \rightleftharpoons H^+ + HSO_4^-,$ $K = (H^+)(HSO_4^-)/(H_2SO_4)$

is strong, but the second (6.32) is weak (Eichler and Rabideau, 1955). As a result of electrostriction, however, the extent of dissociation of bisulfate ion increases with increasing pressure (Horne, Courant, and Frysinger, 1964). We have already seen that, at pH 8, HSO_4^- is negligible compared to SO_4^{2-}. Slightly more than half of the SO_4^{2-} in seawater is in the form of the free ion (Table 6.1). The remainder is tied up with Mg^{2+}, Na^+, and, to a lesser extent, Ca^{2+}.

Morris and Riley (1966), in agreement with the values summarized in Table 5.2, have found a very steady $SO_4^{2-}/\%_0$ Cl ratio of 0.1400 ± 0.0002 in the major oceans and seas of the world except for the Baltic Sea. In these waters the average $SO_4^{2-}/\%_0$ Cl is somewhat high, 0.1413, and varies geographically and with depth (Figure 6.10). Unlike most of the species examined hitherto, the SO_4^{2-} concentration is subject to perturbation by biological activity, in particular, reduction to H_2S by heterotrophic bacteria under anaerobic conditions. Thus in deep, stagnant, bottom water, the $SO_4^{2-}/\%_0$ Cl ratio can decrease (Kwiecinski, 1965; Richards, 1965).

Phosphate. The nature of the phosphate species in seawater is very complex because phosphoric acid dissociates in three steps

(6.56) $H_3PO_4 \rightleftharpoons H^+ + H_2PO_4^-,$ $K_1 = (H^+)(HPO_4^-)/(H_3PO_4)$

(6.57) $H_2PO_4^- \rightleftharpoons H^+ + HPO_4^{2-},$ $K_2 = (H^+)(HPO_4^{2-})/(H_2PO_4^-)$

(6.58) $HPO_4^{2-} \rightleftharpoons H^+ + PO_4^{2-},$ $K_3 = (H^+)(PO_4^{3-})/(HPO_4^{2-})$

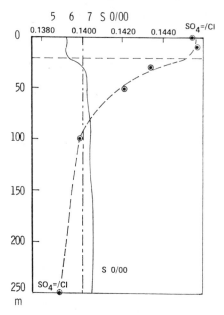

Figure 6.10   Sulfate to Chlorinity Profile in the Gulf of Bothnia, Baltic Sea. From Kwiecinski (1965), with permission of Pergamon Press, Ltd.

It is subject to biological perturbation, and the formation constants for phosphate complexes are, in general, not so well known as those for the analogous chloride, hydroxide, and sulfate complexes. Because of its importance as a nutrient the phosphate content of the seas has been an object of much attention, and the subject has been recently reviewed in some detail by McGill (1964) for the Atlantic Ocean and by Armstrong (1965). We return to the topic in Chapter 9 on the biochemistry of the seas. Here our concern is principally equilibria (6.56) to (6.58) and the other species which may contribute to the inorganic phosphorus content of seawater.

The successive dissociation constants of phosphoric acid in seawater is the subject of a recent thesis by Kester (1966), subsequently published by Kester and Pytkowicz (1967); it reports the values summarized in Table 6.3. Notice that these values are what the author calls an "apparent dissociation constant," denoted by K', which differs from what we have called the "practical equilibrium constant" inasmuch as the activity of the proton has been substituted for its concentration. Using these values in seawater of pH 8, $(PO_4^{3-})/$ (HPO_4^{2-}), $(HPO_4^{2-})/(H_2PO_4^-)$, and $(H_2PO_4^-)/(H_3PO_4)$ are roughly 10^{-1}, 10^{-2}, and 10^{+6}, respectively, from which the conclusion follows that the dominant species, in agreement with Sillen (1961), is HPO_4^{2-}, but that PO_4^{2-} and $H_2PO_4^-$ are still important, and H_3PO_4 is negligible (cf. Table 5.3). The

Table 6.3   Apparent Dissociation Constants for Phosphoric Acid in Seawater
(From Kester, 1966)

Temperature, °C	Salinity		
	30‰	33‰	36‰
K_1'			
5	—	2.68×10^{-2}	2.75×10^{-2}
10	—	2.51×10^{-2}	2.00×10^{-2}
15	2.72×10^{-2}	2.83×10^{-2}	3.05×10^{-2}
20	2.58×10^{-2}	2.35×10^{-2}	2.26×10^{-2}
25	2.36×10^{-2}	2.52×10^{-2}	2.37×10^{-2}
K_2'			
5	—	6.2×10^{-7}	6.7×10^{-7}
10	—	6.8×10^{-7}	7.3×10^{-7}
15	6.6×10^{-7}	7.2×10^{-7}	7.5×10^{-7}
20	7.1×10^{-7}	8.8×10^{-7}	10.3×10^{-7}
25	7.4×10^{-7}	7.9×10^{-7}	9.3×10^{-7}
K_3'			
5	—	0.46×10^{-9}	0.40×10^{-9}
10	—	0.61×10^{-9}	0.58×10^{-9}
15	0.52×10^{-9}	0.95×10^{-9}	0.66×10^{-9}
20	1.26×10^{-9}	1.37×10^{-9}	1.20×10^{-9}
25	1.79×10^{-9}	2.50×10^{-9}	2.80×10^{-9}

various phosphate species probably complex with cations in seawater, but virtually no reliable quantitative information exists to enable even crude estimates of the relative importance of such complexes to be made.

In addition to and as a result of its role as a nutrient in the sea, one very interesting aspect of the phosphate concentration should be mentioned here. Thus far we have given the impression that the oceans and seas of the world are very similar. This is not exactly true, and very luckily so for, in order to supplement the present method of characterizing water by TS diagrams, we would really like to find chemical differences in the waters of the different oceans which could be said to characterize those waters. Like us, seawater grows old. It contains two clocks, one biological and one isotopic (see Chapter 12). On the average, as a consequence of the processes of material exchange and circulation in the great oceans of the world, the waters of the three major oceans—Atlantic, Pacific, and Indian—are of different ages. If we could

Table 6.4   Concentration Ratios of Inorganic Nutrient Anions in the Deep Waters of the Major Oceans (From Chow and Mantyla, 1965)

Ocean	Silicate/Phosphate	Silicate/Nitrate	Nitrate/Phosphate
S.E. Pacific	55–65	3–5	13–14
Equatorial Indian	40–50	3	15
N. Atlantic	20–40	1–2	12–16

measure their ages, we could characterize their waters. This can be done by radioisotope studies, and it can also be done by studying the concentration of nutrient species in the sea for, as biological processes continue in seawater over a period of time, they modify the concentrations and the relations between the concentrations of nutrient substances. As a consequence of such processes the waters of the oceans are chemically identifiable on the basis of their silicate-to-nitrate and, even more spectacularly, their silicate-to-phosphate ratios (Table 6.4) (Chow and Mantyla, 1965).

In Chapter 9 we return to the subject of the phosphorus content of seawater.

Other Species. The CO_2-H_2CO_3-HCO_3^--CO_3^{2-}-$CaCO_3$ system is treated in Chapter 7 (dissolved gases) and silica and silicate species in Chapter 8 (suspended matter). In addition to dissolved N_2 (g), inorganic nitrogen is found in seawater as nitrate, NO_3^-, nitrite, NO_2^-, and ammonium ion NH_4^+ (Vaccaro, 1965). Nitrogenous material, like phosphorus, is an important nutrient in the seas. HNO_3, like HCl, is a strong electrolyte; hence most of the nitrate present should be in the form of the free ion. The concentrations of nitrate and ammonia in seawater are too small to influence inorganic equilibria appreciably (Sillén, 1961). Ammonia is a product of biological activity, and it is not surprising therefore to find its concentration highly variable (Beers and Kelly, 1965).

5   Summary

In order to determine the chemical form or forms of an elemental constituent in seawater it is necessary to consider both simple chemical equilibria and redox potentials. Solutes may be present as hydrated ions, complex ions, or ion pairs. Quantitatively the equilibria can be represented by a true thermodynamic equilibrium constant in terms of activities or by practical equilibrium constants expressed in terms of the species' concentrations.

Thermodynamics tells us whether a reaction in seawater should go. It remains for kinetics to tell us whether in fact it does go and at what rate.

Thus the kinetics of marine chemical reactions represents undoubtedly a very important, yet largely unexplored, field.

The hydrogen ion concentration in the sea, although small, is very important because other equilibria, and certain biochemical processes, are strongly dependent on it. Seawater is a buffered, very slightly alkaline solution. The pH of seawater shows a complex depth profile and a strong dependence on biological activity in the sea.

Sodium and potassium in seawater are almost entirely in the form of the Na^+ and K^+ ions, but magnesium is present mostly as the weak electrolyte, $MgSO_4$. Iron is present in its higher $+3$ valence state, probably in extensively hydrolyzed form.

Of the anions, chloride is present as Cl^-, sulfate as SO_4^{2-} rather than HSO_4^-, and the principal forms of phosphate are unclear.

REFERENCES

Adams, L. H., and R. G. Hall, *J. Phys. Chem.*, **35**, 2145 (1931).

Armstrong, F. A. J., "Phosphorus," in J. P. Riley and G. S. Skirrow, eds., *Chemical Oceanography*, Adademic Press, London, 1965, Vol. I, Chap. 8.

Beers, J. R., and A. C. Kelly, *Deep-Sea Res.*, **12**, 21 (1965).

Berner, R. A., *Geochim., Cosmochim. Acta*, **29**, 947 (1965).

Berner, R. A., *Science*, **153**, 188 (1966).

Bianucci, G., *Ann. Idrol.*, **1**, 72 (1963).

Bjerrum, J., G. Schwarzenbach, and L. G. Sillen, "Stability Constants," *Chem. Soc. (London), Sp. Publ.*, No. 7 (2 vols.: I, organic ligands; II, inorganic ligands) (1958).

Bodanszky, A., and W. Kauzmann, *J. Phys. Chem.*, **66**, 177 (1962).

Boström, K., in W. Stumm, ed., *Equilibrium Concepts in Natural Water Systems*, *Advan. Chem. Ser.*, No. 67 (1967).

Chow, T. J., and A. W. Mantyla, *Nature*, **206**, 383 (1965).

Chow, T. J., and T. G. Thompson, *Anal. Chem.*, **27**, 18 (1955).

Cooper, L. H. N., *J. Marine Biol. Assoc. U.K.*, **27**, 314 (1948).

Culkin, F., and R. A. Cox, *Deep-Sea Res.*, **13**, 789 (1966).

David, H. G., and S. D. Hamann, *Trans. Faraday. Soc.*, **95**, 72 (1959).

Dietrich, G., *General Oceanography*, Interscience, New York, 1963.

Disteche, A., *J. Electrochem. Soc.*, **109**, 1084 (1962).

Eichler, E., and S. Rabideau, *J. Am. Chem. Soc.*, **77**, 5501 (1955).

Fisher, F. H., *J. Phys. Chem.*, **66**, 1607 (1962).

Fisher, F. H., *J. Phys. Chem.*, **69**, 695 (1965).

Folsom, T. R., C. Feldman, and T. C. Rains, *Science*, **144**, 538 (1964).

Garrels, R. M., *Science*, **148**, 69 (1965).

Garrels, R. M., and M. E. Thompson, *Am. J. Sci.*, **260**, 57 (1962).

Glasstone, S., K. J., Laidler, and H. Eyring, *The Theory of Rate Processes*, McGraw-Hill Book Co., New York, 1941.

Hamann, S. D., *J. Phys. Chem.*, **67**, 2233 (1963).

Hamann, S. D., and M. Linton, *Trans. Faraday Soc.*, **62**, 2234 (1966).

Harned, H. S., and W. J. Hamer, *J. Am. Chem. Soc.*, **55**, 2194, 4496 (1933).

Horne, R. A., R. A. Courant, and G. R. Frysinger, *J. Chem. Soc.*, 1515 (1964).

Horne, R. A., and G. R. Frysinger, "Physical Chemistry in the Ocean Depths: The Effect of Pressure in Ionic Transport Processes and Equilibria," Arthur D. Little, Inc. Project Trident Tech. Rept. No. 1320763 (March 1963), Bur. Ships Contract No. brs-81564 S-7001-0307 (Unclass.)

Horne, R. A., B. R. Myers, and G. R. Frysinger, Inorg. Chem., 3, 452 (1964).

Jones, H. W., and C. B. Monk, Trans. Faraday Soc., 48, 929 (1952).

Kester, D. R., M.S. Thesis, Oregon State Univ., 1966.

Kester, D. R., and R. M. Pytkowicz, Limnol. Oceanog., 12, 243 (1967).

Kwiecinski, B., Deep-Sea Res., 12, 797 (1965).

Latimer, W. M., The Oxidation States of the Elements and Their Potentials in Aqueous Solutions, 2nd ed., Prentice-Hall, Englewood Cliffs, N.J., 1952.

Lewis, G. J., Jr., and E. D. Goldberg, J. Marine Res. (Sears Found. Marine Res.), 13, 183 (1954).

McGill, D. A., Progr. Oceanog., 2, 127 (1964).

Mackenzie, F. T., Science, 146, 517 (1964).

Mackenzie, F. T., and R. M. Garrels, Science, 150, 57 (1965).

Martin, D. F., in W. Stumm, ed., Equilibrium Concepts in Natural Water Systems, Advan. Chem. Ser., No. 67 (1967).

Morris, A. W., and J. P. Riley, Deep-Sea Res. Oceanog. Abstr., 13, 699 (1966).

Owen, B. B., "Electrochemical Constants," U.S. Natl. Bur. Std. Cir., No. 524, 193–204 (1953).

Owen, B. B., and S. R. Brinkley, Chem. Rev., 29, 461 (1941).

Park, K., Science, 154, 1540 (1966).

Park, K., D. W. Hood, and H. T. Odum, Inst. Marine Sci., 5, 47 (1958).

Platford, R. F., J. Fisheries Res., Board Can., 22, 113 (1965).

Platford, R. F., J. Marine Res., 23, 55 (1965a). J. Fisheries Res. Board Can., 22, 885 (1965b).

Platford, R. F., and R. Defoe, J. Marine Res., 23, 63 (1965).

Postma, H., Netherlands J. Sea Res., 2, 258 (1964).

Pytkowicz, R. M., I. W. Duedall, and D. N. Connors, Science, 152, 640 (1966).

Richards, F. A., "Anoxic Basins and Fjords," in J. P. Riley and G. Skirrow, eds., Chemical Oceanography, Academic Press, London, 1965, Vol. I, Chap. 13.

Righellato, E. C., and C. W. Davies, Trans. Faraday Soc., 26, 592 (1930).

Riley, J. P., and M. Tongudai, Deep-Sea Res., 11, 563 (1964).

Sillén, L. G., The Physical Chemistry of Sea Water, in M. Sears, ed., "Oceanography," Am. Assoc. Advan. Sci. Publ., No. 67 (1961).

Sillén, L. G., Science, 156, 1189 (1967); see also Prof. Sillén's two chapters in Stumm (1967).

Sillén, L. G., and A. E. Martell, "Stability Constants of Metal-Ion Complexes," Chem. Soc. (London), Sp. Publ., No. 17 (1964).

Smales, A. A., and L. Salmon, Analyst, 80, 37 (1955).

Smith, R. C., K. C. Pillai, T. J. Chow, and T. R. Folsom, Limnol. Oceanog., 10, 226 (1965).

Stokes, R. H., Trans. Faraday Soc., 41, 642 (1945).

Stumm, W., ed., "Equilibrium Concepts in Natural Water Systems," Advan. Chem. Soc., No. 67 (1967).

Thompson, M. E., Science, 153, 866 (1966).

Vaccaro, R. F., "Inorganic Nitrogen in Sea-Water," in J. P. Riley and G. Skirrow, eds., Chemical Oceanography, Academic Press, London, 1965, Vol. 1, Chap. 9.

7 Dissolved Gases and the Carbonate System

1 The Earth's Atmosphere

The gases dissolved in seawater originate from the Earth's atmosphere, from volcanic activity beneath the sea, and from chemical and other processes occurring in the seawater itself. The latter include biological activity, in particular photosynthesis, the decomposition of organic material, and radioactive decay.

This planet may be considered to be blanketed by two concentric mantles: the atmosphere and the oceans or the so-called hydrosphere. These regions also interact with two other zones: the superficial regions of the solid Earth itself, the lithosphere; and a thin scum sometimes designated the biosphere. The nature of these interactions is explored in greater detail: the hydrosphere-biosphere in Chapter 9, the atmosphere-hydrosphere here and in Chapter 11, and the lithosphere-hydrosphere in Chapter 12. The atmosphere and the hydrosphere exhibit many important similarities; in a sense the latter differs from the former principally by virtue of being simply a more viscous medium. Table 7.1 gives the composition of the Earth's atmosphere. Because of its lower

Table 7.1 Composition of the Earth's Atmosphere (Dry) up to 90 km (From U.S.A.F., 1961, with permission of The Macmillan Co.)

Constituent	Volume %	Weight %
Nitrogen, N_2	78.088	75.527
Oxygen, O_2	20.949	23.143
Argon, A	0.93	1.282
Carbon dioxide, CO_2	0.03	0.0456
Neon, Ne	1.8×10^{-3}	1.25×10^{-3}
Helium, He	5.24×10^{-4}	7.24×10^{-5}
Methane, CH_4	1.4×10^{-4}	7.75×10^{-5}
Krypton, Kr	1.14×10^{-4}	3.30×10^{-4}
Nitrous oxide, N_2O	5×10^{-5}	7.6×10^{-5}
Hydrogen, H_2	5×10^{-5}	3.48×10^{-6}
Xenon, Xe	8.6×10^{-6}	3.90×10^{-5}

viscosity the air is more rapidly and thoroughly mixed than the seas. The concentrations of the major constituents of the atmosphere are nearly constant. Other constituents such as ozone, O_3, from ultraviolet radiation, and SO_2, NO_2, I_2, CO, NH_3, from human pollution can be highly variable. Electrical storms can produce ozone and oxides of nitrogen, but the amounts involved are much too small to affect marine chemistry. The formation of organic molecules from inorganic systems by electrical discharge may have been important in biogenesis in the seas (see Part IV).

Although there is no unequivocal evidence that such is in fact the case, the viewpoint is commonly adopted that in the course of oceanic circulation, and as a result of advective and diffusive mixing process, every specimen of seawater has at some time in its history been at the surface and in equilibrium with the gases of the atmosphere, and that the concentration of dissolved atmospheric gases is subsequently modified by processes in the sea.

2   Nitrogen and the Noble Gases

Nitrogen and the noble gases such as argon are chemically very inert. They thus provide us with a good starting point for our examination of dissolved gases in seawater uncomplicated by chemical reactions (as in the case of CO_2) or biological dependence (as in the case of O_2). The equilibrium solubility of a gas in a liquid is given by Henry's law,

(7.1) $$m_g = Cp_g$$

where m_g is the molal concentration of the gas in solution, C is a constant, and p_g is the partial pressure of the gas above the solution. In other words, the solubility of a gas in a liquid is directly proportional to its partial pressure. Just as Raoult's law for the solvent (1.1) is strictly applicable only to ideal solutions, so Henry's law for the solute holds only for ideal or very dilute solutions. The deviations from Henry's law for many gases, however, are not large, and up to pressures of 1 atm it holds to within 1–3%. At elevated pressures the deviations from Henry's law become more appreciable, amounting to more than 10% at 100 atm (Klotz, 1963; Enns et al., 1965). The solubilities of gases, unlike those of most electrolytes, increase with decreasing temperature (Table 7.2 and Figure 7.1). And in an aqueous electrolytic solution they decrease with increasing electrolyte concentration or salinity (Table 7.2 and Figure 7.1). Discussion of the causes of these phenomena and of the rates of gas dissolution are deferred to Chapter 11. Here we concern ourselves only with the *concentrations* of dissolved gases in the sea.

Table 7.2   Nitrogen Solubility in Seawater, ml N_2/ml H_2O at STP
(From Douglas, 1965, with permission of American Chemical Society)

Temp., °C	Chlorinity						
	15	16	17	18	19	20	21
0	0.01931	0.01904	0.01877	0.01850	0.01824	0.01797	0.01770
1	0.01866	0.01860	0.01835	0.01809	0.01783	0.01758	0.01732
2	0.01845	0.01820	0.01795	0.01770	0.01745	0.01720	0.01696
3	0.01804	0.01780	0.01756	0.01732	0.01708	0.01684	0.01660
4	0.01766	0.01742	0.01719	0.01696	0.01673	0.01650	0.01626
5	0.01727	0.01705	0.01683	0.01661	0.01639	0.01616	0.01594
6	0.01691	0.01669	0.01648	0.01627	0.01606	0.01584	0.01563
7	0.01656	0.01635	0.01615	0.01595	0.01574	0.01554	0.01533
8	0.01622	0.01602	0.01582	0.01562	0.01542	0.01522	0.01502
9	0.01587	0.01568	0.01549	0.01530	0.01511	0.01492	0.01473
10	0.01554	0.01536	0.01518	0.01500	0.01481	0.01463	0.01445
11	0.01523	0.01506	0.01488	0.01471	0.01453	0.01436	0.01418
12	0.01496	0.01478	0.01461	0.01444	0.01426	0.01409	0.01392
13	0.01469	0.01452	0.01435	0.01418	0.01401	0.01384	0.01367
14	0.01440	0.01424	0.01408	0.01392	0.01376	0.01360	0.01344
15	0.01419	0.01403	0.01386	0.01370	0.01354	0.01338	0.01321
16	0.01396	0.01380	0.01363	0.01346	0.01329	0.01312	0.01295
17	0.01373	0.01357	0.01341	0.01325	0.01309	0.01293	0.01277
18	0.01351	0.01335	0.01319	0.01303	0.01287	0.01271	0.01255
19	0.01329	0.01314	0.01298	0.01282	0.01266	0.01250	0.01235
20	0.01309	0.01293	0.01278	0.01263	0.01248	0.01332	0.01217
21	0.01288	0.01273	0.01259	0.01244	0.01229	0.01214	0.01199
22	0.01268	0.01254	0.01240	0.01226	0.01211	0.01197	0.01183
23	0.01249	0.01235	0.01222	0.01208	0.01194	0.01180	0.01166
24	0.01231	0.01218	0.01205	0.01192	0.01179	0.01165	0.01152
25	0.01214	0.01202	0.01189	0.01176	0.01164	0.01151	0.01138
26	0.01199	0.01187	0.01174	0.01162	0.01150	0.01137	0.01125
27	0.01185	0.01173	0.01160	0.01148	0.01136	0.01124	0.01112
28	0.01173	0.01160	0.01148	0.01136	0.01124	0.01112	0.01100
29	0.01159	0.01147	0.01135	0.01124	0.01112	0.01101	0.01089
30	0.01146	0.01134	0.01123	0.01111	0.01100	0.01088	0.01077

The most recent detailed studies of the solubilities of N_2 (Table 7.2) and A (Table 7.3) in seawater are those of Douglas (1965) whose results for the former are somewhat less than the earlier findings of Rakestraw and Emmel, and for the latter in poor agreement with König but good agreement with Rakestraw and Emmel (1938) at higher salinities. Table 7.4 lists König's (1963) values for the solubilities of argon and the other noble gases in 34.54‰ S seawater. In each case both the experimental value of the pure gas (at 760 mm) and the calculated value corresponding to the gas pressure in air are

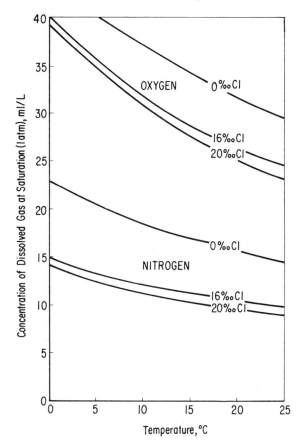

Figure 7.1 Solubility of the Major Atmospheric Gases in Seawater. (This figure was drawn on the basis of older data and is therefore only in approximate agreement with the recent results summarized in Tables 7.2 and 7.6.)

listed. The latter pressures are 5.24×10^{-6}, 1.82×10^{-5}, 9.32×10^{-3}, 1.14×10^{-6}, and 8.6×10^{-8} atm for He, Ne, Ar, Kr, and Xe, respectively.

In view of its chemical and biological inertness the N_2 concentration in the seas should be relatively invariant. Rakestraw and Emmel (1938) found that the concentration of N_2 in Atlantic Ocean waters varied from 101% to 108% of its saturation values, while Hamm and Thompson (1941) found a range of 91–102% saturation for combined N_2 and Ar in Pacific waters. Craig, Weiss, and Clarke (1967) also found samples systematically about 2% supersaturated with N_2. These variations can be attributed, at least in part, to changes in temperature and barometric pressure (Carritt, 1954). The possible

Table 7.3   Argon Solubility in Seawater ml A/ml H_2O at STP
(From Douglas, 1965, with permission of American Chemical Society)

Temp., °C.	Chlorinity						
	15	16	17	18	19	20	21
0	0.04443	0.04387	0.04331	0.04276	0.04220	0.04165	0.04110
1	0.04326	0.04274	0.04222	0.04170	0.04118	0.04066	0.04014
2	0.04212	0.04164	0.04115	0.04066	0.04017	0.03968	0.03920
3	0.04117	0.04069	0.04021	0.03974	0.03926	0.03878	0.03830
4	0.04022	0.03974	0.03928	0.03882	0.03836	0.03789	0.03742
5	0.03931	0.03885	0.03840	0.03794	0.03748	0.03703	0.03657
6	0.03840	0.03795	0.03751	0.03707	0.03663	0.03618	0.03574
7	0.03752	0.03709	0.03666	0.03623	0.03580	0.03536	0.03494
8	0.03665	0.03624	0.03582	0.03540	0.03500	0.03458	0.03416
9	0.03581	0.03541	0.03502	0.03462	0.03423	0.03383	0.03344
10	0.03504	0.03465	0.03427	0.03389	0.03350	0.03312	0.03273
11	0.03430	0.03393	0.03355	0.03318	0.03281	0.03243	0.03206
12	0.03362	0.03325	0.03288	0.03252	0.03215	0.03178	0.03141
13	0.03295	0.03259	0.03223	0.03187	0.03152	0.03116	0.03080
14	0.03230	0.03195	0.03160	0.03126	0.03092	0.03058	0.03023
15	0.03169	0.03135	0.03102	0.03068	0.03035	0.03002	0.02969
16	0.03112	0.03079	0.03046	0.03013	0.02980	0.02948	0.02915
17	0.03059	0.03025	0.02991	0.02958	0.02924	0.02890	0.02856
18	0.03001	0.02970	0.02938	0.02908	0.02876	0.02846	0.02815
19	0.02949	0.02919	0.02889	0.02858	0.02828	0.02799	0.02768
20	0.02900	0.02870	0.02840	0.02811	0.02782	0.02751	0.02722
21	0.02851	0.02822	0.02793	0.02763	0.03734	0.02706	0.02677
22	0.02806	0.02776	0.02748	0.02718	0.02690	0.02661	0.02632
23	0.02760	0.02732	0.02703	0.02674	0.02645	0.02617	0.02589
24	0.02717	0.02689	0.02661	0.02633	0.02605	0.02576	0.02549
25	0.02675	0.02648	0.02620	0.02593	0.02566	0.02538	0.02511
26	0.02636	0.02608	0.02582	0.02554	0.02527	0.02500	0.02474
27	0.02597	0.02570	0.02544	0.02518	0.02492	0.02466	0.02440
28	0.02560	0.02534	0.02509	0.02483	0.02458	0.02432	0.02407
29	0.02521	0.02496	0.02472	0.02447	0.02422	0.02398	0.02373
30	0.02485	0.02461	0.02437	0.02414	0.02390	0.02366	0.02343

perturbation of the N_2 content of seawater by nitrogen fixation and denitrification by organisms has been suggested (see Richards, 1965, and Vaccaro, 1965), and Benson and Parker(1961)have measured N_2/Ar ratios in Atlantic Ocean waters in an attempt to uncover evidence for such perturbations. The ratios they obtained differed only to within 1% of those expected simply from the dissolution of air in seawater. Richards and Benson (1965), however, have observed a significant increase in the N_2/Ar ratio in the deep anaerobic waters of the Cariaco Trench and Dramsfjord. Assuming that Ar is biologically inert,

Table 7.4 Solubilities of the Noble Gases in Seawater, ml of Gas at STP/kg Seawater (From von König, 1963, with permission of the publisher)

Temp., °C	He		Ne		Ar		Kr		Xe	
	Pure	Air	Pure	Air	Pure	Air	Pure	Air	Pure	Air
0	—	—	9.37	171×10^{-6}	—	—	71.5	81.5×10^{-6}	136	11.70×10^{-6}
1	7.91	41.4×10^{-6}	—	—	38.5	0.359	—	—	—	—
5	—	—	9.02	164×10^{-6}	35.3	0.329	63.9	72.8×10^{-6}	115	9.89×10^{-6}
10	7.40	38.8×10^{-6}	8.67	158×10^{-6}	32.8	0.306	58.2	66.3×10^{-6}	103	8.86×10^{-6}
15	6.95	36.4×10^{-6}	8.35	152×10^{-6}	30.2	0.282	51.6	58.5×10^{-6}	90.0	7.74×10^{-6}
17.5	—	—	—	—	—	—	50.2	57.2×10^{-6}	—	—
20	7.00	36.7×10^{-6}	8.20	149×10^{-6}	26.3	0.245	44.8	51.1×10^{-6}	80.0	6.88×10^{-6}
22.8	—	—	—	—	—	—	44.2	50.4×10^{-6}	—	—
24	—	—	—	—	—	—	42.9	48.9×10^{-6}	—	—
25	—	—	8.07	147×10^{-6}	—	—	—	—	70.2	6.04×10^{-6}

they attribute the N_2 in excess of the amount expected of that dissolved from the atmosphere to nitrogen release from the decomposition of organic material.

Mazor, Wasserburg, and Craig (1964) measured the He, Ne, Ar, Kr, and Xe concentrations in South Pacific waters and conclude that, to within $\pm 10\%$, the concentrations of these gases, except He, is as expected from their solubility from the atmosphere. Similar results were also obtained by König and co-workers (1964). Table 7.5 gives the noble gas contents, measured by mass spectrometry, of Pacific waters taken off the California coast. The helium concentration is greater than what one expects on the basis of the other noble gases. This excess increases with increasing depth and is believed to originate from He production from the radioactive decay of isotopes of the U and Th series in marine sediments (Bieri, Koide, and Goldberg, 1964). More recent and detailed studies of the noble gas content of Pacific seawater (Bieri, Koide, and Goldberg, 1966) confirm the He enrichment with increasing depth and have yielded further interesting information concerning the other gases. Contrary to earlier results (Mazor, Wasserburg, and Craig, 1964), the concentrations are *not* what one would expect from thermodynamic equilibrium; rather, on the average, supersaturation occurs for all gases at all depths (Figure 7.2). The concentrations of the gases, therefore, must be modified by physical processes, such as diffusion and advection in the oceans. The supersaturation in the mixed layer can be attributed to any one or a combination of factors, including high atmospheric pressures at the time of dissolution from the atmosphere, bubbles, temperature effects, etc. The intermediate waters appear to be more amenable to theoretic analysis, and the observed saturation maxima can be qualitatively reproduced by a differential equation

Table 7.5   Concentration of Noble Gases (in cm^3/l) in Pacific Seawater (Selected values from Bieri, Koide, and Goldberg, 1966)

Temp, °C	Depth M	He	Ne	Ar	Kr	Xe
23.4	6	4.17×10^{-5}	1.65×10^{-4}	2.64×10^{-1}	5.8×10^{-5}	
23.4	45	4.10	1.64	2.62	5.7	(7.7×10^{-6})
17.4	244	4.32	1.74	2.98	6.6	
4.2	981	4.60	1.87	3.60	8.5	(4.4×10^{-6})
3.0	1477	4.58	1.91	3.78	9.0	(3.7×10^{-6})
2.2	1978	4.57	1.93	3.76	9.0	(12.5×10^{-6})
2.1	2021	4.60	1.91	3.92	9.6	
1.8	2520	4.07	1.67	3.38	8.2	
1.8	3517	4.55	1.93	3.90	9.7	(11.0×10^{-6})

[a] From Biere, Koide, and Goldberg (1964).

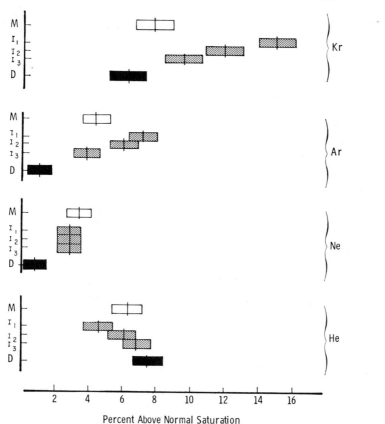

Figure 7.2   Excess Noble Gas Content of Pacific Seawater. M = mixed layer; I_1, I_2, I_3 = intermediate layers; D = deep water. From Bieri, Koide, and Goldberg (1966), with permission of American Geophysical Union and the authors.

analogous to the one developed by Wyrtki (1961) for the vertical temperature distribution in the ocean and which takes diffusion and advection into account.

Nitrous oxide has been found in surface and in deep South Pacific waters (Craig and Gordon, 1963).

3   Oxygen

The oxygen solubilities quoted in the literature are in rather poor agreement, and these discrepancies have led Green (1965) to redetermine the equilibrium solubility of O_2 in seawater at 1 atm as a function of temperature and salinity (Table 7.6) (Green's thesis has recently been published, Green and

Table 7.6 Solubility of Oxygen in Seawater, ml/1 Seawater from a Dry Atmosphere 0.2094 mole fraction O_2
(Selected values from Green, 1965)

Temp., °C	Chlorinity, ‰						
	0	5	10	15	20	25	30
0							
5	10.35	9.72	9.11	8.56	8.03	7.52	7.07
10	9.08	8.54	8.04	7.57	7.13	6.73	6.34
15	7.20	6.83	6.47	6.11	5.80	5.48	5.20
20	6.53	6.18	5.88	5.56	5.29	5.02	4.76
25	5.95	5.65	5.38	5.11	4.84	4.61	4.38
30	5.49	5.23	4.97	4.71	4.49	4.25	4.04
35	5.12	4.85	4.61	4.38	4.18	3.97	3.76

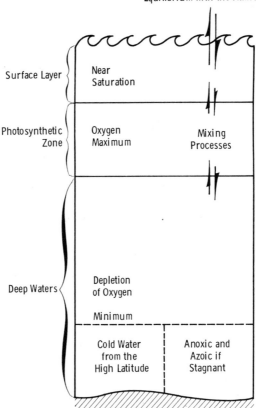

Figure 7.3 Diagrammatic Representation of the Several Conditions of Dissolved Oxygen Content.

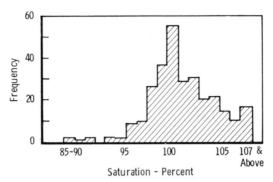

Figure 7.4   Oxygen Saturation of South Atlantic Waters. From Richards (1965), with permission of Academic Press, Inc.

Carritt, 1967). The solubility of O_2 is roughly comparable to the solubility of A (Table 7.3), and the same trends are exemplified, namely, the decrease in solubility with increasing temperature and increasing salt concentration.

The concentration of dissolved oxygen in the oceans is both of great biological importance and highly variable and has, accordingly, been the subject of many investigations. Yet, despite its highly variable nature, certain useful generalizations about the oxygen content of the seas can be made. For this purpose it is useful to divide the vertical water column into four regions (Figure 7.3). At the sea's surface is a layer well-mixed by the wind and other processes in which the oxygen concentration is fairly uniform down to the thermocline. The value of the concentration in this zone tends to be the saturation value (Figure 7.4). established by the sea-air interface at the ambient temperature and barometric pressure. If the water column is relatively stable, a subsurface maximum in the concentration of dissolved oxygen is frequently observed somewhere in the first 50 m as a result of O_2 production by photosynthesis (see Chapter 9). Below the photosynthetic zone, *in situ* processes such as the oxidation of organic matter tend to reduce the oxygen content. In regions such as the tropical North Pacific, where the organic content is high and/or the waters are warmer (thus speeding up the rates of oxidation), the decrease can be as much as 5 ml/l per 100 *m* and, where it is lower and the water colder, as in the Antarctic Convergence, the decrease can even be less than 1 ml/l per 100 m (Richards, 1965). Within this zone itself, a certain amount of fine structure in the oxygen profile in the form of maxima and minima may be real (Figures 7.5 and 7.6) and the profile can vary from year to year (Figure 7.5) and even from month to month (Figure 7.6). If the oceans were not well-stirred, we might expect the O_2 concentration to continue to fall off and the very deep waters to be anoxic and azoic. This is just what happens in certain

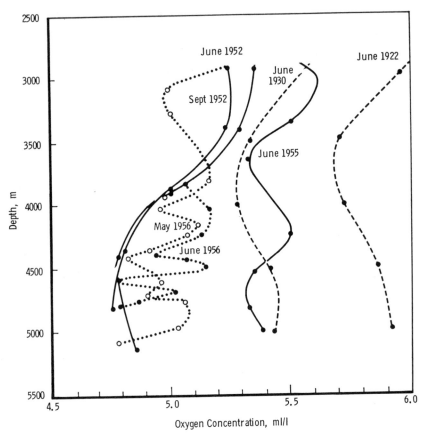

Figure 7.5 Yearly Variations in the Oxygen Profile in Bay of Biscay Waters. From Cooper (1961), with permission of AAAS and the author.

stagnant environments (Figure 7.3) (Richards, 1965), but such circumstances are the exception rather than the rule. Beneath the Earth's major oceans run great masses of young, dense, cold, oxygen-rich water that has sunk down from the polar regions (Taft, 1963). Hence the O_2 concentration does not continue to decrease with increasing depth, but rather goes through a minimum and begins to increase (Wyrtki, 1962; Richards, 1957).

Like the phosphate concentration (see above), the concentration of dissolved oxygen represents a biological clock which can be used to characterize and determine the age of the several oceans. The three great oceans of the world tend to have characteristic dissolved oxygen profiles (Figure 7.7). The Atlantic enjoys an influx of new, cold, oxygen-rich water from both polar

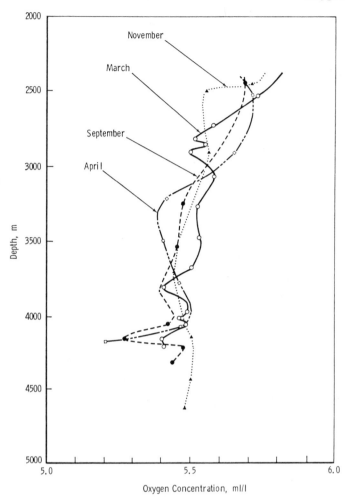

Figure 7.6 Monthly Variations in the Oxygen Profile of Bay of Biscay Waters, 1958. From Cooper (1961), with permission of AAAS and the author.

regions, but the northern boundaries of the Pacific Ocean and, more markedly, the Indian ocean are landlocked. Their waters are older, more tired and worn out, if you will, than the Atlantic's younger waters (shown in cross section in Figure 7.8); and they are relatively depleted in oxygen, not entirely because their circulation is poorer, but because more organic material has been oxidized in them for a longer period of time. Generally speaking, where the concentration of nutrients such as phosphate and nitrate is high, the oxygen content tends to be low (Figure 7.9).

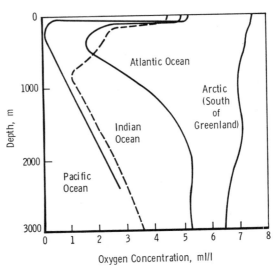

Figure 7.7 Characteristic Oxygen Profiles of the Atlantic, Pacific, and Indian Oceans and of the Area of Formation of the Deep Atlantic Waters. From Dietrich (1963), with permission of John Wiley & Sons.

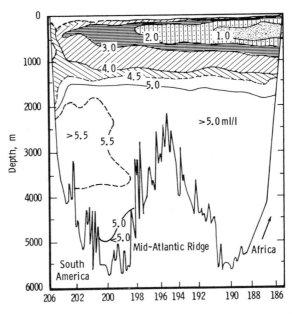

Figure 7.8 Dissolved Oxygen Cross Section for the South Atlantic. From Dietrich (1963), with permission of John Wiley & Sons.

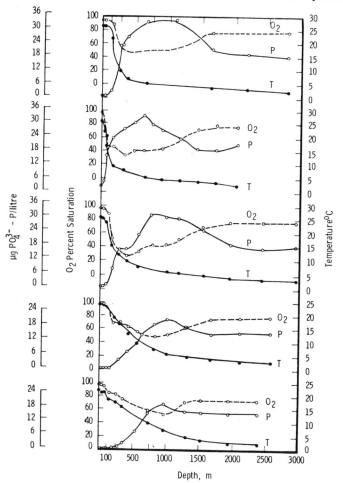

Figure 7.9 Inverse Relation between the Phosphate and Dissolved Oxygen Concentration in Seawater. From Seiwell (1935).

4 Carbon Dioxide and the Carbonate System

With the exception only of the chemistry of the water itself, by far the most important and certainly one of the most complex equilibria systems in the sea is that of the carbon dioxide-carbonate system (shown simplified and schematically in Figure 7.10). It touches upon every aspect of the Earth sciences—meterology, geology, and oceanography—and on ecology as well. Within oceanography it plays an important role in each of the three principal reactive zones of the oceans: in the atmosphere-sea interaction (Bolin, 1960),

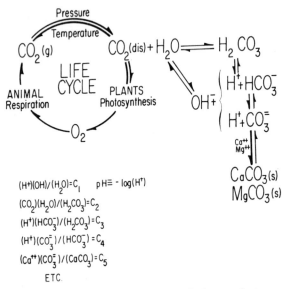

Figure 7.10   The Carbon Dioxide-Carbonate System.

in the chemistry of seawater, and in the deposition of marine sediments. It is largely responsible for controlling the pH of seawater and thus affects directly many of the chemical equilibria in the seas, and it is thus one of the factors responsible for the formation and maintenance of an environment conducive to the genesis and sustenance of life. The carbon dioxide-carbonate system was a major participant in the evolutionary history of this planet's atmosphere, hydrosphere, and lithosphere, and it may play a role of particular importance in the future, for the fear has been expressed that, if the oceans prove inadequate to the task of removing the CO_2 pollution being added to the atmosphere in ever-increasing amounts by human activity, the surface temperature of the planet may increase and man will pay the price of extinction for his inability to control his fertility (Rakestraw, 1965).

The distribution of carbon on Earth is shown in Figure 7.11 in terms of weight per square centimeter of Earth's surface. The oceanic content is in the upper right corner and consists of 0.20 g/cm² of inorganic carbon above the thermocline and 7.25 g/cm² below. The dissolved organic carbon amounts to some 0.533 g/cm², while the carbon content of the marine biosphere is 0.002 g/cm²—less than that of the land biosphere (0.06 g/cm²), and much less than the 0.19 g/cm² of humus (Skirrow, 1965).

Weibe (1941) has reviewed the composition of the two phases in the CO_2^- pure H_2O system, yet despite its obvious importance the solubility of carbon dioxide in seawater does not appear to be known with any high degree of

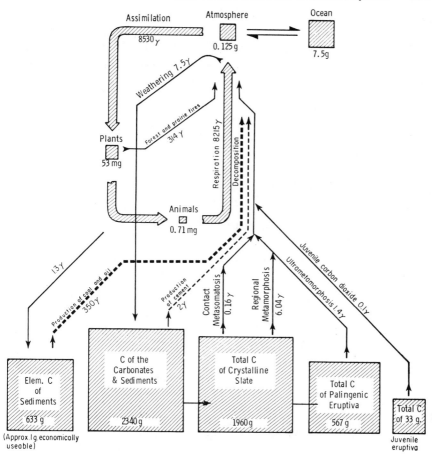

Figure 7.11   The Carbon Distribution and Carbon Cycles on Earth. From Dietrich (1963), with permission of John Wiley & Sons.

certainty. The values usually quoted (Table 7.7) are based on old measurements on NaCl solutions, and there is reason to believe that they are slightly too high for seawater (Skirrow, 1965). Perhaps more reliable data, but again for NaCl solutions rather than seawater, are the Henry's law constants given in Table A.18 of the Appendix. At 0°C and 20‰ Cl, 1434 ml of CO_2 (Table 7.7) dissolve in 1 liter compared to about 19 ml of N_2 (Table 7.2), and CO_2 deviates from Henry's law more widely than N_2, O_2, or A. The reason for this different behavior is that CO_2, unlike these other atmospheric gases, is not chemically inert with respect to water but, in addition to the usual inescapable, hydration phenomena,

(7.2) $$CO_2 \rightleftharpoons CO_2 \text{ (solution)}$$

Table 7.7   Solubility of Carbon Dioxide in "Seawater"
(moles/l) $\times 10^4$, STP)
(From Riley and Skirrow, 1965, with permission of Academic Press, Inc.)

Cl‰	T, °C							
	0	2	4	6	8	10	12	14
0	770	712	662	619	576	536	502	472×10^{-4}
15	674	623	578	538	504	472	442	416
16	667	617	573	533	499	468	438	413
17	660	611	567	528	495	464	434	410
18	653	605	562	524	490	460	431	406
19	646	599	557	519	486	456	428	403
20	640	593	551	514	482	452	424	400
21	633	587	546	509	477	448	421	396

Cl‰	T, °C							
	16	18	20	22	24	26	28	30
0	442	417	394	372	351	332	314	299×10^{-4}
15	393	371	351	331	314	299	284	270
16	390	368	348	329	312	297	281	268
17	387	365	346	327	310	294	279	266
18	384	362	343	324	307	292	277	264
19	381	359	340	321	304	289	275	262
20	377	356	337	319	302	287	273	260
21	374	354	335	317	300	285	271	258

reacts with water to form carbonic acid

(7.3) CO_2 (solution) + $H_2O \rightleftharpoons H_2CO_3$, $K = (H_2CO_3)/(CO_2)$

This acid then dissociates in two steps:

(7.4) $H_2CO_3 \rightleftharpoons H^+ + HCO_3^-$, $K_1' = (H^+)(HCO_3^-)/(H_2CO_3)$

(7.5) $HCO_3^- \rightleftharpoons H^+ + CO_3^{2-}$ $K_2 = (H^+)(CO_3^{2-})/(HCO_3^-)$

The dissolution steps, (7.2) and (7.3), are relatively slow, but the ionization steps are very rapid. Nothing is known concerning equilibrium (7.3) since CO_2 (solution) and H_2CO_3 are operationally indistinguishable. For this reason (7.3) and (7.4) are usually combined to give

(7.6) CO_2 (solution) + $H_2O \rightleftharpoons H^+ + HCO_3^-$,

$$K_1 = (H^+)(HCO_3^-)/(CO_2)$$

The thermodynamic equilibrium constants

(7.7)
$$K_1{}^\circ = \frac{a_{H^+} \, a_{HCO_3^-}}{a_{CO_2}}$$

and

(7.8)
$$K_2{}^\circ = \frac{a_{H^+} \, a_{CO_3^{2-}}}{a_{HCO_3^-}}$$

as determined by Harned and Davis (1943) and Harned and Scholes (1941) are given in Table 7.8 for 1 atm, and Table 7.9 shows the ionic strength dependence (in NaCl solutions) of $K_2{}^\circ$ at 50°C and 1 atm and corrected for hydrolysis. $K_2{}^\circ$ decreases with increasing ionic strength. The effect of ionic strength tends to diminish with decreasing temperature. Constants in terms of concentrations, K_1 and K_2, can be calculated from $K_1{}^\circ$ and $K_2{}^\circ$ using the

Table 7.8   Thermodynamic Dissociation Constants of Carbonic Acid at 1 atm

Temperature, °C	$K_1{}^\circ$	$K_2{}^\circ$
0	2.64×10^7	2.36×10^{-11}
5	3.04	2.77
10	3.44	3.27
15	3.81	3.71
20	4.16	4.20
25	4.44	4.69
30	4.71	5.13

Table 7.9   Ionic Strength Dependence of the Second Thermodynamic Dissociation Constant of Carbonic Acid in Aqueous NaCl at 50°C and 1 atm

I	$K_2{}^\circ$
0.029	6.68×10^{-11}
0.058	6.65
0.083	6.50
0.112	6.47
0.131	6.44

activity coefficients in Table A.15. Commonly oceanographers use mixed constants defined as

$$(7.9) \qquad K_1'' \equiv a_{H^+}(HCO_3^-)/a_{CO_2}$$

$$(7.10) \qquad K_2'' \equiv a_{H^+}(CO_3^{2-})/(HCO_3^-)$$

In his thesis Lyman (1956) defines a third constant

$$(7.11) \qquad K_L'' = K_1''(p/p_0)(\alpha_0/\alpha)$$

where p and p_0 are the vapor pressures and α and α_0 the CO_2 solubility coefficients, i.e., Henry's law constants, of seawater and pure water, respectively. His values of $\log K_L''$ and $\log K_2''$ are given in Table 7.10. Inasmuch as they contain directly measurable terms, these mixed constants, unlike the thermodynamic constants, have the advantage of greater practicality, but the price of a great deal of needless confusion has been paid for this convenience.

Table 7.10   Apparent Dissociation Constants of Carbonic Acid in Seawater

	Temperature, °C						
	$-\log K_L''$						
$Cl, \%_{00}$	0	5	10	15	20	25	30
0	6.58	6.52	6.47	6.42	6.38	6.35	6.33
1	6.47	6.42	6.37	6.33	6.29	6.26	6.24
4	6.36	6.32	6.28	6.24	6.21	6.18	6.16
9	6.27	6.23	6.19	6.15	6.13	6.10	6.08
16	6.18	6.14	6.11	6.07	6.05	6.03	6.01
17	6.17	6.13	6.10	6.06	6.04	6.02	6.00
18	6.16	6.12	6.09	6.06	6.03	6.01	5.99
19	6.15	6.11	6.08	6.05	6.02	6.00	5.98
20	6.14	6.10	6.07	6.04	6.01	5.99	5.97
21	6.13	6.09	6.06	6.03	6.00	5.98	5.96
25	6.09	6.05	6.02	6.00	5.97	5.95	5.93
	$-\log K_2''$						
0	10.62	10.55	10.49	10.43	10.38	10.33	10.29
1	10.06	9.99	9.93	9.87	9.81	9.76	9.71
4	9.78	9.72	9.67	9.61	9.54	9.49	9.43
9	9.64	9.58	9.52	9.46	9.40	9.34	9.27
16	9.46	9.40	9.35	9.29	9.23	9.17	9.10
17	9.44	9.38	9.32	9.27	9.21	9.15	9.08
18	9.42	9.36	9.30	9.25	9.19	9.12	9.06
19	9.40	9.34	9.28	9.23	9.17	9.10	9.02
20	9.38	9.32	9.26	9.21	9.15	9.08	9.01
21	9.36	9.30	9.25	9.19	9.13	9.06	8.98
25	9.29	9.23	9.17	9.11	9.05	8.98	8.91

Douglas (1967) has recently measured the solubility of carbon monoxide in seawater, obtaining the values summarized in Table A.20.

The most straightforward means of achieving an understanding of the CO_2-HCO_3-CO_3^{2-} system is a diagram such as that in Figure 7.12 in which the percentage of the total inorganic carbon represented by each of the carbon species is plotted as a function of pH. The dissociation constants decrease with increasing ionic strength (Table 7.9), shifting the curves to the left in Figure 7.12 as we go from pure water to seawater. From such a diagram we can see at a glance which species is the most important in the seawater pH range, and clearly it is bicarbonate ion, HCO_3^-. At the lower end of the oceanographic scale at pH 7, more than 80% of the carbon is in the form of HCO_3^-, CO_2 representing the remainder, while at the upper end at pH 8.5 even more than 80% of the carbon is HCO_3^-, but now the remainder is almost entirely in the form of CO_3^{2-}.

Deffeyes (1965) has published a series of graphs of "alkalinity" versus total carbonate carbon as a function of pH, HCO_3^- and CO_3^{2-} concentrations, and CO_2 pressure.

Both K_1° and K_2° increase with increasing temperature. The effect of increased temperature then is to shift the curves in Figure 7.12 to the right. At the same time, however, since temperature favors the left side of equilibria (7.2) quite strongly, the total amount of inorganic carbon in solution decreases.

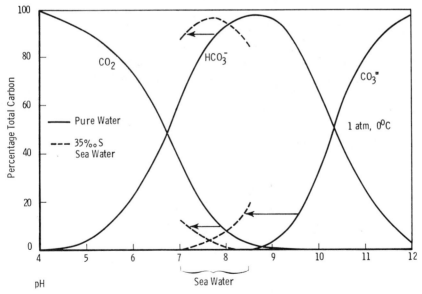

Figure 7.12   Distribution of the CO_2-HCO_3-CO_3^{2-} System in Pure Water and Seawater at 1 atm, as a Function of pH.

The application of hydrostatic pressure, as we have seen, favors the side of a chemical equilibrium of smaller volume, the change in equilibrium constant being given by

(7.12)
$$\left(\frac{\partial \ln K}{\partial P} \right)_{c,T} = \frac{-\Delta V^\circ}{RT}$$

where ΔV° is the partial molal volume change. Pressure shifts equilibria in favor of condensed phases and greater charge; thus, with increasing depth in the sea, equilibria (7.2), (7.4), and (7.5) are all shifted to the right. ΔV° is -25.4 cm^3/mole and -25.6 cm^3/mole for equilibria (7.4) and (7.5), respectively (Disteche and Disteche, 1967).

Values of the logarithms of K_1', K_2, K_D'', and K_2'', where K_D'' is again a mixed constant given by

(7.13)
$$K_D'' \equiv \frac{a_{H^+}(HCO_3^-)}{(H_2CO_3)}$$

for 20‰ Cl seawater at 22°C, are given in Table 7.11 (Disteche and Disteche, 1967). The changes in the pH of seawater with pressure can be attributed to these shifts of equilibria together with the increased dissociation of $CaCO_3$ and $MgCO_3$, a sulfate effect, and some contribution from the boric acid, a weak electrolyte and minor seawater constituent (Buch and Grienberg, 1932). More recently Culberson (1968) has studied the effect of pressure on the dissociation of carbonic and boric acids in seawater. Earlier Buch and Grienberg estimated that for a surface pH of 7.5 the corresponding value at a depth of 1000 m is 0.035 unit less, whereas for a surface pH of 8.5 the corresponding increase at 100 m is 0.020 unit, but Pytkowicz (1963) has criticized the oversimplified assumptions on which these estimates were based, and recently published measurements on the effect of pressure on the dissociation of carbonic acid (Culberson et al., 1967) indicate that these changes are larger, especially at higher pH's.

The air-equilibrium concentration of CO_2 gas in surface seawater, determined by equilibrating seawater with a closed volume of air, is compared

Table 7.11   The Effect of Pressure on
the Dissociation of Carbonic Acid
in Seawater at 22°C

	1 atm	1000 kg/cm^2
$-\log K_1'$	5.89	5.55
$-\log K_2$	9.49	9.19
$-\log K_D''$	6.00	5.66
$-\log K_2''$	9.13	8.93

with the *in situ* CO_2 content of the atmosphere for the Pacific, Indian, and Atlantic oceans in Figure 7.13. The Pacific is markedly undersaturated with CO_2, and the Indian supersaturated near the equator. Earlier measurements revealed an area of high CO_2 concentration, relative to the atmosphere, in the equatorial Pacific Ocean approaching the coast of South America, whereas elsewhere the concentration is close to what one would expect from equilibrium with the atmosphere (Keeling, Rakestraw, and Waterman, 1965).

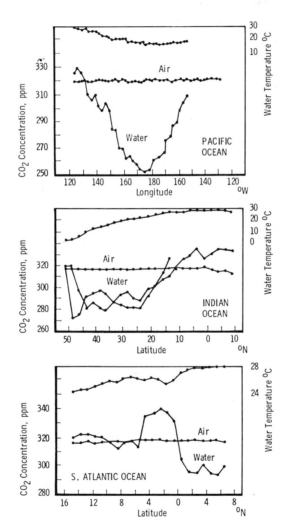

Figure 7.13 The Air-Equilibrium Dissolved Carbon Dioxide Content of Surface Waters. From Waterman (1965), with permission of Macmillan, Ltd., and the author.

The CO_2 content of the surface waters with respect to that of the atmosphere above the sea rarely behaves as expected. Notice, for example, in addition to the under- and supersaturation shown in Figure 7.13, the lack of the expected temperature correlation. Although CO_2 is more soluble in colder water, the dissolved CO_2 content of Indian Ocean waters increases with increasing temperature. All of which leads to the conclusion that many complex factors contribute to the discrepancies between predicted and observed CO_2 content, and one of these factors may be the slowness of the establishment of equilibria (7.2) and (7.3) (Skirrow, 1965). Inasmuch as CO_2 is utilized by the photosynthetic process, we might expect diurnal variations of the dissolved CO_2 content to exist. And indeed they do. But again they are complex and may be exactly the opposite of what we might have expected. When the intensity of solar energy is at its peak, rather than a minimum, Takahasi (1961) commonly observed a peak in the CO_2 content of surface Atlantic waters (Figure 7.14) indicating that temperature-dependent physical-chemical processes may be more important than photosynthetic processes. The effects tend to diminish markedly in the first 10 m (Figure 7.14) as the effect of solar warming diminishes. But here, again, the interpretation is not straightforward: equilibria (7.4) and (7.5) are shifted to the right with increased temperature, but (7.2) and possibly (7.3) are shifted to the left. The increase, rather than decrease, in the amount of dissolved CO_2 with increasing pressure may depend more on rate than on equilibrium phenomena. Temperature may increase the rate of (7.2) and (7.3), and it may also increase the rate of production of CO_2 *in* the surface sea water by acceleration of the kinetics of organic decomposition.

The only conclusion which seems to emerge from all this is that the dis-

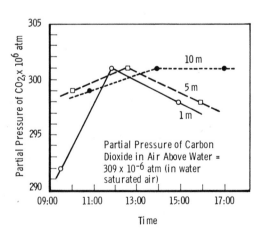

Figure 7.14 Diurnal Variation of the CO_2 Partial pressure in Atlantic Seawater. From Takahasi (1961), with permission of American Geophysical Union and the author.

solved CO_2 content of seawater is a subject worthy of much further experimental and theoretical effort by virtue of the patently incomplete nature of our knowledge and its very great importance to the ecology of this planet in general and chemical oceanography in particular.

The vertical distribution of dissolved CO_2 in the Pacific Ocean (Figure 7.15), shows a rapid increase in the first 500 m due to production from the decomposition of dead organisms and the greater solubility of the gas with increased pressure.

The boric acid system is analogous to the carbonic acid system, and the equilibria

(7.14) $H_3BO_3 \rightleftharpoons H^+ + H_2BO_3^-$, $K_1 = (H^+)(H_2BO_3^-)/(H_3BO_3)$

(7.15) $H_2BO_3^- \rightleftharpoons H^+ + HBO_3^{2-}$, $K_2 = (H^+)(HBO_3^{2-})/(H_2BO_3^-)$

(7.16) $HBP_3^{2-} \rightleftharpoons H^+ + BO_3^{3-}$, $K_3 = (H^+)(BO_3^{3-})/(HBO_3^{2-})$

are reminiscent of equilibria (7.4) and (7.5). Values of the first and second mixed dissociation constants for boric acid are given in Table A.19. For 34.8‰ seawater at 22°C, Culberson and co-workers (1967) have found that the value of K_1 [but where a_{H^+} is substituted for (H^+)] increases by a factor of 1.94 in going from 1 to 654 atm. At seawater pH's, boron is mostly in the form of H_3BO_3 and $H_2BO_3^-$. Because of its smaller concentration the borate system in seawater is far less important than the carbonate system, but its contribution is not entirely negligible.

Over the years the boron content of seawater has been the subject of several investigations (Foote, 1932; Wilcox, 1932; Moberg and Harding, 1933;

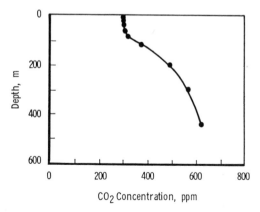

Figure 7.15 Average Dissolved Carbon Dioxide Content of Some Subsurface Pacific Waters. From Keeling, Rakestraw, and Waterman (1965), with permission of American Geophysical Union and the authors.
Keeling, C. D., N.W. Rakestraw, and L. S. Waterman, *J. Geophys. Res.*, **70**, 6089 (1965).

Buch, 1933; Rakestraw and Mahnecke, 1935; Igelsrud, Thompson, and Zwicker, 1938; Gast and Thompson, 1958; and Noakes and Hood, 1961.

In the previous discussion we were concerned largely with equilibria relating to the air-sea interface. Now we turn our attention to equilibria involved for the most part with the solid-sea interface (Figure 7.10). Just as we confined our earlier treatment to equilibria, deferring a more detailed examination of the air-sea interaction to Chapter 11, so again we limit our present attention to a few selected equilibria, reserving a more detailed discussion of the solid-sea interactions to Chapter 12. These equilibria are

(7.17) $CaCO_3 \text{ (s, calcite)} \rightleftharpoons Ca^{2+} + CO_3^{2-}, \qquad K_{sp,0}^{2-}(Ca^{2+})(CO_3^{2-})$

(7.18) $CaCO_3 \text{ (s, aragonite)} \rightleftharpoons Ca^{2+} + CO_3^{2-}, \qquad K_{sp,A} = (Ca^{2+})(CO_3^{2-})$

(7.19) $CaCO_3 \text{ (s, aragonite)} \rightleftharpoons CaCO_3 \text{ (s, calcite)}$

(7.20) $MgCO_3 \text{ (s)} \rightleftharpoons Mg^{2+} + CO_3^{2-}, \qquad K_{sp} = (Mg^{2+})(CO_3^{2-})$

To dispose of the least important aspects of carbonate precipitation first, $MgCO_3$ and dolomite, $CaMg(CO_3)_2$, for reasons unclear, are not important marine sediments despite their estimated thermodynamic relationships and their geological abundance (Cloud, 1965). Theoretical efforts to approximate the equilibria (Garrels, Thompson, and Siever, 1960) indicate that dolomite should be an important precipitate and that sedimented $CaCO_3$ should react

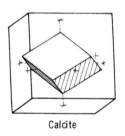

Calcite

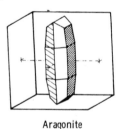

Aragonite

Figure 7.16 Crystalline Forms of Calcite and Aragonite. From Berry and Mason (1959), with permission of W. H. Freeman and Co.

with seawater to form dolomite. Not only is this rarely observed in nature (Berner, 1965, has reported dolomitization of reef $CaCO_3$ material), but attempts to precipitate $MgCO_3$ and $CaMg(CO_3)_2$ under simulated marine conditions have not been successful. We might note in passing that $MgCO_3$ is quite a bit more soluble than $CaCO_3$, K_{sp}° for $MgCO_3$ (7.20) being about 10^{-4} in comparison to 10^{-8} for $CaCO_3$ (see Table A.13).

The equilibria involving $CaCO_3$ in the sea are complicated by the existence of two crystalline polymorphs of this substance, calcite and aragonite (Figure 7.16). The two forms are by no means identical chemically; they have different free energies of formation, and the apparent solubility product of calcite in 36‰ S seawater at 25°C appears to be about half the corresponding value for aragonite (Cloud, 1965). Therefore equilibrium (7.19) should lie somewhat to the right. Both forms are found in the sea, yet calcite is the more important. Although aragonite is formed by some precipitation processes, notably by organisms, it is metastable with respect to calcite at moderate pressures, except possibly at low temperatures (Jamieson, 1953; Simmons and Bell, 1963) and is converted to calcite (7.19). Similarly, attempts to simulate the precipitation of $CaCO_3$ in cavern environments yield predominantly calcite, although aragonite is found in many cave deposits (Siegel and Reams, 1966). The relative paucity of aragonite in the marine environment cannot be attributed to pressure effects in the deep sea, for aragonite is more dense than calcite (specific gravity of 2.85–2.94 compared to 2.7). Thus the application of pressure should force equilibrium (7.19) to the left, and this is in harmony with Jamieson's (1953) characterization of aragonite as the high-pressure modification. In our subsequent discussion, $CaCO_3$, unless specified otherwise, will mean calcite.

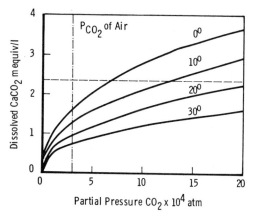

Figure 7.17 Solubility of $CaCO_3$ in 19‰ Cl Seawater as a Function of CO_2 Partial Pressure. From Dietrich (1963), with permission of John Wiley & Sons.

The principal features of the solubility of $CaCO_3$ in seawater are shown in Figures 7.17 and 7.18. Most striking is the supersaturation of seawater with respect to $CaCO_3$. In the surface waters of the equatorial Atlantic Ocean, saturation values between 200% and 300% are the rule (Figure 7.19).

Studies in the waters off Bermuda (Schmalz and Chave, 1963) indicated that only the deepest water was in equilibrium with the sediment present, in this case an ooze containing aragonite, but that the bulk of the water was supersaturated with respect to calcite. Factors such as the particle size and the Mg content of the sediments appear to be important in controlling the degree of saturation (Schmalz and Chave, 1963), and in this connection it is interesting to note that the dissolution of skeletal carbonates in pure water and seawater is an irreversible, i.e., nonequilibrium, process strongly dependent on the mineral content, low Mg calcite being the least soluble, calcite with 20–30% $MgCO_3$ the most soluble, and aragonite intermediate (Chave, Deffeyes, Weyl, Garrels, and Thompson, 1962).

In the Pacific, except for surface water, the water mass is undersaturated with respect to $CaCO_3$ (Peterson, 1966). Kramer (1964) has found that above 14°C Pacific waters tend to be saturated with calcite and above 20°C with aragonite (Figure 7.20).

$CaCO_3$ is not the only inorganic substance that is sometimes supersaturated in seawater. Pacific waters, for example, are slightly supersaturated with hydroxyapatite, $Ca_{10}(PO_4)_6(OH)_2$, and carbonate fluorapatite, $Na_{0.29}Ca_{9.56}$-$(PO_4)_{5.37}(SO_4)_{0.30}(CO_3)_{0.33}F_{2.04}$, at all temperatures (Kramer, 1964). The majority of the inorganic substances in seawater, however, are undersaturated

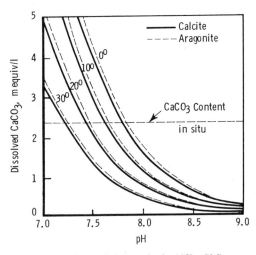

Figure 7.18 Solubilities of Calcite and Aragonite in 19‰ Cl Seawater as a Function of pH. From Dietrich (1963), with permission of John Wiley & Sons.

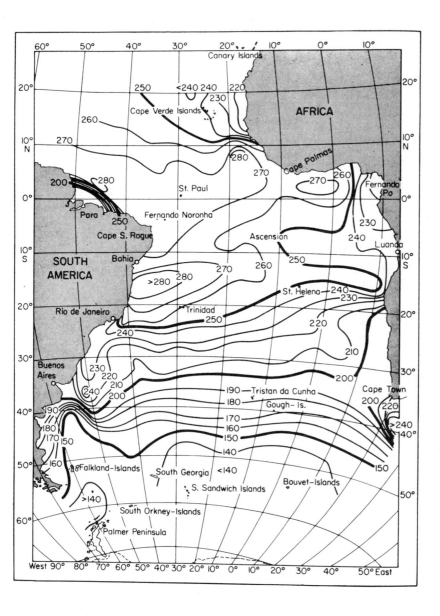

Figure 7.19 Percentage of Saturation of South Atlantic Surface Waters with $CaCO_3$ ($P_{CO_2} = 3 \times 10^{-4}$ atm). From Dietrich (1963), with permission of John Wiley & Sons.

217

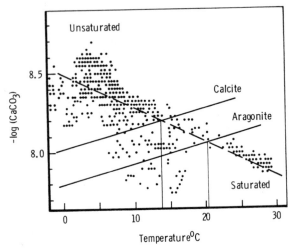

Figure 7.20   Saturation of Pacific Waters (0–3000 m) with Calcite and Aragonite. From Kramer (1964), with permission of AAAS and the author.

(Table 7.12). From Table 7.12, Goldberg (1963) draws the important conclusion that the concentrations of the elements listed, except Ca, Sr, and possibly Ba in deep waters, "cannot be controlled by solubility equilibria." Equilibrium chemical models of seawater (Garrels and Thompson, 1962; Sillén, 1961), while they can be most helpful in summarizing the chemistry of that medium, must therefore always be employed with a certain amount of circumspection. Goldberg also calls attention to the decrease in unsaturation with increasing residence time.

At 25°C and 1 atm, $K_{sp}°$ for calcite and aragonite are $10^{-8.34}$ and $10^{-8.0}$, respectively. Owen and Brinkley (1941) calculated the effect of pressure on the solubility of $CaCO_3$ in pure water and aqueous NaCl solutions of the same ionic strength as seawater, and their results indicate that the product of $\gamma_{Ca^{2+}}$ $\gamma_{CO_3^{2-}}$ increases by less than 10% under 500 atm. On the basis of partial molal volume data of Zen (1957), Berner (1965) has estimated that $\gamma_{Ca^{2+}}$ $\gamma_{CO_3^{2-}}$ is 1.04, 1.08, 1.12, 1.16, and 1.20 times its 1 atm value at 100, 200, 300, 400, and 500 atm or 1000, 2000, 3000, 4000, and 5000 m depth, respectively. A considerably greater pressure effect on the solubility of aragonitic oolites has been reported by Pytkowicz and Conners (1964) who found that the $(Ca^{2+})(CO_3^{2-})$ product at 21°C in 20.5‰ Cl seawater increases some 2.35–2.95 times its 1 atm value at 1000 atm. On the basis of these findings they expected the deep ocean waters to be undersaturated in $CaCO_3$, an expectation subsequently confirmed by measurements on South Pacific Ocean waters (Pytkowicz, 1965) that clearly show undersaturation below the first

Table 7.12 Comparison of Calculated and observed Concentrations of Metals at Saturation with the Observed Concentrations in Seawater (From Goldberg, 1963, with permission of John Wiley & Son)

Metal ion	Insoluble Compound	Solubility Product	Concentration in Saturated Solution Calcd., mg/l	Concentration in Saturated Solution Measured, mg/l	Seawater Concentration, mg/l	Measured Concentration to seawater Concentration	Residence Time, years
Pb	$PbCO_3$	1.5×10^{-13}	0.01	0.3–0.7	0.00003	4000–10,000	2.0×10^3
Ni	$Ni(OH)_2$	1.6×10^{-16}	150	20–450	0.002	10,000–225,000	1.8×10^4
Co	$CoCO_3$	8×10^{-13}	0.02	25–200	0.0005	50,000–400,000	1.8×10^4
Cu	$CuCO_3$	2.5×10^{-10}	5.7	0.4–0.8	0.003	133–266	5.0×10^4
Ba	$BaSO_4$	1×10^{-10}	0.03	0.11	0.03	3.7	8.4×10^4
Zn	$ZnCO_3$	2×10^{-10}	4.6	1.2–2.5	0.01	120–250	1.8×10^5
Cd	$Cd(OH)Cl$	3.2×10^{-11}	105	4–1000	0.0001	40,000–10,000,000	5.0×10^5
Ca	$CaCO_3$	5×10^{-9}	70	100–480	400	0.25–1.2	8.0×10^6
Sr	$SrCO_3$	$3–16 \times 10^{-10}$	9–44	22	8	2.75	1.9×10^7
Mg	$MgCO_3 \cdot H_2O$	1×10^{-5}	84,000	36,000	1350	27	4.5×10^7

Table 7.13   CaCO₃ Saturation Ratios in South Pacific Waters

Depth, m	0	33	94	141	188	236	378	668	955	1710	4252	627(
Saturation ratio[a]	2.13	1.29	1.09	0.94	0.94	0.85	0.65	0.45	0.58	0.73	0.45	0.29

[a] The saturation ratio is defined as the ratio of the observed carbonate concentration to t concentration if the sample had been equilibrated with calcite at the *in situ* pressure.

few hundred meters (Table 7.13) (see also above). Both carbonates and silicates buffer the oceans, the former being by far the more important (Pytkowicz, 1967), and elucidation of the mechanism of *in situ* pH control involves the temperature and pressure dependence of the dissociation constants of carbonic acid and the solubility product of $CaCO_3$ (Pytkowicz, 1963). But, in any event, I think that equilibria are clearly inadequate for accounting for the observed concentrations of carbonate species in the sea and that the rates of dissolution and precipitation processes play crucial roles. This is an area in which research has only just begun. De Groot (1965) has studied the precipitation of $CaCO_3$ (mostly as aragonite) by removal of dissolved CO_2 and found it much too slow to account for the observed rapid "whitings" formation in Persian Gulf waters. The rate of dissolution of calcite specimens in Pacific waters increases abruptly between 3500 and 4000 *m* (Figure 7.21). Similarly

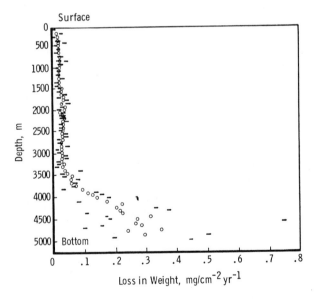

Figure 7.21   Rates of Dissolution of Calcite Spheres in Central Pacific Waters. From Peterson (1966), with permission of AAAS and the author.

the rate of dissolution of foraminiferal ooze also increases with increasing depth (Berger, 1967). These accelerated rates are not necessarily related to the degree of unsaturation (Peterson, 1966), but the depths at which it occurs does seem to correspond roughly to the depths at which calcite removal from marine sediments occurs (Bramlette, 1961).

Absorbed organic coatings may impede the dissolution of suspended carbonate mineral particles in surface waters (Chave, 1965).

5 Summary

The gases of the Earth's atmosphere are well mixed and are principally N_2, O_2, A, and CO_2. An assumption crucial in the interpretation of atmosphere-hydrosphere interactions is that at some time in its history each specimen of seawater was once at the surface and in equilibrium with the gases of the atmosphere.

The solubility of inert gases such as N_2, O_2, and A in seawater decreases with increasing temperature and increasing salinity. The O_2 content of seawater depends strongly on biological activity and is thus highly variable in the seas. The surface waters are nearly saturated, and oxygen maximum appears in the photosynthetic zone, while the deep waters tend to be depleted of oxygen. The very deep waters can be oxygen-rich or anoxic, depending on whether they are new cold waters from high latitudes or stagnant waters. The oxygen profile can be used to characterize the waters of the several oceans.

The chemistry of CO_2 in the seas is very complex and involves interactions between the hydrosphere and the atmosphere, lithosphere, and biosphere. Also, unlike the other atmospheric gases, CO_2 reacts chemically with the seawater to form carbonic acid, H_2CO_3, which in turn can dissociate to form CO_3^{2-} and HCO_3^{-}. At marine pH's, most of the dissolved carbonate in seawater is in the latter form. CO_2, like O_2, is involved in photosynthesis and thus exhibits systematic diurnal and seasonal cycles.

$CaCO_3$ is present in the sea's lithospheric boundaries as calcite or aragonite, the former, for reasons unclear, being far the more common. For all their importance, the many equilibria involved in the CO_2-CO_3^{2-} cycle in the sea are still imperfectly understood, and even less is known concerning their kinetics.

REFERENCES

Benson, B. B., and P. D. M. Parker, *Deep-Sea Res.*, **7**, 237 (1961).
Benson, B. B., and P. D. M. Parker, *J. Phys. Chem.*, **65**, 1489 (1961).
Berger, W. K., *Science*, **156**, 383 (1967).
Berner, R. A., *Geochim. Cosmochim. Acta*, **29**, 947 (1965).

Berner, R. A., *Science*, **147**, 1297 (1965).

Berry, L. G., and B. Mason, *Mineralogy*, W. H. Freeman and Co., San Francisco, Calif., 1959.

Bieri, R. H., M. Koide, and E. D. Goldberg, *J. Geophys. Res.*, **71**, 5243 (1966).

Bieri, R. H., M. Koide, and E. D. Goldberg, *Science*, **146**, 1035 (1964).

Bolin, B., *Tellus*, **12**, 274 (1960).

Bramlette, M. N., "Pelagic Sediments," in M. Sears, ed., *Oceanography*, Am. Assoc. Advan. Sci. Publ. No. 67 Washington, D.C., 1961.

Buch, K., *J. Conseil, Perm. Exploration Mer.*, **8**, 309 (1933).

Buch, K., and S. Grienberg, *J. Conseil Conseil Intern. Exploration Mer*, **7**, 233 (1932).

Carritt, D. E., *Deep-Sea Res.*, **2**, 59 (1954).

Chave, K. E., *Science*, **148**, 1723 (1965).

Chave, K. E., K. S. Deffeyes, P. K. Weyl, R. M. Garrels, and M. E. Thompson, *Science*, **137**, 33 (1962).

Cloud, P. E., Jr., "Carbonate Precipitation and Dissolution in the Marine Environment," in J. P. Riley and G. Skirrow, eds., *Chemical Oceanography*, Academic Press, London, 1965, Vol. I, Chap. 17.

Cooper, L. H. N., "Vertical and Horizontal Movements in the Oceans, in M. Sears ed., *Oceanography*, Am. Assoc. Advan. Sci. Publ. No. 67 (1961).

Craig, H., and L. I. Gordon, *Geochim. Cosmochim. Acta*, **27**, 949 (1963).

Craig, H., R. F. Weiss, and W. B. Clarke, *J. Geophys. Res.*, **72**, 6165 (1967).

Culberson, C. H., M. S. Thesis, Oregon State Univ., 1968.

Culberson, C., D. R. Kester, and R. M. Pytkowicz, *Science*, **157**, 59 (1967).

Deffeyes, K. S., *Limnol. Oceanog.*, **10**, 412 (1965).

de Groot, K., *Nature*, **207**, 404 (1965).

Dietrich, G., *General Oceanography*, Interscience, New York, 1963.

Disteche, A., and S. Disteche, *J. Electrochem. Soc.*, **114**, 330 (1967).

Douglas, E., *J. Phys. Chem.*, **69**, 2608 (1965).

Douglas, E., *J. Phys. Chem.*, **71**, 1931 (1967).

Enns, T., P. F. Scholander, and E. D. Bradstreet, *J. Phys. Chem.* **69**, 389 (1965).

Foote, F. J., *Ind. Eng. Chem. Anal. Ed.*, **4**, 39 (1932).

Garrels, R. M., and M. E. Thompson, *Am. J. Sci.*, **260**, 57 (1962).

Garrels, R. M., M. E. Thompson, and R. Siever, *Am. J. Sci.*, **258**, 402 (1960).

Gast, J. A., and T. G. Thompson, *Anal. Chem.*, **30**, 1549 (1958).

Goldberg, E. D., "The Oceans as a Chemical System," in M. N. Hill, ed., *The Sea*, Interscience, New York, 1963, Vol. II, Chap. 1.

Green, E. J., Ph.D. Thesis, *Massachusetts Inst. Technol.*, 1965.

Green, E. J., and D. E. Carritt, *J. Marine Res.* (*Sears Found. Marine Res.*), **25**, 140 (1967).

Hamm, R. E., and T. G. Thompson, *J. Marine Res.* (*Sears Found. Marine Res.*), **4**, 11 (1941).

Harned, H. S., and R. Davis, Jr., *J. Am. Chem. Soc.*, **65**, 2030 (1943).

Harned, H. S., and S. R. Scholes, Jr., *J. Am. Chem. Soc.*, **63**, 1706 (1941).

Igelsrud, I., T. G. Thompson, and B. M. G. Zwicker, *Am. J. Sci.*, **35**, 47 (1938).

Jamieson, J. C., *J. Chem. Phys.*, **21**, 1385 (1953).

Keeling, C. D., N. W. Rakestraw, and L. S. Waterman, *J. Geophys. Res.*, **70**, 6087 (1965).

Klotz, I. M., *Limnol. Oceanog.*, **8**, 149 (1963).

König, H., H. Wänke, G. S. Bien, N. W. Rakestraw, and H. E. Suess, *Deep-Sea Res.*, **11**, 243 (1964).

Kramer, J. R., *Science*, **146**, 637 (1964).

Lyman, J., Ph.D. Thesis, Univ. of Calif., 1956.

Mazor, E., G. I. Wasserburg, and H. Craig, *Deep-Sea Res.*, **11**, 929 (1964).

Moberg, E. G., and M. W. Harding, *Science*, **77**, 510 (1933).

Noakes, J. E., and D. W. Hood, *Deep-Sea Res.*, **8**, 121 (1961).

Owen, B. B., and S. R. Brinkley, *Chem. Rev.*, **29**, 461 (1941).

Peterson, M. N. A., *Science*, **154**, 1542 (1966).

Pytkowicz, R. M., *Deep-Sea Res.*, **10**, 633 (1963).

Pytkowicz, R. M., *Geochim. Cosmochim. Acta*, **31**, 63 (1967).

Pytkowicz, R. M., *Limnol. Oceanog.*, **10**, 220 (1965).

Pytkowicz, R. M., and D. N. Conners, *Science*, **144**, 840 (1964).

Rakestraw, N. W., *Chemistry*, **38**, 15 (1965).

Rakestraw, N. W., and U. M. Emmel, *J. Marine Res.* (*Sears Found. Marine Res.*), **1**, 207 (1938).

Rakestraw, N. W., and V. M. Emmel, *J. Phys. Chem.*, **42**, 1211 (1938).

Rakestraw, N. W., and H. E. Mahncke, *Ind. Eng. Chem. Anal. Ed.*, **7**, 425 (1935).

Richards, F. A., "Anoxic Basins and Fjords," in J. P. Riley and G. Skirrow, eds., *Chemical Oceanography*, Academic Press, London, 1965, Vol. I, Chap. 13.

Richards, F. A., "Dissolved Gases," in J. P. Riley and G. Skirrow, eds., *Chemical Oceanography*, Academic Press, London, 1965, Vol. I, Chap. 6.

Richards, F. A., in J. W. Hedgpeth, ed., *Treatise on Marine Ecology and Paleoecology*, Geol. Soc. Am., New York, 1957, Vol. I.

Richards, F. A., and B. B. Benson, *Deep-Sea Res.*, **7**, 254 (1965).

Riley, J. P., and G. Skirrow, *Chemical Oceanography*, Academic Press, London, 1965.

Schmalz, R. F., and K. E. Chave, *Science*, **139**, 1206 (1963).

Seiwell, H. R., *Papers Phys. Oceanog. Meteorol.*, *Mass. Inst. Technol.*, Woods Hole Oceanog. Inst., **3**, 1 (1935).

Siegel, F. R., and M. W. Reams, *Sedimentology*, **7**, 241 (1966).

Sillén, L. G., "The Physical Chemistry of Sea Water," in M. Sears, ed., *Oceanography*, Am. Assoc. Advan. Sci. Publ., No. 67, Washington, D.C., 1961.

Simmons, G., and P. Bell, *Science*, **139**, 1197 (1963).

Skirrow, G., "The Dissolved Gases–Carbon Dioxide," in J. P. Riley and G. Skirrow, eds., *Chemical Oceanography*. Academic Press, London, 1965, Vol. I, Chap. 7.

Taft, B. A., *Marine Res.* (*Sears Found. Marine Res.*), **21**, 129 (1963).

Takahashi, T., *J. Geophys. Res.*, **66**, 477 (1961).

Vaccaro, R. F., "Inorganic Nitrogen in Sea Water," in J. P. Riley and G. Skirrow, eds., *Chemical Oceanography*, Academic Press, London, 1965, Vol. 1, Chap. 9.

von König, H., *Z. Naturforsch.*, **18a**, 363 (1963).

U.S.A.F., *Handbook of Geophysics*, The Macmillan Co., New York, 1961.

Waterman, L. S., *Nature*, **205**, 1099 (1965).

Wiebe, R., *Chem. Rev.*, **29**, 475 (1941).

Wilcox, L. V., *Ind. Eng. Chem. Anal. Ed.*, **4**, 38 (1932).

Wyrtki, K., *Deep-Sea Res.*, **8**, 39 (1961).

Wyrtki, K., *Deep-Sea Res.*, **9**, 11 (1962).

Zen, E. A., *Geochim. Cosmochim. Acta*, **12**, 103 (1957).

8 Suspended Particulate Matter in Seawater

1 Introduction

In addition to dissolved substances, a seawater sample contains materials present as a second phase (Figure I.2 of the Introduction). The nature and both the relative and total concentrations of the dissolved constituents of seawater are remarkably invariant; in contrast, the nature and concentration of suspended material in the sea can be highly variable. The dissolved constituents give rise to such important seawater properties as the freezing point depression, the density, and the electrical conduction of seawater, while the suspended materials affect for the most part the optical and acoustical properties of seawater, giving rise to absorption and scattering of sound and being responsible for the color and turbidity of the ocean water.

The total amount of the dissolved substances in seawater is, of course, very much greater than of the suspended material. A liter of seawater contains between 30 and 40 g of dissolved salts. A sample taken at a depth of 4030 m in the Atlantic (160 m off the bottom) and described as having a large concentration of suspended material contained less than 0.003 g of solid in a liter (Groot and Ewing, 1963); centrifugation of another sample, taken in the Caribbean Sea, yielded about 0.0003 g of solid stuff in a liter, and of this residue only about 16% was of a mineral nature (Jacobs and Ewing, 1965). The clear blue-green coastal waters off Plymouth (England) contain 0.0004–0.002 g of suspended particulate matter per liter, while the blue Atlantic Ocean waters some 160 km beyond the continental shelf contain roughly half these amounts (Harvey, 1966). In deep Atlantic waters, Ewing and Thorndike (1965) have found a zone of cloudy or nepheloid water consisting of a layer of suspended lutite in concentrations which may be sufficient to induce downslope flow.

2 State of Division

Although the various filtration procedures qualitatively classify particle sizes, no careful, quantitative investigations of the distribution of the sizes of the suspended heterogeneities in seawater has yet come to my attention.

Table 8.1   Increase in Surface Area with
Subdivision of a 1-cm Cube

Edge, cm	Surface Area, cm²
1	6
0.5	12
0.1	60
0.0001	60,000
0.0000001	60,000,000

This is surprising, for the particle size is an important property and is crucial in a number of phenomena ranging from optical and acoustic scattering to surface chemical activity and the rate of sedimentation.

If a cube, 1 cm on an edge, is repeatedly subdivided into smaller cubes, the surface area increases rapidly (Table 8.1). As a consequence, highly divided or porous materials can have greatly enhanced chemical activity, either participating in reactions as primary reactants or as catalytic substances. This applies to biological as well as to strictly inorganic processes. The concentration of bacteria in the open sea is low, for example, but, if an object is left in the sea, it accumulates a flourishing population, presumably because of the concentrating of nutrients by absorption on the surface.

The transport of suspended material in the seas does not fall within the scope of chemical oceanography. Suffice it to say here that the settling rate depends directly on particle size (Table 8.2), very largely determined by Stokes' law (3.12). Differences in settling rate can give rise to interesting chemical consequences. Dobrzanskaya and Kovalevsky (1959) have concluded that the vertical distribution of iron in the Black Sea is for the most part determined by the size of the aggregates containing this element. A density gradient in the sea acts as a "barrier." The heavier particles fall through, but the smaller, more slowly sinking particles are retained, resulting in a layer of increased iron concentration.

The sizes of the heterogeneities in the sea range from the blue whale down to objects so minute that it ceases to be operationally meaningful to call them heterogeneities and they pass imperceptibly into the category of dissolved substances. The classification of substances as dissolved, colloidal, or particulate (Table 8.3) is therefore inescapably arbitrary.

The particulate matter in seawater eventually settles out. It is a transient constituent. Colloidal matter, however, in principle (but see below) could remain dispersed indefinitely and must be considered a seawater constituent almost in the same sense as the dissolved solutes. Colloids are produced by dispersion processes in which a coarser material is subdivided into finer particles or by condensation processes in which subcolloidal species aggregate.

Table 8.2 Settling Velocity of Quartz Spheres in Distilled Water at 20°C
(From Sverdrup, Johnson, and Fleming, 1942, with permission of Prentice-Hall, Inc.)

	Diameter		Settling Velocity		Time To Fall 10 cm				Settling Velocity
	mm	microns	Stokes' Law	Wadell	Days	Hours	Min	Sec	m/day
Boulder	256	256,000	—	—	—	—	—	—	—
Cobble	64	64,000	—	—	—	—	—	—	—
Pebble	4	4,000	—	—	—	—	—	—	—
Granule	2	2,000	—	—	—	—	—	—	—
Very coarse sand	1	1,000	(89.2 cm/sec)	16.0 cm/sec	—	—	—	0.6	—
Coarse sand	1/2	500	(22.3 cm/sec)	7.7 cm/sec	—	—	—	1.2	—
Medium sand	1/4	250	(5.58 cm/sec)	3.4 cm/sec	—	—	—	2.7	—
Fine sand	1/8	125	(1.39 cm/sec)	1.2 cm/sec	—	—	—	8.3	1040
Very fine sand	1/16	62.5	3482 microns/sec	—	—	—	—	29	301
Silt	1/32	31.2	870	—	—	—	1	55	75.2
	1/64	15.6	218	—	—	—	7	40	18.8
	1/128	7.8	54.4	—	—	—	30	38	4.7
	1/256	3.9	13.6	—	—	2	2	32	1.2
Clay	1/512	1.95	3.4	—	—	8	10	—	0.3
	1/1024	0.98	0.85	—	1	8	41	—	0.074
	1/2048	0.49	0.21	—	5	10	42	—	0.018
	1/4096	0.25	0.052	—	21	18	50	—	0.004
	1/8192	0.12	0.013	—	87	3	19	—	0.001

Table 8.3   Classification of Substances in Seawater

Type	Particle Size	Example Constituents	
Solution	$<10^{-7}$ cm	Inorganic simple and complex ions, molecules, and polymeric species; polyelectrolytes; organic molecules; undissociated solutes, small aggregates.	
Colloidal dispersion (submicroscopic)	10^{-7} to 10^{-5} cm	Mineral substances, hydrolysis and precipitation products, macromolecules, biopolymers, detritus	
Fine particulate dispersion (visible under the microscope)	10^{-5} to 5×10^{-3} cm	Mineral substances, precipitated and coagulated particles, detritus, bacteria, plankton, and other microorganisms	
Coarse dispersion (visible to the naked eye)	$>5 \times 10^{-3}$ cm	Mineral substances, precipitated and coagulated particles, detritus, macroplankton, organisms of all types including the largest fish, cephalopods, and mammals	
Seston (turbidity)		Mineral substances (carried into the sea by rivers and winds); detritus (fine inorganic and organic remains from the decomposition of dead organisms); nanno (plankton and other organisms)	Passively transported
		Nekton (larger organisms such as fish)	Actively transported

They are classified into hydrophilic and hydrophobic colloids, depending on whether or not they are attracted by and associated with water. Colloidal particles are small enough to be subject to the forces producing Brownian motion as they diffuse through the liquid much as dissolved solutes diffuse. Like dissolved substances, they are capable of lowering the freezing point of water but the effect is very much smaller. In part, colloids owe their stability to their electrical charge. Coulombic repulsion of like-charged particles keeps them separate, thus preventing their further aggregation and subsequent precipitation. Destruction of the charge results in destruction of the colloid (Parks, 1967), and a very effective way of doing this is to add an electrolyte. Very little appears to be known about colloidal substances in the sea, and the reason may be that they are relatively unimportant in that electrolytic solution. In particular, the hydroxides formed on the hydrolysis of Fe(III) in the sea are believed to go through a colloidal state, but the $Fe(OH)_3$ can then be flocculated by electrolytes in seawater, yet the rate of the latter process can be exceedingly slow because of the very small concentrations involved (Harvey, 1966).

We might expect the precipitation of colloidal materials in river waters where they are mixed with the electrolyte-containing water of the sea. Appreciable quantities of mineral colloidal material are carried into Chesapeake Bay by rivers. At the mouth of their estuaries, the dispersed material has decreased to about 0.005 mg/l from about 0.2 mg/l at their head, but the decrease in concentration is due to dilution rather than flocculation or settling (Nelson, 1959).

3   Inorganic Suspended Material

Not unexpectedly, the chemical composition of the inorganic suspended material in the sea resembles that of the marine sediment beneath. The sediment once was suspended material; the suspended material will be added to the accumulation of sediment. In addition, water movement may stir up the bottom so that sediment may find itself again suspended. Groot and Ewing (1963) found no mineral matter in the surface Atlantic waters, but 160 m off the bottom they found quantities of suspended clay. Wangersky and Gordon (1965), however, found greater values of particulate carbonate in surface waters and constancy below 100 m. X-ray diffraction studies of the mineral content of the suspended material have helped to establish the similarity of the suspended and sediment compositions. Jacobs and Ewing (1965) found that in the Caribbean Sea the suspended matter contained the same minerals —illite, kaolinite, chlorite, talc, quartz, feldspar, and amphibole—as the sediments in that area with kaolinite and chlorite being more abundant in the sediment relative to illite. Evidence for feldspar, quartz, calcite, and kaolinite was found in Pacific waters off the coast of Japan, and turbid China sea waters showed calcite and quartz but no montmorillonite, a mineral absent from the sediments in that region (Ishii and Ishikawa, 1965). However, the situation can be complicated by the transmutation of some mineral substances such as silica and $CaCO_3$ which dissolve and are reprecipitated by organisms. Thus some authors claim that all the suspended silica in the world's oceans are remnants of organisms and that even the small amount of colloidal SiO_2 is biogenic (Lisitsyn, Belyaev, Bogdanov, and Bogayavlenskii, 1966).

4   Organic Suspended Material

The quantity of organic matter suspended in seawater can be comparable to and even in excess of the quantity of inorganic particulate material (Table 8.4). According to the results summarized by Parsons (1963), the amount of sus-

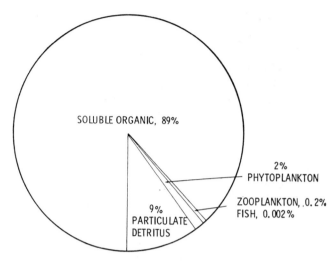

SOLUBLE ORGANIC, 89%

2%
PHYTOPLANKTON

ZOOPLANKTON, .0. 2%
FISH, 0. 002%

9%
PARTICULATE
DETRITUS

Figure 8.1   Distribution of Organic Material in the Euphotic Zone of the Hydrosphere.

pended particulate organic material is considerably less than the amount of dissolved organic substance (Figure 8.1). He is careful, however, to point out that the distinction between the two categories is contingent on the filtering procedures employed and consists mostly of detritus (Figure 8.1). Menzel (1964) found that the waters of the Arabian Sea contained about 1300 g of dissolved organic carbon under 1 m^2 of surface but only 120 g of particulate, the concentration of both forms decreasing with increasing depth. The dissolved organic material and selected aspects of the biochemistry of the seas form the topic of the next chapter, and here the discussion will be restricted

Table 8.4   Distribution of Organic and Inorganic Fractions
in Suspended Particulate Matter

Area	Total Suspended Matter, mg/l	Organic Matter, %	Reference
Offshore, Pacific	10.5	62	Fox et al. (1953)
Offshore, Pacific	3.8	29	Fox et al. (1953)
North Sea	6.0	27	Postma (1954)
Wadden Sea	18.0	14	Postma (1954)
Oceanic average	0.8–2.5	20–60[a]	Lisitzen (1959)
Bering Sea	2–4	—	Lisitzen (1959)
Indian Ocean	—	6–36[a]	Lisitzen (1959)
Long Island Sound	2–7	20–45	Riley (1959)

to the detritus. The organic detritus content of seawater varies widely in horizontal and vertical space and in time (Table 8.5). The detritus content is particularly variable in coastal areas. The amount of organic detritus is much reduced, yet fairly constant, with increasing depth below the euphotic zone (Riley, van Hemert and Wangersky, 1965). Below the euphotic zone the detritus content appears to be related to the salinity (Jerlov, 1959). The distribution of particles also sometimes appears to depend on the position of the phosphate and CO_2 maxima and the O_2 minima.

Microscopic examination of the detritus reveals a rich variety of debris, including fragments of phytoplankton cells and the exoskeletons of zooplankton and fecal pellets from zooplankton and, in coastal areas, sponge spicules, seaweed and even wood fragments, pollen grains, and evidence of human pollution. The size of the particles seems to decrease with increasing depth (Parsons, 1963). As mentioned earlier, the inorganic particulate material can be very effective in concentrating dissolved organic species from seawater by absorption, thereby providing an environment conducive to bacterial growth (Bader, Hood, and Smith, 1960).

Chemical analyses of the detritus have indicated the presence of chlorophyll and carotenoids (Yentsch and Ryther, 1959; Steele and Baird, 1965); vitamin B_{12} (Burkholder and Burkholder, 1956); monosaccharides including glucose, galactose, mannose, arabinose, and xylose (Parsons and Strickland, 1962); and amino acids including glutamic acid, aspartic acid, lysine, arginine, serine, proline, and especially alanine and glycine. (See Table 9.5 and compare with Tables 9.6, 9.7, and 9.8.)

Riley (1963) has proposed that the 5×10^{-4} to 3×10^{-1} cm inanimate organic-inorganic particles frequently found in seawater might originate from the absorption of dissolved materials on bubbles and other surfaces. Particulate organic matter has been produced in the laboratory by bubbling air through seawater and has been used to feed shrimp cultures (Baylor and Sutcliffe, 1963). The material aggregated on bubble surfaces has a higher carbon content than living phytoplankton (Riley, Wangersky, and van Hemert, 1964).

5   Silicon

The concentration of silicon in seawater is affected by both geological and biological processes. Armstrong (1965) describes it as the most variable element in the sea. It can be undetectable in surface waters, yet its concentration in the deep waters of the Pacific and Indian oceans are in the 2 to 3×10^{-3} g/l range, or roughly twice the Atlantic Ocean concentration. Silicon is found both as dissolved silicate ion and as suspended silica.

Table 8.5 Distribution of Organic Detritus

Area	Date	Depth (m)	Organic Detritus, dry wt, mg/m³	Reference
		Pacific Ocean		
St "P" 50°N 145°W	July 1959	0–50	250[a, b]	McAllister et al. (1960)
St "P" 50°N 145°W	August 1959	0–50	250[a, b]	McAllister et al. (1960)
St "P" 50°N 145°W	July 1961	10	362[a, b]	Parsons and Strickland (1962)
48°57′N 132°30′W	July 1961	15–35	478[a, b]	Parsons and Strickland (1962)
48°57′N 132°30′W	July 1961	500	92[a]	Parsons and Strickland (1962)
48°57′N 132°30′W	July 1961	1500	64[a]	Parsons and Strickland (1962)
48°57′N 132°30′W	July 1961	3000	88[a]	Parsons and Strickland (1962)
51°27′N 133°20′W	July 1961	500	152[a]	Parsons and Strickland (1962)
Coastal, La Jolla	—	6	6500	Fox et al. (1953)
Coastal, La Jolla	—	6	1100	Fox et al. (1953)
Coastal, Nanaimo	September	0–21	1062[a]	Parsons (1960)
Coastal, Nanaimo	January	5	400[a]	Parsons (1960)
Coastal, Nanaimo	April	0–6	80[a]	Parsons (1960)
Coastal, Nanaimo	July	5	590[a]	Parsons (1960)
Coastal, Nanaimo	August	5	868[a]	Parsons (1960)
		Atlantic Ocean and North Sea		
Sargasso Sea	February–March	surface	180[c]	Ryther and Menzel (1960)
North Sea (coastal)	—	—	830	Postma (1954)
Waddensee (coastal)	—	—	750	Postma (1954)
Reykjanes Ridge	May–June	0–50	275[c]	Hagmeier (1960)
Reykjanes Ridge	May–June	150	85[c]	Hagmeier (1960)
Reykjanes Ridge	May–June	650	30[c]	Hagmeier (1960)
North Sea 58°01′N 6°44′E	August	0	160[c]	Hagmeier (1960)
	August	60	120[c]	Hagmeier (1960)
Fladen Ground, 57°51′N 1°26′E	August	0	174[c]	Hagmeier (1960)
	August	50	145[c]	Hagmeier (1960)

alculated from carbon ×2 = dry organic matter.
orrected for an average chlorophyll a of 0.5 mg/m³ × 60 = plant carbon.
tal detritus assumed to contain 30% dry organic matter.

The exact form of both the dissolved and suspended silicon species is unclear. Sillén (1961) has pointed out that silicic acid, H_4SiO_4 [or $Si(OH)_4$], is a very weak acid

$$(8.1) \quad H_4SiO_4 \rightleftharpoons H^+ + H_3SiO_4^- \qquad K_1 = (H^+)(H_3SiO_4^-)/(H_4SiO_4)$$

$$(8.2) \quad H_3SiO_4^- \rightleftharpoons H^+ + H_2SiO_4^{2-}, \quad \text{etc.}$$
$$K_2 = (H^+)(H_2SiO_4^{2-})/(H_3SiO_4^-)$$

so that in the marine pH range only about 5% of the dissolved silicon would be in the form $H_3SiO_4^-$. He concludes that the dissolved silicon exists mostly as mononuclear $Si(OH)_4$ and not in colloidal form as earlier believed. This is perhaps surprising, as H_4SiO_4 readily hydrolyzes to form a gel and the absence of a colloid in seawater must be attributed to the concentrations involved and to depolymerization by the other electrolytes present (Armstrong, 1965). The degree of hydration of Si(IV) is variable. Silicic acid is readily and irreversibly dehydrated to form very stable silica, SiO_2:

$$(8.3) \quad H_4SiO_4 \rightarrow 2\,H_2O + SiO_2$$

The form of the suspended silicon is equally mysterious. Lisitsin (1961) found that over 70% of the suspended clays in the Indian Ocean consisted of very minute particles less than 10^{-3} cm in diameter.

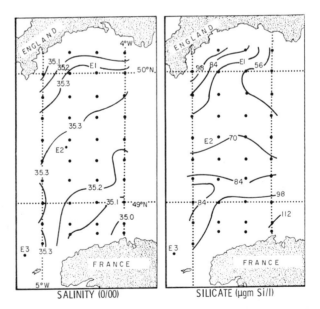

Figure 8.2 Salinity and Silicate Content of Western English Channel Surface Waters in February. From Armstrong (1965), with permission of Academic Press, Inc.

A considerable amount of silicon-containing material washes and blows into the seas from land; hence coastal waters commonly exhibit low salinities and high silicon contents (Figure 8.2). MacKenzie, Garrels, Bricker, and Bickley (1967) suggest that suspended mineral substances carried into the ocean by rivers may be a major factor in determining the silicon content of seawater.

Silicon is a biologically significant element and, as a consequence, like phosphate and nitrate (see Figure 6.7), it exhibits a strong seasonal dependence reflecting the waxing and waning of the life processes (Figures 8.3 and 8.4). The variations in silicate tend to be related to, yet greater and more erratic than, the phosphate content. The silicate and dissolved oxygen contents are also related. The cycle (Figure 8.5) begins in the spring with the uptake of silicon by the growing phytoplankton population, resulting in a depletion of the Si content of the seawater. In the summer the rate of this growth slackens and the silicate is replenished somewhat. But in the fall there may be a second spurt in phytoplankton growth (notice the double minimum in Figure 8.4). With the approach of winter the organisms die and, as their remains slowly sink, the silicon is restored to the seawater by processes of redissolution. Owing to electrostriction about the resultant ions, the increased pressure at

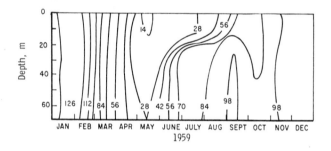

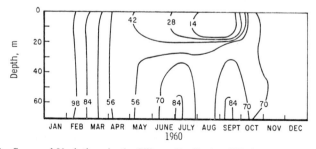

Figure 8.3   Seasonal Variations in the Silicate Profile (μg Si/l) for the English Channel. From Armstrong and Butler (1960, 1962), with permission of Marine Biological Association of the United Kingdom.

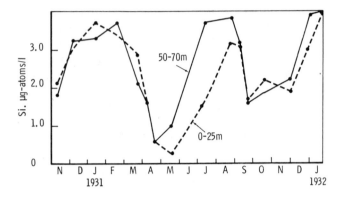

Figure 8.4 Seasonal Variations in the Silicate Content of the English Channel. From Sverdrup, Johnson, and Fleming (1942), with permission of Prentice-Hall, Inc.

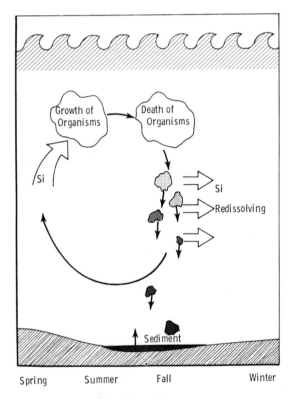

Figure 8.5 Silicon Cycle in the Sea.

greater depths undoubtedly facilitates this process (Berger, 1967). However, although seawater is undersaturated with respect to silicon, the dissolution process is unexpectedly slow (Armstrong, 1965), possibly because of absorbed cations (Lewin, 1959), at least in part—still another example of the importance of *rates* as well as equilibria in marine chemistry. The silicon which does not dissolve finally settles to the bottom, adding to the siliceous sediments (Figure 8.5). In addition to this biological cycle, the silicon content of seawater is further complicated by downward and upward movements of the water mass and diffusional processes.

The vertical silicate distribution can be very complex (Figure 8.6). Earlier (Table 6.4) we saw that the major oceans tend to have characteristic silicate-to-phosphate and silicate-to-nitrate ratios. One of the most fascinating features of the marine chemistry of silicon is that the major oceans of the world

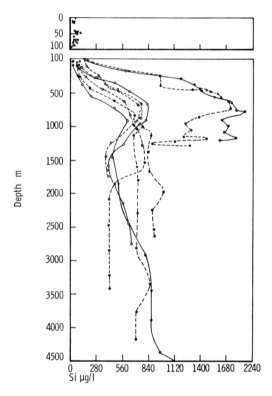

Figure 8.6 Silicate Profiles for Western North Atlantic Stations in the Antilles Arc Region. 5601, 5607, Cariaco Trench; 5282, 5284, Venezuelan Basin; 5276, 5278, Atlantic Ocean, east of Windward Islands; 5289, Puerto Rico Trench; 5286, Jungferm Passage region. From Richards (1958), with permission of Sears Foundation for Marine Research.

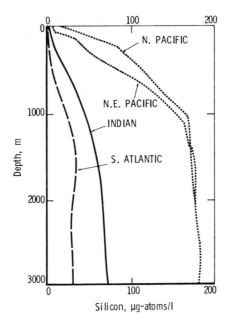

Figure 8.7 Silicate Profiles in the North Pacific, Indian, and South Atlantic Oceans. From Sverdrup, Johnson, and Fleming (1942), with permission of Prentice-Hall, Inc.

appear to have strikingly different silicon profiles (Figure 8.7). The deep waters of the Pacific and Indian oceans are much richer in silicon than those of the Atlantic. Exchange of bottom water in these oceans is poor compared with the freer movement of Atlantic water, and in the Atlantic Ocean the silicate is better dispersed. In the Antarctic Ocean very high concentrations of phosphorus and silicon occur at the ice edge and in deep water. Silicon contents as high as 5200 μg Si/l have been reported in the Antarctic Convergence. In these regions the dissolution of siliceous material, in contrast to the results mentioned above, appears to be quite rapid, possibly because diatoms eaten and excreted by zooplankton have had any protective coating removed by the digestive processes—a possibility compatible with the rapid dissolution of silica minerals in seawater (MacKenzie and Garrels, 1965). Or it may be simply because thin-walled (hence greater reactive surface area in a given weight) species are more numerous in the area (Harvey, 1963; J. C. Lewin, 1961).

As mentioned in the preceding chapter, siliceous materials also play a role in buffering seawater (Garrels, 1965), and Sillen (1967) has reiterated his conviction of their significance, but they are less important than the carbonate system.

6 Summary

In addition to dissolved substances, seawater contains suspended materials ranging through all sizes from colloids to particulate matter. The distribution of this matter is dependent on their settling rates and thus on the size of the aggregates. Colloidal substances can be precipitated by electrolyte addition, and one might therefore expect some interesting chemistry when the colloid-carrying fresh waters of rivers mix with the saline seawater.

The chemical composition of the inorganic suspended matter, not unexpectedly, is closely related to that of marine sediments. The organic material is largely detritus.

Silicon is an important nutrient in the oceans, and the silicon content of seawater, both dissolved and particulate, is closely linked to the growth and decay of organisms. The silicate profile, like the O_2 profile which is also closely related to biopopulations, tends to be characteristic of a given ocean.

REFERENCES

Armstrong, F. A. J., "Silicon," in J. P. Riley and G. Skirrow, eds., *Chemical Oceanography*, Academic Press, London, 1965, Vol. 1, Chap. 10.

Armstrong, F. A. J., and E. I. Butler, *J. Nav. Biol. Assoc. U.K.*, **39**, 525 (1960).

Armstrong, F. A. J., and E. I. Butler, *J. Nav. Biol. Assoc. U.K.*, **42**, 253 (1962).

Bader, R. G., D. W. Hood, and J. B. Smith, *Geochim. Cosmochim. Acta*, **19**, 236 (1960).

Baylor, E. R., and W. H. Sutcliffe, *Limnol. Oceanog.*, **8**, 369 (1963).

Berger, W. H., *Science*, **156**, 383 (1967).

Burkholder, P. R., and L. M. Burkholder, *Limnol. Oceanog.*, **1**, 202 (1956).

Dobrzanskaya, M. A., and A. D. Kovalevsky, *Abstr. Intern. Oceanog. Congr.* 1959, Am. Assoc. Advan. Sci., Washington, D.C., 1959, p. 884.

Ewing, M., and E. M. Thorndike, *Science*, **147**, 1291 (1965).

Fox, D. L., C. M. Oppenheimer, and J. S. Kittredge, *J. Marine Res. (Sears Found. Marine Res.)*, **12**, 233 (1953).

Garrels, R. M., *Science*, **148**, 69 (1965).

Groot, J. J., and M. Ewing, *Science*, **142**, 579, (1963)

Hagmeier, E., Dissertation, Kiel, 1960.

Harvey, H. W., *The Chemistry and Fertility of Sea Waters*, Cambridge University Press, Cambridge, England, 1966.

Ishii, J., and I. Ishikawa, *Studies Oceanog., Collection Papers*, 288 (1965).

Jacobs, M. B., and M. Ewing, *Science*, **149**, 179 (1965).

Jerlov, N. G., *Deep-Sea Res.*, **5**, 173 (1959).

Lewin, J. C., *Abstr. Intern. Oceanog. Congr., 1959*, Am. Assoc. Advan. Sci., Washington, D.C., 1959, p. 950.

Lewin, J. C., *Geochim. Cosmochim. Acta.*, **21**, 182 (1961).

Lisitsin, A. P., *Okeanol. Issled., Akad. Nauk. SSSR*, **3**, 52 (1961).

Lisitsin, A. P., Y. I. Belyaev, Y. A. Bogdanov, and A. N. Bogoyavlenskii, *Geokhim. Kremnezema, Akad. Nauk. SSSR*, 37 (1966).

Lisitsin, A. P., *Abstr. Intern. Oceanog. Congr. 1959*, Am. Assoc. Advan. Sci., Washington, D.C., 1959, p. 240.

McAllister, C. D., T. R. Parsons, and J. D. H. Strickland, *J. Conseil Conseil Perm. Intern. Exploration Mer.*, **25**, 240 (1960).

MacKenzie, F. T., and R. M. Garrels, *Science*, **150**, 57 (1965).

MacKenzie, F. T., R. M. Garrels, O. P. Bricker, and F. Bickley, *Science*, **155**, 1404 (1967).

Menzel, D. W., *Deep-Sea Res.*, **11**, 757 (1964).

Nelson, B. W., *Abstr. Intern. Oceanog. Congr. 1959*, Am. Assoc. Advan. Sci., Washington, D.C., 1959, p. 640.

Parks, G. A., in W. Stumm ed., *Equilibrium Concepts in Natural Water Systems*, Advan. Chem. Ser., No. 67 (1967).

Parson, T. R., *Fisheries Res. Board Can.*, *M.S. Rept.*, No. 81 (1960).

Parsons, T. R., *Progr. Oceanog.*, **1**, 205 (1963).

Parsons, T. R., and J. D. H. Strickland, *Science*, **136**, 313 (1962).

Postma, H., *Arch. Neerl. Zool.*, **10**, 1 (1954).

Richards, F. A., *J. Marine Res. (Sears Found. Marine Res.)*, **17**, 449 (1958).

Riley, G. A., *Bull. Bingham Oceanog., Coll.*, **73**, 83 (1959).

Riley, G. A., *Limnol. Oceanog.*, **8**, 372 (1963).

Riley, G. R., D. van Hemert, and P. J. Wangersky, *Limnol. Oceanog.*, **10**, 354 (1965).

Riley, G. A., P. J. Wangersky, and D. van Hemert, *Limnol. Oceanog.*, **9**, 546 (1964).

Ryther, J. M., and D. W. Menzel, *Bermuda Biol. Sta. Rept.* (1960).

Sillén, L. G., "The Physical Chemistry of Sea Water," in M. Sears, ed., *Oceanography*, Am. Assoc. Advan. Sci., Publ. No. 67 (1961).

Sillén, L. G., *Science*, **156**, 1189 (1967).

Steele, J. H., and I. E. Baird, *Limnol. Oceanog.* **10**, 261 (1965).

Sverdrup, H. U., M. W. Johnson, and R. M. Fleming, *The Oceans*, Prentice-Hall, Englewood Cliffs, N.J., 1942.

Wangersky, P. J., and D. C. Gordon, Jr., *Limnol. Oceanog.*, **10**, 574 (1965).

Yentsch, C. S., and J. H. Ryther, *Deep-Sea Res.*, **6**, 72 (1959).

9 Biochemistry of the Oceans

1 Introduction

Marine biology is, obviously, a very large topic, so large in fact that it continues to dominate oceanography. We are obliged, therefore, in this chapter to restrict our attention narrowly to those aspects of marine biology which directly relate to chemical oceanography as defined in the Introduction.

An organism can be regarded as a heterogeneity in the sea. While viable, it removes chemical constituents from seawater and releases others. On death its materials are returned to the ambient sea. These materials fall into three categories (Figure 9.1): the biomaterial proper, including life cycle nutrients and products and the living biosubstance itself composed of complex organic compounds; the body fluids; and the supporting or containing skeletal structures consisting of relatively simple mineral substances.

In addition to the chemical nature of the substances taken from, formed in, and released into the sea by organisms, we also consider in this chapter selected examples of the chemical reactions responsible for these chemical changes. The principal biochemical processes in the sea (and everywhere

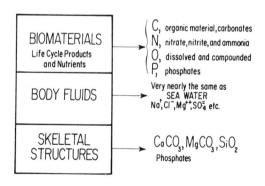

Figure 9.1   The Three Types of Materials Constituting Marine Organisms.

239

else in the biosphere) which result in changes in the environment fall into four types:

Photosynthesis
Metabolism
Decomposition
Mineral fixation

and each is considered in turn.

2 Body Fluids

Living organisms consist mostly of water, ranging from about 50% by weight for trees up to as much as 99.7% water in jellyfish. The importance of things closest to experience is often overlooked. In past times the role of water in living organisms was neglected, the water was thought simply to fill up the empty spaces in cells. Today, however, with the quickening advance of molecular biology, there is a growing appreciation of the profoundly important, active role water plays in living cells (Dick, 1966), not only in determining the exact configuration of the forms of life, but also in those chemical reactions which constitute the very bases of the life processes. The modern interest in the structure of water and aqueous solutions and the nature of transport processes in these media (Chapters 1, 2, and 3) has taken its rise, not from oceanography as it might reasonably have done so, but from molecular biology and physiology. It is no accident, then, that the most detailed treatment of water structure published in the United States was written by a professor of zoology (Kavanau, 1964), as a second volume of a treatise on structure and function in biological membranes.

Life began and evolved in the seas. Even those creatures who left the sea to populate the landmasses brought with them their miniature marine environment. We are constantly reminded of our marine origin. For example, our sexual cycle harks back to tidal rhythm. In the most simplified terms, a man is a bag of seawater. With respect to its chemical composition, the various body fluids of organisms bear a striking resemblance to seawater (Table 9.1). Both human blood and seawater are buffered, NaCl solutions. The pH of the body fluids, including blood, lymph, cerebrospinal fluid, etc., of animals, varies within the narrow range 7.1–7.7, centering somewhat nearer neutrality than the seawater range, 7.5–8.4. Departures from seawater composition in body fluids arise through evolutionary changes or as a consequence of the specific functions the fluids perform. This is especially conspicuous in the greater concentrations of some of the minor constituents in Table 9.1.

The evolutionary history of the chemical composition of body fluids and of the oceans are closely, albeit not always clearly, related. Some authorities have argued that the salt composition of the body fluids of organisms can give clues to the composition of the ancient seas in which they evolved. Others have held that, in view of the extreme sensitivity of many marine organisms to changes in the chemical composition of seawater, the salt content of the Earth's oceans cannot have changed significantly in recent geological times.

Table 9.1   A. Comparison of the Composition of Seawater and Human Blood[a]
(From Dietrich, 1963, with permission of John Wiley and Sons)

	Major Constituents			
	Seawater		Human Serum	
Species	g/kg	%	g/kg	%
Na^+	10.75	30.7	3.00	34.9
K^+	0.39	1.1	0.20	2.3
Ca^{2+}	0.416	1.2	0.10	1.2
Mg^{2+}	1.295	3.7	0.025	0.3
Cl^-	19.345	55.2	3.55	41.3
SO_4^{2-}	2.701	7.7	0.02	0.2
HPO_4^-	0.000185	0.0005	0.10	1.2
HCO_3^-	0.145	0.4	1.60	18.6

	Minor Constituents		
Element	Seawater mg/m^3	Human Serum, mg/kg	Serum/seawater
Fe	50	1	20
Zn	5	3.3	660
Cu	5	1.7	340
		In Total Blood	Blood/Seawater
As	15	0.6	40
Mn	5	0.3–1.5	60–300
Al	120	Up to 2	17
Pb	5	4–7	800–1,400
I	50	0.03–0.1	0.6–2
F	1,400	0.5–1	0.36–0.71

[a] For a very detailed survey of the composition of body fluid see Altman and Dittmer(1961).

B. Relative Composition of the Body Fluids of Some Sea Organisms (From Sverdrup, Johnson, and Fleming, 1942, with permission of Prentice-Hall, Inc.)

Element	Seawater	*Echinus esculentus* (Sea urchin)	*Homarus vulgaris* (Lobster)	*Cancer pagurus* (Crab)
Cl	180	182	156	156
Na	100	100	100	100
Mg	12.1	12.0	1.5	5.7
S in SO_4	8.4	8.5	2.2	6.7
Ca	3.8	3.9	5.0	4.8
K	3.6	3.7	4.7	4.0

Particularly fascinating is the Mg/Ca ratio in the blood serum of animals. Taking Ca to be equal to 100, the value of this ratio for seawater is 311. Primitive marine animals (Table 9.2) have values comparable to those of seawater, and the blood serum of such ancient forms as sharks and rays is isotonic to seawater, but land and fresh water animals exhibit values an order of magnitude smaller. Curiously enough, values for the very large class of bonefish, which includes such common species as cod and pollack, are also low, leading to the conclusion that they were once fresh water forms which returned to the sea in some more recent geological period.

Because the biosphere mass, and thus the total content of body fluids it contains, is so very small compared to the mass of the hydrosphere, the accumulation and/or release of these liquids has no perceptible effect on the composition of seawater.

Table 9.2   Mg/Ca Ratio in the Blood Serum of Selected Animals (From G. Dietrich, 1963, with permission of John Wiley & Sons)

Primitive Marine Animals		Land and Fresh Water Animals		Recent Fish	
Crustacea	130–196	Human	25	*Homaris americanus*	34.7
Echinoderm	240–253	Mammals	18–29	*Gadus collarus* (codfish)	35.9
Mollusk	214–275	Chicken	19	*Pollachius virens*	
		Crocodile	33	(pollack)	47.0
		Tench	34		
		Edible snail	11		

3   The Biomaterial

The chemical elements comprising the biomaterial can be conveniently divided into two more classifications: those present in relatively large amounts in the organic material which forms the structure of the organism, and those present, sometimes in only the most minute traces, in the substances that catalyze biochemical processes. Table 9.3 compares the contents of structural and catalytic elements of plant cells with seawater, and Table 9.4 classifies the chemical elements with respect to their role in the life processes. The major structural elements, H, C, N, and O, are all low atomic weight elements restricted to the first two horizontal rows of the periodic table (Table A.2). All of the structural and most of the catalytic elements are found in the first few rows. With the exception of lead, the high atomic weight elements in the lower part of the periodic table appear to play insignificant roles in the life processes.

The organic matter of living cells is largely composed of three types of substances: proteins, carbohydrates, and a sort of chemical catchall referred to as lipids and distinguished by their solubility in organic solvents. Table 9.4B gives the average composition of these materials.

Table 9.3   Elements Present in Plant Cells and the Sea
(From Dietrich, with permission of John Wiley & Sons)

	Structural Elements			
	Content per 100 g Dry Organism	Moles	Content per m^3 Seawater (35‰ S)	Content Seawater / Content Organism
Hydrogen	7 g	7	—	—
Sodium	3 g	0.27	10.75 kg	3600
Potassium	1 g	0.053	390 g	390
Magnesium	0.4 g	0.033	1.3 kg	3300
Calcium	0.5 g	0.025	416 g	830
Carbon	30 g	5	28 g	1
Silicon (a)	0.5 g	0.036	500 mg	1
Silicon (b)	10 g	0.71	500 mg	0.05
Nitrogen	5 g	0.71	300 mg	0.06
Phosphorus	0.6 g	0.04	30 mg	0.05
Oxygen, as O_2 and CO_2	47 g	—	90 g	2
Sulfur	1 g	0.063	900 g	900
Chlorine	4 g	0.23	19.3 kg	4800

Table 9.3 (Continued)

	Catalytic Elements		
	Content per 100 g Dry Organism	Content per m³ Seawater (35‰ S)	Content Seawater / Content Organism
Copper	5 mg	10 mg	2
Zinc	20 mg	5 mg	4
Boron	2 mg	5 g	2500
Vanadium	3 mg	0.3 mg	0.1
Arsenic	0.1 mg	15 mg	150
Manganese	2 mg	5 mg	2.5
Fluorine	1 mg	1.4 g.	1400
Bromine	2.5 mg	66 g	26,000
Iron (a)	1 g	50 mg	0.05
Iron (b)	40 mg	50 mg	1.3
Cobalt	0.05 mg	0.1 mg	2
Aluminum	1 mg	120 mg	120
Titanium	100 mg	—	—
Radium	4×10^{-12} g	10^{-10} g	25

Proteins occupy a position of central importance both in the microscopic structural organization and in the dynamic processes of life. Life can be said to be characterized by the presence of (in addition to water) protein material, and the only system in nature where protein synthesis occurs is the living cell. With respect to their chemical composition, proteins contain 50–55 w % C, 6–7% H, 20–23% O, and 12–19% N. The presence of nitrogen is characteristic of proteins, and it is a common practice to determine the protein content of plant or animal material by analyzing for this element and multiplying the weights by a factor, 100/16, based on an average protein nitrogen content of

Table 9.4   A. The Role of the Chemical Elements in the Life Process

Elements necessary for the life processes
  H, B, C, N, O, F, Na, Mg, Si, P, S, Cl, K, Ca, V, Mn, Fe, Co, Ni, Cu, Zn, Br, I

Elements probably necessary for life
  Al, Ti, As, Sn, Pb

Elements probably not necessary for life
  He, Li, Be, Ne, A, Sc, Cr, Ga, Ge, Se, Kr, Rb, Sr, Y, Zr, Nb, Tc, Ru, Rh, Pd, Ag, Cd, In, Sb, Te, Xe, Cs, Ba, La, Rare earths, Hf, Ta, W, Re, Os, Ir, Pt, Au, Hg, Tl, Bi, Po, At, Rn, Fr, Ac, Th, Pa, U

B. Average Composition of Organic Materials
(From Sverdrup, Johnson, and Fleming, 1942, with permission of Prentice-Hall, Inc.)

	Percentage Composition				Relative Proportions by Weight, $C = 100$		
Element	Carbohydrates	Lipids	Proteins	Element	Seawater	Lipids	Proteins
O	49.38	17.90	22.4	C	100	100	100
C	44.44	69.056	51.3	P	0.05	3.1	1.4
H	6.18	10.00	6.9	N	0.5	0.88	34.7
P	—	2.13	0.7	S	3150	0.45	1.6
N	—	0.61	17.8	Fe	0.07	—	0.2
S	—	0.31	0.8				
Fe	—	—	0.1				

C. Percentage Composition of Skeletal Material
(From Sverdrup, Johnson and Fleming, 1942, with permission of Prentice-Hall, Inc.)

Substance	Foraminifera (*Orbitolites marginatis*)	Coral (*Oculina diffusa*)	Calcareous alga (*Lithophyllum antillarum*)	Lobster (*Homarus* sp.)	Phosphatic brachiopod (*Discinisca lamellosa*)	Siliceous sponge (*Euplectella speciosa*)
Ca	34.90	38.50	31.00	16.80	26.18	0.16
Mg	2.97	0.11	4.36	1.08	1.45	0.00
CO_3	59.70	58.00	62.50	22.40	7.31	0.24
SO_4	—	—	0.68	0.52	4.43	0.00
PO_4	tr	tr	tr	5.45	34.55	0.00
SiO_2	0.03	0.07	0.04	} 0.30	{ 0.64	88.56
$(Al,Fe)_2O_3$	0.13	0.05	0.10		{ 0.44	0.32
Organic matter, etc.	2.27	3.27	1.32	53.45	25.00	10.72

16%. Proteins are very large macromolecules. Egg albumin has a molecular weight of about 34,000, for example, and hemoglobin 64,000. The milk protein, β-lactoglobulin, has the approximate formula $C_{1864}H_{3012}O_{506}N_{468}S_{21}$.

When a protein is hydrolyzed, by warming with acid for example, it breaks up into a number of much simpler amino acid units. Amino acids have the general form $R—CH(NH_2)COOH$; some of the principal ones and their formulas are listed in Table 9.5. In the protein molecule the amino acid segments are joined together to form long polypeptide chains:

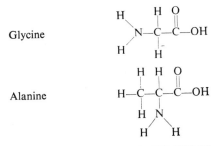

Table 9.5   Principal Amino Acids Derived from Proteins

I. Aliphatic Amino Acids

A. Monoaminiomoncarboxylic acids

Glycine

$$\underset{H}{\overset{H}{N}}-\underset{H}{\overset{H}{C}}-\overset{O}{C}-OH$$

Alanine

$$H-\underset{H}{\overset{H}{C}}-\underset{\underset{H}{N}\!\!\diagup\diagdown H}{\overset{H}{C}}-\overset{O}{C}-OH$$

Valine	$(CH_3)_2CH \cdot CH(NH_2)COOH$
Leucine	$(CH_3)_2CH \cdot CH_2CH(NH_2)COOH$
Isoleucine	$CH_3 \cdot CH_2CH(CH_3) \cdot CH(NH_2)COOH$
Serine	$HOCH_2 \cdot CH(NH_2)COOH$
Threonine	$CH_3 \cdot CH(OH) \cdot CH(NH_2)COOH$

B. Sulfur-containing Amino Acids

Cysteine	$HSCH_2 \cdot CH(NH_2)COOH$
Methionine	$CH_3SCH_2 \cdot CH_2 \cdot CH(NH_2)COOH$

C. Monoaminodicarboxylic acids and their amides

Aspartic acid	$HOOC \cdot CH_2 \cdot CH(NH_2)COOH$
Asparagine	$NH_2CO \cdot CH_2 \cdot CH(NH_2)COOH$
Glutamic acid	$HOOC \cdot CH_2 \cdot CH_2 \cdot CH(NH_2)COOH$
Glutamine	$NH_2CO \cdot CH_2 \cdot CH_2 \cdot CH(NH_2)COOH$

D. Basic amino acids

Lysine	$NH_2 \cdot CH_2 \cdot CH_2 \cdot CH_2 \cdot CH_2 \cdot CH(NH_2)COOH$
Hydroxylysine	$NH_2 \cdot CH \cdot CH(OH) \cdot CH_2 \cdot CH_2 \cdot CH(OH_2)COOH$
Arginine	$NH_2 \cdot C \cdot NH \cdot CH_2 \cdot CH_2 \cdot CH_2 \cdot CH(NH_2)COOH$

Histidine

$$H-\underset{\underset{N-CH}{\diagdown}}{\overset{\overset{H}{N}-C-CH_2 \cdot CH(NH_2)COOH}{\diagup}}{C}$$

II. Aromatic Amino Acids

Phenylalanine

$$HC \underset{\underset{H}{C} - \underset{H}{C}}{\overset{\overset{H}{C} = \overset{H}{C}}{\Big|}} C - CH_2 \cdot CH(NH_2)COOH$$

Tyrosine

$HO\langle \bigcirc \rangle CH_2CH(NH_2)COOH$

Diiodotryrosine

$HO\langle \bigcirc \rangle CH_2 \cdot CH(NH_2)COOH$

Thyroxine

$HO\langle \bigcirc \rangle - O - \langle \bigcirc \rangle CH_2 \cdot CH(NH_2)COOH$

III. Heterocyclic Amino Acids

Tryptophan

$CH_2 \cdot CH(NH_2)COOH$

Proline

$$\begin{array}{c} H_2C-\!\!\!-CH_2 \\ H_2C \underset{\underset{H}{N}}{} CH\cdot COOH \end{array}$$

Hydroxyproline

$$\begin{array}{c} HO-HC-\!\!\!-CH_2 \\ H_2C \underset{\underset{H}{N}}{} CH\cdot COOH \end{array}$$

Histidine (see above)

The overall form of the enormous macromolecule thus made can be fibrous or globular. The chains can be cross-linked and folded and spiraled in various ways. Figure 9.2 shows a single segment of the typical α-helix of the protein molecule. Table 9.6 gives the approximate amino acid content in 100 g of some familiar proteins (the numbers are free amino acid and thus the total is is greater that 100 g). Table 9.7 tabulates the amino acid composition of algae cells. In addition to the

$$\begin{array}{c} O \\ \| \\ -C-NH- \end{array}$$

peptide bonds, conjugated proteins can also be linked by nonamino acid units known as nucleic acids.

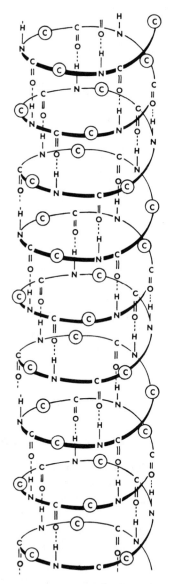

Figure 9.2 The α-Helix. Hydrogen bonds which hold the form of the helix are indicated by broken lines.

Table 9.6 Weight of Free Amino Acid in 100 g of Some Proteins
(From Fruton and Simmonds, 1953, with permission of John Wiley & Sons)

	β-Lactoglobulin	Hemoglobin (Horse)	Insulin	Edestin	Egg Albumin	Chymotrypsinogen	Silk Fibroin	Gelatin	Salmine
Glycine	1.5	5.6	4.5	—	3.1	5.3	43.6	26.9	2.9
Alanine	6.6	7.4	4.4	4.3	6.7	—	29.7	9.3	1.1
Valine	5.7	8.4	7.5	5.7	7.1	10.1	3.6	3.3	3.1
Leucine	15.5	16.0	12.9	7.4	9.2	10.4	0.9	5.2[a]	1.6
Isoleucine	5.9	—	2.8	4.7	7.0	5.7	1.1	—	—
Serine	4.1	5.8	5.2	6.3	8.2	11.4	16.2	3.2	9.1
Threonine	5.8	4.4	2.1	3.9	4.0	11.4	1.6	2.2	—
Cystine	3.4	1.0	12.5	1.4	1.0	6.6	—	—	—
Cysteine	1.1	0.4	—	—	1.4	1.3	—	—	—
Methionine	3.2	1.5	—	2.3	5.2	1.2	2.8	0.9	—
Aspartic acid	11.2	10.4	7.5	12.8	9.3	11.3	2.2	5.6	—
Glutamic acid	21.5	8.5	18.6	19.3	16.5	9.0	—	11.2	—
Amide NH$_3$	1.3	1.1	1.7	2.2	1.2	1.9	0.7	0.1	—
Lysine	11.2	8.6	2.5	2.4	6.3	8.0	1.1	4.6	—
Arginine	2.8	3.6	3.4	16.7	5.7	2.8	0.4	8.6	85.2
Histidine	1.6	7.9	4.9	2.9	2.4	1.2	—	0.7	—
Phenylalanine	3.8	7.9	8.0	5.7	7.7	3.6	3.4	2.6	—
Tyrosine	3.8	3.0	12.2	4.5	3.7	3.0	12.8	1.0	—
Tryptophan	1.9	—	—	1.4	1.2	5.6	—	—	—
Proline	5.2	3.9	2.6	4.6	3.6	5.9	0.7	14.8	5.8
Hydroxyproline	—	—	—	—	—	—	—	14.5	—
Total	117.1	105.4	113.3	108.5	110.5	115.7	120.8	114.7	108.8

[a] Includes both leucine and isoleucine.

Table 9.7 Amino Acid Composition of Algae Cells
(From Parsons, 1963, with permission of Pergamon Press, Ltd.)

Amino Acids	Tetra-selmis maculata	Duna-liella salina	Mono-chrysis lutheri	Syracos phaera carterae	Skeleto-nema costatum	Coscino-discus sp.	Phaeodac-tylum tricor-nutum	Exuviella sp.	Amphi-dinium carteri	Agmenellum quadru-plicatum
	Amino Acid Nitrogen as a % of the Total Amino Acid Nitrogen									
Aspartic acid	26.0	16.5	25.3	35.5	28.0	15.6	14.8	20.0	15.4	17.0
Glutamic acid	8.1	13.6	5.9	5.1	13.5	15.6	12.9	13.1	11.0	17.7
Glycine	11.1	16.8	8.6	11.4	16.5	12.8	15.6	17.6	21.2	16.4
Alanine	15.0	19.0	26.0	11.4	7.5	13.8	17.0	13.1	2.85	15.7
Threonine	—	6.3	5.3	5.1	5.0	7.3	8.7	6.0	+	3.9
Lysine	21.0	12.0	19.6	17.8	+	12.9	7.2	16.6	+	13.8
Valine	5.1	4.2	3.7	5.1	8.5	7.3	6.4	4.0	7.1	2.6
Phenylalanine	7.5	4.5	1.5	2.5	8.0	7.3	11.0	3.5	8.6	5.2
Leucine ⎫ Isoleucine ⎬ Methionine ⎭	6.0	6.9	4.0	6.3	13.0	7.3	6.4	6.0	8.1	7.9
	Occurrence of Other Amino Acids[a]									
Serine	—	—	+	+	+	+	+	+	+	+
Proline	+	+	+	+	+	—	+	+	+	+
Tyrosine	—	—	+	+	—	+	—	—	+	+
Histidine	—	—	—	—	+	—	—	—	—	+
Arginine	—	—	+	—	+	—	—	—	—	—
Cysteic acid	—	—	—	—	—	—	+	—	—	—
Methionine sulphoxide	—	+	+	+	+	—	—	+	+	—

a + Amino acid detected but not estimated; — amino acid not detected.

A group of proteins of special importance in their function as the catalysts of biochemical reactions are the enzymes.

Carbohydrates, as the name implies, are compounds of C, H, and O, although they may also contain N or S. They are polyhydroxyl compounds. D-Glucose, a very important monosaccharide found in the blood of animals as well as in the sap of plants, exists in two forms:

(9.1)

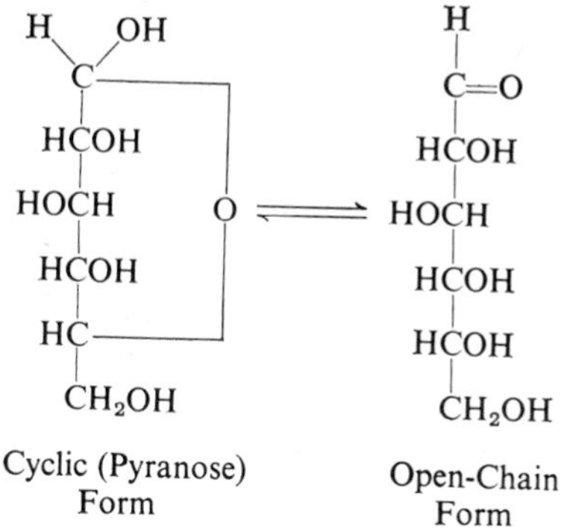

Cyclic (Pyranose)
Form

Open-Chain
Form

Glucose

Oligosaccharides are formed by the condensation of the hydroxyl group of one monosaccharide molecule with the reducing group of another. Ordinary sugar (sucrose) serves as an example:

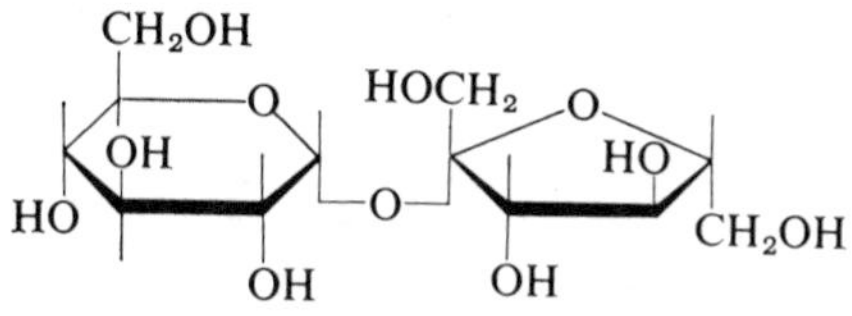

Sucrose

Polysaccharides, containing many units, fall into two groups: the skeletal or structural polysaccharides which serve as rigid mechanical, internal, or external frameworks for organisms; and the nutrient polysaccharides, such as starch, which function as metabolic reserves of monosaccharides in animals and plants. An example of an important invertebrate structural polysaccharide is chitin, found in large amounts in the shells of lobsters and crabs. Table 9.8 summarizes the carbohydrate composition of algae cells.

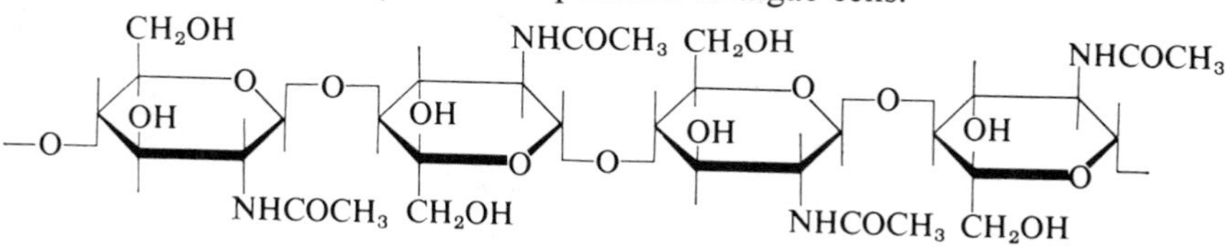

Chitin

Table 9.8 Carbohydrate Composition of Algae Cells as % Dry Weight
(From Parsons, 1963, with permission of Pergamon Press, Ltd.)

Species	Glucose	Galactose	Mannose	Ribose	Xylose	Arabinose	Rhamnose	Fucose	Fructose	Hexosamine	Hexuronic Acids
Chlorophyceae											
Tetraselmis maculata	11.9	2.3	–	0.95	–	–	–	–	–	–	+
Dunaliella salina	17.2	11.8	–	1.7	–	–	–	–	–	–	+
Chrysophyceae											
Monochrysis lutheri	22.1	4.4	–	1.3	3.5	–	–	–	–	–	+
Syracosphaera carterae	9.2	7.1	–	1.5	0.8	1.9	–	–	–	–	+
Bacillariophyceae											
Chaetoceros sp.	3.3	1.5	0.79	0.71	0.4	–	2.8	+	–	–	+
Skeletonema costatum	16.4	1.8	0.87	1.2	–	–	1.0	0.9	–	–	+
Coscinodiscus sp.	2.1	0.4	0.41	+	–	–	0.7	0.5	–	–	+
Phaeodactylum tricornutum	10.7	2.7	3.7	0.72	0.7	–	1.5	–	–	–	+
Dinophyceae											
Amphidinium carteri	19.0	8.4	–	0.9	–	–	+	–	–	–	–
Exuviella sp.	26.8	8.3	–	+	+	+	+	–	–	–	–
Myxophyceae											
Agmenellum quadruplicatum	17.4	3.2	–	1.5	–	–	–	–	3.5	0.3	+

– Sugars not detected; + sugars detected but not estimated.

4   Skeletal Structures

Despite their obvious importance, relatively little appears to be known concerning the details of the chemical processes whereby organisms extract elements from seawater and precipitate the mineral materials that form their supporting structures. The oceans tend to be supersaturated with respect to $CaCO_3$ (Figure 7.19); thus, in a sense, the organism can be considered to serve simply as a catalyst for its precipitation. However, Cloud (1965) is careful to point out that "skeletal construction may take place even where surrounding waters are barely saturated or actually undersaturated."

The secreted skeletal $CaCO_3$ can be either calcite or aragonite; if calcite, it can have a high or low magnesium content depending on the organism involved and/or the environmental conditions. The important environmental factors in the biological precipitation of $CaCO_3$ appear to be temperature, degree of $CaCO_3$ saturation, and pressure. The characteristic calcareous organisms, the corals and coccolithophoridae, are most conspicuous in the tropical oceans where $CaCO_3$ precipitation is facilitated by high super-saturation (Figure 7.19), and its kinetics hastened by the warm temperatures. The corals flourish only in waters about $20°C$; thus reef formation occurs only in the shallow waters of tropical seas. The presence of coral skeletal material in colder or deeper water hints of profound changes in the geological past, such as the sinking of the land or the rising of the sea level. In cold and/or deep waters the biological precipitation of $CaCO_3$ is more difficult. Thus, in cold polar seas, the population of calcareous organisms is diminished, and diatoms, whose skeletons are composed of silica, dominate. Similarly, in deep waters, $CaCO_3$ skeletons and shells tend to be nonexistent or at best skimpy affairs. The shells of deep crustacea tend to have little lime, the bones of deep-sea fish are fragile and poorly calcified, and at a depth of only 300–400 m calcareous sponges tend to be replaced by siliceous forms (Sverdrup, Johnson, and Fleming, 1942). The increased solubility of $CaCO_3$ with increasing pressure contributes to the difficulty of calcareous skeletal formation at great depths.

The details of the chemical processes involved in silicon metabolism and the precipitation of skeletal silica remain equally mysterious. The extent of silicification depends on the amount of silica available in the seawater, and heavily silicified diatoms may contain as much as 25% silica (Jørgensen, 1953, 1955; Lewin, 1955, 1957).

5   Photosynthesis

The variety and intricacy of the chemistry of life on this planet stagger the imagination. Yet the whole of this enormously complex system has as its foundation a single type of chemical process—photosynthesis. Life represents a transitory and tenuous advance in the direction of organization in the face of a cosmical flood toward disintegration and disorder. While photosynthesis may have come relatively late in the genesis and early evolution of life, in the continuation of existing life it is the first synthetic step in the direction of organization, making large organic molecules from small inorganic ones.

The overall photosynthetic process can be represented by

$$(9.2) \qquad n\,CO_2 + 2_n\,H_2A \rightarrow (CH_2O)_n + n\,H_2O + 2n\,A$$

where H_2A is a hydrogen donor such as H_2O, H_2S, or H_2. Of particular importance is the process

$$(9.3) \qquad 6\,CO_2 + 6\,H_2O \xrightarrow[\text{Chlorophyll}]{\text{Light}} \underset{\text{Glucose}}{C_6H_{12}O_6} + 6\,O_2$$

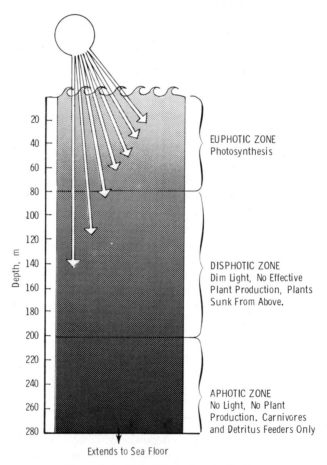

Figure 9.3   Light Penetration in the Sea.

Life's struggle against disorder requires energy, some 690 kcal/mole of glucose in (9.3), and this energy is supplied by sunlight.

In the seas, only a little light penetrates below 80 m (Figure 9.3; Table A.21 of the Appendix contains information on the optical properties of seawater, including refractive index and light absorption). Thus the life contained in the marine biosphere ultimately derives from a shallow surface layer. In this euphotic zone (Figure 9.3), phytosynthesis by phytoplankton forms the first and crucial step in the production of biomaterials. The creatures of the sea feed on phytoplankton or on creatures which feed on phytoplankton, or on creatures which feed on creatures which feed on phytoplankton. The remnants of this carnage sink slowly down into the deeper, lightless waters where it is eaten by detritus feeders who are in turn the prey of carnivores. And, in the words of the Flemish proverb, the big fish eat the little fish.

In recent years modern science has made progress in uncovering the secrets of the photosynthetic process (Franck and Loomis, 1949; Rabinowitch, 1959; McElroy and Glass, 1961). Thus more recent evidence indicates that (9.3) is a gross oversimplification, that the primary photosynthetic step is not the production of carbohydrates from CO_2 but the conversion of light energy to chemical energy, and that the cells utilize this chemical energy in various ways. The green pigment, Chlorophyll, the essential catalyst in the photosynthetic process, absorbs a photon of light (Figure 9.4), thereby raising its electrons to an excited state. This enables the chlorophyll to function as an electron source for the production of adenosine triphosphate (ATP).

The water molecule serves as a source of protons (Figure 9.4). The electrons, protons, and triphosphopyridine nucleotide (TPN) form the reduced triphosphopyridine nucleotide ($TPNH_2$), which is in turn responsible for the reduction of CO_2 and the production of glucose. Carbohydrates are

Chlorophyll

Adenosine Diphosphate(ADP)

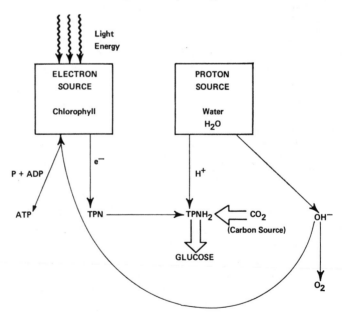

Adenosine Triphosphate(ATP)

Reduced Triphosphopyridine
nucleotide (TPNH₂)

Figure 9.4   The Photosynthetic Process.

Table 9.9 Oceanic Distribution[a] of Chlorophyll a
(From Parsons, 1963, with permission of Pergamon Press, Ltd.)

Area		Date	Depth	Chlorophyll a mg/m³	Reference
Atlantic Ocean					
Continental shelf off the coast of New York	<100 m	February	Euphotic av.	1.5–3.5	Ketchum et al. (1958)
	<200 m	February	Euphotic av.	0.3–1.0	Ketchum et al. (1958)
	<2000 m	February	Euphotic av.	ca. 0.2	Ketchum et al. (1958)
	>2000 m	February	Euphotic av.	<0.2	Ketchum et al. (1958)
Sargasso Sea ca. 40°N		Seasonal av.	Surface	0.3	Hulburt et al. (1960)
		October	Surface	0.1–0.4	U.S., Agric. Mech. Coll., Texas (1961)
ca. 50–60°N		Summer	Surface	0.3–2.0	Hansen (1961)
ca. 60–70°N		July	Surface	0.3–2.8	Steemann Nielsen and Hansen (1961)
Pacific Ocean					
Northern Peru current		—	Surface	0.7	Holmes (1958)
Approx. average over 5° square of lat. and long. regardless of date (East Pacific)	ca. 10°N	—	Surface	0.1–0.8	Holmes (1958)
	15–40°N	—	Surface	<0.3	Holmes (1958)
	45–55°N		Surface	0.2–2.0	Holmes (1958)
Northwest Pacific ca. 160°E	30–40°N	Summer	Euphotic av.	0.2 or less	Saijo and Ichimura (1960)
	40–50°N	Summer	Euphotic av.	0.2–0.6	
Southwest Pacific	20–30°S	April	Surface	ca. 0.1	Humphrey (1959)
	30–35°S	December	Surface	<0.2	Humphrey (1960)
	35–40°S	June	Surface	0.4–0.6	Humphrey (1959)
	30–35°S	April	Surface	0.2–0.6	Humphrey (1960)
Chukchi Sea		August	Surface to 50 m	ca. 0.2–10	U.S., Agric. Mech. Coll., Texas (1961)

[a] More recently the chlorophyll a content of particulate organic matter in the North Sea was reported by Steel and Baird (1965).

Table 9.10   Coastal Variations in Chlorophyll a
(From Parsons, 1963, with permission of Pergamon Press, Ltd.)

Region	Date	Depth, m	Chlorophyll a, mg/m³	Reference
Departure Bay, British Columbia	January	5	ca. 0.6	Parsons (1960)
	March	5	0.3–14	Parsons (1960)
	June	5	4–28	Parsons (1960)
	August	5	2–3	Parsons (1960)
	October	5	ca. 1.0	Parsons (1960)
Gerlache and Bransfield Straits, Antarctica	Summer	Surface	0.5–17.8	Burkholder and Sieburth (1961)
Long Island Sound	January	Surface	5–10	Riley (1959)
	February	Surface	10–20	Riley (1959)
	May	Surface	4–5	Riley (1959)
	August	Surface	10–25	Riley (1959)
	October	Surface	3–5	Riley (1959)
West coast, North Island, New Zealand	August	Surface	1–91	Cassie and Cassie (1960)

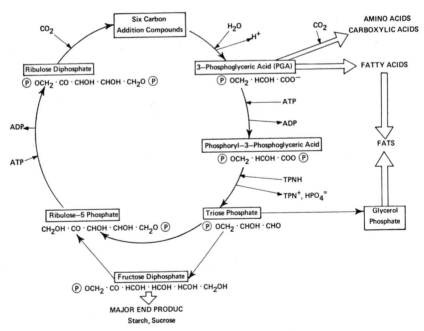

Figure 9.5   Products of the Photosynthetic Process.

not the only product of photosynthesis; fatty acids, fats, and carboxylic and amino acids are also produced, as shown in the highly oversimplified Figure 9.5.

The concentration of chlorophyll a is readily measured spectrophotometrically and, despite certain difficulties, such as distinguishing whether the pigment is in live or dead cells, can be used as an indication of the size of the phytoplankton crop in a particular ocean area or at a given time. The chlorophyll a content of seawater decreases with increasing depth (Table 9.9), changes with season (Tables 9.9 and 9.10), and is high and variable in coastal areas. Holm-Hansen and Booth (1966) have reported results of the measurement of ATP in the ocean.

6 Animal Metabolism

The photosynthetic process in plants has many interesting similarities to and differences from the respiratory metabolism in animals. In photosynthesis, as we have seen, energy is absorbed and simple materials are chemically reduced and converted into more complex substances. In the respiratory metabolic system, foodstuffs, such as carbohydrates, are oxidized,

(9.4) $C_6H_{12}O_6 + 6 O_2 \rightarrow 6 CO_2 + 6 H_2O + energy$

and the energy release utilized by the organism in numerous ways, ranging from the maintenance of body temperature in the case of homeothermic creatures to muscular movement. Thus, in Nature's great scheme of things, photosynthesis in plants and respiration in animals are chemically opposite and complementary processes which maintain the delicate balance of the life cycle:

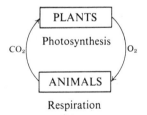

Respiration

Many (but not all) plants are characterized by the green catalytic pigment chlorophyll; many (but not all) animals are characterized by the (often) red respiratory pigments, the hematin compounds (Wyman, 1948; Lemberg and Legge, 1949). The respiratory pigments in the higher animals are extremely complex proteinous compounds. Figure 9.6 shows a *part* of the myoglobin molecule, a red pigment in the muscle tissue of the whale. The resolution of this structure by three-dimensional X-ray analysis by Kendrew, Perutz, and their co-workers at Cambridge University was a superb piece of work which represents the zenith of modern structural analysis and was quite appropriately rewarded with the Nobel prize in 1962 (Kendrew, et al., 1960; Kendrew, 1963). Hemoglobin, the red pigment in human blood, contains four units similar to that represented in Figure 9.6.

In Fig. 9.6 notice the sphere surrounded by a number of polygons lying in a plane. This is the heme group, the functional site in the catalyst, and it consists of Fe(II) surrounded by a planar porphyrin ring system. The heme group bears an obvious resemblance to chlorophyll (see above), except now the Mg has been replaced by Fe. In free heme the 5th and 6th coordination positions around the iron above and below the plan of the porphyrin ring are occupied

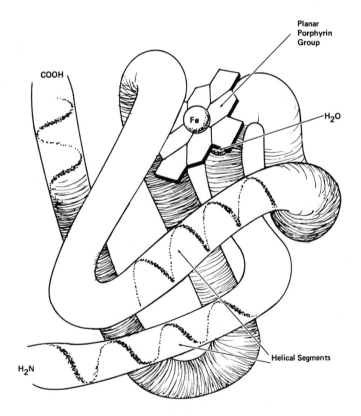

Figure 9.6 A Portion of the Myoglobin Molecule. From Kendrew (1961). Based on a model constructed from three-dimensional Fourier synthesis by Kendrew and Perutz; see Kendrew (1963).

by water molecules, while in hemoglobin for example, one of these water molecules is replaced by a gigantic protein complex. Although iron is the

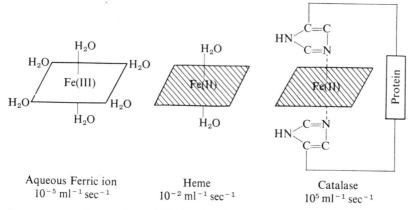

CH₃—C═══C—CH═CH₂

Ferrous Protoporphyrin
(Heme)

most common element, the heme-like substances in certain marine creatures contain other transition elements. The key biological macromolecules have undergone evolutionary development parallel to the evolutionary development of whole organisms, and the tracing of such evolutionary lines has opened up fascinating realms of speculation concerning the history of life on this planet (Ingram, 1961; Zuckerkandl, 1965). The Fe(II)–Fe(III) redox reaction plays a highly varied and tremendously important role in biological systems, and a primary "objective" in the course of the evolutionary development of Fe-containing molecules appears to have been the enhancement of their catalytic potency. Thus Calvin (1959), by way of example, has pointed out the marked increase in catalytic activity with respect to the decomposition of hydrogen peroxide in going from free ferric ion in aqueous solution to the enzyme catalase.

Aqueous Ferric ion
10^{-5} ml^{-1} sec^{-1}

Heme
10^{-2} ml^{-1} sec^{-1}

Catalase
10^{5} ml^{-1} sec^{-1}

In a very small organism, diffusion is adequate for the transport of oxygen to the sites of need, but in a large organism, such as man, the function of hemoglobin is to transport the molecular O_2 required by process (9.4) to the cells throughout the organism where it is needed. In order to achieve this purpose the O_2 combines reversibly with the respiratory pigment, releasing a water molecule (Haurowitz, 1951; Keilin and Hartree, 1952).

Increasing Catalytic Activity $\longrightarrow$
(The cross-hatched areas represent the porphyrin system.)

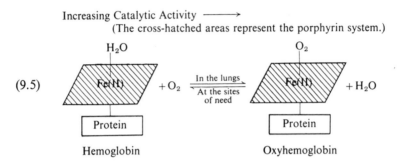

(9.5)

Hemoglobin   Oxyhemoglobin

The details of this equilibrium are imperfectly understood and appear to be very complex: the four hemes in hemoglobin interact with one another so that successive O_2 additions are not equivalent, the failure of the bound O_2 to oxidize Fe(II) is still a matter of controversy, and the equilibrium appears to entail profound structural changes (Perutz, 1963; McGinnety, Doedens, and Ibers, 1967).

7   Decomposition and the Dissolved Organic Content of Seawater

No sooner is life extinguished than the forces of decay begin to undo its work. The fragile, unstable state represented by the living organism disintegrates, the organization that was the creature disappears, and its substance is returned to the cosmic chaos. The complex macromolecules—the proteins, the carbohydrates—so artfully and laboriously built are now fragmented and broken down as if Nature were an irritable child scattering the pieces of a broken toy. The larger shards of life make up the suspended organic content of seawater, the smaller fragments may go into solution and thereby contribute to the dissolved organic content of seawater.

In a subsequent examination of the origin of life in the oceans of Earth we shall consider the likelihood of the formation of organic molecules from smaller inorganic ones, but it is perfectly safe to say here that the overwhelming majority of the dissolved organic content of seawater is the consequence of the forces of destruction rather than construction. Jeffrey and Hood (1958) have proposed that, in addition to the decomposition of dead organisms, excretion from living organisms, soil leaching, and exchange and/or decomposition processes in sediments are also sources of dissolved organic matter in seawater. The bulk of living organic matter in the sea consists of phytoplankton, less than 10% being eventually converted into animal materials. Dead organisms can represent a large fraction of the total particulate matter, and Duursma (1961) has concluded that the dissolved organic substances originate for the greater part from dead phytoplankton and detritus, rather than from excretion by living cells.

Marine algae excrete about 3–6% of their photoassimilated carbon during growth, although a few species excrete as much as 10–25%. Glycolic acid constitutes 9–38% or less of the total excreted carbon, while the amount of protein material ranges from 0.2% to 10% (Hellebust, 1965).

The decomposition scheme for the breakdown of proteins in nature can be represented by

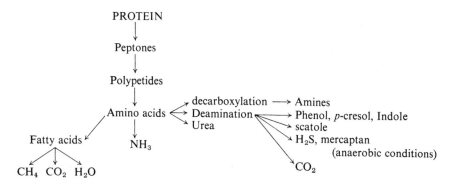

The decomposition process is not simply a progression from complex to simple molecules, although that is the overall result. Bacteria play an important role in decay, and through their agency there is some intermediate synthesis of complex substances from simpler ones in addition to the breakdown of larger molecules.

The heterotrophic microorganisms largely responsible for the decomposition process concentrate in the vicinity of solid surfaces such as provided by particulate matter. Thus, not only do less stable compounds stand a better chance of survival in the marine than in the terrestrial biosphere, but also the amounts of primary dissolved organic substances in seawater should be inversely dependent on the quantity of particulate material in suspension (Duursma, 1965).

Whether the "dissolved" organic material in seawater is in true solution or in the colloidal state in unknown, and, in view of the very low concentrations involved, it will be extremely difficult to resolve this question. The ease with which organic matter is absorbed on iron and aluminum hydroxides suggests that in the sea some of the organic material is adsorbed on colloida and larger particles of these substances.

The soluble organic substances in the sea are more or less what one expects proteinous material, carbohydrates, and their decomposition products. Lipids constitute a minority of the life substance but are more likely to be detected than are primary proteins and carbohydrates because of their greater resistance to decomposition (Duursma, 1959). We should also note in passing that a small fraction of the dissolved organic content of seawater results from decomposition processes on land or of materials from the terrestrial biosphere.

Duursma (1959) has compiled an extensive table of location, concentration, and investigators of the organic compounds that have been identified in seawater, including carbohydrates, proteins and their derivatives, aliphatic, carboxylic, and hydroxy carboxylic acids, and biologically active compounds such as vitamins. Here we confine our attention to the results of some selected recent studies.

Tatsumoto and co-workers (1958, 1961) examined the concentration of dissolved amino acids in surface waters, and this work has been extended to deep water by Park, Williams, et al. (1962) using the technique of ion exchange resin chromatography. The latter investigation showed that the identifiable

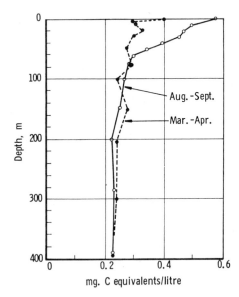

Figure 9.7 Vertical Profile of Readily Oxidized Dissolved Organic Matter in the Irminger Sea. From Gillbricht (1959), with permission of Deutsches Hydrographisches institut.

amino acids were "rather uniformly distributed throughout a wide range of depths" (Table 9.11). This is, of course, not unexpected for a truly soluble constituent in a well-mixed ocean. From Table 9.11, the most abundant amino acids in seawater are glutamic acid, lysine, aspartic acid, serine, alanine, leucine, and valine (see Table 9.5 for formulas). Threonine is also a commonly encountered amino acid, although its relatively high abundance is not represented by Table 9.11.

Williams (1961) has separated the fatty acids from seawater by solvent extraction and has determined the amounts of the acids with various numbers of carbons by gas chromatography with the results summarized in Table 9.12, and Ushakov, Vityuk, et al. have investigated fatty acids in Black Sea waters.

Koyama and Thompson (1959) have analyzed surface Northwest Pacific waters for simple carboxylic acids—acetic, formic, and lactic acids—and have found concentrations as high as 1.4 mg/l. The concentrations decrease with increasing distance from shore.

The dissolved carbohydrate content of water off Cape Cod ranges from 0.40 to 1.00 mg/l (Walsh, 1965). It exhibits seasonal variations, is greater in surface than in deep waters, and is clearly related to phytoplankton activity. On the basis of this and a second study of a eutrophic coastal pond, the accompanying CO_2-carbohydrate cycle was proposed.

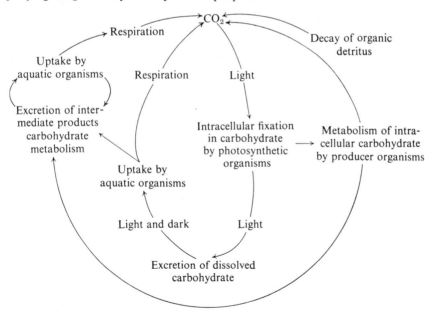

The carbohydrate content and the concentration of the other dissolved organic constituents of seawater generally follow the same pattern: large and

Table 9.11   Dissolved Amino Acids in Seawater[a]
(From Park, Williams, Prescott, and Hood, 1962, with permission of
AAAS and the authors)

		Distribution at Depths Indicated			
10 m	500 m	900 m	1500 m	3000 m	3500 m
		Glutamic Acid			
+++	+++	+++	+++	+++	+++
		Lysine			
+++	+++	+++	+++	+++	+++
		Glycine			
+++	+++	+++	+++	+++	+++
		Aspartic Acid			
++	+++	+++	+++	++	+++
		Serine			
++	+++	+++	++	+	++
		Alanine			
++	++	++	++	++	++
		Leucine			
++	++	++	++	++	++
		Valine plus Eystine			
++	++	++	++	+	++
		Isoleucine			
++	++	+	++	+	+
		Ornithine			
+	++	+	+	+	+
		Methionine Sulfoxide			
+	+	+	+	+	+
		Threonine			
+	+	+	+	+	+
		Tyrosine and Phenylalanine			
+	+	+	+	+	+
		Histidine			
+	+	+	+	+	+
		Arginine			
+	+	+	+	+	+
		Proline			
+		+	+	+	+
		Methionine			
+	+		+	+	+

[a] In order of abundance: +++ high, > 1 mg/m^3; ++ medium, 1–0.5 mg/m^3; + low,
< 0.5 mg/m^3.

Table 9.12 Fatty Acids Isolated from Seawater
(From Williams, with permission of Macmillan, Ltd., and the author)

Location	Depth Collected, m	Depth to Bottom, m	Volume Extracted, l	Extraction, pH	Saturated						Unsaturated				Unknown			Miscellaneous	No. of Major Peaks	Total, µg/l
											Mono-		Di-							
					C_{12}	C_{14}	C_{16}	C_{16}	C_{20}	C_{22}	C_{16} (C=C)	C_{16} (C=C)	C_{16} (C=C)	C_z	C_y	C_z			
31° 40′ N 119° 35′ W	1000	3290	19.0	1.30	2.3	4.3	20.9	14.2	1.8	1.4	0.3	8.3	1.4	—	—	—	1.6	13	56.5
32° 40′ N 118° 09′ W	1200	2020	18.1	3.10	0.4	1.8	14.6	22.2	8.1	Tr	—	—	—	6.6	15.5	14.1	6.4	14	89.7
32° 40′ N 118° 09′ W	1500	2020	19.0	2.88	0.2	1.3	7.9	14.0	7.0	Tr	—	—	—	3.3	10.1	9.2	7.4	14	60.4
25° 19′ N 114° 05′ W	1957	4070	20.0	2.30	0.3	2.5	12.2	17.7	5.4	—	—	—	—	6.2	13.1	10.1	4.2	14	71.7
46° 46′ S 123° 57′ W	2427	4330	15.4	2.23	1.5	1.6	3.2	7.1	Tr	—	—	—	—	Tr	Tr	Tr	6.7	10	20.1
26° 45′ N 123° 00′ W	987	4280	20.0	7.50	0.7	2.9	11.1	8.9	2.5	—	0.8	1.4	1.0	—	—	—	3.6	10	32.9
31° 41′ N 119° 35′ W	1020	3570	19.0	7.50	0.5	1.9	8.3	6.5	1.5	—	—	3.0	2.9	—	—	—	0.1	10	32.2
46° 46′ S 123° 57′ W	2427	4330	15.3	7.58	0.3	0.4	2.7	—	—	—	—	—	—	—	—	—	—	7	3.4
32° 40′ N 118° 09′ W	1500	2020	56	4.5	0.3	0.5	1.5	3.7	2.1	Tr	—	—	—	1.0	2.0	2.5	2.7	10	16.3

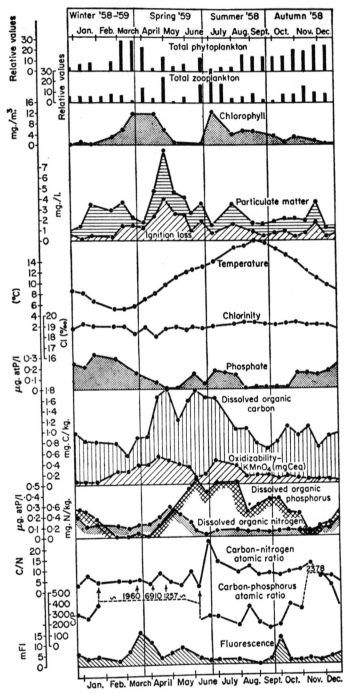

Figure 9.8 Seasonal Variations in the Distribution of Dissolved Organic Materials and Related Parameters in the North Sea. From Duursma (1965), with permission of Academic Press, Inc.

variable concentrations in shallow, especially coastal, waters; small and uniform concentrations in deep water; seasonal fluctuations (Figure 9.7), and a strong dependence on biological activity (Figure 9.8). This biological activity, in particular the factors affecting and accompanying phytoplankton population growth and utilization, have been examined in detail by Strickland (1965) and Provasoli (1963), and are to some extent evident in Figure 9.8.

A number of results have been reported for the vitamin content of seawater, and they are summarized in the compilation of Duursma (1965) mentioned above. A recent bioassay of thiamin (vitamin B_1) in North Pacific and Gulf of Alaska waters (Natarajan and Dugdale, 1966) showed a relatively high and variable content in shallow, especially coastal waters, and low contents in deep water. In particular, a high concentration of the substance was recorded in a thermally stratified layer above the thermocline. (For further information about the bioassay of vitamins and other constituents of seawater see the review of Belser, 1963.)

8 Phosphate, Nitrate, and Nitrogen Substances

The "inorganic" chemical constituents of seawater most intimately involved in the life processes are water, oxygen, carbonate, silicate, phosphate, nitrite, and nitrate. We have mentioned these substances before and now we return to them. In the case of carbonate, so great are the reserves of this material in the sea's extensive boundaries, CO_2 in the air and carbonates in rocks and sediments, that the perturbation of the carbonate content of seawater by biological activity appears to be negligible. However, the concentration of phosphates and nitrogen compounds in seawater is sufficiently small, relatively speaking, that the biological perturbation is very evident.

Our earlier examination of the chemistry of photosynthesis (Figures 9.4 and 9.5) clearly revealed the importance of phosphates in that fundamental process. Not surprisingly, then, the particulate phosphate profile closely parallels the chlorophyll content profile (Figure 9.9). Harvey (1966) gives the phosphorus cycle in the seas as shown in the accompanying diagram, and to this scheme I should like to add only that the inorganic, dissolved phosphate is probably HPO_4^{3-} rather than orthophosphate, PO_4^{3-}. Phosphorus is present in virtually every form we have examined, dissolved inorganic, dissolved and particulate organic and, quite probably, by absorption, particulate inorganic. Usually dissolved orthophosphate is the predominating form, but the partitioning of the phosphorus content is seasonally dependent. Thus the concentration of dissolved organic phosphorus and, to a lesser extent, the phosphorus of detritus, phytoplankton, and zooplankton exhibit a prominent maximum in late spring and early summer at a depth of 70 m in

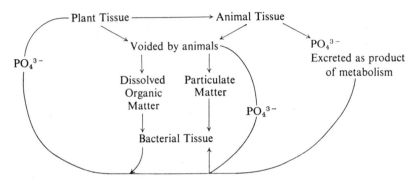

the English Channel. Table 9.13, based on results of Correll (1965), gives the analysis of the particulate phosphorus. In surface waters of the Bering Sea and Gulf of Alaska, organic phosphorus may represent up to 47% of the total, whereas below 200 m inorganic phosphorus predominates (Wardani, 1960). The phosphorus cycle is not strictly a closed one, as shown above. There is, for example, an exchange of this element in the seawater with deposits on the ocean floor, and in the case of suspended sediments the phosphate exchange can be significant (Pomeroy, Smith, and Grant, 1965). The overall result of the cycle is a general downward movement of phosphorus

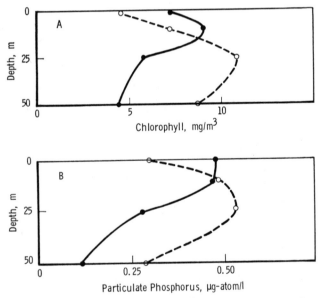

Figure 9.9 Comparison of the Chlorophyll and Particulate Phosphorus Content in Gulf of Maine Waters. From Ketchum and Corwin (1965), with permission of *Limnology and Oceanography*.

Table 9.13   Types of Phosphorus in Particulate Form in Antarctic Waters

Dissolved P, more than 90%		
Particulate P, 3–7.5%	Acid-soluble organic P	3–15%
	Phospholipid	4–29%
	Orthophosphate	12–59%
	Oligopolyanions	3–11%
	RNA-poly-P	15–74%

compounds in the sea. The phosphorus content of seawater is also highly dependent on ocean currents. In the sinking Antarctic waters the phosphorus content is about 85 mg/m³. As the Antarctic deep current flows north, the content is as low as 60 mg/m³ and lower (Harvey, 1966).

Phosphorus can be considered a primary nutrient of phytoplankton, and phytoplankton provide food for zooplankton. Thus, as the population of each step in this chain grows, the concentration of its predecessor diminishes (Figure 9.10). The concentrations of the biologically permutable constituents of seawater are intimately interconnected, and many interesting and important correlations can be discerned among them (Redfield, Ketchum, and Richards, 1963). The photosynthetic process utilizes phosphate and releases oxygen; thus these two constituents are inversely related (Figure 9.11). This

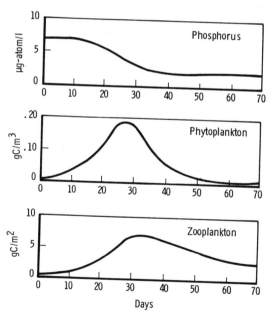

Figure 9.10   Idealized Sequence of Events in the Euphotic Zone during a Spring Plankton Bloom. From Steele (1961), with permission of AAAS and the author.

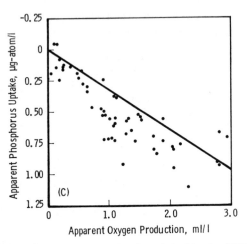

Figure 9.11 The Relation between Apparent Phosphate Uptake (APU, μg atom/l) and Apparent Oxygen Production (AOP, ml/l) for Near Shore Stations off the Washington-Oregon Coast in June. From Stefansson and Richards (1964), with permission of Pergamon Press, Ltd.

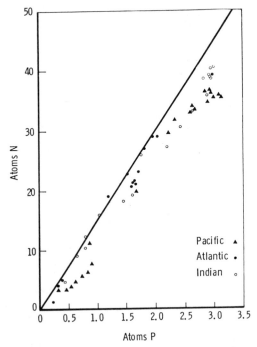

Figure 9.12 The N/P Ratio in the Atlantic, Pacific and Indian Oceans. From Sverdrup, Johnson, and Fleming (1942), with permission of Prentice-Hall, Inc.

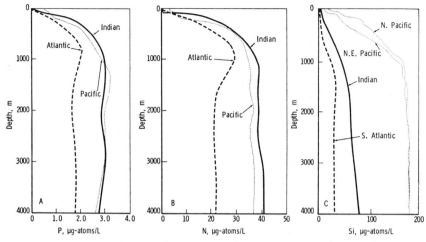

Figure 9.13   Vertical Distributions of Phosphate (A), Nitrate (B), and Silicate (C) in the Atlantic, Pacific, and Indian Oceans. From Sverdrup, Johnson, and Fleming (1942), with permission of Prentice-Hall, Inc.

reciprocal relationship was found to obtain in a detailed study of the phosphorus and oxygen content of the Atlantic as a part of the 1957–1958 I.G.Y. (McGill, 1964). Many years ago Harvey (1926) pointed out that phytoplankton simultaneously utilize phosphate and nitrate and that, therefore, the content of these two nutrients should vary in the same way and the N/P ratio should remain constant in the seas. The ratio does, in fact, remain remarkably constant (and on an atom basis it has a value of 15:1; see Table 6.4) although exceptions do exist, especially in restricted and/or surface waters (Armstrong 1965). Figure 9.12 shows the N/P ratio in the Atlantic, Pacific, and Indian oceans, and a comparison of Figures 9.13A and 9.13B shows how strikingly similar the P and N profiles are in the oceans. Analogous correlations appear between the phosphate and silicate and the nitrate and silicate concentrations, but they are less straightforward (Figure 9.13C), perhaps because the silicate concentration is subject to appreciable geological perturbation.

The profiles taken off California by Holm-Hansen, Strickland, and Williams (1966) and shown in Figure 9.14 serve as a summary of the relationships discussed above. Again notice the similarity of the nitrate (G) and phosphate (E) curves and their inverse relation to the O_2 curve (D). Vitamin B_{12} and ammonia plus amino acid profiles were also measured.

In the seas the nitrogen is released from the decomposition of the biomaterial mostly as ammonia, NH_3, which is then oxidized first to nitrite, NO_2^-, and finally to nitrate, NO_3^- (von Brand and Rakestraw, 1941) (Figure 9.15). In the seas the oxidation to NO_3^- appears to be activated by

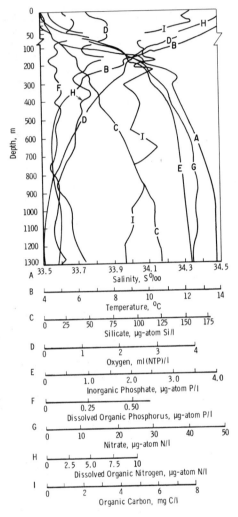

Figure 9.14 Profiles of Biologically Important Substances off Southern California. Compiled from Holm-Hansen, Strickland, and Williams (1966).

bacteria, and it takes place for the most part immediately below the photosynthetic zone in the open seas. Oxidation may also occur photochemically in the first meter or so of the surface water, and by means of the free oxygen dissolved in seawater in the presence of catalytic surfaces, although the relative importance of such process does not appear to be known with any degree of clarity. In addition to decomposition, nitrogenous material, largely as polypeptides and, in lesser quantities, as amides, is released by excretion. During

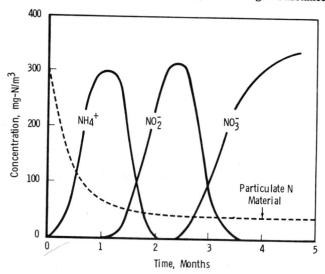

Figure 9.15 Production of Nitrogenous Material from the Decomposition of Phytoplankton in Aerated Seawater Stored in the Dark.

early growth of the organism as much as 50% of the total nitrogen taken up can be excreted, but this rate falls off to a minimum of about 10–20% at half-growth (Provalsoli, 1963).

If I may digress a moment, there is presently a great deal of poorly informed and highly optimistic talk that the seas may provide a solution to the world's overpopulation problem. Ryther (1963) makes a most interesting and germane point when he calls attention to the fact that the ocean waters richest in nutrients "contain only 60μg-atoms/1 or 0.00005% nitrogen, or four orders of magnitude less than fertile land. A cubic meter of this seawater could supply a crop of no more than about 5g of dry organic matter." When we remember that primary food production in the seas is confined to a very thin surface layer, clearly the oceans are no easy panacea which will enable man to avoid the terrible consequences of his reproductive irresponsibility.

To return to our topic, Harvey (1966) gives the nitrogen cycle in the seas as shown in the accompanying diagram, which is reminiscent of the phosphorus cycle discussed earlier (Vaccaro, 1965, gives a more detailed cycle). And again, as for the phosphorus cycle, the system is not a closed one, but in the case of nitrogen the air-sea rather than the sea-bottom interface is involved. Rivers and rain add small quantities of both NH_4^+ and NO_3^- to the seas. Harvey (1966) quotes Clark's estimate that rain annually brings about 28 mg of nitrate-N and 56–240 mg of ammonium-N to each square meter of

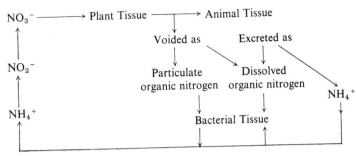

the sea's surface (see also Vaccaro, 1965) and Riley's analysis of the average nitrogen content of Mississippi River waters give:

NH_4^+-nitrogen 20 mg/m^3 NO_3^--nitrogen 350 mg/m^3
NO_2^-- nitrogen 5 mg/m^3 Albuminoid-nitrogen 350 mg/m^3

The concentrations of NH_4^+, NO_2^-, and NO_3^- in seawater are as highly variable in depth, location, and season (Figure 9.16) as are the biopopulations responsible for them. The ammonia content also exhibits diurnal variations. The euphotic zone of the Sargasso Sea contains 0.37–0.97 μg atoms of NH_3-N per liter, and the level is high in mid and late morning and low near midnight (Beers and Kelly, 1965).

Bacteria play an important role in the nitrogen cycle (see above). In addition to oxidizing NH_4^+ to NO_2^- and NO_3^-, bacteria are also capable of fixing atmospheric nitrogen and of reducing NO_3^- and NO_2^- to NH_4^+, and even to free N_2 (see below). ZoBell and Budge (1965) have studied the effect of hydrostatic pressure on NO_3^- reduction by marine bacteria. Harvey (1966)

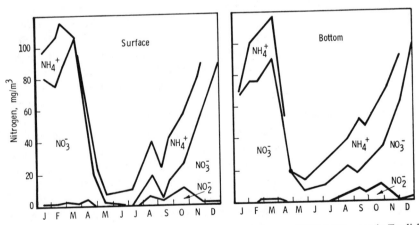

Figure 9.16 Variation of NH_4^+ Nitrogen, NO_2^- Nitrogen, and NO_3^- Nitrogen in English Channel Waters. From Cooper (1933).

devotes a chapter of his thin, but very valuable book to the changes in the composition of seawater brought about by bacterial action. These changes are the decomposition of chitin, cellulose, lignin, urea, and petroleum hydrocarbons; the oxidation of ammonium to nitrite, nitrite to nitrate, and sulfide to sulfate; the utilization of gaseous nitrogen, ammonium, urea, nitrate, nitrite, and phosphate in the bacterial cell and their release as ammonium and phosphate after death; and the formation of iron and manganese oxide concretions.

As for the utilization of nitrogenous material by organisms, virtually every type of nitrogen bond seems susceptible to attack, and all three inorganic nitrogen sources, NH_4^+ (or NH_3), NO_2^-, and NO_3^-, are taken up with comparable rates by many marine algae. However, generally speaking, ammonia seems to be the preferred form, and nitrate is first converted to NO_2^-, and then NH_3, when assimilated by algae (Vaccaro, 1965) since the valence of N in the amino acid linkage in proteins is of course -3 as in ammonia rather than $+3$ or $+5$ as in NO_2^- and NO_3^-, respectively.

(9.6) $$NO_3^- + 2\,H^+ + 2\,e^- \rightarrow NO_2^- + H_2O$$

(9.7) $$2\,NO_2^- + 4\,H^+ + 4\,e^- \rightarrow N_2O_2^{2-} + 2\,H_2O$$

(9.8) $$N_2O_2^{2-} + 6\,H^+ + 4\,e^- \rightarrow 2\,NH_2OH$$

(9.9) $$NH_2OH + 2\,H^+ + 2\,e^- \rightarrow NH_3 + H_2O$$

9 Anoxic Conditions

Below the photosynthetic zone in the sea there is a net consumption of oxygen by the respiration of animals and the decomposition and biochemical oxidation of organic material. Usually circulation of the water mass is adequate to ensure that the deeper waters are never entirely depleted of their supply of dissolved oxygen. However, there exists some particular areas where circulation is restricted, oxygen consumption exceeds oxygen removal, and the waters are anoxic or devoid of dissolved oxygen (Richards, 1965). The largest and best known of these areas is the Black Sea. Anoxic conditions are also encountered in the fjords and inlets of the coast of Norway, western British Columbia, and presumably Chile, and in basins such as the Gulf of Cariaco in the Caribbean Sea and off the California coast (Figure 9.17).

The chain of animal life in the deeper waters, as we have seen, has as its crucial links the rain of organic material from above for food and the dissolved oxygen for the respiratory processes. We might expect, therefore, the stagnant anoxic waters to be azoic or devoid of life. And so they might seem. They are graveyards, regions of death and decay. Literally marine deserts. The fish and other microscopic forms of life which teem in waters elsewhere

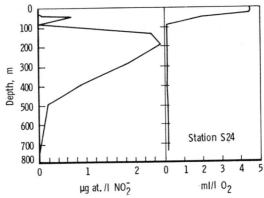

Figure 9.17 Oxygen and Nitrite Profiles for Anoxic Waters in the East Tropical Pacific Ocean. From Richards (1961), with permission of Academic Press, Inc.

are conspicuously absent here. To the casual traveler a terrestrial desert seems a place devoid of life; nevertheless the trained observer can detect myriad plants and creatures which have adapted themselves to this forbidding environment and flourish there. Similarly, although the familiar forms are missing, life persists even in these most inhospitable regions of the sea.

The phenomenon of life confronts us with many paradoxes, and one which never ceases to amaze me is the extreme fragility of life on the one hand and its extreme persistence and resilience on the other. Many marine creatures are so sensitive to environmental conditions, such as temperature and salinity, that their entire lives must be spent in very narrowly defined zones in the oceans; for them to stray out of these zones is fatal. Yet, at the opposite extreme, life forms are capable of flourishing in the seemingly most hostile environments—the desert, extremes of cold, hot springs, and anoxic waters. The persistence and ability of life to adapt to such environments is truly marvelous.

In anoxic waters, in the absence of oxygen, life turns to that element's closest relative in the periodic table (Table A.2), sulfur. Anoxic waters are characterized by a chemistry quite different from that of oxygenated waters and, in many of the fundamental metabolic processes of the anaerobic life forms that dwell in such waters, sulfur plays the role more familiarly taken by oxygen. Richards (1965) lists three important chemical features of the anoxic environment:

1. Concurrent with the disappearance of oxygen, denitrification and disappearance of NO_2^- and NO_3^-.

2. Reduction of sulfate ions and the production of H_2S (Figure 9.18).

3. Lowering of the redox potential with, consequently, reduced removal of organic materials.

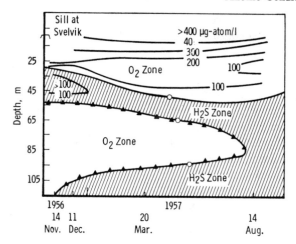

Figure 9.18 Dissolved Oxygen and Hydrogen Sulfide in Dramsfjord, Norway. From Richards and Benson (1961), with permission of Pergamon Press, Ltd.

If we formulate the organic materials as $(CH_2O)_{106}(NH_3)_{16}H_3PO_4$, then the normal oxidation of organic material in seawater by oxygen can be represented by the equation

$$(9.10) \quad (CH_2O)_{106}(NH_3)_{16}H_3PO_4 + 138 \ O_2$$
$$\rightarrow 106 \ CO_2 + 122 \ H_2O + 16 \ HNO_3 + H_3PO_4$$

When the dissolved oxygen is exhausted, the system turns to the next most abundant source of free energy for the oxidation of organic material, NO_3^-. The overall dentrification process can be represented by

$$(9.11) \quad (CH_2O)_{106}(NH_3)_{16}H_3PO_4 + 84.8 \ HNO_3$$
$$\rightarrow 106 \ CO_2 + 42.4 \ N_2 + 148.4 \ H_2O + 16 \ NH_3 + H_3PO_4$$

Nitrite ion, NO_2^-, is an intermediate, and considerable concentrations of this species can accumulate (Figure 9.17). In addition the NH_3 can be oxidized:

$$(9.12) \quad 5 \ NH_3 + 3 \ HNO_3 \rightarrow 4 \ N_2 + 9 \ H_2O$$

In deep anoxic waters the N_2 concentration may exceed that expected from equilibrium solubility. In the northeastern tropical Pacific Ocean the NO_2^- concentration exhibits a maximum (and the NO_3^- a minimum) in the oxygen minimum layer below the thermocline (300–400 m). Although no NH_3 was found, there does appear to be further denitrification of NO_2^- to N_2O and N_2 (Thomas, 1966). The rate of the conversion of NO_3^- to N_2 in low O_2, high NO_2^- content water was recently examined by Goering and Dugdale (1966).

After the dissolved oxygen level has fallen below 0.11 ml/l and all the NO_3^- and NO_2^- has been consumed, sulfate bacterial reduction occurs:

$$(9.13) \quad (CH_2O)_{106}(NH_3)_{16}H_3PO_4 + 53\ SO_4^{2-}$$
$$\rightarrow 106\ CO_2 + 53\ S^{2-} + 16\ NH_3 + 106\ H_2O + H_3PO_4$$

with consequent production of hydrogen sulfide and ammonia; in fact, an impressive correlation between the H_2S and NH_3 concentrations has been observed (Figure 9.19). Under anoxic conditions the principal source of sulfide is sulfate reduction and not the sulfur of decomposing organic matter (Redfield, Ketchum, and Richards, 1963). I might add parenthetically here the recent curious finding of Lloyd (1967) that the O^{18} in SO_4^{2-} is *not* in isotopic equilibrium with seawater, leading to the conclusion that the sulfur cycle in the sea must be so rapid that equilibrium is not established.

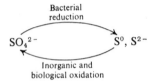

In addition to denitrification, the peculiar chemistry of anoxic water gives rise to some further interesting consequences. The first one is the lowering of the redox potential (Table 9.14); as a result of this reduced oxidizing capability of the water, there tends to be a greater accumulation of organic material in anoxic waters and in the sediments beneath them. The lowered redox potential also affects the oxidation states of metal ions; for example, iron is generally

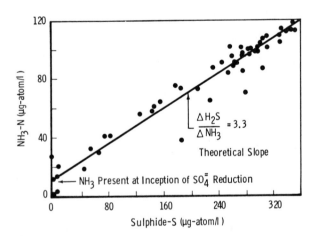

Figure 9.19 Correlation between the Ammonia and Sulfide Contents of the Waters of Lake Nitinat. From Richards (1961), with permission of Academic Press, Inc.

Table 9.14   Some Properties of Anoxic Black Sea Water
[From Richards (1965) (based on data of Skopintsev), with permission of
Academic Press, Inc.]

Depth m	Redox potential, mV	Oxygen Concentration ml/l	Oxygen Concentration μg at./l	Sulphide Concentration ml/l	Sulphide Concentration μg at./l	$SO_4^{2-}/Cl‰$
0	395	5.60	498	—	—	0.1408
25	408	7.06	628	—	—	—
50	404	6.35	565	—	—	0.1418
100	340	1.08	96	—	—	0.1408
150	−26	0.25	22	0.02	2	0.1402
200	−88	0.08	7	0.67	60	0.1400
300	−139	0	—	1.74	155	0.1394
500	−170	0	—	3.60	320	0.1387
750	−152	0	—	5.29	471	0.1377
1000	−144	0	—	6.15	547	0.1380
1500	−129	0	—	6.34	653	0.1378
1750	—	—	—	—	—	0.1378
2000	—	—	—	—	—	0.1361

reduced from +3 to +2 in anoxic conditions. Then, too, sulfide ion is a very
strong complexing ligand, and many metal sulfides are highly insoluble. Even
under such circumstances, the concentrations involved in seawater still may
be so very low that the solubility products are not approached (Krauskopf,
1956). Finally, the nutrients ammonia, phosphate, and silicate tend to accumulate in anoxic waters and are thus found in greater concentrations there
than in surface waters in the same area.

In fresh water, CO_2 is known to be reduced to methane, CH_4 ("marsh
gas"), under anoxic conditions. Methane has been detected in an anoxic

Table 9.15   Concentration of Hydrocarbons ($\times 10^{-7}$ ml/l) in Gulf of
Mexico (G) and North Atlantic (A) Waters
(From Swinnerton and Linnenbom, 1967, with permission of AAAS and
the authors)

Depth m	Methane G	Methane A	Ethane and Ethylene G	Ethane and Ethylene A	Propane G	Propane A	Propylene G	Propylene A	iso-Butane G	iso-Butane A
0	676	474	45.0	99	2.4	1.9	0.3	5.9	5.9	2.9
30	2830	600	102	51.3	6.6	3.4	6.3	8.5	6.9	2.4
500	246	488	5.4	12.3	1.7	0.8	1.6	3.1	5.9	1.4

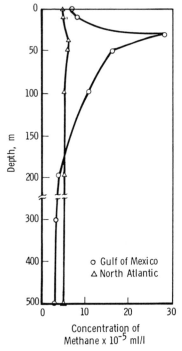

Figure 9.20 Methane Profile in Gulf of Mexico and North Atlantic Waters. From Swinnerton and Linnenbom (1967), with permission of AAAS and the authors.

fjord along with H_2S (Richards, Cline, et al. 1965). Recently the results of the first measurements of the hydrocarbon content of the open ocean have been reported (Swinnerton and Linnenbom, 1967). However, the presence of hydrocarbons other than methane (Table 9.15) and an obvious maximum near the surface (Figure 9.20) in certainly nonanoxic waters indicate a source other than CO_2 reduction.

10   The Concentration of Chemical Elements by Organisms

The several natural processes in the ocean of mixing and diffusion all tend to randomize the distribution of the chemical elements in seawater. Earlier we characterized life as a metastable derandomization, and one aspect of organisms is their ability to permute and, in particular, increase the concentrations of some elements within their bodies. In the skeletal material and such major elemental components of the biomaterial as carbon, nitrogen, and phosphorus, this is perfectly obvious (Table 9.16). Other examples of the concentration of less familiar elements have also been known for some time, such as the high iodine content of kelp; but only recently has there been an appreciation of the prevalence of the phenomenon of the concentration of the trace seawater constituents by organisms and of the tremendous concentration factors sometimes involved. Quite different organisms may enrich the primary bioelements, C, N, and P, in comparable relative amounts, and the lesser bioelements, Fe, Ca, and Si, at the same time in conspicuously different amounts (Table 9.16). Additional composition and enrichment data are given in Table 9.17. Brooks and Rumsby (1965) have published the results of a detailed study of trace element uptake by three species of New Zealand bivalves (Table 9.18). The observed enrichment factors compared with the marine environment differed for scallops, oysters, and mussels, and some of the enrichment factors were very high, such as 2,260,000 for Cd in the scallops (Table 9.19). In addition to the organism as a whole, they also analyzed the concentrations of trace elements in individual organs (Table 9.20). The results

Table 9.16   Relative Composition of Plankton Organisms and
Enrichment Factors Relative to Seawater
(From Dietrich, 1957, with permission of John Wiley & Sons)

Element	Relative Content (C = 100)			Enrichment Factor (C = 2100)		
	Diatom	Peridinium	Copepod	Diatom	Peridinium	Copepod
C	100	100	100	2,100	2,100	2,100
N	18.2	13.8	25.0	34,000	28,000	50,000
P	2.7	1.7	2.1	55,000	35,000	43,000
Fe	9.6	3.4	0.11	120,000	42,000	1,400
Ca	12.5	2.7	0.66	18	4	1
Si	93.0	6.6	0.11	110,000	8,000	130

Table 9.17   A. Composition of the Copepod *Calanus finmarchicus*
(From Dietrich, 1957, with permission of John Wiley & Sons)

Element	Content, %	Enrichment Factor	Element	Content, %	Enrichment Factor
O	79.99	—	P	0.13	43,000
H	10.21	—	Ca	0.04	1
C	6.10	2,100	Mg	0.03	0.2
N	1.52	50,000	Fe	0.007	1,400
Cl	1.05	0.5	Si	0.007	140
Na	0.54	0.5	Br	0.0009	0.14
K	0.29	7.4	I	0.0002	40
S	0.14	1.6			

B. Relative Composition of Marine Animals (Na = 100)
(From Sverdrup, Johnson, and Fleming, 1942, with permission of Prentice-Hall, Inc.)

Element	Calanus (Copepod)	Fish (average)	Archidoris britannica (Nudibranch)	Seawater	Concentration Factors		
					Copepod	Fish	Nudibranch
Cl	194	—	180	180	1.1	—	1.0
Na	100	100	100	100	1.0	1.0	1.0
Mg	5.6	36	156	12.1	0.46	3.0	12.9
S	25.9	259	7.1	8.4	3.1	31	0.85
Ca	7.4	52	262	3.8	1.9	13.7	69
K	53.7	383	20	3.6	15	109	5.5
Br	1.7	—	—	0.6	—	—	3
C	1113	ca. 4100	ca. 480	0.26	4,300	15,800	1,850
Sr	—	—	11	0.12	—	—	92
Si	1.3	—	—	0.001	13,000	—	—
F	—	—	69	0.01	—	—	6,900
N	280	1276	107	0.001	280,000	1,276,000	107,000
P	24.1	256	6	0.0001	241,000	2,560,000	60,000
I	0.04	—	—	0.0005	80	—	—
Fe	1.3	1.3	0.23	0.0002	6,000	6,000	—
Mn	—	0.0008	—	0.0001	—	8	—
Cu	—	0.008	0.43	0.0001	—	80	4,300

C. Some Further Enrichment Factors
(From Dietrich, 1957, with permission of John Wiley & Sons)

Organism	Element	Enrichment Factor
Ascidiacea	Vanadium	280,000
Gasteropoda, jellyfish, sea anemones	Zinc	32,000
	Tin	2,700
	Lead	2,600
Algae, sponges	Iodine	—
Sea anemones	Molybdenum	6,000
Jellyfish, lip-fish	Cobalt	21,000
Lip-fish	Silver	22,000

Table 9.18   Concentrations of Trace Elements in the Soft Portions of Whole Shellfish[a]
(From Brooks and Rumsby, 1965, with permission of *Limnology and Oceanography*)

Description	Mn	Pb	Cr	Mo	V	Cu	Ag	Cd	Zn	Ni	Fe
Scallop	161	23	23	2.3	14	12	2.3	299	368	17	6,900
	98	13	4	0.8	7	7	0.5	256	288	4	3,200
	63	12	8	0.6	6	7	0.5	225	293	2	1,950
	28	10	3	1.1	8	14	0.2	210	195	2	1,140
	12	14	7	0.5	5	7	0.2	216	288	4	1,242
	306	23	12	0.1	11	2	0.3	272	268	5	3,060
Oyster	10	6	6	0.3	2	53	6.3	43	1,500	3	630
	11	6	5	0.3	2	49	7.3	43	1,050	3	700
	1	7	4	0.1	2	49	6.3	38	1,204	2	688
	9	13	2[b]	0.4	3	32	4.8	10[b]	850	1[b]	646
	8	12	2[b]	0.1	3	39	4.6	42	1,040	1[b]	676
	8	14	6	0.3	4	21	4.5	34	975	1[b]	750
Mussel	18	18	9	1.0	4	5	0.1	<10	50[b]	5	1,610
	34	25	18	0.6	6	10	0.1[b]	<10	50[b]	17	2,520
	38	12	14	0.5	5	10	0.1[b]	<10	168	13	2,640
	30	8	17	0.1[b]	8	11	0.1[b]	<10	50[b]	3	2,430
	28	6	11	1.0	2	8	0.3	<10	180	5	1,600
	12	3[b]	24	0.1[b]	3	9	0.1[b]	<10	50[b]	1[b]	960
Av. scallop	111	16	10	0.9	9	9	0.7	249	283	6	2,915
Av. oyster	8	10	3	0.3	3	41	5.6	35	1,103	2	682
Av. mussel	27	12	16	0.6	5	9	0.1	<10	91	7	1,960

[a] All concentrations in ppm on material dried at 110°C.
[b] Below spectrographic detection limit, estimated value based on 50% of detection limit.

Table 9.19   Element Enrichment Factors in Shellfish
Compared with the Marine Environment
(From Brooks and Rumsby, 1965, with permission
of *Limnology and Oceanography*)

Element	Enrichment Factors		
	Scallop	Oyster	Mussel
Ag	2,300	18,700	330
Cd	2,260,000	318,000	100,000
Cr	200,000	60,000	320,000
Cu	3,000	13,700	3,000
Fe	291,500	68,200	196,000
Mn	55,500	4,000	13,500
Mo	90	30	60
Ni	12,000	4,000	14,000
Pb	5,300	3,300	4,000
V	4,500	1,500	2,500
Zn	28,000	110,300	9,100

show that the elements are concentrated in particular organs, such as Fe in the gills of the scallop. The authors note that the blood system of the oysters studied contain no hemoglobin-like pigment, and that, in general, porphyrin-bound Fe has not been found in mussels or scallops.

The ability of organisms to concentrate trace elements is a highly fascinating topic and one of very great complexity (Goldberg, 1965). There can be little doubt that, in many instances, specific, but as yet unidentified, biological functions are served by these elements. Nevertheless Goldberg's (1957) observation that the enrichment factors of the metals parallel the stability of their complexes with ligands could form the basis of a general theory of trace metal enrichment in marine organisms.

11 Summary

The biomaterials, body fluids, and skeletal structures of organisms are formed from chemicals abstracted from seawater by growing organisms, and they and their decomposition products are released to the sea on death, thus forming, together with mineral fixation, a great cycle of biochemical processes in the sea—photosynthesis, metabolism, and decomposition.

(From Brooks and Rumsby, 1965, with permission of *Limnology and Oceanography*)

Sample	% of Whole Animal	Mn	Pb	Cr	Mo	V	Cu	Ag	Cd	Zn	Ni	Fe	Sb
Scallop													
Mantle	13	45	<5	<3	1.8	7	15	0.2	<20	<100	<2	1540	<30
Gills	10	353	52	145	3.1	3	36	1.0	<20	<100	68	21,600	<30
Muscle	24	2	<5	<3	<0.1	<2	1	<0.1	<20	108	<2	34	<30
Visceral mass[b]	17	24	8	8	2.0	30	24	1.8	2000	400	2	2200	<30
Intestine[b]	1	435	28	24	3.6	16	131	2.9	<20	392	52	6090	<30
Kidney	1	2660	137	17	3.4	4	78	4.8	<20	2630	106	2470	<30
Foot	1	27	14	8	0.4	4	17	1.1	<20	210	22	2380	<30
Gonads	20	5	78	<3	<0.1	32	9	0.2	<20	256	<2	228	<30
Shell	—	1	<5	<3	<0.1	130	2	<0.1	<20	<100	<2	2000	<30
Oyster													
Mantle	7	4	<5	<3	<0.1	12	34	4.8	207	4760	<2	840	<30
Gills	10	11	<5	<3	0.4	3	100	4.6	<20	3300	<2	620	<30
Muscle	7	2	<5	<3	<0.1	<2	23	1.8	<20	400	8	274	80
Striated muscle	20	1	<5	<3	<0.1	4	6	0.5	97	369	<2	2880	<30
Visceral mass[b]	33	2	<5	<3	0.5	<2	49	68	61	1122	12	1156	<30
Kidney	4	2	<5	<3	<0.1	4	22	1.0	118	1248	<2	384	<30
Heart	1	2	15	9	1.3	4	59	48	154	2090	2	247	<30
Shell	—	1	<5	<3	<0.1	49	<1	<0.1	<20	<100	<2	570	<30
Mussel													
Mantle	15	2	<5	<3	<0.1	3	11	0.4	<20	192	<2	180	<30
Gills	14	28	36	10	0.6	5	20	1.0	<20	336	8	2940	<30
Muscle	18	<1	<5	<3	<0.1	2	1	<0.1	<20	<100	<2	82	<30
Visceral mass[b]	14	105	26	29	1.9	23	19	0.2	<20	525	32	13,650	<30
Intestine[b]	2	65	69	<3	<0.1	25	15	0.3	<20	<100	42	570	<30
Foot	15	<1	<5	<3	<0.1	44	13	<0.1	<20	<100	<2	<10	<30
Gonads	6	6	7	<3	0.1	<2	6	0.1	<20	260	5	93	<30
Shell	—	<1	<5	<3	11	110	3	<0.1	<20	<100	<2	<10	<30
Sediment	—	693	<5	307	1.5	84	102	0.3	<20	10	219	73,000	<30
Seawater[c] (ppb)	—	2	3	0.05	10	2	3	0.3	0.11	10	0.5	10	0.5

[a] All concentrations in ppm on material dried at 110°C.   [b] Includes the gut content.   [c] Data of Goldberg (1957).

287

The composition of the body fluids tends to resemble closely that of the seawater itself and, in fact, can give clues to the evolutionary history of the organisms.

Proteins and their constituent amino acids and carbohydrates constitute the biomaterial proper, but many other specialized organic compounds such as lipids and biological catalysts or enzymes are also present.

The foundation of life in the seas is photosynthesis—the formation of complex biocompounds such as carbohydrates from the simple inorganic substances CO_2 and H_2O using solar energy. The catalyst for this process is chlorophyll; thus the chlorophyll content of seawater provides a measure of the biopopulation of the waters.

Animal metabolism represents, in a sense, the reversal of photosynthesis, that is, the combustion of carbohydrates to yield back CO_2, H_2O, and especially the energy needed for the animal functions.

The inorganic solutes, phosphate, nitrate, nitrite, and ammonia, are nutrients and/or products of life processes, and their concentrations in seawater correlate with biological activity and its products. In oxygen-depleted waters, organisms tend to utilize NO_3^-, NO_2^-, and even SO_4^{2-} as oxidants, yielding in the last instance, H_2S as a product. Such anoxic waters have a chemistry quite different from that of "normal" ocean waters. In them the elements tend to occur in reduced valence states, and the removal of lifeless organic material is greatly slowed.

Some organisms are able to concentrate particular trace chemical elements so that those elements appear in their organs in concentrations sometimes many orders of magnitude greater than in seawater.

REFERENCES

Altman, P. L., and D. S. Dittmer, "Blood and Other Body Fluids," U.S. Air Force, Aeronautical Systems Division Tech. Rept. 61–199 (June 1961).

Armstrong, F. A. J., "Phosphorus," in J. P. Riley and G. Skirrow, eds., Chemical Oceanography, Academic Press, London, 1965, Chap. 8.

Beers, J. R., and A. C. Kelly, Deep-Sea Res., 12, 21 (1965).

Belser, W. L., "Bioassay of Trace Substances," in M. N. Hill, ed., The Sea, Interscience, New York, 1963, Vol. 2, Chap. 9.

Brooks, R. R., and M. G. Rumsby, Limnol. Oceanog., 10, 521 (1965).

Burkholder, P. R., and J. H. Sieburth, Limnol. Oceanog., 6, 45 (1961).

Calvin, M., Science, 130, 1170 (1959).

Cassie, R. M., and V. Cassie, New Zealand J. Sci., 3, 173 (1960).

Cloud, P. E., Jr., "Carbonate Precipitation and Dissolution," in J. P. Riley and G. Skirrow, eds., Chemical Oceanography, Academic Press, London, 1965, Vol. 2, Chap. 17.

Cooper, L. H. N., J. Marine Biol. Assoc. U.K., 18, 677 (1933).

Correll, D. L., *Limnol. Oceanog.*, **10**, 364 (1965).

Dick, D. A. T., *Cell Water*, Butterworths, Washington, D.C., 1966.

Dietrich, G., *General Oceanography*, Interscience, New York, 1963.

Duursma, E. K., *Nethe. J. Sea Res.*, **1**, 1 (1961).

Duursma, E. K., "The Dissolved Organic Constituents of Sea Water," in J. P. Riley and G. Skirrow, eds., *Chemical Oceanography*, Academic Press, London, 1965, Vol. 1, Chap. 11.

El Wardani S. A., *Deep-Sea Res.*, **7**, 201 (1960).

Franck, J., and W. E. Loomis, *Photosynthesis in Plants*, Iowa State College Press, Ames., Iowa, 1949.

Fruton, J. S., and S. Simmonds, *General Biochemistry*, John Wiley & Sons, New York, 1953.

Gillbricht M., *Deutsch. Hydrograph. Z.*, No. 3B, 90 (1959).

Goering, J. J., and R. C. Dugdale, *Science*, **154**, 505 (1966).

Goldberg, E. D., *Mem. Geol. Soc. Am.*, **67**, 345 (1957).

Goldberg, E. D., "Minor Elements in Sea-Water," in J. P. Riley and G. Skirrow, eds., *Chemical Oceanography*, Academic Press, London, 1965, Vol. 1, Chap. 5.

Hansen, V. K., *Rapp. Proces. Verbaux Reunions, Conseil Perm. Intern. Exploration Mer*, **149**, 160 (1961).

Harvey, H. W., *The Chemistry and Fertility of Sea Water*, Cambridge University Press, Cambridge, England, 1966.

Harvey, H. W., *J. Marine Biol. Assoc. U.K.*, **14**, 71 (1926).

Haurowitz, F., *J. Biol. Chem.*, **193**, 443 (1951).

Hellebust, J. A., *Limnol. Oceanog.*, **10**, 192 (1965).

Holm-Hansen, O., and C. R. Booth, *Limnol. Oceanog.*, **11**, 510 (1966).

Holm-Hansen, O., J. D. H. Strickland, and P. M. Williams, *Limnol. Oceanog.*, **11**, 548 (1966).

Holmes, R. W., *Rapp. Proces-Verbaux Reunions, Conseil, Perm. Intern. Exploration Mer*, **144**, 109 (1958).

Hulburt, E. M., J. H. Ryther, and R. R. L. Guillard, *J. Conseil, Conseil Perm. Intern. Exploration, Mer.*, **25**, 115 (1960).

Humphrey, G. F., *C.S.I.R.O. Div. Fisheries Oceanog. Rept.*, No. 24 (1959).

Humphrey, G. F., *C.S.I.R.O. Div. Fisheries Oceanog. Rept.*, No. 30 (1960).

Ingram, V. M., *Nature*, **189**, 704 (1961).

Jeffrey, L. M., and D. W. Hood, *J. Marine Res. (Sears Found. Marine Res.)*, **17**, 242 (1958).

Jørgensen, E. G., *Physiol. Plantarum*, **6**, 301 (1953); **8**, 840 (1955).

Kavanau, J. L., *Water and Solute-Water Interactions*, Holden-Day, Inc., San Francisco, 1964.

Keilin, D., and E. F. Hartree, *Nature*, **170**, 161 (1952).

Kendrew, J. D., *Science*, **139**, 1259 (1963).

Kendrew, J. D., *Sci. Amer.*, **205**, No. 6, 96 (1961).

Kendrew, J. D., R. E. Dickerson, B. E. Strandberh, R. G. Hart, D. R. Davies, D. C. Phillips, and V. C. Shore, *Nature*, **185**, 422 (1960).

Ketchum, B. H., and N. Corwin, *Limnol. Oceanog.*, **10**, R148 (1965).

Ketchum, B. H., J. H. Ryther, C. S. Yentsch, and N. Corwin, *Rapp. Procces-Vervaux. Reunions, Conseil Perm. Intern. Exploration Mer*, **144**, 132 (1958).

Koyama, T., and T. G. Thompson, *Abstr. Intern. Oceanog. Congr.*, *1959*, Am. Assoc. Advan. Sci., Washington, D.C., 1959, p. 925.

Krauskopf, K. B., *Geochim. Cosmochim. Acta*, **9**, 1 (1956).

290   Biochemistry of the Oceans

Lemberg, R., and J. W. Legge, *Hematin Compounds and Bile Pigments*, Interscience, New York, 1949.

Lewin, J. C., *Can. J. Microbiol.*, **3**, 427 (1957).

Lewin, J. C., *J. Gen. Physiol.*, **39**, 1 (1955).

Lewin, J. C., *Plant Physiol.*, **30**, 129 (1955).

Lloyd, R. M., *Science*, **156**, 1228 (1967).

McElroy, W. D., and B. Glass, eds., *Light and Life*, Johns Hopkins Press, Baltimore, Ind., 1961.

McGill, D. A., *Progr. Oceanog.*, **2**, 129 (1964).

McGinnety, J. A., R. J. Doedens, and J. S. Ibers, *Science*, **155**, 709 (1967).

Natarajan, K. V., and R. C. Dugdale, *Limnol. Oceanog.*, **11**, 621 (1966).

Nielsen, E. S., and V. K. Hansen, *Rapp. Proces-Verbaux Reunions, Conseil Perm. Intern. Exploration Mer.*, **149**, 158 (1961).

Park, K., W. T. Williams, J. M. Prescott, and D. W. Hood, *Science*, **138**, 531 (1962).

Parsons, T. R., *Fisheries Res. Board Can. M.S. Rept.* No. 81 (1960).

Parsons, T. R., *Progr. Oceanog.* **1**, 205 (1963).

Perutz, M. F., *Science*, **140**, 863 (1963).

Pomeroy, L. R., E. E. Smith, and D. M. Grant, *Limnol. Oceanog.*, **10**, 167 (1967).

Provasoli, L., "Organic-Regulation of Phytoplankton Fertility," in M. N. Hill, ed., *The Sea*, Interscience, New York, 1963, Vol. 2, Chap. 8.

Rabinowitch, E. I., *Photosynthesis*, Interscience, New York, 1951.

Redfield, A. C., B. H. Ketchum, and F. A. Richards, "The Influence of Organisms on the Composition of Sea-water," in M. N. Hill, ed., *The Sea*, Interscience, New York, 1963, Vol. 2, Chap. 2.

Richards, F. A., and B. B. Benson, *Deep-Sea Res.*, **7**, 254 (1961).

Richards, F. A., "Anoxic Basins and Fjords," in J. P. Riley and G. Skirrow, eds., *Chemical Oceanography*, Academic Press London, 1965, Vol. 1, Chap. 13.

Richards, F. A., J. D. Cline, W. W. Broenkow, and L. P. Atkinson, *Limnol. Oceanog.*, **10**, R158 (1965).

Riley, G. A., *Bull. Bingham Oceanog. Coll.*, **17**, 83 (1959).

Ryther, J. H., "Geographic Variations in Productivity," in M. N. Hill, ed., *The Sea*, Interscience, New York, 1963, Vol. 2, Chap. 17.

Saijo, Y., and S. Ichimura, *J. Oceanog. Soc. Japan*, **16**, 29 (1960).

Steele, J. H., and I. E. Baird, *Limnol. Oceanog.*, **10**, 261 (1965).

Steele, J. H., "Primary Production" in M. Sears, ed., *Oceanography*, Am. Assc. Advan. Sci. Publ., No. 67, Washington, D.C., 1961.

Stefansson, U., and F. A. Richards, *Deep-Sea Res.*, **11**, 355 (1964).

Strickland, J. D. H., "Production of Organic Matter in the Primary Stages of the Marine Food Chain," in J. P. Riley and G. Skirrow, eds., *Chemical Oceanography*, Academic Press, London, 1965, Vol. 1, Chap. 12.

Sverdrup, H. U., M. W. Johnson, and R. H. Fleming, *The Oceans*, Prentice-Hall, Englewood Cliffs, N. J., 1942.

Swinnerton, J. W., and V. J. Linnenbom, *Science*, **156**, 1119 (1967).

Tatsumoto, M., W. T. Williams, J. M. Prescott, and D. W. Hood, *J. Marine Res. (Sears Found. Marine Res.*, **17**, 247 (1958); **19**, 89 (1961).

Thomas, W. H., *Deep-Sea Res.*, **13**, 1109 (1966).

Ushakov, A. N., D. M. Vityuk, V. A. Vaver, and L. D. Bergelson, *Okeanoligiya*, **6**, 891 (1966).

Vaccaro, R. F., Inorganic Nitrogen in the Sea," in J. P. Riley and G. Skirrow, eds., *Chemical Oceanography*, Academic Press, London, 1965, Vol. 1, Chap. 9.

von Brand, T., and N. W. Rakestraw, *Biol. Bull.*, **81**, 63 (1941).
Walsh, G. E., *Limnol. Oceanog.* **10**, 570, 577 (1965).
Williams, P. M., *Nature*, **189**, 219 (1961).
Wyman, J., Jr., *Advan. Protein Chem.*, **4**, 407 (1948).
ZoBell, C. E., and K. M. Budge, *Limnol. Oceanog.*, **10**, 207 (1965).
Zuckerkandl, E., *Sci. Am.*, **212**, No. 5, 110 (1965).

10 Radiochemistry of the Seas

1 Introduction

The atomic weight given in Table A.2 of the Appendix for any given element is the average of that element's isotopes; for example, the abundances of the three isotopes of potassium, K^{39}, K^{40}, and K^{41}, are 93.3%, 0.012%, and 6.7%, respectively. The isotopes may be either stable or unstable and subject to radioactive decay. Each radionuclide decays, emitting characteristic radiation with a characteristic half-life. Radioactive decay is a statistical process, and the probability of a given atom of a radioactive element decaying is, for our purposes, independent of that atom's past history or its circumstances such as temperature, pressure, chemical state. The number of atoms, N, remaining unchanged at time t is given by the familiar rate law

$$(10.1) \qquad N = N_0 e^{-\lambda t}$$

where λ is a constant characteristic of the decaying species and N_0 is the number of atoms that were present at $t = 0$. Equation (10.1) can be rewritten

$$(10.2) \qquad \ln \frac{N}{N_0} = \frac{-1}{t}$$

When half of the atoms have decayed,

$$(10.3) \qquad \ln \frac{1}{2} = -\lambda t_{\frac{1}{2}}, \qquad t_{\frac{1}{2}} = \frac{0.69315}{\lambda}$$

where the useful quantity $t_{1/2}$ is called the half-life.

Figure 10.1 shows the decay curves of four radionuclides used for dating pelagic sediments. At the end of one-half life (8×10^4 years), one-half of the Th^{230} remains; at the end of two half-lives (or $16 \times 10_4$ years), only one-fourth of the radionuclide remains; at the end of 24×10^4 years, only one-eighth; and so on.

The radioisotopes present in the marine environment fall into three classes (Burton, 1965):

1. Natural, very long, half-lived nuclides which have persisted since the formation of the planet and their shorter-lived daughter nuclides which are continually renewed by decay.

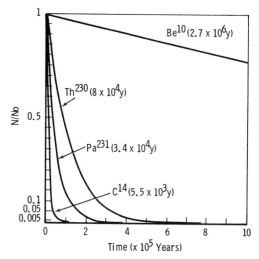

Figure 10.1 Decay Curves of Four Radionuclides. From Picciotto (1961), with permission of AAAS and the author.

2. Natural, relatively short-lived nuclides, continually formed by processes such as cosmic radiation in the atmosphere.
3. Artificial nuclides resulting from human pollution.

2 Natural Radioactive Elements in the Sea

Table 10.1 summarizes the concentration of radionuclides in surface ocean waters before the 1953 thermonuclear tests. Notice that K^{40} is easily the principal contributor to the natural radioactivity of the seas, being responsible for more than 90% of the total radioactivity of seawater. K^{40} decays by β-emission and K-electron capture to yield Ca^{40} and A^{40} as products. About 28% of the rubidium in seawater is present as the long-lived nuclide Rb^{87}, but the radioactivity from this isotope is less than 1% of that from K^{40}. The three geologically important radioactive series involving nuclides of class 1 above are:

Uranium Series

$$U^{238} \xrightarrow[\alpha]{4.51 \times 10^9 \text{ y}} Th^{234} \xrightarrow[\beta^-]{24.10 \text{ days}} Pa^{239} \xrightarrow[\beta^-]{1.18 \text{ mo}} U^{234}$$

(99.274% of U) $\qquad\qquad\qquad\qquad$ (0.056% of U

$$2.48 \times 10^5 \text{ y} \Big| \alpha$$

$$\leftarrow Pb^{210} \leftarrow\leftarrow Po^{218} \xleftarrow[\alpha]{3.825 \text{ days}} Rn^{222} \xleftarrow[\alpha]{1622 \text{ yr}} Ra^{226} \xleftarrow[\alpha]{8.0 \times 10^4} Th^{230}$$

Table 10.1   Natural Radionuclides Present in the Ocean
(From Picciotto, 1961, with permission of AAAS and the author)

Nuclide	Half-life, yr	Concentration, g/ml	Isotopic Abundance, %	Disintegrations, per sec and per ml
H^3	1.2×10^1	3.2×10^{-21}	1.0×10^{-16}	$1.1 \times 10^{-6} \beta$
C^{14}	5.5×10^3	3.1×10^{-17}	1.3×10^{-10}	$5.2 \times 10^{-6} \beta$
Be^{10}	2.7×10^6	1×10^{-16}		$7 \times 10^{-8} \beta$
K^{40}	1.3×10^9	4.5×10^{-8}	1.2×10^{-2}	$1.1 \times 10^{-2} \beta+\gamma$
Rb^{87}	5.0×10^{10}	3.4×10^{-8}	27.8	$1.0 \times 10^{-4} \beta$
U^{238}	4.5×10^9	2×10^{-9}	99.3	$2.5 \times 10^{-6} \alpha$
Th^{230}	8.0×10^4	6×10^{-16}	$>3 \times 10^{-3}$	$4 \times 10^{-7} \alpha$
Ra^{226}	1.6×10^3	8×10^{-17}	~ 100	$2.9 \times 10^{-6} \alpha$
U^{235}	7.1×10^8	1.4×10^{-11}	0.7	$1.1 \times 10^{-6} \alpha$
Pa^{231}	3.4×10^4	5×10^{-17}	~ 100	$8 \times 10^{-8} \alpha$
Th^{227} (RdAc)	—	7×10^{-23}	—	$8 \times 10^{-8} \alpha$
Th^{232}	1.4×10^{10}	2×10^{-11}	~ 100	$8 \times 10^{-8} \alpha$
Th^{228} (RdTh)	1.9	4.0×10^{-21}	—	$1.2 \times 10^{-7} \alpha$
Ra^{228} (MsTh)	6.7	1.4×10^{-20}	$\sim 1 \times 10^{-2}$	$1.2 \times 10^{-7} \beta$

Thorium Series

$$Th^{232} \xrightarrow[\alpha]{1.39 \times 10^{10} \text{ yr}} Ra^{228} \xrightarrow[\beta-]{6.7 \text{ yr}} Ac^{228} \xrightarrow[\beta-]{6.13 \text{ hr}} Th^{228} \longrightarrow$$
(100% of Th)

Actinouranium Series

$$U^{235} \xrightarrow[\alpha]{7.1 \times 10^8 \text{ yr}} Th^{231} \xrightarrow[\beta-]{95.64 \text{ hr}} Pa^{231} \xrightarrow[\alpha]{3.43 \times 10^4 \text{ y}} Ac^{227}$$
(0.720% of U)

$$21.89 \downarrow \beta-$$

$$\longleftarrow Th^{227}$$

In the marine environment the concentrations of the members of this series (Table 10.1 and 10.2) are, in general, much lower than in the Earth's crust, and their ratios are also strikingly dissimilar. The Th/U ratio, for example, is only 0.01 in the ocean compared to 3 in rocks. Clearly, then, in the sea there is "complete disruption of the radioactive equilibria by geochemical processes" (Burton, 1965). Table 10.3 summarizes some of the more important among these processes.

Burton (1965) gives a detailed table summarizing determinations of uranium in seawater. In well-mixed ocean waters the concentration of U is 0.003 mg/l (Table 5.3), and it is proportional to the salinity, although

Table 10.2   Oceanic Concentrations of the Radioactive Isotopes Belonging
to the Three Natural Radioactive Series

Isotope	Half-Life	Mode of Decay	Estimated Average Concentration in Seawater, g/l	Estimated Average Concentration Surface Sediments, g/g	Range of Concentration in Surface Sediments, g/g
U^{238}	4.5×10^9 yr	α	3.0×10^{-6}	1×10^{-6}	$(0.4 - 80) \times 10^{-6}$
U^{235}	7.13×10^8 yr	α	2.1×10^{-8}	7.1×10^{-9}	
U^{234}	2.48×10^5 yr	α	1.6×10^{-10}	8.1×10^{-11}	
Pa^{234}	1.14 min	β	1.4×10^{-19}	4.7×10^{-20}	
Pa^{231}	3.43×10^4 yr	α	$<2 \times 10^{-12}$	1×10^{-11}	$(0.08 - 9) \times 10^{-11}$
Th^{234}	24.1 days	β	4.3×10^{-17}	1.4×10^{-17}	
Th^{232}	1.42×10^{10} yr	α	$<2 \times 10^{-8}$	5.0×10^{-6}	$(1 - 16) \times 10^{-6}$
Th^{231}	25.6 hr	β	8.6×10^{-20}	2.9×10^{-20}	
Th^{230}	8.0×10^4 yr	α	$<3 \times 10^{-13}$	2.0×10^{-10}	$(0.3 - 20) \times 10^{-10}$
Th^{228}	1.91 yr	α	4.0×10^{-18}	7×10^{-16}	
Th^{227}	18.17 days	α	$<7.0 \times 10^{-20}$	1.3×10^{-17}	
Ac^{228}	6.13 hr	β	1.5×10^{-21}	2.4×10^{-19}	
Ac^{227}	21.6 yr	β, α	$<1 \times 10^{-15}$	5.9×10^{-15}	
Ra^{228}	6.7 yr	β	1.4×10^{-17}	2.3×10^{-15}	
Ra^{226}	1.622×10^3 yr	α	1.0×10^{-13}	4.0×10^{-12}	$(0.3 - 40) \times 10^{-12}$
Ra^{224}	3.64 days	α	2.1×10^{-20}	3.4×10^{-18}	
Ra^{223}	11.68 days	α	$<4.4 \times 10^{-20}$	8.5×10^{-18}	
Fr^{223}	22 min	β	$<7.0 \times 10^{-24}$	1.4×10^{-21}	
Rn^{222}	3.823 days	α	6.3×10^{-19}	2.5×10^{-17}	
Rn^{220}	51.5 sec	α	3.3×10^{-24}	5.4×10^{-22}	
Rn^{219}	3.92 sec	α	$<1.7 \times 10^{-25}$	3.1×10^{-23}	
Po^{218}	3.05 min	α	3.4×10^{-22}	1.4×10^{-20}	
Po^{216}	0.158 sec	α	1.0×10^{-26}	1.7×10^{-24}	
Po^{215}	1.83×10^{-3} sec	α	$<8.1 \times 10^{-29}$	1.4×10^{-26}	
Po^{214}	1.64×10^{-4} sec	α	3.0×10^{-28}	1.1×10^{-27}	
Po^{212}	3.04×10^{-7} sec	α	1.2×10^{-32}	2.4×10^{-29}	
Po^{211}	0.52 sec	α	$<6.8 \times 10^{-29}$	1.2×10^{-26}	
Po^{210}	138.4 days	α	2.2×10^{-17}	8.8×10^{-16}	
Bi^{214}	19.7 min	β	2.1×10^{-21}	8.8×10^{-20}	
Bi^{212}	60.5 min	β, α	2.2×10^{-22}	3.7×10^{-24}	
Bi^{211}	2.16 min	α, β	$<5.6 \times 10^{-24}$	1.0×10^{-21}	
Bi^{210}	5.01 days	β	7.8×10^{-19}	3.1×10^{-17}	
Pb^{214}	26.8 min	β	2.9×10^{-21}	1.2×10^{-19}	
Pb^{212}	10.6 hr	β	2.4×10^{-21}	3.9×10^{-19}	
Pb^{211}	36.1 min	β	$<9.0 \times 10^{-23}$	1.6×10^{-20}	
Pb^{210}	19.4 yr	β	1.1×10^{-15}	4.5×10^{-14}	
Tl^{208}	3.10 min	β	4.1×10^{-24}	6.7×10^{-22}	
Tl^{207}	4.79 min	β	$<1.2 \times 10^{-23}$	2.1×10^{-21}	

exceptions can occur in isolated basins. At pH 8 the element is believed to be in the form of the complex $[UO_2(CO_3)]^{4-}$.

The thorium content of seawater is very depleted and the concentrations are consequently very small (Table 5.3); in coastal waters they appear to be

Table 10.3   Geochemical Balance of Some Radionuclides in the Ocean (From Koczy, Picciotto, Poulaert, and Wilgain, 1957, with permission of Pergamon Press, Ltd.)

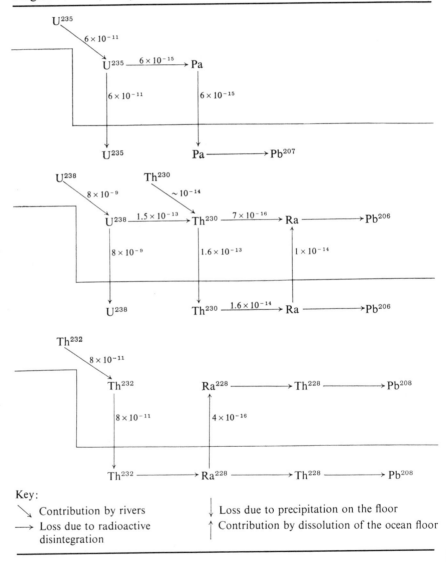

Key:

↘ Contribution by rivers   ↓ Loss due to precipitation on the floor

⟶ Loss due to radioactive   ↑ Contribution by dissolution of the ocean floor
   disintegration

appreciably affected by geological processes. Rn^{222} in seawater appears to be in excess of the concentrations expected from equilibrium with Ra^{226}, possibly because of release of Rn into near-bottom waters from sediments. Pb^{210} is produced in the oceans as a decay product but is also added from the atmosphere from the decay of Rn^{222}. Its concentration in the seas is variable and exhibits depth dependence as yet poorly understood. The concentration of Pa^{231} is also very depleted in seawater. Both Th^{230} and Pa^{231} are quantitatively precipitated in the sea, as they (or one of their parent nuclides) are formed from uranium so that less than 0.05% and 0.2%, respectively, remain dissolved in seawater (Moore and Sackett, 1964).

The concentration of Ra^{226} in seawater is variable and tends to increase with depth. This has been attributed to the fact that this nuclide, unlike those

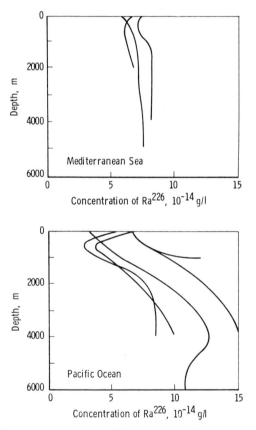

Figure 10.2   Vertical Distribution of Ra^{226}. From Koczy (1958). A similar distribution of Ra^{226} in Pacific waters has been obtained by Broecker and Li and reported in Broecker (1966).

Table 10.4 Steady State Fractional Inventories and Decay Rates of Cosmic-Ray-Produced Radioisotopes in the Exchange Reservoirs (From Lal (1967) with permission of the International Atomic Energy Agency)

Exchange Reservoir	Radioisotope												
	Be^{10}	Al^{26}	Cl^{36}	C^{14}	Si^{32}	Ar^{39}	H^3	Na^{22}	S^{35}	Be^7	Ar^{37}	P^{33}	P^{32}
Stratosphere	3.7×10^{-7}	1.3×10^{-6}	10^{-6}	3×10^{-3}	1.9×10^{-3}	0.16	6.8×10^{-2}	0.25	0.57	0.60	0.63	0.64	0.60
Troposphere	2.3×10^{-8}	7.7×10^{-8}	6×10^{-8}	6×10^{-2a}	1.1×10^{-4}	0.83	4×10^{-3}	1.7×10^{-2}	8×10^{-2}	0.11	0.37	0.16	0.24
Mixed oceanic layer	8×10^{-6}	2×10^{-5}	2×10^{-2}	2.3×10^{-2}	5×10^{-3}	5×10^{-4}	0.50^b	0.62	0.34	0.28	0	0.19	0.16
Deep oceanic layer	1.4×10^{-4}	10^{-4}	0.98	0.91	0.96	3×10^{-3}	0.43	0.11	6×10^{-3}	3×10^{-3}	0	10^{-3}	1.5×10^{-4}
Oceanic sediments	0.999	0.999	0	10^{-2}	4×10^{-2}	0	0	0	0	0	0	0	0
λ (yr^{-1})	2.8×10^{-7}	9.4×10^{-7}	2.2×10^{-6}	1.24×10^{-4}	1.4×10^{-3}	2.3×10^{-3}	5.6×10^{-2}	0.27	2.9	4.8	7.2	10	17.7
Normalizing factor for obtaining disintegrations/ min in cm² column in different reservoirs	2.7	1.4×10^{-2}	0.21	108	9.6×10^{-3}	0.34	15	3.4×10^{-3}	8.4×10^{-2}	4.9	0.05	4.1×10^{-2}	4.9×10^{-2}

a Includes amount present in biosphere and humus.
b Includes amount present in the continental hydrosphere.

originating from cosmic rays, enters the ocean largely at its bottom rather than at the surface. The hypothesis that Ra^{226} dissolves from bottom sediments and is transported upward by eddy diffusion is nicely supported by the marked contrast in the Ra^{226} profiles in Pacific waters, where sedimentation is slow, and in Mediterranean waters, where sedimentation is so rapid that Ra^{226} has little opportunity to leach out (Figure 10.2).

We shall return to the important topic of the radionuclide content of marine sediments in Part III.

Still other radionuclides enter the water from the earth's atmosphere where they are produced by the action of cosmic radiation. The more important of these nuclides include:

C^{14} produced in the atmosphere at a rate of about 2 atoms/cm^2/sec by capture of neutrons by nitrogen.

H^3 ("tritium") produced at a rate of about 0.14–1.3 atoms/cm^2/sec by the spallation of air molecules and the interaction of secondary neutrons with nitrogen.

Table 10.5   Oceanic Concentration of Cosmic-Ray-Produced and Artificial Isotopes

Isotope	Half-life	Estimated Concentration in Surface Seawater, g/l	Estimated Concentration in Surface Sediments, g/g
		Cosmic-Ray-Produced Isotopes	
Cl^{39}	1 hr	—	—
S^{35}	87 days	$<1.8 \times 10^{-18}$	—
P^{32}	14.3 days	$<1.5 \times 10^{-18}$	—
P^{33}	25 days	$<3.1 \times 10^{-18}$	—
Si^{32}	710 yr	5×10^{-19}	$(0–2) \times 10^{-16}$
Na^{22}	2.6 yr	—	—
C^{14}	5570 yr	$(2–3) \times 10^{-14}$	$(0.1–1) \times 10^{-13}$
Be^{10}	2.5×10^6 yr	$(0.7–8) \times 10^{-17}$	$(1–3) \times 10^{-13}$
Be^7	53 days	$<4.9 \times 10^{-17}$	—
H^3	12.26 yr	$(0.7–5) \times 10^{-16}$	—
		Artificial Isotopes	
Pm^{147}	2.8 yr	$(0.2–3) \times 10^{-17}$	—
Ce^{144}	285 days	$(0.1–2.5) \times 10^{-17}$	—
Cs^{137}	28 yr	$(0.5–1.2) \times 10^{-15}$	—
Sr^{90}	28 yr	$(0.6–7) \times 10^{-15}$	—
Cl^{36}	3.1×10^5 yr	$<5 \times 10^{-18}$	—
C^{14}	5570 yr	$\sim 3 \times 10^{-16}$	—
H^3	12.26 yr	$(0.1–1) \times 10^{-15}$	—

Si^{32} produced at the rate of about 2×10^{-4} atoms/cm^2/sec, probably by spallation of argon.

Be^{10} produced at a rate of about 0.06–0.08 atoms/cm^2/sec from atmospheric nitrogen and oxygen.

The half-lives of these nuclides are given in Table 10.5, and their average specific activities in the oceans are 10, 3.3×10^{-4}, 8×10^{-3}, and $1.6 \times 10^{-}$ dpm/g, respectively. Table 10.4 summarizes the Earth's inventories of these and other cosmic-ray-produced radioisotopes, and Table 10.5 gives further information about their concentration in seawater and in sediments.

Of these isotopes, C^{14} and H^3 are of particular interest. Atoms of the former become oxidized and enter the carbon cycle (Figures 7.10 and 7.11) as $C^{14}O_2$. The removal of both from the oceans is slow compared to their decay rates and the time scale of oceanic circulation; hence they are very useful tags for following water masses and determining rates of mixing processes in the sea.

3 Radioactive Contamination of the Seas

With his characteristic and rather considerable nest-fouling capabilities and inclinations, man has managed to contaminate the entire atmosphere, lithosphere, biosphere, and hydrosphere of Earth with radioactivity—a not insignificant achievement. The half-lives and seawater concentrations of some of these contaminants are listed in Table 10.4. The radioactivity legacy in the seas left by the detonation of nuclear weapons now consists mainly of the fission products of U^{235}, Pu^{239}, and U^{238}. Radioactive waste disposal makes an additional, allegedly small and localized, contribution. The source of the radioactivity from the slow neutron fission products of U^{235} after 20 days and 1 year is shown in Table 10.6. After 20 years, Sr^{90}-Y^{90} and Cs^{137}-Ba^{137} account for 48% and 45% respectively, of the remaining total activity, the remainder coming from Pm^{147} and Sm^{157}. In addition to the fission products themselves as sources of radioactivity, activities are induced by neutron capture and other nuclear reactions, so that H^3, C^{14}, S^{35}, Co^{57}, Co^{58}, Co^{60}, Zn^{65}, Mn^{54}, Fe^{59}, and Fe^{55} must be added to the list of contaminants. Their persistence makes Sr^{90} and Cs^{137} especially important contaminants and much attention has been focused on them. The sinister tendency of Sr^{90} to concentrate in bone material, especially of growing children, has been another factor responsible for the concern with this nuclide. Burton (1965) gives tables summarizing determinations of these two radionuclides in seawater. Fortunately, the penetration of these nuclides appears to be slow (Figure 10.3)—a depth in the seas of less than 500 m as of 1963—in contrast with the more rapid penetration suggested by earlier work (Bowen and Sugihara, 1960; Miyake et al., 1962). Bowen, Noshkin, and Sugihara (1966) have examined Sr^{90}

Table 10.6   The Percentage of Activity in Curies of the Principal Radioisotopes from Slow Neutron Fission of ^{235}U (1 kg) (From Miyake, 1963, with permission of John Wiley & Son)

Twenty days	9.8×10^6 curies		One year	3.1×10^4 curies	
Nuclide	Activity curies (10^5)	%	Nuclide	Activity, curies (10^3)	%
^{140}La	13.6	13.9	$^{144}Ce\text{-}^{144}Pr$	164	52.8
^{140}Ba	11.8	12.0	^{95}Nb	45.6	14.7
^{143}Pr	11.8	12.0	^{95}Zr	22.4	7.2
^{141}Ce	9.5	9.7	^{147}Pm	17.7	5.7
^{133}Xe	6.2	6.3	^{91}Y	11.8	3.8
^{95}Zr	5.8	5.9	^{89}Sr	8.4	2.7
^{91}Y	5.5	5.6	$^{106}Ru\text{-}^{106}Rh$	15.2	4.9
^{131}I	5.5	5.6	$^{90}Sr\text{-}^{90}Y$	11.5	3.7
^{89}Sr	4.9	5.0	$^{137}Cs\text{-}^{137}Ba$	9.0	2.9
^{147}Nd	4.9	5.0	^{103}Ru	2.5	0.8
^{103}Ru	4.3	4.4	^{103}Rh	2.5	0.8
^{95}Nb	4.1	4.2			
^{144}Ce	2.3	2.3			
^{144}Pr	2.6	2.6			
^{90}Mo	1.3	1.3			
^{131}I	1.0	1.05			

transport in the Atlantic Ocean. There is some evidence that cesium tends to concentrate in the 500–1500 m zone (Folsom, Feldman, and Rains, 1964; Mackenzie, 1964). Broecker (1966) has devised a six-reservoir model for mixing across the main thermocline in the absence of particulate transport which enables the distributions of Sr^{90} and Cs^{137} to be predicted.

By concentrating elements in their substance far in excess of their concentrations in seawater (Tables 9.14, through 9.20), organisms at the same time concentrate radionuclides. Activities of Zn^{65} and Fe^{55} as high as 3.8 and 7.7 $m\mu c/g$, respectively, have been detected in the liver of the albacore. Other nuclides observed to be concentrated by organisms include Fe^{59}, Mn^{54}, Co^{57}, Co^{58}, Co^{60}, Sr^{90}, Cd^{113}, and Cs^{137}. The upward motions of living, and the settling of dead, organisms containing concentrations of radioisotopes may be an important factor in determining the mobility and distribution of radionuclides in the sea.

In addition to radionuclides, an underwater nuclear explosion produces radiolysis of water. The radiolysis products, such as H_2 and H_2O_2, can then in turn affect the oxidation states of the radionuclide content of the seawater (Pestaner, 1967).

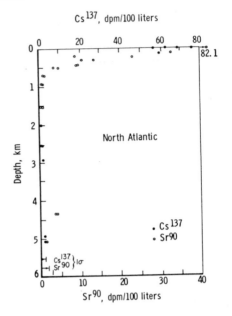

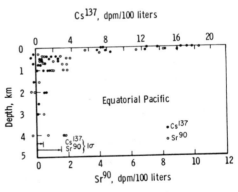

Figure 10.3   Vertical Profiles of Sr^{90} and Cs^{137} in 1963. Based on Broecker, Bonebakker, and Rocco (1966), with permission of American Geophysical Union and the authors.

4   Residence Times and Mixing in the Seas

The application of radiochemical techniques to marine geochemistry and to problems of ocean circulation, mixing, and residence times in recent years has been extraordinarily fruitful and has contributed massively to our understanding of these problem areas. Earlier (Chapters 6 and 9) we saw how the

ratios of nutrients such as phosphate, nitrate, and silicate can both characterize a water mass and provide a biological clock which enables us to make very
qualitative estimates of the age of the water. Similarly, isotopic decay can
form the basis of a radiological clock which, in principle, should enable us to
make quantitative estimates of the age of a water sample in addition to
providing a "tag" for following the course of a water sample through
oceanic circulation and mixing processes.

Perhaps the best known of these radiological clocks is the technique of
carbon-dating which has been applied with great success in archaeology.
The mixing in the Earth's atmosphere is sufficiently rapid, compared to the
decay rate, that the C^{14} formed by cosmic rays is evenly distributed in the
oxidized form $C^{14}O_2$. The mean residence time of CO_2 in the atmosphere is
estimated to range from 2 to 10 years. C^{14} decays by β-emission with a half-
life of 5600 years. The clock "starts" when the $C^{14}O_2$ dissolves at the sea's
surface. The "older" a seawater sample is, that is, the longer it has been since
it was last at the sea's surface in contact with the air, the lower is its C^{14}
content, compared with, for example, the content of normal stable C^{12}
(for a review see Broecker, 1963).

The results of C^{14} measurements on a seawater sample can be and often
are expressed in terms of the "age" of the sample; however, this is a derived
quality and is contingent on a number of questionable assumptions. A quantity
not subject to such ambiguities, directly related to measured parameters, and
therefore probably preferable, is ΔC^{14}, or the per mille difference with reference to a standard. Rate processes depend on the masses of the atoms involved. The resulting isotope effect is usually negligible for the heavier elements (although the Atomic Energy Commission separates $U^{238}F_6$ from
$U^{235}F_6$, using gas diffusion) but can be appreciable in the case of lighter atoms
such as carbon. Fortunately the isotopic fractionation resulting from the
isotope effect can be readily compensated for on the basis of the ratio of the
stable isotopes C^{13} and C^{12}. Broecker and Olson (1959) have proposed the
following normalized expression for ΔC^{14}.

$$(10.4) \qquad \Delta C^{14} \equiv \delta C^{14} - 2\,\delta C^{13}\left(1 + \frac{\delta C^{14}}{1000}\right) - 50$$

where

$$(10.5) \qquad \delta C^{13} \equiv \frac{(C^{13}/C^{12})_s - (C^{13}/C^{12})_o}{(C^{13}/C^{12})_o}\,1000$$

and

$$(10.6) \qquad \delta C^{14} \equiv \frac{(C^{14}/C^{12})_s - 0.95(C^{14}/C^{12})_o}{0.95(C^{14}/C^{12})_o}\,1000$$

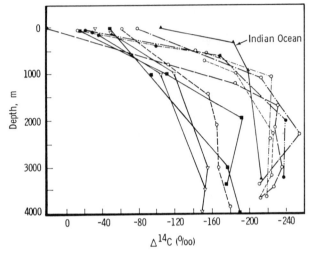

Figure 10.4 Vertical Profiles of C^{14} in Pacific Waters. From Bien, Rakestraw and Suess (1963), with permission of National Academy of Sciences.

The subscripts s and o indicate sample and reference standard, respectively; in (10.5) the latter is a belemnite and in (10.6) a National Bureau of Standards oxalic acid specimen.

As the origin of the C^{14} is at the sea's surface, we might expect the C^{14} activity to decrease with depth (Figure 10.4). Despite the great differences in time scale, Figures 10.4 and 10.3 point to the same conclusion—a rapid falloff in the concentration of surface-originating radionuclides in the first 1000 m, relatively constant concentration in deeper water—in short, surprisingly slow mixing between surface and deep waters.

Broecker, Gerard, et al. (1961) have found that, in the Atlantic, waters originating from the Antarctic surface have lower ΔC^{14} values than those originating in the northern Atlantic waters (Figure 10.5). When the ΔC^{14} values are plotted on a TS diagram (Figure 10.6), they are seen to decrease as temperature and salinity decrease. The pattern described by the ΔC^{15} values (Figure 10.5) also describes the boundaries and movements of the water masses. Measurements of ΔC^{14} in the Pacific and Indian oceans reveal a similar pattern (Figure 10.7). The surface waters (open points) exhibit a decrease in ΔC^{14} in going from north to south, whereas ΔC^{14} in the deep water (closed points) is not only much lower but also increases in this direction. That is, these waters are older the farther north they are, and the water mass appears to be moving northward at a velocity of 0.06 cm/sec. This velocity is at best only a rough guess, for there may be radiocarbon enrichment from the rain of organic material from above. As mentioned before, the

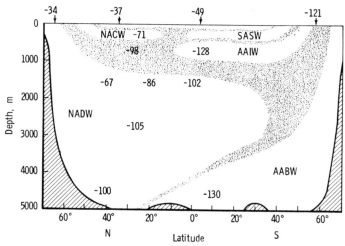

Figure 10.5 Idealized Vertical ΔC^{14} Profile in the Western Basin of the Atlantic. From Broecker, Gerard, Ewing, and Heezen (1961), with permission of AAAS and the authors. NACW = North Atlantic central waters; SASW = South Atlantic surface waters; AAIW = Antarctic intermediate waters; NADW = North Atlantic deep waters; AABW = Antarctic bottom waters.

Indian and Pacific oceans, unlike the Atlantic, are closed systems to the north, so the deep waters flow from the south with virtually no northern influx. The deep waters of the Indian Ocean are more radioactive, thus younger, than the Pacific waters, and the west to east decrease in C^{14} corresponds to a water flow of 0.03 cm/sec.

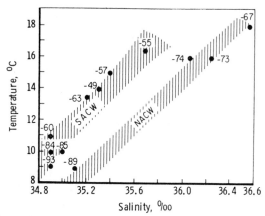

Figure 10.6 The Relationship among Temperature, Salinity, and ΔC^{14} in South Atlantic Central Waters (SACW) and North Atlantic Central Waters (NACW). From Broecker, Gerard, Ewing, and Heezen (1961), with permission of AAAS and the authors.

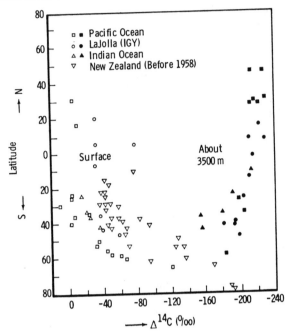

Figure 10.7   Values of ΔC^{14} in Pacific and Indian Ocean Waters. From Bien, Rakestraw, and Suess (1963), with permission of AAAS and the authors.

Estimating the age of waters from C^{14} content is beset with certain difficulties and is correspondingly uncertain. Among the more important of these difficulties are:

1. The inconstancy of the C^{14} content of the atmosphere.
2. Inadequate sea-air contact time to ensure equilibrium.
3. Perturbation by C^{14} containing organic material.
4. Mixing of water masses in the sea.
5. The possible interaction with sea-bed carbonate.

The rain of C^{14}-rich organic material into the old, deep waters can make them appear younger (see above), and the mixing of water masses of different origins and C^{14} activity results in intermediate activities, but the uncertainty which has received most attention is the inconstancy of the C^{14} content of the atmosphere. Man's use of fossil carbonaceous fuels pours great quantities of very old carbon, with all of its C^{14} long since vanished, into the atmosphere. Fortunately, in the wood of trees we have a very good record of man's pollution of atmospheric carbon and an appropriate correction can be made. The factors 50 in (10.4) and 0.95 in (10.6) are introduced to normalize the scale in ΔC^{14} and δC^{14} in terms of nineteenth century wood. Table 10.7 summarizes

Table 10.7   Industrial CO_2 Effect Correction
(From Broecker, Gerard, Ewing and Heezen, 1961, with
permission of AAAS and the authors)

Location	ΔC^{14} 1955	ΔC^{14} 1890
North Atlantic (60–80°N)	−40	−34
North Atlantic (15–40°N)	−51	−37
Caribbean	−57	−43
South Atlantic (0–42°S)	−63	−49
Antarctic (< 3°C)	−124	−121
South Pacific (15–42°S)	−51	−37
Atmosphere	−21	0
Average ocean	−58	−46

the effects of the introduction of fossil fuel carbon into the atmosphere by human activity [the Suess (1955) effect]. The depletion of atmospheric C^{14} by the Suess effect may be as great as 4%, and it appears to be greater in the northern hemisphere, presumably because of the great concentration of industry there (for a brief review see Skirrow, 1965). Fergusson (1958) has estimated, on the basis of tree rings, that in 1955 the average Suess effect on the Earth's atmosphere was about 2%.

But, although, on the one hand man has been diluting the atmospheric C^{14} content by the combustion of fossil fuels, on the other he has more than made amends of this diminution by the injection of large quantities of C^{14} into the atmosphere from nuclear devices. By May 1957 the activity of tropospheric CO_2 in the Southern Hemisphere had increased 7%, by March 1959 by 16%, while that in the Northern Hemisphere, where most of these devices are detonated, had increased by 25–30%. Much nuclear explosion activity is injected into the stratosphere, and it mixes slowly with the troposphere, thus delaying somewhat the contamination. The troposphere contamination from explosions before 1958 began to decrease by late 1960, but subsequent nuclear blasts have since sent activity climbing again. With every minor nation developing its own bomb, no end of this contamination is in sight. Fortunately, at least from a scientific if not a humanitarian point of view, the amount of this contamination is well-known and, as in the case of the Suess effect, a correction can be applied. Table 10.8 summarizes the effect of atomic bomb detonations on the C^{14} content of surface waters as of 1958–1960.

Nevertheless, despite the uncertainties and difficulties mentioned above, crude estimates can be made of the ages and residence times of the water masses. For example, the age of deep Atlantic water can be estimated by comparison of its C^{14} content with that of its presumed source region.

Table 10.8   Effects of Nuclear Bomb Detonation on the C^{14} of Seawater (From Broecker, Gerard, Ewing, and Heezen, 1961, with permission of AAAS and the authors)

Location	ΔC^{14} Uncorrected for Bomb C^{14}	ΔC^{14} Pre-1956	ΔC^{14} Corrected for Bomb C^{14}
North Atlantic, (60–80°N)	−35	—	−40
North Atlantic, 15–40°N	−49	−47	−54
Caribbean	−56	−46	−57
South Atlantic, 0–40°S	−57	—	−63
Falkland Current	−77	—	−80
Antarctic (< 3°C)	−120	—	−124
South Pacific, 15–42°S	−41	−54	−51

Table 10.9   Estimates of the Age of Atlantic Waters from C^{14} Content (From Broecker, Gerard, Ewing, and Heezen, 1961, with permission of AAAS and the authors)

Deep Water of North Atlantic Origin

Depth, m	ΔC^{14}, %	Difference from Source, %	Crude Age, yr
Surface	−35	—	—
200–400	−71	36	300
800–1100	−98	63	600
122–2500	−102	67	650
1200–2500	−75	40	350
Western boundary 2500–4000	−105	70	700
> 4000	−100	65	650

Deep Water of South Atlantic Origin

Surface 1–3°	−121	—	—
Surface 3–7°	−90	—	—
Surface 12°	−49	—	—
200–400 m	−64	15	120
600–1200 m	−128	38	300
1200–2500 m	−130	9	75
> 500 m	−144	< 23	< 200

Table 10.10   Reservoirs for an Ocean mixing Model
(From Broecker, Gerard, Ewing, and Heezen, 1961, with permission of
AAAS and the authors)

Reservoir[a]	Latitude Range	Depth Range, m	Surface Area, Volume, m²	moles C
"Arctic" North Atlantic surface water	> 55°N	Entire	15×10^{12}	40×10^{15}
(NASW) South Atlantic surface water	55°N–5°N	< 100	40×10^{12}	9×10^{15}
(SASW) North Atlantic deep water	5°N–50°S	< 100	40×10^{12}	9×10^{15}
(NADW)	55°N–50°S	> 100	—	655×10^{15}
Antarctic	> 50°S	Entire	45×10^{12}	420×10^{15}
Pacific + Indian Ocean surface water (P + IOSW)	< 55°S	< 100	220×10^{12}	50×10^{15}
Pacific + Indian Ocean deep water (P + IODW)	< 55°S	> 100	—	1980×10^{15}
Atmosphere	Entire	Entire	360×10^{12}	57×10^{15}

[a] Assume uniform ΔC^{14} in each reservoir.

Values so obtained (Table 10.9) range around 500 years for the North
Atlantic and 200 years for the South Atlantic. Broecker, Gerard, Ewing, and
Heezen (1961) interpret their C^{14} content and age determinations on the basis
of "a grossly simplified model" of the ocean-atmosphere system divided into
eight internally well-mixed reservoirs (Table 10.10 and Figure 10.8). On
the basis of this model they give the mean water residence times listed in
Table 10.11 and conclude that "the deep oceans are being renewed with a
time scale of the order of magnitude of 1000 years."

Table 10.11   Mean Residence Times
(From Broecker, Gerard, Ewing, and
Heezen, 1961, with permission of
AAAS and the authors)

Location	Time, yr
Arctic	45
NASW	10
SASW	10
NADW	600
Antarctic	100
P + IOSW	25
P + IODW	1300

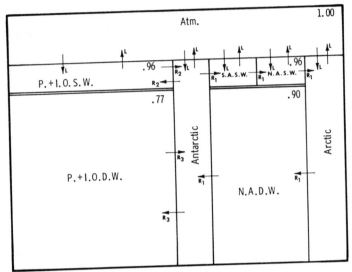

Figure 10.8 Large-Scale Oceanic Mixing Model. Arrows refer to allowed modes of transfer between reservoirs (Table 10.10), and the numbers refer to relative C^{14}/C^{12} ratios (atm = 1). From Broecker, Gerard, Ewing, and Heezen (1961), with permission of AAAS and the authors.

5   Summary

Radionuclides in the sea fall into three classes: natural, long-lived nuclides present since the birth of the planet, and their decay products; natural, relatively short-lived nuclides continually being formed by radiation in the Earth's atmosphere; and artificial nuclides from weapons, testing, and other human pollution.

The isotope K^{40} is responsible for most of the natural radioactivity of seawater.

Of the natural short-lived nuclides, C^{14} is of particular interest because it can be used to date water and thus estimate residence times and mixing in the seas.

REFERENCES

Bien, G. S., N. W. Rakestraw, and H. E. Suess, "Nuclear Geophysics," *Natl. Acad. Sci. Natl. Res. Council Pub.*, No. **1075**, 152 (1963). Bien, G. S., N. W. Rakestraw, and H. E. Seuss, *Limnol. Oceanog.*, **10**, R25 (1965).

Bien, G. S., N. W. Rakestraw, and H. E. Suess, *Nuclear Geophysics* (Nucl. Sci. Ser. Rept. No. 38), Committee on Nucl. Sci., Natl. Acad. Sci., Natl. Res. Council, Washington, D.C., 1963.

Bowen, V. T., V. E. Noshkin, and T. T. Sugihara, *Nature*, **212**, 383 (1966).

Bowen, V. T., and T. T. Sugihara, *Nature*, **186**, 71 (1960).

Broecker, W. S., E. R. Bonebakker, and G. G. Rocco, *J. Geophys. Res.*, **71**, 1999 (1966).

Broecker, W. S., *J. Geophys. Res.*, **71**, 5827 (1966).

Broecker, W. S., R. D. Gerard, M. Ewing, and B. C. Heezen, "Geochemistry and Physics of Ocean Circulation," in M. Sears, ed., *Oceanography*, Am. Assoc. Advan. Sci. Pub. No. 67, Washington, D. C., 1961.

Broecker, W. S., *J. Geophys. Res.*, **71**, 5827 (1966).

Broecker, W., in M. N. Hill, ed., *The Sea*, Interscience, New York, 1963, Vol. 2.

Broecker, W. S., and E. A. Olson, *Am. J. Sci., Radiocarbon Suppl.*, **1**, 111 (1959).

Burton, J. D., "Radioactive Nuclides in Sea-Water, Marine Sediments and Marine Organisms," in J. P. Riley and G. Skirrow, eds., *Chemical Oceanography*, Academic Press, London, 1965, Vol. 2, Chap. 22.

Fergusson, G. J., *Proc. Roy. Soc. (London)*, **243A**, 561 (1958).

Folsom, T. R., C. Feldman, and T. S. Rains, *Science*, **144**, 538 (1964).

Koczy, F. F., *Proc. 2nd U.N. Internat. Conf. Peaceful Uses of Atomic Energy*, **18**, 351 (1958).

Koczy, F. F., E. Picciotto, G. Poulaert, and S. Wilgain, *Geochim. Cosmochim. Acta.*, **11**, 103 (1957).

Lal. D., in *Radioactive Dating*, International Atomic Energy Agency, Vienna, 1967, p. 149.

Mackenzie, F. T., *Science*, **146**, 517 (1964).

Miyake, Y., "Artificial Radiation Intensity in the Sea" in M. N. Hill, ed., *The Sea*, Interscience, New York, 1963, Vol. 2.

Miyake, Y., K. Saruhashi, Y. Katsuragi, and T. Kanazawa, *J. Rad. Res.*, **3**, 141 (1962).

Moore, W. S., and W. M. Sackett, *J. Geophys. Res.*, **69**, 5401 (1964).

Pestaner, J. F., "The Radiolysis of Sea Water," U.S.N. Rad. Def. Lab. Rept. No. USNRDL-TR-67-44, 3 April 1967.

Picciotto, E. E., "Geochemistry of Radioactive Elements in the Ocean and the Chronology of Deep-Sea Sediments," in M. Sears, ed., *Oceanography*, Am. Assoc. Advan. Sci. Pub., No. 67, Washington, D.C., 1961.

Skirrow, G., "The Dissolved Gases—Carbon Dioxide," in J. P. Riley and G. Skirrow, eds., *Chemical Oceanography*, Academic Press, London, 1965, Vol. 1, Chap. 7.

Suess, H. E., *Science*, **122**, 415 (1955).

Part III
Chemistry of the Marine
Interfaces

I I The Air-Sea Interface

1 Introduction

The first chapter of this book was devoted to making amends for an enormous oversight—the failure of oceanographers to recognize the most obvious fact of chemical oceanography, namely, that the seas consist of water. The second most obvious fact of marine chemistry has also largely escaped the attention it deserves—the fact that marine chemistry is almost entirely interface chemistry. "Chemical reactions in the ocean . . . are largely determined by phenomena which occur at interfaces . . . seawater is bounded by two of the most extensive interfaces on earth—the one where it meets the air above, the other where it mingles with the sediments below" (Koczy, 1966). In addition to interactions with the atmospheric boundary above and with the lithospheric boundary below and at the surface of suspended particulate material, the marine biosphere represents a third interfacial reaction zone. In fact, with the exception of the adjustment of ionic equilibrium to changes in temperature, pressure, and electrolyte concentration, we are rather hard-pressed to find chemical reactions in the sea which do not occur at an interface of some kind or another. Inasmuch as that is where the bulk of the difficulty, lack of knowledge, and interesting chemistry lies, I feel it safe to predict that the future course of chemical oceanography will be directed strongly toward the investigation of phenomena occurring at the sea's boundaries. An ocean is not an isolated or isolatable phenomenon. It is a system, a phenomenon in a box. And in the long run I strongly suspect that this box will prove more interesting than the phenomenon. Interfaces in Nature correspond to boundaries between the disciplines. In the Introduction, oceanography was described, not as a science, but as a collection of sciences, and already the stimulation, the excitement, and the new advances in chemical oceanography are largely concentrated in the interfacial, the interdisciplinary areas where the geologist, the meteorologist, and the biologist meet the sea.

On the assumption that an understanding of the nature of water; of its structure; of the effects of temperature, pressure, and solute addition on that structure; and of the mechanism of transport processes occurring therein was a necessary prerequisite to a genuine appreciation of the chemistry of a representative element of bulk seawater (Figure I.1), Chapter 1 was devoted to a description, for the most part qualitative, of the nature of the water substance on an atomic-molecular scale. But, as we shall see presently, the water near an interface is quite different from the bulk or "normal" water a safe distance from the boundary. So in a sense, in order to lay the foundation for an understanding of the all-important chemistry of marine interfaces, we must go back to where we started and reexamine the question of the structure of water and solutions, now near interfaces, in as microscopic terms as we are able.

The prospect of revisiting Chapter 1 may fill you with some trepidation for, although we said much there, we have concluded little. You may retain little from that chapter except the impression that the whole question of the nature of water and aqueous solutions is extremely controversial, unsettled, and, above all, complex. But this at least is a realistic impression. The state of affairs with respect to the nature of water and aqueous solutions near interfaces is, if you can imagine it, worse. The amount of controversy may be a bit smaller, but only because our ignorance is that much more profound. Some of the ideas concerning the nature of water near interfaces that I will discuss, I should warn you, are speculations of my own which are not necessarily shared by the majority of the scientific community and which time may even prove to be quite wrong.

The discussion of interfacial water structure is divided into two parts. We now discuss the problem in connection with the air-sea interface, and we return to it again in the following chapter on the solid-seawater interface.

2   Surface Tension and Water Structure at the Gas–Water Interface

Comparison with a "normal" liquid (Table 11.1) illustrates the extraordinary nature of water. The excellent solvent capabilities of water, so important in marine chemistry, are in large part a consequence of its very high dielectric constant, while the high density, specific heat, and viscosity all are evidence for the tendency of water molecules to associate with one another and resist separation. Earlier we saw that the boiling point of water is very much higher than those of its sister compounds (Figure 1.4), and we see in Table 11.1 that the heat of evaporation is also anomalously great. Evidently it is relatively difficult for a water molecule to escape from the liquid into the gaseous phase across the interface. Notice that I have said relatively difficult, for the rapid establishment of vapor-liquid equilibrium in a closed system indicates that many molecules do manage to escape (and return), and the interface is the scene of turmoil with "immense turbulence and disturbance on the molecular scale" (Drost-Hansen, 1965).

Table 11.1   Comparison of Physical Properties of Water and n-Heptane (From Dietrich, 1963, with permission of John Wiley & Sons)

	Water $(H_2O)_n$	Normal Heptane (C_7H_{16})
Molecular weight	$(18)_n$	100
Dipole moment, e.s.n.	1.84×10^{-18}	$>0.2 \times 10^{-18}$
Dielectric constant	80	1.97
Density g/cm³	1.0	0.73
Boiling point, °C	100	98.4
Melting point, °C	0	-97
Specific heat, cal/g/°C	1.0	0.5
Heat of evaporation, cal/g	540	76
Melting heat, cal/g	79	34
Surface tension at 20°C, dyne/cm	73	25
Viscosity at 20°C, poise	0.01	0.005

Most of our knowledge concerning the gas-aqueous electrolytic solution interface derives from surface potential and surface tension measurements (Randles, 1957, 1963). The surface tension of water is strikingly high compared to that of other liquids (Figure 11.1). If, as in the case of the boiling points (Figure 1.4), we compare water's surface tension and those of a series

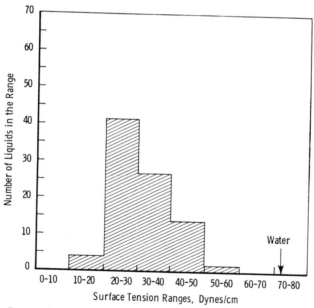

Figure 11.1   Range of Surface Tension of Some Ninety Representative Liquids at 20°C.

of similar compounds (Table 11.2), its anomalously high surface tension stands out clearly. As the molecular weight decreases, the surface tension of a series of alcohols is about 23 dynes/cm and decreases slightly, but for water, the smallest "alcohol," the value abruptly soars to about 72 dynes/cm. The analogy of a crowded dance hall has been frequently used successfully elsewhere to illustrate the dynamic nature of chemical equilibria, and I think we can put it to good purpose here: in order to "leave," a water molecule must exert itself, not once, but twice. First it must disengage itself from the partner, then it must make its way through the crowd at the door. The difficulty in the escape from liquid to gaseous phases, as reflected in the high boiling point and heat of evaporation of water, is a consequence both of molecular association or structure in the bulk phase and some kind of barrier or structural enhancement at the interface. Much of our information concerning the gas-solution interface is thermodynamic (Randles, 1963) and, as usual, thermodynamics tells us relatively little about what we would really like to know, namely, the spatial orientations of the molecules on a microscopic scale. However, the entropy is a measure of the relative amount of order-disorder obtaining and is thus of some help. On the basis of surface tension measurements, the average molar surface entropy of nonpolar liquids is about 24 joules/deg, whereas that of water is only about 10 joules/deg. In other words, the water surface is more ordered, and the entropy difference of about 14 joules/deg is attributed to the considerable particular orientation of the near-surface molecules (Good, 1957). There remains little doubt that a zone or layer of molecules oriented in some particular way exists at the surface of liquids, especially polar liquids such as water (Henniker, 1949; Frenkel, 1955), but the all-important questions about the nature of these layers in terms of molecular arrangements and their thickness are surrounded by controversy and remain largely unanswered.

Guastulla (1947) and more recently Claussen (1967) have theorized that the surface layer consists of a plane of puckered hexagons, about 19.5×10^{-16} cm^2 each in area, similar to the planes in Ice-I$_h$ (Figure 1.7), and the latter has had some success in calculating the surface energy as function of temperature, but, perhaps fortunately, the calculation is more dependent on the

Table 11.2 Surface Tension of Water and
Related Compounds at 20°C

Water	HOH	73 dynes/cm
Methanol	CH_3OH	22.6 dynes/cm
Ethanol	C_2H_5OH	22.7 dynes/cm
n-Propanol	C_3H_7OH	23.8 dynes/cm
n-Butanol	C_4H_9OH	24.6 dynes/cm

assumption that the thermal expansion of the surface and interior water behave similarly than on the particular ice-like model of the surface layer. Earlier (Chapter I) we rejected the notion that the Frank-Wen clusters in the bulk water might be ice-like in the light of the phenomenon of supercooling. Fascinating results from measurements on water in very thin capillaries (Schufle and Venugupalan, 1967, and the references cited therein; to be discussed in greater detail in the next chapter) indicate that interfacial water is especially loath to freeze. Thus our earlier argument applies with added force to the surface water, and I believe we can safely assert that the structure of interfacial water is *not* ice-like.

Furthermore, I think it probable that we can say that the structure of the interfacial water is not the same as the structure of the Frank-Wen clusters— whatever that may be. The subject of the effect of pressure on surface tension deserves a great deal of further experimental study. The few data available (Slowinski, Gates, and Waring, 1947) appear to indicate that the effects are small and are thus in contrast to the profound effect of pressure on the structure of bulk water (Chapter 1). The thermodynamic theory of surface tension under pressure (Eriksson, 1962) again tells us virtually nothing about the structural details of the surface layer. The effects of solutes on the surface tension of water are complex and varied (Drost-Hansen, 1965) and, although important relationships with bulk effects undoubtedly exist, it is far from clear at this time what they may be. The most straightforward evidence for the conclusion that the structure of the surface layer and the bulk clusters are different is the difference in their stability with respect to temperatures. The relative viscosity of water, dependent on the bulk structure (and, inasmuch as it is the shear viscosity, probably dependent at least in part on surface structure), falls off much more rapidly with increasing temperature than does the relative surface tension (Figure 11.2), which is dependent on the surface structure, and the difference is much greater for water than for a "normal" liquid such as *n*-octane.

Wicke (1966) points out that we might expect a breaking-off or loosening of the water structure at the interface. That (quite the reverse) we should observe an apparent strengthening or consolidation of the structure means that any disruption must be masked by a greater increase in short-range order. He further suggests that this order may be analogous to the "hydration of the second kind" such as encountered in the hydration of the tetraalkylammonium cations (Chapter 3). At the present time my own prejudices on this matter lie very much in the same direction. Perhaps even more germane than the hydration of the tetraalkylammonium cations is the thermodynamics of gas solubilities in water and aqueous solutions (Chapter 2). There, we found that the large entropy effect indicated extensive water structure enhancement around the dissolved alien molecules, and similar effects are also encountered

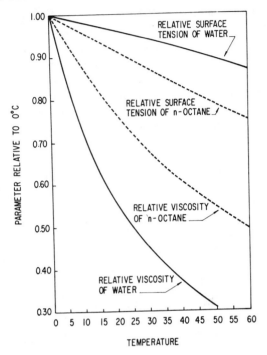

Figure 11.2   Comparison of Surface and Bulk Properties of Water and a "Normal" Liquid. From Horne, Day, Young, and Yu (1968).

in the dissolution of hydrocarbons. If a single dissolved gas molecule is capable of profoundly perturbing the structure of liquid water, no great stretch of the imagination is required to imagine a similar perturbation of the water structure as the gas-water interface is approached. Figure 11.3 represents a very crude attempt to represent the alteration of water structure as the interface is approached. This figure shows the H_2O's ranged in a highly random manner and the density of the near-surface water less than that of the more distant water, both of which may or may not be, and probably are not, true. Two features probably correctly represented in Figure 11.3 are the increased H bonding as the interface is approached, and the H_2O's at the interface aligned with their oxygen pointing to the air and their hydrogens to the water phase. With respect to the latter, Fletcher (1962) has calculated that the energy gained as a consequence of the water dipole orienting itself at the surface with the negative (oxygen) vertex outward is about 10^{-12} erg/molecule.

Unfortunately the proponents of the various theories of water structure in the bulk phase (Chapter 1) have been slow to apply their favorite models to

Surface

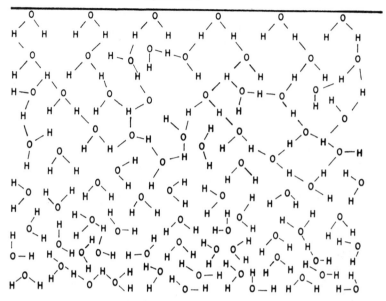

Figure 11.3 Perturbation of the Water Structure as an Interface Is Approached. From Horne (1968), with permission of Academic Press, Inc.

the structure near the interface. An exception is Lu, John, et al. (1967), who have attempted to apply the significant structure theory to the phenomenon of surface tension.

From classical electrostatic theory, electrolyte ions should be repulsed from the low-dielectric-constant air phase to the high-dielectric-constant water phase and be crowded away from the interface into the interior of the solution (Onsager and Samaras, 1934), but the actual situation may be considerably more complex.

As for the very important question of the depth of the surface area, opinion seems to be divided. The older, more established point of view appears to be that it is relatively shallow, comparable in thickness to the electrical double layer, that is, a few, but not much more than, 10 molecules thick. In the paper mentioned above, Fletcher (1962) points out that the preferential orientation of the water dipoles decays exponentially, and he estimates the thickness of the surface zone to be 13 molecular layers or about 26 Å. The so-called "Dutch School," on the other hand, believes that the range of surface van der Waals and London forces in liquids is very much greater than for individual molecules and may extend as far as thousands of angstroms (see Henniker's, 1949, old but excellent review). The direction of some recent experimental evidence appears to substantiate the second viewpoint and indicates that the perturbation of the water structure near an interface is quite extensive.

We shall return to interfacial water structure in somewhat greater detail in the next chapter in connection with the aqueous solution-solid surface interface. Suffice it here to summarize our conclusions thus far:

1. Liquid water becomes more highly structured, i.e., H-bonded, as the interface is approached.

2. The structure of this water is probably not that of Ice-I_h or the same as that of the Frank-Wen clusters.

3. At the surface the H_2O's are oriented with their oxygen negative charge centers pointing away from the liquid water phase.

4. Electrolyte ions tend to be excluded from this structured surface zone.

5. The perturbation of water structure may extend quite deeply into the solution.

Below about 0.002 M, electrolytes reduce the surface tension of water, but above this concentration they increase it, whereas a solute such as alcohol has just the opposite effect, increasing the surface tension below 2.5×10^{-4} mole fraction and decreasing it above (Drost-Hansen, 1965). Over the range of oceanographic salinities the surface tension or capillarity increases by only about 1% over that for pure water. Hence surface capillary waves, whose restoring force is surface tension, behave similarly in all oceans and in pure water. At any given temperature the surface tension of seawater, γ, of chlorinity, Cl, is related to that of pure water, γ_0 (Figure 11.2 and Table 11.3) by the expression

$$(11.1) \qquad \gamma = \gamma_{0,0°C} - 0.144T + 0.0399\ Cl, \quad \gamma_{0,0°C} = 75.64 \text{ dyne/cm}$$

where the temperature is in centigrade degrees (Fleming and Revelle, 1939).

Table 11.3   A. Surface Tension of Pure Water
(From Dorsey, 1940, with permission of Reinhold Pub. Corp.)

Temp., °C	γ, dyne/cm	Temp., °C	γ, dyne/cm
−8	76.96	+18	73.05
−5	76.42	+20	72.75
0	75.64	+22	72.44
+5	74.92	+24	72.13
+10	74.22	+26	71.82
+12	73.93	+28	71.50
+14	73.64	+30	71.18
+16	73.34	+35	70.38

B. Surface Tension of Clean Seawater
(From Fleming and Revelle, 1939, with permission of
American Association of Petroleum Geologists)

S, ‰	Temperature, °C			
	0	10	20	30
0	75.64	74.20	72.76	71.32
10	75.86	74.42	72.98	71.54
20	76.08	74.64	73.20	71.76
30	76.30	74.86	73.42	71.98
35	76.41	74.97	73.53	72.09
40	76.52	75.08	73.64	72.20

Surface potential studies on aqueous solutions (Randles, 1957) indicate that most ions (but not tetraalkylammonium ions) are repelled from the surface, anions less strongly than cations. The order of repulsion appears to be related with the size and hydration energy of the ions.

3 Surface-Active Agents and Surface Films

The surface tension and other interfacial properties of water and aqueous solutions are very sensitive to impurities. The behavior of surface-active material or "surfactants" and of surface films has expanded into a science in its own right and, thanks to practical applications such as the reduction of evaporation from reservoirs by surface films, has even achieved a certain measure of public attention (LaMer and Healy, 1965). Many film-forming molecules are bifunctional; that is, they have hydrophobic and hydrophilic segments, and at the surface they align themselves nicely with the former part directed out of, and the latter part directed into, the water phase.

Extensive areas of the sea's surface have been observed to be covered with organic films or slicks from shipping and other human pollution near shore and from natural biological processes (plankton) in the open sea. A scum sample formed by film collapse collected near the Scripps pier contained about 27% organic material (Ewing, 1950). In addition to measuring film pressures, surface tensions, and surface viscosities of slicks, Jarvis, Garrett, Scheiman, and Timmons (1967) found that the films were composed of the same type of compounds reported by Garrett (1964), namely, fatty esters, fatty acids (of 8–12 carbon atoms), fatty alcohols, and both saturated and unsaturated hydrocarbons.

Analyses of the water-insoluble organic components of film materials collected from the Atlantic and Pacific oceans, gulfs of California and Mexico, and the Bay of Panama revealed fatty esters, acids, and alcohols, and hydrocarbons, the relative amounts being dependent on the meteorological and oceanographic conditions prevalent at the location when a sample was taken (Garrett, 1964). These films affect several of the surface properties of the sea, including the surface potential and viscosity and the damping of capillary waves (Jarvis, Garrett, et al., 1967). In addition, by reducing evaporation, these films can also give rise to an excess warming of the surface water (Jarvis, 1962). In the open sea, much of the organic material of the surface film may come from dissolved organic matter in seawater (Parsons, 1963) which is concentrated by adsorption on bubble surfaces; carried upward, and dispersed at the air-sea interface when the bubble bursts. Phosphate, ammonia, and nitrite concentrations also appear to be larger at the sea's surface (Goering and Menzel, 1965). Surface films have been troublesome to some investigators of the surface properties of seawater, and they have often taken elaborate precautions in their laboratories to avoid them and obtain "clean" seawater. While the importance of surface films on the sea must not be underestimated —they can affect such large-scale phenomena as the surface wave state— their effect on the physical chemistry of the marine interface may not be over-riding. Blanchard (1963) found that, although the addition of the surfactant oleic acid decreases the surface tension of seawater, the property is restored to its original "clean" seawater value by passing through a stream of small bubbles. In the open sea the action of wind, waves, spray, precipitation, and bubble bursting is probably quite effective in continually breaking up films and freshening the surface. So effective is wind and water motion in collapsing films that Goldacre (1949) raised doubts that an extended film can exist if a wind is blowing. In a second study Blanchard (1964) found evidence that spray from surf may carry a highly compressed surface-active film.

4 Energy Transfer across the Air-Sea Boundary

At any given point the fluxes of energy and matter in either direction across the air-sea interface are usually modest, but so vast is the surface of the sea that the total quantities are enormous. A number of transport processes occur across the air-sea boundary; some of the more obvious ones are represented pictorially in Figure 11.4. These processes may be divided into two categories:

ENERGY TRANSFER
  Solar energy
  Heat
  Mechanical energy (wind, waves)
  Gravitational energy (tides)

MASS TRANSFER
Water transport
   evaporation
   precipitation
Chemical transport
   salt spray
   rain and dust

In addition there are processes which entail the simultaneous transport of both energy and mass. As we shall see, the bursting of bubbles at the surface ejects both salt and charge (electric energy) into the atmosphere. Freezing also can produce electrification. Rain, of course, transports water into the sea, but may also transfer sufficient mechanical energy (momentum) to alter the characteristics of the surface waves. And rivers introduce large quantities of water and dissolved and particulate substances into the sea along with sufficient mechanical energy to locally permeate oceanic circulation near their estuaries.

Because in this book our attention is restricted to mass transfer, I mention here briefly only those aspects of energy transfer which relate directly to chemical processes. The principal heat sources and losses in the sea are given in Table 11.4, while the total heat budget (Table 11.5) gives us some idea of

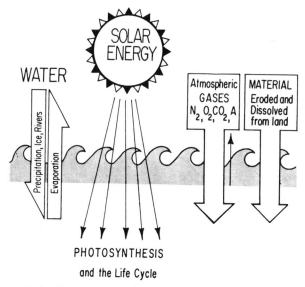

THE AIR-SEA INTERFACE

Figure 11.4   Some Transport Processes across the Air-Sea Interface.

Table 11.4   Principal Heat Sources and Losses in the Oceans
(From Defant, 1961, with permission of Pergamon Press, Ltd.)

Heat Sources	Heat Losses
1. Absorption of solar and sky radiation.	1. Radiation from the sea surface.
2. Convection of sensible heat from atmosphere to sea.	2. Convection of sensible heat from sea to atmosphere.
3. Conduction of heat through the sea bottom from the interior of the Earth.	3. Evaporation from the sea surface.
4. Conversion of kinetic energy into heat.	
5. Heat produced by chemical and biological processes.	
6. Condensation of water vapor on the sea surface.	
7. Radioactive disintegration in the seawater.	

Table 11.5   Heat Budget of the Total Ocean
(From Defant, 1961, with permission of Pergamon Press, Ltd.)

	Latitude									
	0°	10°	20°	30°	40°	50°	60°	70°	80°	90°
Heat Gain, gcal/cm²/day										
Direct solar radiation after allowing for cloudiness	202	255	267	233	171	107	80	58	44	39
Diffuse radiation	166	129	106	99	98	95	73	54	41	36
Total heat gain	368	384	373	332	269	202	153	112	85	75
Heat Loss, gcal/cm²/day										
Effective back radiation	118	134	144	143	133	116	121	126	131	137
Evaporation heat	164	170	176	160	125	78	36	13	6	0
Convection	45	45	40	35	20	20	20	20	20	20
Total heat loss	327	349	360	338	278	214	177	159	157	157
Gains–losses	+41	+35	+13	−6	−9	−12	−24	−47	−72	−82

their relative importance. Of these several processes, it is clear that solar radiation, either direct or diffuse, is easily the most important. The entire surface of the Earth receives on the average some 0.485 gcal/min cm² or about 700 gcal/cm² in the course of a day. Table 11.6 gives more detailed information about the flux of incident solar energy, and Figure 11.5 shows the average annual hours of sunshine over the oceans of the world. For a given ray of light of, for example, intensity 100, some fraction will be reflected and some fraction will penetrate the sea, in accordance with Figure 11.6. The index of refraction of pure water, n_o, is give in Table 11.7, and that of seawater, n_s, can be calculated from n_o and the expression

(11.2) $$n_s = n_o + C \times 10^{-6}$$

Values of the correction factor C are listed in Table 11.8, and the wavelength dependence is small. The light which penetrates the sea's surface then is attenuated by absorption and scattering, the latter either by the water itself or by suspended impurities in it. If I_o is the intensity of the entering light, the intensity I at depth x in meters is given by

(11.3) $$I = I e_o^{-\kappa x}$$

where κ is the attenuation coefficient. Figure 11.7 shows the dependence of κ on wavelength in seawater.

The nature and penetration of the light is important in marine chemistry for two reasons: most of it eventually is degraded into thermal energy, warming the surface waters and giving rise to evaporation and subsequent perturbations of the surface salinity; and some of it is, of course, utilized in the all-important process of photosynthesis. The warming of the surface water undoubtedly shifts chemical equilibria and accelerates the rates of chemical reactions, including biological processes, but so far as I know no systematic study of the problem has yet been undertaken. Neither does much appear to be known concerning the possibility of direct photochemical reactions, other than photosynthesis, in the air-sea interface region, although the iodine enrichment of the atmosphere above the oceans has been attributed to photochemical oxidation of iodide at the sea's surface (Duce et al., 1963, 1965; Miyake and Tsunogai, 1963).

Earlier I mentioned that mass transport in the seas by diffusion is negligible compared to the transport by macroscopic mixing processes; similarly heat transport by thermal conductivity, comparatively speaking, plays a very minor role. Defant (1961) has calculated that it would take more than 2000 years for a thermal disturbance to penetrate 100 m by thermal conduction alone. Dorsey (1940) gives the thermal conductivity of seawater at 17.5°C (listed in Table 11.9). The conductivity of pure water at 30°C increases by 5.8% in going from 1 to 1000 atm. But we must not forget that, in circum-

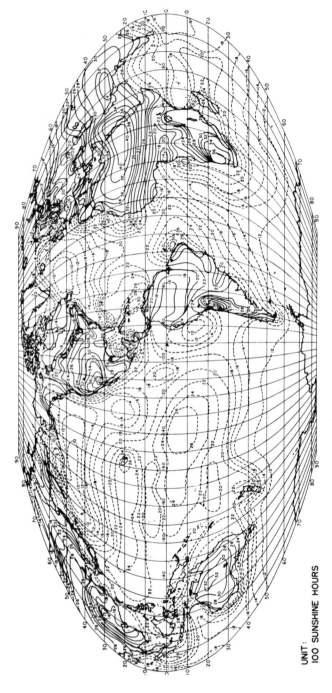

Figure 11.5 Total Annual Hours of Sunshine. From U.S.A.F. (1961), with permission of Macmillan Co. U.S. Air Force, *Handbook of Geophysics*, The Macmillan Co., New York, 1961, Chap. 16.

UNIT:
100 SUNSHINE HOURS

328

Table 11.6 Average Amounts of Radiation from Sun and Sky,[a] which
Every Month Reaches the Sea Surface in the Stated Localities
(From Sverdrup, Johnson, and Fleming, 1942, with permission of Prentice-Hall, Inc.)

| Locality | | Month | | | | | | | | | | | |
Latitude	Longitude	Jan.	Feb.	Mar.	Apr.	May	June	July	Aug.	Sept.	Oct.	Nov.	Dec.
60°N	7°E–56°W	0.002	0.053	0.125	0.207	0.272	0.292	0.267	0.212	0.147	0.074	0.006	0
60°N	135–170°W	0.005	0.078	0.155	0.208	0.269	0.260	0.242	0.185	0.127	0.077	0.015	0
52°N	10°W	0.048	0.089	0.148	0.219	0.258	0.267	0.251	0.211	0.160	0.104	0.062	0.041
52°N	129°W	0.053	0.091	0.135	0.185	0.246	0.250	0.230	0.214	0.158	0.097	0.058	0.039
42°N	66–70°W	0.094	0.138	0.212	0.272	0.306	0.329	0.302	0.267	0.230	0.174	0.115	0.086
42°N	124°W	0.100	0.151	0.210	0.286	0.331	0.360	0.320	0.274	0.231	0.174	0.113	0.029
30°N	65–77°W	0.146	0.165	0.238	0.285	0.317	0.310	0.301	0.282	0.239	0.188	0.169	0.142
30°N	128–130°E	0.141	0.153	0.199	0.241	0.258	0.238	0.256	0.260	0.219	0.178	0.153	0.135
10°N	61–69°W	0.254	0.276	0.299	0.305	0.272	0.276	0.285	0.292	0.287	0.269	0.248	0.239
10°N	116°E–80°W	0.226	0.257	0.292	0.278	0.255	0.239	0.240	0.242	0.247	0.237	0.224	0.219
0	7–12°E	0.239	0.248	0.244	0.230	0.210	0.196	0.188	0.194	0.220	0.240	0.239	0.235
0	48°W & 170°E	0.261	0.265	0.282	0.297	0.309	0.300	0.300	0.340	0.366	0.362	0.339	0.278
10°S	14°E; 36–38°W	0.329	0.328	0.301	0.254	0.219	0.206	0.232	0.278	0.312	0.324	0.317	0.320
10°S	72–171°E	0.290	0.308	0.315	0.289	0.266	0.253	0.269	0.306	0.332	0.313	0.301	0.303
30°S	17 and 116°E	0.452	0.406	0.340	0.254	0.186	0.148	0.166	0.214	0.274	0.362	0.401	0.430
30°S	110°W	0.380	0.330	0.260	0.209	0.162	0.130	0.145	0.176	0.237	0.321	0.340	0.390
42°S	73°W; 147°E	0.343	0.297	0.223	0.154	0.104	0.085	0.092	0.135	0.187	0.264	0.310	0.348
52°S	58°W	0.289	0.237	0.167	0.112	0.062	0.039	0.049	0.097	0.150	0.222	0.273	0.302
60°S	45°W	0.213	0.171	0.105	0.056	0.011	0	0.003	0.054	0.111	0.156	0.204	0.221

[a] Expressed in gcal/cm^2/min.

329

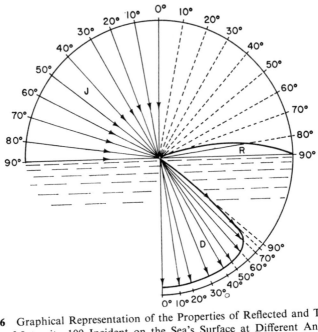

Figure 11.6 Graphical Representation of the Properties of Reflected and Transmitted Radiation of Intensity 100 Incident on the Sea's Surface at Different Angles. From Defant (1963), with permission of Pergamon Press.

Table 11.7 Index of Refraction of Pure Water (From Dorsey, 1940, with permission of Reinhold Pub. Corp.)

		T		
	10	20	30	40
λ		n_0		
0.70652	1.330704	1.330019	1.328993	1.327685
0.66781	1567	0876	9843	8528
0.65628	1843	1151	1.330116	8798
0.58926	3690	2988	1940	1.330610
0.58756	3744	3041	1993	0662
0.57696	4085	3380	2331	0998
0.54607	5176	4466	3411	2071
0.50157	7070	6353	5289	3939
0.48613	7842	7123	6055	4702
0.47131	8653	7931	6860	5504
0.44715	1.340149	9423	8347	6984
0.43583	0938	1.340210	9131	7765
0.40466	3476	2742	1.341656	1.340280

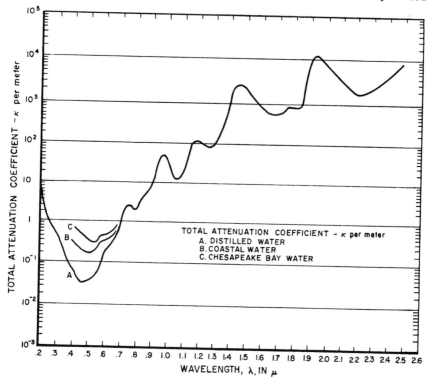

Figure 11.7 Total Attenuation Coefficients as a Function of Wavelength. From Tyler and Preisendorfer (1962), with permission of John Wiley & Sons.

stances where water circulation is restricted, as in organisms and marine sediments, all of the mass transfer can be by diffusion and all of the heat transfer by thermal conduction.

The mechanical energy (wind-wave) transported across the sea-air interface also has its chemical consequences. Not only is it partially dissipated as

Table 11.8   Refraction Index Correction for Seawater: $C = (n_s - n_o) \times 10^6$

S, ‰	Temperature, °C				
	0	10	20	30	40
20	4001	3814	3697	3621	3571
25	4989	4759	4617	4524	4463
30	5977	5708	5538	5429	5357
35	6966	6657	6463	6337	6254
40	7956	7610	7391	7250	7157

Table 11.9   Thermal Conductivity of Seawater at 17.5°C at
1 atm (in watt/cm °C)
(From Dorsey, 1940)

$S, \%_0$				
0	10	20	30	40
0.00583	0.00569	0.00563	0.00560	0.00558

thermal energy, but, more important, as every chemist knows, stirring helps most reactions along. Presently we shall examine three of the most obvious cases in which much agitation plays an important role: evaporation, the dissolution of atmospheric gases, and aerosol formation.

5   Water Transport Across the Air-Sea Boundary

Evaporation and Precipitation. In view of the tightly knit structure at the water interface and the tenaciousness with which all water molecules, and especially water molecules threatened by an alien intruder such as an interface, cling to one another, I sometimes am amazed that any water molecules manage to escape into the gaseous phase. Yet they do escape, incessantly and in great numbers, but only by dint of enormous effort, some 540 cal/g at a very high temperature of 100°C (Table 11.1) in contrast to only 132 cal/g at −61.4°C for their closest relative, H_2S.

In a closed, isothermal, static system the partial pressure of water vapor above pure water depends directly on the temperature (see Table A.26 in the Appendix) and is lowered by the addition of electrolyte, in the case of dilute solutions, by Raoult's law (Eq. 1.1). For seawater the following useful relation holds:

(11.4) $P = P_0 (1 - 0.000537 \, S)$

where P_0 is the vapor pressure of pure water. Values of the vapor pressure of water of $35\%_0$ S seawater at different temperatures are given in Table 11.10. If the initial vapor pressure in the gas phase exceeds the equilibrium vapor pressure at that temperature, then water condenses out into the liquid phase; if less, then water evaporates out of the liquid into the gaseous phase. In the oceans, evaporation is more commonly encountered than condensation at the interface. In addition to occurring at the surface of the sea, condensation can occur far above and away from the sea (rain) or near the sea's surface (fog.) The rate of attainment of equilibrium in a closed, isothermal system is a somewhat more complicated question but still amenable to straightforward

Table 11.10   Vapor Pressure (in millibars) over 35‰ S Seawater
(From Sverdrup, Johnson, and Fleming, 1942, with
permission of Prentice-Hall, Inc.)

Temperature, °C	Vapor Pressure, mb	Temperature, °C	Vapor Pressure, mb
−2	5.19	16	17.85
−1	5.57	17	19.02
0	5.99	18	20.26
1	6.44	19	21.57
2	6.92	20	22.96
3	7.43	21	24.42
4	7.98	22	25.96
5	8.56	23	27.59
6	9.17	24	29.30
7	9.83	25	31.12
8	10.52	26	33.01
9	11.26	27	35.02
10	12.05	28	37.13
11	12.88	29	39.33
12	13.76	30	41.68
13	14.70	31	44.13
14	15.69	32	46.71
15	16.74		

theoretical treatment, being determined only by the temperature, the quantities present, gaseous diffusion, and thermal conduction. The real situation at the sea's surface is extremely complex. The details of the theory of evaporation from the sea's surface lie outside the compass of this book and are reviewed elsewhere (Anderson, Anderson, and Marciano, 1950; Sverdrup, 1951; and Schrage, 1953; see above Deacon and Webb, 1962). Figure 11.8, which is sort of a rexamination of Figure 11.4 in detail, represents an attempt to summarize the heat transfer (to the left), mass transfer (to the right), and structural phenomena (in the center), all of which play an important role in the evaporation process. The wind is especially effective in hastening evaporation from the seas. Attempts to treat the effect of wind rigorously lead to very complicated formulas, but Sverdrup, Johnson, and Fleming (1942) give the following simple rule of thumb for calculating the mean annual evaporation E (in centimeters):

$$(11.5) \qquad E = 3.7 \, (\bar{p} - \bar{p}_a) \, \bar{u}$$

where $\bar{p}$ is the average vapor pressure, in millibars, at the sea's surface and is dependent on temperatures and salinity, $\bar{p}_a$ is the average water vapor pressure

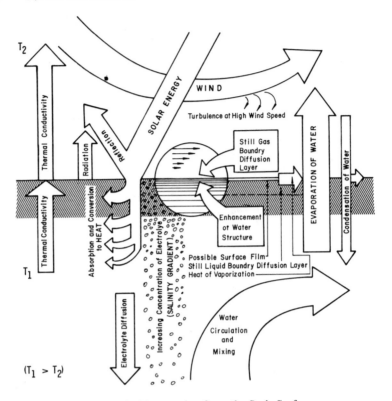

Figure 11.8   Evaporation from the Sea's Surface.

in the air 6 m above the sea's surface, and $\bar{u}$ is the average wind velocity, in meters per second at the same height.

In Figure 11.8 we have taken the sea's surface to be smooth. Needless to say, if the surface is roughened by the winds, the situation is further complicated. In the figure we have taken the temperature of the surface seawater, T_1, to be greater than that of the air above. Under such conditions, which are common in the winter months in the middle and higher latitudes, evaporation is at a maximum.

The evaporation can also be estimated from a consideration of the heat balance between atmosphere and hydrosphere, but this approach, which is basically a thermodynamic one, gives very little insight into the detailed nature of the microscopic processes involved.

Average values of evaporation, precipitation, and their difference are tabulated in Table 11.11. A plot of these values (thin line in Figure 5.7) shows the close relationship between evaporation and the salinity of the surface

Table 11.11 Average Values of Salinity, S, Evaporation, E, and Precipitation, P, and the Difference, $E - P$, for Every Fifth Parallel of Latitude Between 40°N and 50°S (From Sverdrup, Johnson, and Fleming, 1942, with permission of Prentice-Hall, Inc.)

Latitude	Atlantic Ocean				Indian Ocean				Pacific Ocean				All Oceans			
	S, ‰	E, cm/yr	P, cm/yr	$E-P$, cm/yr	S, ‰	E, cm/yr	P, cm/yr	$E-P$, cm/yr	S, ‰	E, cm/yr	P, cm/yr	$E-P$, cm/yr	S, ‰	E, cm/yr	P, cm/yr	$E-P$, cm/yr
40°N	35.80	94	76	18	—	—	—	—	33.64	94	93	1	34.54	94	93	1
35	36.46	107	64	43	—	—	—	—	34.10	106	79	27	35.05	106	79	27
30	36.79	121	54	67	—	—	—	—	34.77	116	65	51	35.56	120	65	55
25	36.87	140	42	98	—	—	—	—	35.00	127	55	72	35.79	129	55	74
20	36.47	149	40	110	(35.05)	(125)	(74)	(51)	34.88	130	62	68	35.44	133	65	68
15	35.92	145	62	83	(35.07)	(125)	(73)	(52)	34.67	128	82	46	35.09	130	82	48
10	35.62	132	101	31	(34.92)	(125)	(88)	(37)	34.29	123	127	−4	34.72	129	127	2
5	34.98	105	144	−39	(34.82)	(125)	(107)	(18)	34.29	102	(177)	(−75)	34.54	110	177	−67
0	35.67	116	96	20	34.15	125	131	−6	34.85	116	98	18	35.08	119	102	17
5°S	35.77	141	42	99	34.93	121	167	−46	35.11	131	91	40	35.20	124	91	33
10	36.45	143	22	121	34.57	99	156	−57	35.38	131	96	35	35.34	130	96	34
15	36.79	138	19	119	34.75	121	83	38	35.57	125	85	40	35.54	134	85	49
20	36.54	132	30	102	35.15	143	59	84	35.70	121	70	51	35.69	134	70	64
25	36.20	124	40	84	35.45	145	46	99	35.62	116	61	55	35.69	124	62	62
30	35.72	116	45	71	35.89	134	58	76	35.40	110	64	46	35.62	111	64	47
35	35.35	99	55	44	35.60	121	60	61	35.00	97	64	33	35.32	99	64	35
40	35.65	81	72	9	35.10	83	73	10	34.61	81	84	−3	34.79	81	84	−3
45	34.19	64	73	−9	34.25	64	79	−15	34.32	64	85	−21	34.14	64	85	−21
50	33.94	43	72	−29	33.87	43	79	−36	34.16	43	84	−41	33.99	43	84	−41

Table 11.12   Water Budget of the Earth
(From Defant, 1961, with permission of Pergamon Press, Ltd.)

	Precipitation		Evaporation	
	km³/yr	cm/yr	km³/yr	cm/yr
Ocean	3.24×10^5	90	3.61×10^5	100
Continent	0.99×10^5	67	0.62×10^5	42
Entire earth	4.23×10^5	83	4.23×10^5	83

water. Evaporation also varies in the course of the day, thereby giving rise to diurnal variations in salinity (Figure 5.4).

The Earth's total water budget is given in Table 11.12, and the details of the hydrologic cycle are schematically represented in Figure 11.9.

The percent difference between the masses of the different isotopes of the lighter elements are relatively great and can give rise to appreciable isotope

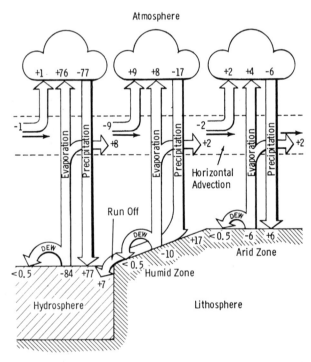

Figure 11.9   The Earth's Hydrologic Cycle (Based on a relative unit of 100).

effects in chemical reactions and in phase changes such as evaporation and freezing. Natural water, including seawater, consists of H_2O^{16}, HDO, and H_2O^{18} roughly in the proportions 1,000,000/320/2000, respectively (Craig and Gordon, 1965). The structure of D_2O is believed to be very similar to that of H_2O, although somewhat more extensive (Kavanau, 1964; Kirshenbaum, 1951; Bhandari and Sisodia, 1964; Horne and Johnson, 1967) because of the greater strength of the D bond compared to the H bond. In natural waters, because the H/D ratio is so enormous, there is no D_2O and the HDO always is surrounded by H_2O neighbors. The effect that HDO might have on either the bulk or surface structure of H_2O is unclear, but in any event the greater mass of HDO and the greater strength of the D bond make it more difficult for this molecule to leave the liquid phase than H_2O. As a consequence the vapor pressure of HDO is lower than that of H_2O, and during the evaporative process HDO tends to be left behind and concentrated in the liquid phase. This isotopic fractionation on evaporation results in higher D/H ratios in equatorial water and in surface waters compared with deep waters (Friedman, 1953; Horibe and Kobayakawa, 1960; Friedman et al., 1964; Redfield and Friedman, 1965). The reverse situation occurs in precipitation: the water containing the heavier isotope, whether D or O^{18}, is preferentially precipitated and the rain thereby enriched (Kirshenbaum, 1951). Isotopic fractionation also occurs in freezing and melting. Arctic ice contains about 2% more D

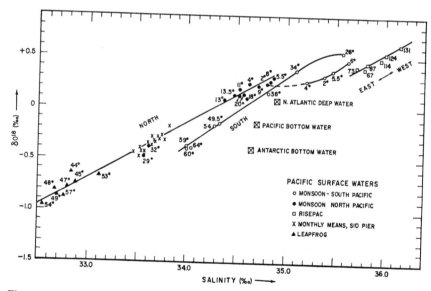

Figure 11.10 The O^{18} Isotope Effect-Salinity Relationship in Pacific Waters. From Craig and Gordon (1965), with permission of the authors.

than the seawater from which it was formed (Friedman, Harris, and Schoen, 1961; Weston, 1955).

The isotope O^{18} is also fractionated as a consequence of phase transitions in nature but, inasmuch as the relative mass difference between H_2O^{18} and H_2O^{16} is much less than between HDO and H_2O, the isotope effects are correspondingly smaller. The incremental isotope effect, δ, for O^{18}, defined by

$$(11.6) \qquad \frac{R}{R_s} = 1 + \delta$$

where R is D/H or O^{18}/O^{16} and the subscript s indicates standard mean ocean water (SMOW), increases with increasing salinity (Figure 11.10). Like D, O^{18} tends to be concentrated in surface waters by evaporation, and the O^{18} content is highly dependent on fresh water sources; for example, the O^{18} content of deep ocean waters appears to be largely determined by dilution from sinking polar waters (Epstein and Mayeda, 1953; Craig and Gordon, 1965). The theory of isotopic fractionation in the sea, with emphasis on O^{18}, is treated by Craig and Gordon (1965); they present the evaporation-precipitation-mixing model shown in Figure 11.11 for the trade winds region. The temperature of ancient seas can be estimated from the O^{18} ratio in water and carbonates (Urey, 1947; Epstein, Buchsbaum, Lowenstam, and Urey, 1951).

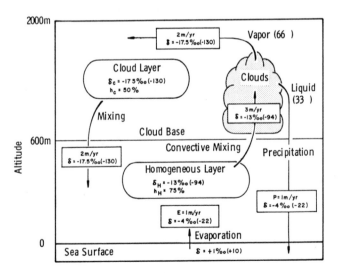

Figure 11.11 Evaporation-Precipitation-Mixing Model for the Isotope Effects in Trade Wind Regions. (The first isotopic δ value is for O^{18}, the second in parenthesis for D.) From Craig and Gordon (1965), with permission of the authors.

6   Gas Exchange Across the Air-Sea Interface

In the preceding section we examined the transport of water vapor from the sea to the atmosphere by evaporation. In a subsequent section on bubbles we shall turn our attention to the return of dissolved gases from the sea to the atmosphere. Therefore here we are largely concerned with the dissolution of atmospheric gases in the sea and, inasmuch as the thermodynamics of such processes were treated in Chapter 2 and the equilibria and solubilities in Chapter 7, we concentrate on the kinetics or rates of gas solution in the sea.

If the partial pressure of the gas in the atmosphere is greater than the equilibrium partial pressure of the gas in the sea (for a given temperature and salinity), then there will be mass transport of gas from the gaseous to the liquid phase; if less, then the seawater will release gas to the atmosphere. Inasmuch as many of the same factors are involved in gas transfer as in evaporation (Figure 11.8), the situation may on first inspection seem too complex to be amenable to theoretical attack. Fortunately, however, the rate of gas transfer is commonly determined by gas diffusion through a surface boundary layer. Winds and ocean currents ensure relatively rapid mixing in the bulk phases, so that diffusion through the quiet zones immediately adjacent to the interface becomes the slow, rate-limiting steps. The residence time of water in the atmosphere is only about 10 days, whereas the residence time for CO_2 in the sea is about 5–10 years. Bolin (1960) interprets this to mean that, in the case of CO_2, it is the aqueous, rather than the atmospheric, boundary layer which is rate-determining, a not unexpected finding because the gaseous diffusion constant is much greater than the diffusion constant in the liquid phase, 10^{-2} cm²/sec compared to 3×10^{-5} cm⁻²/sec (Skirrow, 1965).

We must remember that the rate of solution of CO_2, unlike that of O_2, N_2, and the noble gases, may be complicated by chemical reactions such as equilibria (7.2) and (7.3). However, Kanwisher (1963) has presented arguments that the reactivity of CO_2 does not appreciably alter its exchange rate across the sea's surface.

Eddy diffusion and advection bring the gas molecules to the surface boundaries relatively rapidly; then within the layers the gas molecules are transported by molecular diffusion. Kanwisher (1962) gives the gas flux, dG/dt, through the layer as

(11.7) $$ \text{Flux} = \frac{dG}{dt} = \frac{\Delta C D}{t} $$

where the "driving force" is the gas concentration difference between the bottom and top of the layer of thickness t, and D is the diffusibility, which at 20°C has very nearly the same value of 2×10^{-5} cm/sec for all the atmospheric gases. The concentration difference depends on the partial pressure and the solubilities of the gases, and these vary considerably for the several atmospheric gases; thus, for a given difference in partial pressure, the (dG/dt_{O_2}) should be twice as great as $(dG/dt)_{N_2}$.

Depending on the state of agitation, the surface layer is estimated to be between 5×10^{-3} and 0.1 cm thick with 0.01 cm, corresponding to 10^5 molecular diameters, being a representative average. Agitation reduces the effective layer thickness: strong stirring easily halves the film thickness, and its reduction becomes evident for wind velocities in excess of about 2m/sec (Figure 11.12).

Downing and Truesdale (1955) also found that the rate of solution of O_2 in pure water and saline water increases with increasing wind velocity. The rate also increased with stirring, wave height, and temperature. Oil films have little effect on the rate until their thickness exceeds about 10^{-4} cm, but certain soluble surfactants can reduce the rate of solution.

While I have no doubt that the foregoing theory of gas transfer is qualitatively correct and, despite its simplicity, applicable to many real situations at sea for all their endless complexity, I do have some reservations about the meaningfulness of the film thicknesses that have been calculated. If, as we have suggested, liquid water becomes more highly structured as the interface is approached, then we might expect the diffusion of gas molecules to be slower in the surface layer than in the bulk liquid phase. The diffusion coefficient could be reduced still more if the surface film were cooled by evapora-

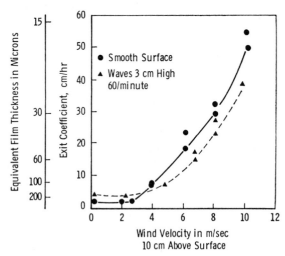

Figure 11.12 Effect of Wind Velocity of Liquid-Air Gas Exchange. From Kanwisher (1963), with permission of Pergamon Press, Ltd.

tion (Ewing and McAllister, 1960). Still, on the other hand, in the structured surface layer the gas molecules may diffuse, analogously to gas diffusion in solids, divested of all or part of their bulk phase hydration envelope. These are interesting questions which deserve further study; gas diffusion rates through the boundary layer might, in fact, provide a useful research technique for investigating the extent and nature of surface water structure. In any event, I suspect that the gas diffusion rates in the surface layer have been overestimated and that, as a consequence, the estimated layer thicknesses reported are too great, perhaps by as much as a factor of 10.

Table 11.13 summarizes the results of investigations of gas transfer across the liquid surface reported in the recent literature. In addition to the oxygen flux at the surface as given in Table 11.13, Pytkowicz (1964) has also estimated average diffusivities of oxygen in waters off the Oregon coast for the June-September period as a function of depth. The oxygen flux decreases with increasing depth: 3.88×10^{-7}, 2.51×10^{-7}, 0.55×10^{-7}, and 0.05×10^{-7} ml/cm² sec for depths of 10, 20, 30, and 40 m, respectively, hence over these months the oxygen loss decreases with increasing depth. (See Table A.27 for

Table 11.13   Gas Transfer across the Gas-Liquid Interface

Gas	Transport Rate across the Air-Sea Interface	Conditions	Reference
O_2, N_2	—	Solution of bubbles	Wyman, Scholander, Edwards, and Irving (1952)
O_2	30×10^4 ml O_2/m²/yr	Left surface of Gulf of Maine, October–March	Redfield (1948)
CO_2	24 moles/m² yr	Ocean average	Craig (1957)
O_2	1–60 cm/hr exit coefficient	Dependent on stirring wind, velocity, wave height film, etc.	Downing and Truesdale (1955)
Air	2–60 cm/hr exit coefficient	Strongly dependent on wind velocity	Kanwisher (1963)
CO_2	0.32 mg/cm²/min/atm	—	Suguira, Ibert, and Hood (1963)
CO_2	0.020 mg/cm²/min/atm	—	Miyake and Hamanda (1960)
CO_2	3–5 moles/m²/yr	Lakes	Broecker and Walton (1959)
O_2	1.23×10^4 ml O_2/m²/month	*In situ*, Oregon coast	Pytkowicz (1964)
CO_2	18 moles/cm²/atm/yr	Pacific Ocean	Keeling (1965)
CO_2	8×10^{-3} cm/sec	Gulf of Mexico	Park and Hood (1963)

some recent values of gas diffussion coefficients in pure water.) As mentioned earlier, the rate of dissolution of CO_2 in the seas may be crucial in the removal of industrial pollution from the Earth's atmosphere (Revelle and Seuss, 1957). The rate of CO_2 exchange between sea and atmosphere is decreased by the finite rate of hydration of that gas (Bolin, 1960).

7 Chemical Mass Transfer into the Sea from the Atmosphere

Although the solution of atmospheric gases and rainwater represents the most obvious mass transport from the air to the sea, other chemicals also cross the interface in the seaward direction. The total worldwide quantity of these chemicals can be considerable, but, relative to the chemical content of the seawater itself, they are literally only a "drop in the ocean," Nevertheless, despite the minute relative amounts, their importance must not be overlooked, for they provide us with some of the most useful clues to the nature of the atmosphere-lithosphere-hydrosphere interaction.

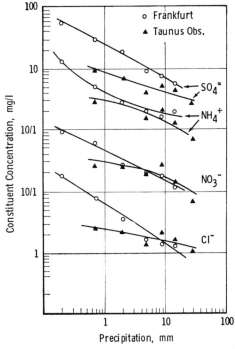

Figure 11.13 Concentration of Rainwater Constituents in Polluted Frankfurt and Unpolluted Taunus Observatory Atmospheres. From Junge (1963), with permission of Academic Press, Inc.

Some of these chemicals fall directly into the sea in solid form, others are washed in by rain. The chemical content of rainwater is highly variable. It depends strongly on the origin and history of the liquid particles and is subject to very marked perturbation by human pollution, both industrial pollution and radioactive fallout. A great deal of very detailed chemical data for rainwater constituents exists, but it is for the most part for rain collected over the landmasses. The most abundant, and frequently determined, constituents are Cl^-, SO_4^{2-}, NO_3^-, NO_2^-, HCO_3^-, Na^+, K^+, Ca^{2+}, Mg^{2+}, and NH_4^+ (Figure 11.13). In addition, rainwater may also contain more exotic compounds—P_2O_5, I_2, H_2O_2, and formaldehyde, and even organic and insoluble material, which may have served as the nucleation center for raindrop formation. Organic carbon in precipitation over Sweden amounted to 1.7–3.4 mg/l, and the concentration of insoluble matter can be comparable to that of the soluble material (Junge, 1963). Some of the rainwater constituents originate in the sea (see the next section) and are returned on precipitation. Thus it is not surprising to find that the rainwater content of these constituents of marine origin falls off with increasing distance from the coast (Figure 11.14).

The winds which sweep over the continental landmasses carry quantities of fine dust out over the oceans for great distances. This dust falls into the sea

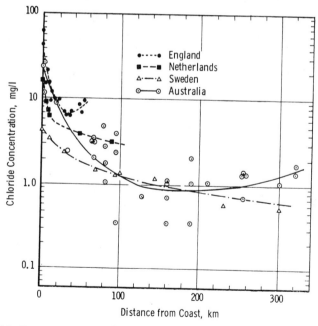

Figure 11.14   Decrease in the Chloride Content of Rainwater with Increasing Distance from the Coast. From Junge (1963), with permission of Academic Press, Inc.

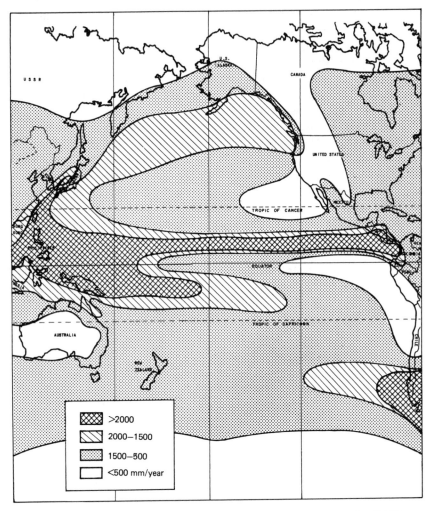

Figure 11.15   Annual Rainfall over the Pacific Ocean. From Arrhenius (1966).

and settles to the bottom, where it adds to the accumulation of the sediment (Arrhenius, 1961). Figure 11.15 shows the rate of rainfall over the Pacific; Figure 11.16 shows the biologically produced opaline silica in the same area; and, in contrast, Figure 11.17 shows the distribution of mica from the land-masses in the present sediment surface on the Pacific Ocean floor. Of particular interest is the Southern Pacific where, in the absence of complication by river-transported continental minerals, we can clearly trace the dust carried

OPALINE SILICA

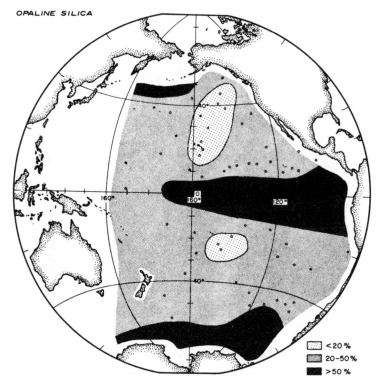

Figure 11.16   Present-Day Organic Productivity of Opaline Silica in the Pacific Ocean. From Bonatti and Arrhenius (in press), with permission of Prof. Arrhenius.

eastward in the "roaring forties" from Australia almost across the entire breadth of the ocean. The basaltic rocks of the Pacific basin contain little quartz; hence the quartz found in surface sediment samples, like the mica, is distributed over the waters by the wind (Rex and Goldberg, 1958).

Finally, two other interesting sources of material that enter the seas via the atmosphere are meteoric and volcanic substances (Rex and Goldberg, 1962; Arrhenius, 1961). The spherules of extraterrestrial origin are for the most part iron mixed with some cobalt and nickel (Castaing and Fredriksson, 1958; Petterson and Fredriksson, 1958), and their distribution both on the landmasses and in the sea is a subject which currently enjoys considerable publicity. "Tektites" (see O'Keefe, 1963) are glass-like particles. They are found both on land and in marine sediments (Glass and Heezen, 1967), but their origin is controversial. They may be molten terrestrial substance, splashed into the atmosphere on meteoric impact with Earth, or they may be material spallated from the surface of the Moon.

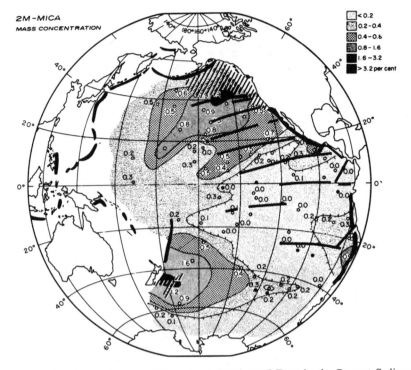

Figure 11.17 Concentration of Mica from Continental Dust in the Present Sediment Surface of the Pacific Ocean Floor. From Bonatti and Arrhenius (in press), with permission of Prof. Arrhenius.

Volcanoes spew solutes, insoluble particulate material, and gases into the sea, either directly if they are subsurface or via the atmosphere and the air-sea interface. Ash deposits represent a scarcely explored field that holds considerable promise of increasing our understanding of the transport and distribution of material in sea by both atmospheric and oceanic processes. In some instances archeological evidence and even historic testimony enable the exact date of the deposition to be established.

8   Bubbles and Chemical Mass Transport from the Sea into the Atmosphere

Figure 11.14 shows that the chloride content of rainwater is highest near the sea and falls off abruptly as the distance from the coast increases. Clearly, then, not only do sediment and rain carry chemicals into the sea, but also other processes must be at work which result in the effective transport of seawater constituents in the opposite direction into the atmosphere. On first thought, we might conclude that the violent action of waves and surf simply splashes seawater into the atmosphere to form the ion-containing marine aerosol. The salt content of the air above the sea remains nearly constant for Beaufort wind strengths of 0–3. Then, at about 3, the wind conditions when whitecaps appear, it increases sharply and goes through a maximum at about wind force 6 (Waite, 1929; Moore and Mason, 1954). But more subtle processes also seem to be at work in the formation of the marine aerosol, and bubble collapse at the sea's surface over vast expanses of open ocean probably is more important than the more spectacular but less frequent and localized violent storm conditions in contributing to the salt content of the atmosphere (Blanchard, 1963; Blanchard and Woodstock, 1957). Therefore we conclude this chapter on the air-sea interface with a brief consideration of the physical chemistry of bubbles in the ocean.

The origin of the free, visible bubbles in the sea is somewhat mysterious. Some bubbles originate from air entrapment by waves, but under "normal" circumstances the majority appear to result from the growth of microbubbles of gas already existing in the water column. The existence of such bubble nuclei is necessary, for in their absence very high gas partial pressures are needed to produce bubbles, pressures representing extremes of gas super-saturation far in excess of those occurring in the sea. There is no direct proof of the existence of invisible microbubbles, yet the ease of onset of cavitation in seawater would seem to testify to their presence (Liebermann, 1957). But the postulation of microbubbles only defers the question. What, then, is the origin of the microbubbles? The majority of guesses seem to center on micro-

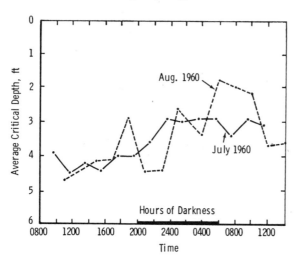

Figure 11.18   Hourly Changes in the Critical Depth for Bubble Growth off Mission Beach, California. From Ramsey (1962), with permission of *Limnology and Oceanography*.

bubble formation by gas sorption on the surface of particulate material, or by biological processes such as photosynthesis and respiration by marine organisms.

Taking the CO_2 contribution to be negligible because of that gas's greater solubility in seawater and the ratio of the N_2 content to the saturation solubility to be constant, Ramsey (1962) has derived an expression for the excess oxygen concentration C_{O_2} necessary to produce bubble formation:

$$(11.8) \qquad C_{O_2} = S_{O_2} (1.0062 + 0.14105d - 4.78P_{H_2O})$$

where S_{O_2} is the saturation value of O_2 from a dry atmosphere of normal composition at the given temperature, d is the depth in feet, and P_{H_2O} is the water vapor partial pressure. This expression (11.8) indicates that extraordinary O_2 supersaturation (in excess of 150% at 20°C) is necessary for bubble growth below about 5 ft. However, water motion, as in vortices, might be capable of creating local pressure reductions, and hence bubble growth, even in deeper water. The hourly variations of the average critical depth for bubble growth, calculated from O_2 saturation measurements, are shown in Figure 11.18.

The growing bubble rises to the sea's surface with an accelerated velocity, for its volume is increasing, not only by virtue of gas expansion as the hydrostatic pressure decreases, but also by virtue of increased gas diffusion into

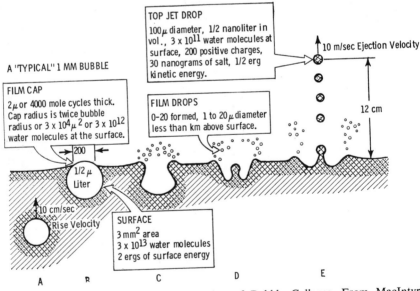

Figure 11.19 Diagrammatic Representation of Bubble Collapse. From MacIntyre (1965), with permission of the author.

the bubble across its enlarging surface area (Wyman, Scholander, et al., 1952).

Although the origin and motions of bubbles in the sea have been somewhat neglected, the bursting of bubbles on the sea's surface has now been investigated in rather considerable detail, and the subject has been reviewed by Blanchard (1963) whose immediate concern was the electrification of the atmosphere. When a bubble bursts, a fascinating and complex sequence of events takes place in rapid succession; see the diagrammatic representation in Figure 11.19. This figure summarizes the experimental results obtained with the apparatus shown in Figure 4.7 and also the findings of high-speed photography. Bubble rupture ejects two types of droplets from the sea's surface into the atmosphere: small droplets representing the remains of the actual atmosphere exposed film droplets of the bubble, and large jet drops which rise rapidly from the bottom of the bubble.

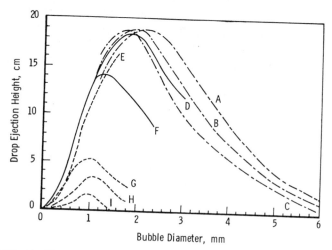

Figure 11.20 Jet Drop Height as a Function of Bubble Diameter. From Blanchard (1963), with permission of the author.

Curve	Drop Position on Jet	Water Temperature, °C	Water Type
A	Top	4	Sea
B	Top	16	Sea
C	Top	30	Sea
D	Top	22–26	Sea
E	Top	4	Sea
F	Top	21	Dist.
G	Second	22–26	Sea
H	Third	22–26	Sea
I	Fourth	22–26	Sea

Because of their greater mass the jet drops tend to fall rapidly back into the sea, whereas the much smaller film droplets are caught up by the wind and are thus the more important contributors to the marine aerosol (Mason, 1954). Figure 11.20 shows the dependence of the height reached by the ejected jet drops on bubble diameter. For bubbles of diameter less than 2 mm the ejection height increases with temperature, whereas for larger bubbles it decreases. Salinity has little effect on ejection height. In Figure 11.21 we see the dependence of droplet size and salt content on bubble size. The total amount of salt transported into the atmosphere from bubble bursting is not inconsequential; at a rate of 10^{-12} g/cm^2/day it amounts to some 10^9 to 10^{10} metric tons per year (Eriksson, 1957; Blanchard, 1963).

The sea salt content of the air is shown in Figure 11.22 as a function of altitude, while Figure 11.23 compares the dependencies of drop production and sea salt fallout on wind speed.

In addition to salt, electric charge is also transported into the atmosphere over the sea by bubble bursting. The oceanic charge separation reaches a maximum of 3.2×10^{-7} esu cm^{-2} sec^{-1} at about 50°S latitude in June–August, and Blanchard (1963) has attributed the ejection of positive electrification to the faster flow of the positive region of the electrical double layer at the

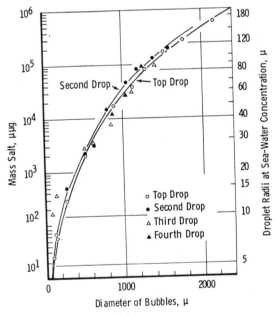

Figure 11.21 Size and Salt Content of Droplets Ejected by Bursting Bubbles in Seawater. From Woodcock (1962), with permission of John Wiley & Sons.

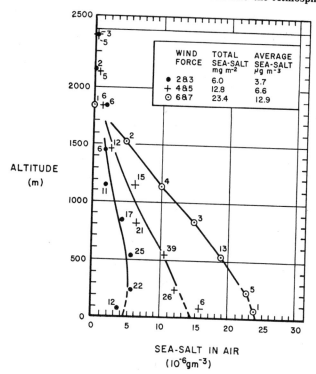

Figure 11.22   Sea Salt Content as a Function of Altitude. From Woodcock (1962), with permission of John Wiley & Sons,

surface (see next chapter) into the rising column from which the jet drops separate (Figure 11.24).

Rossby (1959) has noted that the chemical constituents of seawater "leave the sea in different proportions than those characteristic of seawater," and he has suggested that surface chemistry probably plays an important role. Blanchard (1963) has amplified these remarks: "It must be remembered that the (aerosol) nuclei which originate in the sea came not from the bulk water, but from the surface layer whose chemistry may well be different from that of the bulk water." Table 11.14 lists values of ion ratios which differ appreciably in the aerosol from their sources. In some instances the aerosol values are analyses of precipitation, whereas in others the aerosol was prepared by distillation or by bubbling air through seawater and other solutions in the laboratory. Not included in the table are constituents possibly concentrated in the aerosol by specific chemical processes, such as iodine, mentioned earlier, boron concentrated by evaporation of boric acid (Gast and Thompson, 1959), and sulfate. The SO_4^{2-}/Cl^- ratio is always higher over the sea than in seawater,

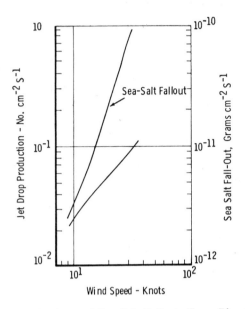

Figure 11.23 Drop Production and Sea Salt Fallout. From Blanchard (1963), with permission of Pergamon Press, Ltd.

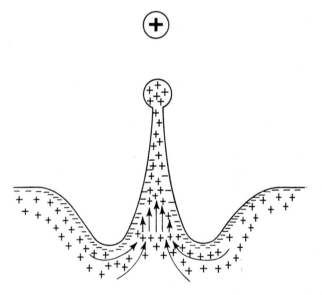

Figure 11.24 Mechanism of Ejection of Positive Electrification from the Electrical Double Layer by Jet Droplets. From Blanchard (1963), with permission of Pergamon Press, Ltd.

352

Table 11.14   Ion Fractionation at the Interface (A = aerosol, 0 = original solution)

Ratio	Value	Comment	Reference
Na^+/K^+	27	Seawater	
Na^+/K^+	18	Rain, southwest coast, Iceland	Eriksson (1960)
Na^+/K^+	16.5	Rain, southwest coast, Norway	Eriksson (1960)
Na^+/K^+	6.3	Rain, westcoast, Ireland	Gorham (1957)
Na^+/K^+	35.2	Rain, coast, Tel-Aviv	Bloch et al. (1966)
Na^+/K^+	6.7	Rain, Jerusalem, Israel	Bloch et al. (1966)
Na^+/K^+	10.4	Rain, Lodd, Israel	Bloch et al. (1966)
Na^+/K^+	9.3	Rain, Berr-sheva, Israel	Bloch et al. (1966)
Na^+/K^+	14.6	Rain, Duphira, Israel	Bloch et al. (1966)
Na^+/K^+	1.11	Snow, Rasman Glacier	Wilson (1959)
Na^+/K^+	2.83	Snow, Mt. Egmont	Wilson (1959)
Na^+/K^+	2.0	Snow, Arthur Pass	Wilson (1959)
Cl^-/Br^-	302	Seawater	—
Cl^-/Br^-	52.9	Rain, Jerusalem	Bloch et al. (1966)
Cl^-/Br^-	44.5	Rain, Jerusalem	Bloch et al. (1966)
Cl^-/Br^-	287	Rain, Tel-Aviv, Israel	Bloch et al. (1966)
Cl^-/Br^-	121	Rain, Haifa-Hadar	Bloch et al. (1966)
Cl^-/Br^-	74.5	Rain, Daphna	Bloch et al. (1966)
Cl^-/Br^-	26.1	Rain, Ein Harod	Bloch et al. (1966)
Cl^-/Br^-	12.3	Rain, Sde Boker	Bloch et al. (1966)
Cl^-/Br^-	48.4	Rain, Nicosia	Bloch et al. (1966)
Cl^-/Br^-	9.48	Rain, Prodronos	Bloch et al. (1966)
Cl^-/Br^-	86.3	Rain, Göttingen	Behne (1953)
Cl^-/Br^-	37.8	Snow, Bolschewo	Selivanoff (1939)
$\dfrac{(Cl^-/Br^-)_A}{(Cl^-/Br^-)_0}$	$\dfrac{25}{304}$	Distillation of seawater	Bloch et al. (1966)
$\dfrac{(Cl^-/Br^-)_A}{(Cl^-/Br^-)_0}$	$\dfrac{12.1}{304}$	Distillation of seawater without spray trap	Bloch et al. (1966)
$\dfrac{(Cl^-/Br^-)_A}{(Cl^-/Br^-)_0}$	$\dfrac{8.2}{114}$	Distillation, Lake of Kinereth water	Bloch et al. (1966)
$\dfrac{(Cl^-/Br^-)_A}{(Cl^-/Br^-)_0}$	$\dfrac{4.6}{114}$	Distillation, Lake of Kinereth water without spray trap	Bloch et al. (1966)
$\dfrac{(Cl^-/Br^-)_A}{(Cl^-/Br^-)_0}$	$\dfrac{11.2}{44}$	Distillation, Dead Sea water	Bloch et al. (1966)
$\dfrac{(Cl^-/Br^-)_A}{(Cl^-/Br^-)_0}$	$\dfrac{11.1}{44}$	Distillation, Dead Sea water without spray trap	Bloch et al. (1966)
$\dfrac{(Cl^-/Br^-)_A}{(Cl^-/Br^-)_0}$	$\dfrac{13}{310}$	Distillation, Jerusalem trap water	Bloch et al. (1966)
$\dfrac{(Cl^-/Br^-)_A}{(Cl^-/Br^-)_0}$	$\dfrac{172}{300}$	Air sweep through sea water	Bloch et al. (1966)
	$\dfrac{151}{300}$	—	—
	$\dfrac{120}{300}$	—	—

Table 11.14 (cont.)

Ratio	Value	Comment	Reference
$\dfrac{(Na^+/K^+)_A}{(Na^+/K^+)_0}$	$\dfrac{5.7}{27}$	Air sweep through seawater	Bloch et al. (1966)
$\dfrac{(Cl^-/Br^-)_A}{(Cl^-/Br^-)_0}$	$\dfrac{98}{364}$	Air sweep through artificial seawater	Bloch et al. (1966)
$\dfrac{(Cl^-/Br^-)_A}{(Cl^-/Br^-)_0}$	$\dfrac{117}{400}$	Air sweep through artificial seawater with added NaCl	Bloch et al. (1966)
$\dfrac{(Cl^-/Br^-)_A}{(Cl^-/Br^-)_0}$	$\dfrac{101}{400}$	Air sweep through artificial seawater with added NaCl	Bloch et al. (1966)
$\dfrac{(Na^+/K^+)_A}{(Na^+/K^+)_0}$	$\dfrac{7.2}{27}$	Air sweep through artificial seawater	Bloch et al. (1966)
$\dfrac{(Mg^{2+}/Pb^{2+})_A}{(Mg^{2+}/Pb^{2+})_0}$	$\dfrac{2.0}{116}$	Air sweep through nitrate solutions	Bloch et al. (1966)
$\dfrac{(Na^+/Ca^{2+})_A}{(Na^+/Ca^{2+})_0}$	$\dfrac{5.4}{10}$	Air sweep through chloride solutions	Bloch et al. (1966)
$\dfrac{(Na^+/Ca^{2+})_A}{(Na^+/Ca^{2+})_0}$	$\dfrac{0.54}{0.10}$	Air sweep through chloride solutions	Bloch et al. (1966)
(Cl^-/Na^+)	1.80	In seawater	—
(Cl^-/Na^+)	1.90–2.30	Rain, Netherlands	Leeflang (1938)
(Cl^-/Na^+)	1.17–1.69	Rain, Shetland Islands	Oddie (1959)
(Cl^-/Na^+)	1.71	Rain, Norway coast	Eriksson (1960)
(Cl^-/Na^+)	1.78	Rain, Iceland coast	Eriksson (1960)
(Cl^-/Na^+)	1.62	Rain, Shetland Islands	Eriksson (1960)
(Cl^-/Na^-)	1.63	Rain, Netherlands coast	Eriksson (1960)
(Cl^-/Na^+)	1.79	Rain, England	Eriksson (1960)
(Cl^-/Na^+)	1.90	Rain, Hawaii	Eriksson (1957)
(Cl^-/Na^+)	1.72	Rain, Bermuda	Junge and Werby (1958)
(Cl^-/Na^+)	1.71	Rain, Newfoundland	Junge and Werby (1958)
(Cl^-/Na^+)	1.75	Rain, Cape Cod	Junge and Werby (1958)
(Cl^-/Na^+)	1.54	Rain, Cape Hatteras	Junge and Werby (1958)
(Cl^-/Na^+)	1.58	Rain, Olympic Peninsula	Junge and Werby (1958)
(Cl^-/Na^+)	1.53	Rain, San Diego	Junge and Werby (1958)
(Cl^-/Na^+)	1.51	Rain, Miami	Junge and Werby (1958)
(Cl^-/Na^+)	1.61	Rain, Boston	Junge and Werby (1958)
(Cl^-/Na^+)	0.61	Rain, Mid-U.S. average	Junge and Werby (1958)
K^+/Na^+	0.036	In seawater	Junge and Werby (1958)
K^+/Na^+	0.050	Rain, Bermuda	Junge and Werby (1958)
K^+/Na^+	0.062	Rain, Newfoundland	Junge and Werby (1958)
K^+/Na^+	0.060	Rain, Cape Cod	Junge and Werby (1958)
K^+/Na^+	0.053	Rain, Cape Hatteras	Junge and Werby (1958)
K^+/Na^+	0.041	Rain, Olympic Peninsula	Junge and Werby (1958)
K^+/Na^+	0.097	Rain, San Diego	Junge and Werby (1958)
K^+/Na^+	0.062	Rain, Miami	Junge and Werby (1958)
K^+/Na^+	0.183	Rain, Boston	Junge and Werby (1958)
K^+/Na^+	0.67	Rain, Mid-U.S. average	Junge and Werby (1958)
(Mg^{2+}/Na^+)	0.12	In seawater	—

Table 11.14 (cont.)

Ratio	Value	Comment	Reference
(Mg^{2+}/Na^+)	0.14	Rain, Iceland	Eriksson (1960)
(Mg^{2+}/Na^+)	0.13	Rain, Shetland Islands	Eriksson (1960)
(Mg^{2+}/Na^+)	0.17	Rain, Norway	Eriksson (1960)
(Mg^{2+}/Na^+)	0.17	Rain, Netherlands	Eriksson (1960)
(Mg^{2+}/Na^+)	0.16	Rain, Ireland	Gorham (1957)
(Ca^{2+}/Na^+)	0.038	In seawater	—
(Ca^{2+}/Na^+)	0.40	Rain, Bermuda	Junge and Werby (1958)
(Ca^{2+}/Na^+)	0.15	Rain, Newfoundland	Junge and Werby (1958)
(Ca^{2+}/Na^+)	0.089	Rain, Cape Cod	Junge and Werby (1958)
(Ca^{2+}/Na^+)	0.098	Rain, Cape Hatteras	Junge and Werby (1958)
(Ca^{2+}/Na^+)	0.051	Rain, Olympic Peninsula	Junge and Werby (1958)
(Ca^{2+}/Na^+)	0.38	Rain, San Diego	Junge and Werby (1958)
(Ca^{2+}/Na^+)	0.35	Rain, Miami	Junge and Werby (1958)
(Ca^{2+}/Na^+)	0.41	Rain, Boston	Junge and Werby (1958)
(Ca^{2+}/Na^+)	0.110	Rain, Iceland	Eriksson (1960)
(Ca^{2+}/Na^+)	0.050	Rain, Shetland Islands	Eriksson (1960)
(Ca^{2+}/Na^+)	0.140	Rain, Norway	Eriksson (1960)
(Ca^{2+}/Na^+)	0.085	Rain, Netherlands	Eriksson (1960)
(Ca^{2+}/Na^+)	0.530	Rain, Iceland	Gorham (1957)
$\dfrac{(PO_4{}^{3-}/Na^+)_A}{(PO_4{}^{3-}/Na^+)_0}$	As great as 700	Bubbling air through	MacIntyre (1965)
$\dfrac{(PO_4{}^{3-})_A}{(PO_4{}^{3-})_0}$	1.41	Bubbling air through	Sutcliffe et al. (1963)

but even higher as one proceeds inland over the continents. Pollution is probably a factor, and the high values sometimes found in coastal areas have led to speculation that its source might be H_2S produced in marine muds and stagnant water (see Junge, 1963). The whole question of ionic fractionation at the sea's surface is further clouded by the unknown, but possibly crucial, role of surface-concentrated organic matter. Yet these difficulties have not deterred the formulation of theories, including differences in ionic absorption at the surface, fractional crystallization during dehydration in the atmosphere, etc. Komabayasi (1962) claims that the enrichment of ions correlates with their weights (Figure 11.25) and has proposed a "thermo-gravitational" mechanism which entails longer residence times of the heavier ions on the bubble film (see also Sugawara and Kawasaki, 1958); others appear to have adopted this hypothesis and its variants (Martin and Lübke, 1964; Bloch et al., 1966). If I may interject my own prejudices at this point, I must say that this model of light ions sliding down the bubble film faster than heavy ones overtaxes my imagination. Remembering Blanchard's comment, quoted above, that the seawater which is ejected into the atmosphere is surface, not bulk, seawater, and in the light of our earlier discussion of the changes in the structure of

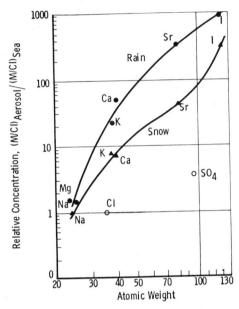

Figure 11.25   Correlation of Ionic Fractionation in the Atmosphere-Sea System with Atomic Weight. From M. Kombayasi (1962), with permission of Meteorological Society of Japan and the author.

liquid water as the interface is approached, it would be most surprising indeed if the ionic composition of the aerosol were the same as the sea from which it originates. While there certainly may be complications accompanying the phenomenon, I feel that its basis is simply the preferential exclusion, or call it selective negative absorption if you prefer, of electrolyte by the more highly structured region near the interface. And I further suspect that the extent to which ions are excluded, and thus their concentration ratios in the aerosol, in turn depend on the degree to which they tend to disrupt the structure that the surface water prefers. Thus one expects a correlation between the ionic fractionation at the interface and those ionic properties which reflect the capability of the ions to alter water structure. In this connection it is interesting to note that, when the K^+/Na^+ ratio in seawater, 0.036, is multiplied by (4.8/3.4), the ratio of the radii of the hydrated cations near a surface as derived from ion-exclusion experiments (McConnell, Williams et al., 1964) and which I have reinterpreted as a direct measure of ion exclusion (Horne, 1966), a value of 0.051 is obtained which is in good agreement with the value of 0.053 for the average K^+/Na^+ ratio in coastal precipitation reported by Junge and Werby (1958). Perhaps this result is fortuitious, but, then again, perhaps it is not. In any event, a systematic investigation of the ionic fractionation of the whole family of alkali metal chlorides is in progress in our laboratories at the time of this writing.

Before bringing this chapter to a close there is one last topic that I should like to treat—the concentration of organic material on bubbles. Dissolved organic substances in seawater appear to sorb on the surface of bubbles in sufficient quantities to form colloidal micelles even particulate forms on bubble collapse (Sutcliffe, Baylor, and Menzel, 1963; Baylor and Sutcliffe, 1963; Riley, 1963; Menzel, 1966; Barber, 1966), and these particles may be an important link in the food chain in the sea. The particles have a high phosphorus content, and their formation by bubbling simultaneously removes phosphate from seawater (Sutcliffe, Baylor, and Menzel, 1963; Baylor, Sutcliffe, and Hirschfeld, 1962; MacIntyre, 1965). However, the sorption of cations such as Zn^{2+}, Mn^{2+}, and Ca^{2+} on bubble-produced organic aggregates does not seem to be appreciable (Siegel and Burke, 1965), nor is the sorption of Sr^{2+} large enough to account for the reported maximum in Sr^{2+}/Cl in the water column (MacKenzie, 1964).

9   Summary

The interfacial physical chemistry occurring at the boundaries of the hydrosphere with the lithosphere and atmosphere form a very large and important part of marine chemistry.

The surface tension and escape energetics of liquid water are extraordinarily high—evidence of a strengthening of the water structure, that is, the intermolecular attractive forces, near the interface. At the interface the waters are oriented with their O's directed away from the bulk liquid phase.

The interfacial water structure appears to be different from that of the Frank-Wen clusters in liquid water, and evidence now appears to indicate that the region of interfacial perturbation penetrates deeply into the bulk phase.

The sea's surface is frequently covered with a surface film of organic material, but, inasmuch as the action of wind and waves is very effective in destroying this film, its importance to the surface chemistry of the sea remains unclear.

Of the several forms of energy transferred across the air-sea interface, by far the most important is solar energy. Water is transported across this boundary by evaporation and precipitation, and these transport processes can give rise to isotopic fractionations both of D/H and O^{18}/O^{16}.

The rate-determining step of gas exchange across the air-sea interface is, ideally, diffusion through the liquid boundary layer. The rate of exchange in either direction is strongly dependent on the state of agitation of the system and then on such parameters as wind velocity and sea state.

The composition of the aerosol above the sea may be appreciably different from that of the parent seawater, presumably because of interfacial ion fractionation processes accompanying the bursting of small bubbles at the sea's surface.

REFERENCES

Anderson, E. R., L. J. Anderson, and J. J. Marciano, "A Review of Evaportaion Theory and Development of Instrumentation," U.S.N. *Electronics Lab. Rept.* No. 159 (1950).

Arrhenius, G., in M. Sears, ed., *Oceanography*, Am. Assoc. Advan. Sci., Publ., No. 67, Washington, D.C., 1961.

Arrhenius, G. "Sedimentary Record of Long Range Phenomena," in P. R. Hurley, ed., *Advances in Earth Science*, M.I.T. Press, Cambridge, Mass., 1966.

Barber, R. T., *Nature*, **211**, 257 (1966).

Baylor, E. R., W. H. Sutcliffe, and D. S. Hirschfeld, *Deep-Sea Res.*, **9**, 120 (1962).

Baylor, E. R., and W. H. Sutcliffe Jr., *Limnol. Oceanog.*, **8**, 369 (1963).

Behne, W., *Geochim. Cosmochim. Acta.*, **3**, 208 (1953).

Bhandari, R. C., and M. L. Sisodia, *Indian J. Pure Appl. Phys.*, **2**, 266 (1964).

Blanchard, D. C., *Progr. Oceanog.*, **1**, 73 (1963).

Blanchard, D. C., *Science*, **146**, 396 (1964).

Blanchard, D. C., and A. H. Woodstock, *Tellus.*, **9**, 145 (1957).

Bloch, M. R., D. Kaplan, V. Kertes, and J. Schnerb, *Nature*, **209**, 802 (1966).

Bolin, B., *Tellus*, **12**, 274 (1960).

Bonatti, E., and G. Arrhenius, *Marine, Geology* in press.

Broecker, W. S., and A. Walton, *Abstr. Intern. Oceanog. Congr.*, *1959*, Am. Assoc. Advan. Sci., Washington D.C., 1959, p. 856.

Castaing, R., and K. Fredriksson, *Geochim. Cosmochim. Acta.*, **14**, 114 (1958).

Claussen, W. F., *Science*, **156**, 1226 (1967).

Craig, H., *Tellus*, **9**, 1 (1957).

Craig, H., and L. I. Gordon, *Proc. Symp. Marine Geochim. Univ. Rhode Island, Publ.*, No. 3, 277 (1965); see also *Proc. Conf. Stable Isotopes in Oceanog. Studies*, E. Tongiorgi, ed., Spoleto, Italy, 1965.

Deacon, E. L., and E. K. Webb, "Small-Scale Interactions," in M. N. Hill, ed., *The Sea*, Interscience, New York, 1962, Vol. 1, Chap. 3.

Defant, A., *Physical Oceanography*, Pergamon Press, New York, 1961, Vol. 1.

Dietrich, G., *General Oceanography*, Interscience, New York, 1963.

Dorsey, N. E., *Properties of Ordinary Water-Substance*, Reinhold Pub.

Downing, A. L., and G. A. Truesdale, *J. Appl. Chem.*, **5**, 570 (1955).

Drost-Hansen, W., *Ind. Eng. Chem.*, **57**, No. 4, 18 (1965).

Duce, R. A., J. T. Wasson, J. W. Winchester, and F. Burns, *J. Geophys. Res.*, **68**, 3943 (1963).

Duce, R. A., J. W. Winchester, and T. W. Van Nahl, *J. Geophys. Res.*, **70**, 1775 (1965).

Epstein, S. R. Buchsbaum, H. A. Lowenstam, and H. C. Urey, *Bull. Geol. Soc. Am.*, **62**, 417 (1953).

Epstein, S., and T. Mayeda, *Geochim. Cosmochim. Acta*, **4**, 213 (1953).

Eriksson, J. C., *Acta, Chem. Scand.*, **16**, 2199 (1962).

Eriksson, E., *Tellus*, **9**, 509 (1957).

Eriksson, E., *Tellus*, **12**, 63 (1960).

Ewing, G., *J. Marine Res. (Sears Found. Marine Res.)*, **9**, 161 (1950).

Ewing, G., and E. D. McAllister, *Science*, **131**, 1374 (1960).

Fleming, R. M., and R. Revelle, "Physical Processes in the Ocean," P. D. Trask, ed., *Recent Marine Sediments*, Am. Assoc. *Petroleum Geologists*, Tulsa, 1939, p. 48.

Fletcher, N. H., *Phil. Mag.*, **7**, 255 (1962).

Frenkel, J., *Kinetic Theory of Liquids*, Dover, Pub, New York, 1955, Chap. 6.

Friedman, I., *Geochim. Cosmochim. Acta*, **4**, 89 (1953).

Friedman, I., J. Harris, and B. Schoen, *J. Geophys. Res.*, **66**, 1861 (1961).

Friedman, I., A. C. Redfield, B. Schoen, and J. Harris, *Rev. Geophys.*, **2**, 177 (1964).

Garrett, W. D., "The Organic Chemical Composition of the Ocean Surface," U.S.N. Res. Lab. Rept. No. 6201 (Dec. 24, 1964) (unclass.).

Garrett, W. D., The Organic Chemical Composition of the Ocean Surface, U.S.N. Res. Lab. Rept. No. 6201 (Dec. 24, 1964).

Gast, J. A., and T. G. Thompson, *Tellus*, **11**, 344 (1959).

Glass, W., and B. C. Heezen, "*Tektites in Deep-Sea Sediments*, Paper 052 presented at 48th Ann. Meeting, Am. Geophys. Union, Washington, D.C., April 1967.

Goering, J. J., and D. W. Menzel, *Deep-Sea Res.*, **12**, 839 (1965).

Goldacre, A. J., *J. Animal Ecology*, **18**, 36 (1949).

Good, R. J., *J. Phys. Chem.*, **61**, 810 (1957).

Gorham, E., *Irish Naturalists' J.*, **12**, 1 (1957).

Guastalla, J., *J. Chim. Phys.*, **44**, 306 (1947).

Henniker, J. C., *Rev. Mod. Phys.*, **21**, 322 (1949).

Horibe, Y., and M. Kobayakawa, *Geochim. Cosmochim. Acta*, **20**, 273 (1960).

Horne, R. A., *J. Phys. Chem.*, **70**, 1335 (1966).

Horne, R. A., *Surv. Progr. Chem.*, 7, 1 (1968).

Horne, R. A., A. F. Day, R. P. Young, and N.-T. Yu, *Electrochem. Octa*, **13**, 397 (1968).

Horne, R. A., *Surv. Progr. Chem.*, in press (1968).

Horne, R. A., and D. S. Johnson, *J. Phys. Chem.*, **71**, 1936 (1967).

Jarvis, N. L., *J. Colloid Sci.*, **17**, 512 (1962).

Jarvis, N. L., W. D. Garrett, M. A. Scheiman, and C. O. Timmons, *Limnol. Oceanog.*, **12**, 88 (1967).

Junge, C. E., *Air Chemistry and Radioactivity*, Academic Press, New York, 1963.

Junge, C. E., and R. T. Werby, *J. Meteorol.*, **15**, 417 (1958).

Kanwisher, J., *Deep-Sea Res.*, **10**, 195 (1963).

Kavanau, J. L., *Water and Solute-Water Interactions*, Holden Day, Inc., San Francisco, Calif., 1964.

Keeling C. D., *J. Geophys. Res.*, **70**, 6099 (1965).

Kirshenbaum, I., *Physical Properties and Analysis of Heavy Water*, McGraw-Hill Book Co., New York, 1951.

Koczy, F. F., *Intern. Sci. Technol.*, No. 60, 52 (Dec. 1966).

Komabayasi, M., *J. Meteorol. Soc. Japan*, **40**, 25 (1962).

La Mer, V. K., and T. W. Healy, *Science*, **148**, 36 (1965).

Leeflang, K. W. M., *Chem. Weekblad*, **35**, 658 (1938).

Liebermann, L., *J. Appl. Phys.*, **28**, 205 (1957).

Lu, W. C., M. S. John, T. Ree, and H. Eyring, *J. Chem. Phys.*, **46**, 1075 (1967).

MacIntyre, F., Ph.D. Thesis, M.I.T., 1965.

MacKenzie, F. T., *Science*, **146**, 517 (1964).

MacIntyre, F., Ph.D. Thesis, M.I.T., 1965. Cambridge, Mass.

Martin, H., and H. J. Lübke, *Z. Naturforsch.*, **19A**, 115 (1964).

Mason, B. J., *Nature*, **174**, 470 (1954).

McConnell, B. L., K. C. Williams, J. L. Daniel, J. H. Stanton, B. N. Irby, D. L. Drugger, and R. W. Maatman, *J. Phys. Chem.*, **68**, 2941 (1964).

Menzel, D. W., *Deep-Sea Res.*, **13**, 963 (1966).

Miyake, Y., and A. Hamada, *Oceanog. Congr.*, *Helsinki Comm.*, K13 (1960).

Miyake, Y., and I. Tsunogai, *J. Geophys. Res.*, **68**, 3989 (1963).

Moore, D. J., and B. J. Mason, *Quart. J. Roy. Meteorol. Soc.*, **80**, 583 (1954).

O'Keefe, J. A., ed., *Tektites*, Univ. Chicago Press, Chicago, 1963.

Oddie, B. C. V., *Quart. J. Roy. Meteorol. Soc.*, **89**, 163 (1959).

Onsager, L., and N. Samaras, *J. Chem. Phys.*, **2**, 528 (1934).

Park, K., and D. W. Hood, *Limnol. Oceanog.*, **8**, 287 (1963).

Parsons, T. R., *Progr. Oceanog.*, **1**, 203 (1964).

Pettersson, H., and K. Fredriksson, *Pacific Sci.*, **12**, 71, (1958).

Pytkowicz, R. M., *Deep-Sea Res.*, **11**, 381 (1964).

Ramsey, W. L., *Limnol. Oceanog.*, **7**, 1 (1962).

Randles, J. E. B., *Discussions Faraday Soc.*, **24**, 194 (1957).

Randles, J. E. B., "The Interface between Aqueous Electrolyte Solutions and the Gas Phase, in P. Delahey, ed., *Electrochemistry*, John Wiley & Sons, New York, 1963.

Redfield, A. C., *J. Mar. Res.* (*Sears Found. Marine Res.*), **7**, 347 (1948).

Redfield, A. C., and I. Friedman, *Proc. Symp. Marine Geochem.*, *Univ. Rhode Island Pub.*, No. 3, 149 (1965).

Revelle, R., and H. E. Suess, *Tellus*, **9**, 18 (1957).

Rex, R. W., and E. D. Goldberg, *Tellus*, **10**, 153 (1958).

Rex, R. W., and E. D. Goldberg, "Insolubles," in M. N. Hill, ed., *The Sea*, Interscience, New York, 1962, Vol. 1, Chap. 5.

Riley, G. A., *Limnol. Oceanog.*, **8**, 372 (1963).

Schrage, R. W., *A Theoretical Study of Interphase Mass Transfer*, Columbia Univ. Press, New York, 1953.

Schufle, J. A., and M. Venugupalan, *J. Geophys. Res.*, **72**, 3271 (1967).

Selivanoff, L. S., *Tr. Biogeokhim. Labor. Akad. Nauk. SSSR.*, **5**, 132 (1939).

Siegel, A., and B. Burke, *Deep-Sea Res.*, **12**, 789 (1965).

Skirrow, G., "The Dissolved Gases–Carbon Dioxide," in J. P. Riley and G. Skirrow, eds., *Chemical Oceanography*, Academic Press, London, 1965, Vol. 1, Chap. 7.

Slowinski, E. J., Jr., E. E. Gates, and C. E. Waring, *J. Phys. Chem.*, **61**, 808 (1957).

Sugawara, K., and N. Kawasaki, *Records Oceanog. Works Japan, Spec. Publ.*, No. 2, 227 (1958).

Suguira, Y., E. I. Ibert, and D. W. Hood, *J. Marine Res.* (*Sears Found. Marine Res.*), **21**, 11 (1963).

Sutcliffe, W. H., Jr., E. R. Baylor, and D. W. Menzel, *Deep-Sea Res.*, **10**, 232 (1963).

Sverdrup, H. U., "Evaporation from the Oceans," in *Compendium of Meteorology*, Am. Met. Soc., 1951.

Sverdrup, H. U., M. W. Johnson, and R. H. Fleming, *The Oceans*, Prentice-Hall, Englewood Cliffs, N.J., 1942.

Tyler, J. E., and R. W. Preisendorfer, "Light," in M. N. Hill, ed., *The Sea*, Interscience, New York, 1962, Vol. 1.

Urey, H. C., *J. Chem. Soc.*, **562** (1947).

Waite, G. R., *Carnegie Inst. Wash. Yearbook*, **28**, 271 (1929).

Weston, R. E., Jr. *Geochim. Cosmochim. Acta.*, **8**, 281 (1955).

Wicke, E., *Angew. Chem.*, **5**, 106 (1966).

Wilson, A. T., *Nature*, **184**, 99 (1959).

Wyman, J., P. F. Scholander, G. A. Edwards, and L. Irving, *J. Marine Res.*, **11**, 47 (1952).

12 The Sea-Ocean Bottom Interface

1 Introduction

The second interface which forms one of the two vast boundaries of the ocean system is that between the Earth's hydrosphere and lithosphere, the aqueous electrolytic solution-solid phase boundary largely at the bottom of the seas. In Figure 11.4 we saw the several energy and mass transfer processes across the air-sea boundary; similarly Figure 12.1 summarizes several of the more important mass transfer processes associated with the hydrosphere-lithosphere interaction. Just as in Chapter 11 we avoided the chemistry of this planet's atmosphere and restricted our discussion to those processes involving mass transport across and chemical changes at the air-sea interface, so here we will avoid the countless and exceedingly complex problems of geochemistry and limit our attention narrowly to those processes directly involved in the interaction of seawater with suspended and sedimented solid material. Also I do not plan to explore the question of mass transport through biomembranes, despite its overwhelming importance in the life process—although life does of course, as I have noted, represent a heterogeneity in the sea—for to do so in any measure of detail would carry us too far afield.

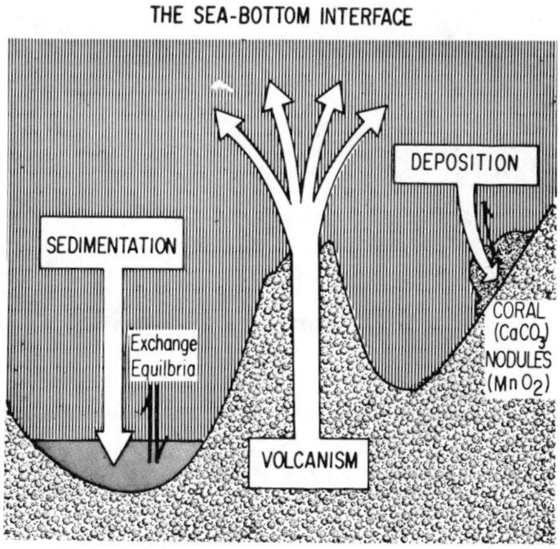

THE SEA-BOTTOM INTERFACE

DEPOSITION

SEDIMENTATION

CORAL
(CaCO$_3$)
NODULES
(MnO$_2$)

Exchange
Equilbria

VOLCANISM

Figure 12.1   Some Important Mass Transfer Processes of the Hydrosphere-Lithosphere Interaction.

361

2   The Electrical Double Layer and the Structure of Water at the Solution-Solid Interface

We begin this chapter by returning a third and final time to that most fundamental problem of chemical oceanography—the structure of liquid water and aqueous electrolytic solutions. In the first chapter we reviewed the current theories of the nature of the bulk water; in the previous chapter on the chemistry of the air-sea boundary we attempted to describe a qualitative model of the structure of water and of solutions near an interface; and now, for the most part on the basis of relatively recent studies of the solution-solid interface, we shall try to add more detail to our picture of interfacial water structure.

Paradoxically, the solution-solid interfacial situation about which we have the most detailed information is a relatively complex one, namely, the distribution of ions in aqueous solutions near a charged metal surface. The situa-

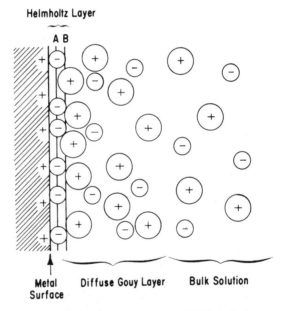

Helmholtz Layer

A B

Metal Surface   Diffuse Gouy Layer   Bulk Solution

A.   Inner Helmholtz or Stern Plane of Anionic Centers

B.   Outer Helmholtz or Limiting Gouy Plane of Cationic

Distance of Closest Approach

Figure 12.2   The Electrical Double Layer. From Horne (1968), with permission of Academic Press, Inc.

tion constitutes the heart of the science of electrochemistry and is of enormous fundamental and applied significance. As a consequence it has been the subject of extensive study, both theoretical and experimental, and from this effort has gradually evolved the theory of the electrical double layer (see Delahay, 1965, Conway, 1965; Devanthan and Tilak, 1965). Qualitatively the picture which has emerged (Figure 12.2) is one of two layers of oppositely charged ions, their order depending on the charge of the solid surface. But, for all the effort expended, our understanding of the electrical double layer is still clearly very incomplete. For example, hitherto attention has been largely focused on the spatial configuration of the ions, and few have had the courage to confront the even more difficult problem of how the water molecules arrange themselves in the double layer. An exception has been Bockris, Devanthan, and Müller (1963) and their effort to say something more specific about what's happening to the water molecules in the double layer (Figure 12.3) represents still, I believe, the deepest penetration of the question.

In addition to studies of surface tension and potential and of the electrical double layer, a third source of clues to the structure of water near interfaces

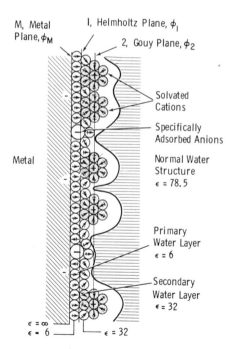

Figure 12.3 Model for the Position and Orientation of Water Molecules in the Electrical Double Layer. From Bockris, Devanathan, Müller (1963), with permission of the Royal Society.

has been the investigation of the properties of liquid water and aqueous solutions in porous materials and very thin capillaries—and very remarkable indeed these properties are! Many years ago Bijl (1927) noted that the temperature of maximum density of a mixture of water and finely divided charcoal is lower than for pure water. This finding was interpreted in terms of the theory of liquid water then current, namely, the concentration of water polymers is different in the layer near the surface from the concentration in the bulk water. Even earlier Duff (1905) reported that liquids flowing through capillaries exhibit abnormally high viscosities, and Wolkowa (1934), working with clay and silica gel, found that the phenomenon was particularly marked in the case of water (for a review see Henniker, 1949). But perhaps the most unexpected property of liquid water in porous materials and thin capillaries is its extreme reluctance to solidify. Such water can be cooled to $-40°C$ before it finally solidifies. This and related phenomena have been the subject of recent studies (Deryagin and Fedyakin, 1962; Chahal and Miller, 1965; Schufle, 1965; Deryagin, Ershova, Telesnyii, and Churajev, 1966). In particular Schufle and Venugopalan (1967) have measured the specific volume of water in fine capillaries and obtained the extraordinary results shown in Figure 12.4.

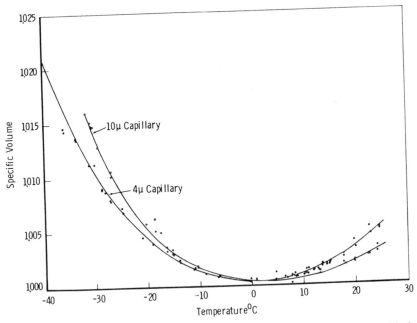

Figure 12.4 Temperature Dependence of the Specific Volume of Water in Capillaries. From Schufle and Venugopalan (1967), with permission of American Geophysical Union and the authors.

At a Discussion of the Faraday Society held at Nottingham in 1966, Professor Derjaguin described in further detail the properties of this peculiar water—the viscosity is 15 times, and the thermal expansion 1.5 times, that of "normal" water, its vapor pressure is lower, and it appears to be in an unusually stable form. He observed peculiar water in glass capillary tubes as wide as 0.03 mm, indicating that the perturbation of the "normal" water structure must extend out from the interface for a very considerable distance. Nuclear magnetic resonance studies (Korringa, 1956) indicate that the interfacial water structure layer is as thick as 700 Å, and Drost-Hansen (1959) has estimated that in a 3 μ diameter capillary approximately 25% of the water is abnormal. He gives 0.1–0.4 μ as the thickness of the layer and states that the thickness depends on the ion exchange capacity (see below) of the porous material and that it decreases slowly with increasing temperature.

NMR studies have also indicated that the water surface layer in, for example, silica gel has considerable crystallinity (Zimmerman and Brittin, 1957), but just what the structure might be remains unclear. There may even be different structures adjacent to different solids; for example, the water structure on kaolinite appears to be different from that on montmorillonite with the waters more rigidly bound in the former instance (Slonimskaya and Raitburd, 1965). Water and silicate minerals have the same crystal lattice parameters; the O—O distance in ice is 4.52 Å and in some silicates 4.51 Å. This has led Macey (1942) to propose that a layer of ice exists on the surface of the solid which extends outward until overcome by thermal agitation and the effect of electrolytes. But, in Chapter 1 we declined to assign an Ice-I_h type of structure to the Frank-Wen clusters in bulk water, and now the great proclivity of water near surfaces to supercool makes our reason even more compelling. Weiss (1966) has pointed out that the dimensions of the water structure on silicate surfaces correspond well with those of the water structure in gas clathrates and, for want of better evidence, I am inclined to favor this view.

To further complicate matters there also can be definite chemisorption of water on the solid material, and this chemisorption can be very strong—the water monolayer on kaolinite is not removed until the temperature reaches 425°C (Brindley and Millhollen, 1966). The most strongly adsorbed H_2O's on silica are chemically bound to solid surface hydroxyl groups (Hambleton, Hockey, and Taylor, 1966), and there may be as many as three chemisorbed layers of immobilized water on the surface (Antoniou, 1964). To summarize, then, in a porous material there can be at least three kinds of water: (a) strongly and rigidly bound chemisorbed water on the solid surface, (b) structured water near the surface, and (c) "normal" water at some greater distance from the surface.

Is the water structure near the water-solid interface the same as that near the water-gas interface? It may very well be. We saw in the previous chapter that the highly structured water in the interfacial layer tends to exclude solutes that disrupt water structure; does the water structure near a solid surface behave in a similar manner? The answer appears to be yes. For example, certain cations are selectively excluded from silica gel. Tien (1965) has attributed this exclusion or negative absorption to the increased difficulty experienced in penetrating the smaller pores in the gel. Maatman (1965), however, has rejected this explanation and attributed the exclusion phenomenon to a generalized "geometric effect" at the water-solid interface (McConnell, Williams, et al., 1964). This theory, I feel, handles ionic hydration in an inadequate way and water structure not at all. Furthermore, inasmuch as ion exclusion by this mechanism can only occur immediately next to the surface, the phenomenon should be observable only when the ratio of surface area to volume in the system is extremely large, which is contrary to observation (see below). Nevertheless, despite these difficulties, if one takes the radii of the hydrated ion in the gel as obtained from this theory and treats them simply as a measure of ion exclusion, plotting them against the corresponding viscosity B coefficient (Chapter 3), a striking correlation obtains with the ions falling mainly into two classes, the structure makers and the structure breakers (Figure 12.5). On the basis of Figure 12.5 the more positive a cation's B coefficient, the more strongly it is excluded, but other factors such as the preference of the ion for the higher dielectric phase (the Onsager-Samaras theory of ion exclusion from the gas-solution interface mentioned in the previous chapter) may also be operative.

High-pressure conductivity and elution experiments also yield evidence for the exclusion of ions from the region near the solution-solid interface (Horne, Day, et al., 1968). The maximum in the relative electrical conductivity versus pressure curve for a "marine sediment" consisting of 80 mesh alumina permeated with $0.10\ M$ aqueous KCl solution is 10% less than that for an $0.10\ M$ aqueous KCl solution alone; in effect, the solution appears to be more concentrated. The fact that any reduction at all is observed at these pressures (about $2{,}000\ kg/cm^2$) indicates that the structural form of water with which we are dealing, unlike the structure in the Frank-Wen clusters and the hydration atmospheres of ions, is relatively stable with respect to pressure. This is nicely complementary to Derjaguin's observation (see above) that the interfacial water structure possesses an unusual thermal stability but is very difficult to reconcile with the large specific volume of the interfacial water (Figure 12.4). When a solution is passed through a long column of granular solid material, the first few emergent aliquots are more concentrated than the initial solution —further evidence of electrolyte exclusion. The qualitative picture (Figure 12.6) is one of enhanced water structure near the solid surfaces with conse-

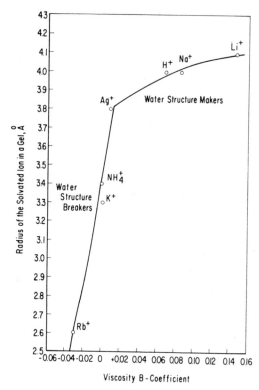

Figure 12.5   Dependence of Ion Exclusion on the Viscosity B Coefficient. From Horne, (1966).

quent exclusion of electrolyte and increased concentration of ions in the interstitial water more removed from the interfaces.

In addition to viscosity, ion transport processes in solution permeated porous materials have also been investigated. Owing to the compensating effect of ionic hydration, the limiting equivalent conductivities of ions in free solution are surprisingly independent of charge—26.5, 31.2, and 34.4 $cm^2/$ ohm-equiv for Na^+ Ca^{2+}, and La^{3+}, respectively but, in an ion exchange resin, a porous medium with fixed charge sites distributed throughout, a strong charge dependence is evident, the corresponding values being 2.82, 0.99, and 0.22 $cm^2/$ohm-equiv (George and Courant, 1967). The Arrhenius activation energies of electrical conductivity tend to be higher for the membrane than for free solution (George and Courant, 1967), and E_a is also higher for protonic conduction in thin capillaries (Yu, 1966) since that process depends on the rotability of water molecules. Similarly, as one might expect,

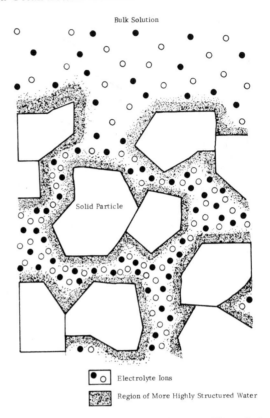

Figure 12.6 Model of Particulate Solids in an Aqueous Electrolytic Solution. From Horne, Day, Young and Yu (1968).

the self-diffusion coefficient for water in zeolites (1.34–1.88×10^{-5} cm^2/sec at $30°C$) is less than for pure bulk water (2.5×10^{-5} cm^2/sec), and its activation energy is 0.6–1.9 kcal/mole higher (Parravano, Baldeschwieler, and Boudart, 1967).

Notice that at the solution-solid interface we can get both positive and negative absorption—the concentration of ions by direct sorption or in the electrical double layer and exclusion, owing to the structuring of the solvent. Which of the two effects dominates appears to depend on the nature of the solid surface. For a conductive highly charged surface such as a metallic electrode, the electrical double layer is very much in evidence whereas, for a nonconducting solid with few charge sites, ion exclusion may be the more important process.

To summarize our ruminations concerning the structure of aqueous electrolytic solutions such as seawater for a final time both in the bulk phase and

Table 12.1   Properties of the Two Structural Forms of Liquid Water
(From Horne, Day, Young, and Yu, 1966)

α Form (present in large amounts of bulk water in the form of Frank-Wen clusters)	β Form (present in small amounts in bulk water)
Coulombic Hydration	*Hydrophobic Hydration*
Surrounds ions and charge sites.	Envelops interfaces (liquid-water, gas-water, and solid-water) hydrophobic materials and hydrophobic portions of molecules.
Structure unknown, may be random. Probably is not Ice I.	Structure unknown, possibly Pauling clathrate type. Probably is not Ice I.
Relatively temperature-stable. Gradually destroyed in going from 0 to 100°C. Gives rise to the density maximum at 4°C.	Greater thermal stability. Gives rise to anomalies in water properties near 40°C.
Completely destroyed by 1000 kg/cm².	Still evident at 5000 kg/cm².

near interfaces, let us say that in addition to unassociated, monomeric water there are *at least* two structural forms of liquid water, call them α and β, with the properties listed in Table 12.1. But above all we must remember that these conclusions are imperfect in the extreme, that Table 12.1 is both tentative and highly speculative, and that the question of the structure of liquid water will remain the most profound, central, and difficult problem in chemical oceanography—and in molecular biology—for the foreseeable future.

3   The Chemical Nature of Recent Marine Sediments

Only rarely does the solid interface at the bottom of the sea consist of the firm rock of the Earth's crust (Figure 12.7). Almost the entire ocean floor is covered with a slowly accumulating deposit of sedimentary material. Pelagic sediments alone (see classification below) cover some 2.68×10^8 km² of the Earth's surface or some 74% of the sea bottom (Table 12.2). The geologically recent marine sediments have been classified in a number of ways. They can, for example, be classified chemically as calcareous ($CaCO_3$–$MgCO_3$), siliceous (SiO_2), etc., but more useful classifications are based on their origin or their distribution, and Table 12.3 represents an effort to combine in a simplified form both of these classifications. Detail is sometimes added to this classification; for example, the calcareous oozes can be divided into globigerina, pteropod, and cocolith oozes, and the siliceous oozes into diatom and radiolarian oozes on the basis of the most prevalent organism remains, while

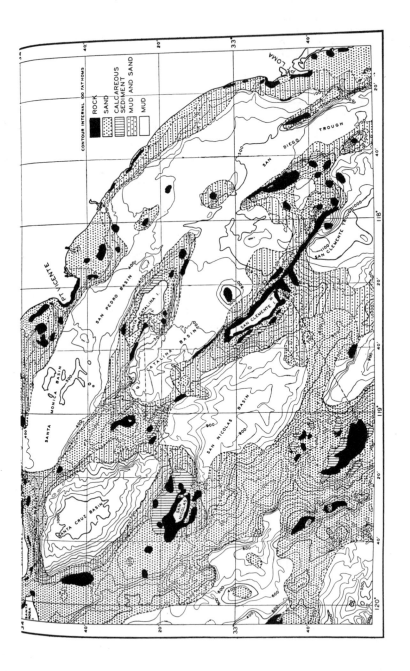

Figure 12.7 Distribution of Rock and Sediments off the Coast of Southern California. From Revelle and Shepard (1939), with permission of American Association of Petroleum Geologists.

Table 12.2   Areas (in 10^6 km^2) Covered by Pelagic Sediments
(From Sverdrup, Johnson, and Fleming, 1942, with permission of
Prentice-Hall, Inc.)

	Atlantic Ocean		Pacific Ocean		Indian Ocean		Total	
	Area	%	Area	%	Area	%	Area	%
Calcareous oozes:								
Globigerina	40.1	—	51.9	—	34.4	—	—	—
Pteropod	1.5	—	—	—	—	—	—	—
Total	41.6	67.5	51.9	36.2	34.4	54.3	127.9	47.7
Siliceous oozes								
Diatom	4.1	—	14.4	—	12.6	—	—	—
Radiolarian	—	—	6.6	—	0.3	—	—	—
Total	4.1	6.7	21.0	14.7	12.9	20.4	38.0	14.2
Red clay	15.9	25.3	70.3	49.1	16.0	25.3	102.2	38.1
	61.6	100.0	143.2	100.0	63.3	100.0	268.1	100.0

terrigenous deposits can be classified on the basis of the coarseness of their particles:

SAND

80% of the particles coarser than 62 μ diameter
Very coarse, 1000–2000 μ
Coarse, 500–1000 μ
Medium, 250–500 μ
Fine, 125–250 μ
Very fine, 62–125 μ

SILTY SAND

50–80% coarser than 62 μ

SANDY SILT

> 50% coarser than 5 μ, > 20% coarser than 62 μ

SILTY MUD

> 50% coarser than 5 μ, > 20% coarser than 62 μ

CLAYEY MUD

> 50% coarser than 5 μ

The origins and means of transportation of sedimentary material in the seas has been reviewed by Kuenen (1965). The finer the particles, the more easily they are carried great distances from their source (Stabaugh and Stump,

Table 12.3   Classification of Marine Sediments

Source	Distribution
TERRIGENOUS From the Lithosphere (the continental landmasses)	
Volcanic	
Clastic (Mech. and/or chem. destruction of rocks) → larger particles	**LITTORAL DEPOSITS** Shallow, coastal waters; Highly variable distribution; Large quantities → Sands, Silts, Muds, Clays
	{ HEMIPELAGIC (Transitional Littoral-Pelagic found on a continental slope) } — Airborne dust, Seaborne turbidity
	PELAGIC DEPOSITS Deep sea; Wide distribution; Largest quantities → $<30\%$ organic origin RED CLAY; $>30\%$ organic origin OOZE
BIOGENOUS From the Biosphere (organic origin)	
Planktonic (Insol. residua from near-surface life forms)	
Benthonic (Bottom-living animals and plants)	More abundant in shallow, coastal water; Relatively small quantities
HALMYROGENOUS From the Hydrosphere (flocculation and chemical precipitation)	Wide and irregular distribution; Relatively small quantities → CONCRETIONS NODULES
COSMOGENOUS Extraterrestial origin	Wide Distribution; Very small quantities → METEORITES

1964), since their settling rates are slower (Table 8.2). Consequently there is a sorting of the terrigineous material, and the littoral deposits tend to be coarser and more highly varied in the distribution of their particle sizes and their chemistry (Figure 12.7), the latter reflecting the complex mineralogical chemistry of their continental sources.

The rock exposed on the ocean floor represents a small potentially active surface area. Thus the chemistry of the sea-ocean bottom interface is largely the surface chemistry of the marine sediments, and the particle sizes, degree of compaction, and water content of the sediment are important, not only because they determine the physical characteristics of the material but also because of the role they play in its chemistry. Coprecipitation and, especially, absorption take trace elements out of seawater (Krauskopf, 1956); the smaller the particles are, the greater the surface area, the more absorption, and the greater the concentration of absorbing trace elements (Table 12.4). On the basis of this finding

Table 12.4   The Dependence of Trace Element Concentration on
Particle Size in a $CaCO_3$ Ooze
(From Turekian, 1965, with permission of Academic Press, Inc.)

Particle Size Range, μ	Parts per million						% $CaCO_2$
	Pb	Sn	Ni	Mn	Cu	Sr	
700–1000	15	9	13	320	54	1400	~100
500–700	15	7	15	400	47	1200	~100
250–500	15	4	15	290	23	1400	~100
180–250	9	10	10	210	28	1200	~100
140–180	15	20	15	190	28	1700	~100
125–140	18	10	15	130	27	1300	~100
100–125	27	<4	13	64	17	1400	~100
89–100	18	<4	5	50	13	1300	~100
63–89	18	9	8	96	31	1400	~100
45–63	12	7	13	170	70	1300	~100
32–45	33	28	40	170	310	1400	~100
22–32	40	27	72	270	130	1200	97.3
16–22	70	44	105	600	410	1400	96.1
11–16	54	44	120	1000	330	1600	94.2
8–11	42	66	220	>1000	730	2280	91.8
5.6–8	230	67	900	>1000	1650	2700	90.3
4.0–5.6	110	42	550	≫1000	1650	2940	86.0
2.8–4.0	94	46	510	≫1000	1650	>3000	86.9
2.0–2.8	82	54	490	≫1000	1100	2820	88.2
1.4–2.0	82	42	390	≫1000	1300	1950	87.9
1.0–1.4	88	58	230	>1000	620	2150	88.5
0.7–1.0	150	62	170	>1000	1300	2100	88.0
0.5–0.7	15	10	15	210	43	1500	91.1

Table 12.5 The Distribution of Trace Elements in Marine Environments and Igneous Rocks (Concentrations in ppm) (From Chester, 1965, with permission of Academic Press, Inc.)

Trace Element	Igneous Rocks	Near-Shore Sediments	Deep-Sea Argillaceous Clays (Pacific)	Manganese Nodules	Ratio, near-shore sediments/ igneous rocks	Ratio, deep-sea clays/ near-shore sediments	Ratio, manganese nodules/ deep-sea clays
Cr	65	100	77	<10	1.5	0.77	<0.13
V	127	130	330	590	1.0	2.5	1.8
Cu	42	48	570	3300	1.1	11.9	5.8
Pb	13	20	162	1500	1.5	8.1	9.2
Ni	50	55	293	5700	1.1	5.3	19.4
Co	19	13	116	3400	0.69	8.9	29.2
Sn	2	21	20	300	10.5	0.95	15.0
Ba	530	750	2237	3100	1.4	3.0	1.4
Sr	335	<250	587	1000	0.75	2.3	1.7
Zr	152	160	145	340	1.0	0.91	2.3
Ga	17	19	20	17	1.1	1.1	0.85

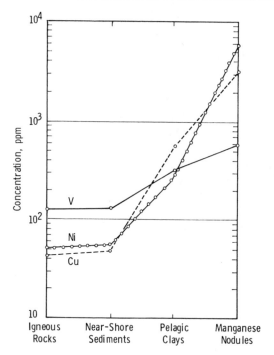

Figure 12.8   Concentration of Some Trace Elements in Rocks, Sediments, and Nodules. From Chester (1965), with permission of Academic Press, Inc.

Table 12.6   Water Content of
Freshly Settled Sediment
(From Sverdrup, Johnson, and
Fleming, 1947, with permission of
Prentice-Hall, Inc.)

Size Group, μ	Water Content, Vol. %
250–500	45.0
125–250	45.4
64–125	46.9
16–64	51.6
4–16	66.2
1–4	85.8
< 1	98.2

we might expect the finer deep-sea sediments to be richer in trace elements than the coarser near-shore materials, and such does indeed seem to be the case (Table 12.5 and Figure 12.8). Organisms, as we have seen, are capable of building up very large relative concentrations of trace elements. But, inasmuch as these materials are concentrated largely in the soft tissues (Table 9.18), my own suspicion is that the processes of dissolution and decay return these elements to the seawater long before the skeletal remains settle finally to the bottom, and that as a consequence Table 12.5 and Figure 12.8 represent

Table 12.7   Computed Consolidation of Marine Sediments (From Hamilton, 1959, with permission of Geological Society of America and the author)

A. Consolidation of Clay and Shale

Assumption: 300 m of clay overlies shale with 100% hydrostatic uplift effective. Properties: density of solids, 2.31 g/cm^3; water density, 1.05 g/cm^3; initial porosity, 72%.

Depth in Sediment, m	Interval Pressure, kg/cm^2	Cumulative Pressure, kg/cm^2	Void ratio, e	Density saturated, g/cm	Submerged Weight g/cm^2	Porosity, n, %
0	0	0	2.58	1.40	0.35	72
10	0.35	0.35	1.73	1.52	0.47	63
20	0.48	0.83	1.48	1.55	0.50	60
30	0.51	1.34	1.35	1.58	0.53	58
40	0.54	1.88	1.25	1.60	0.55	56
50	0.56	2.44	1.18	1.63	0.58	54
60	0.59	3.03	1.13	1.64	0.59	53
70	0.60	3.63	1.08	1.66	0.61	52
80	0.62	4.25	1.05	1.67	0.62	51
90	0.63	4.88	1.02	1.68	0.63	50
100	0.63	5.51	0.98	1.69	0.64	49
150	3.3	8.8	0.87	1.72	0.67	47
200	3.5	12.3	0.80	1.76	0.71	44
250	3.6	15.9	0.75	1.77	0.72	43
300	3.7	19.6	0.70	1.80	0.75	41
400	7.7	27.2	0.58	1.85	0.80	37
500	8.1	35.3	0.54	1.87	0.82	35
600	8.3	43.6	0.50	1.90	0.85	33
700	8.6	52.2	0.47	1.91	0.86	32
800	8.7	60.9	0.43	1.93	0.88	30
900	8.9	69.8	0.40	1.95	0.90	29
1000	9.1	78.9	0.37	1.97	0.92	27
1500	48.4	127.3	0.25	2.06	1.01	20
2000	52.8	180.1	0.15	2.15	1.10	13

B. Consolidation of a *Globigerina* Ooze

Assumptions: A progressive porosity decrease under pressure with no lithification, which is probably a rare situation. Properties: density of solids, 2.69 g/cm³; water density, 1.05 g/cm³ initial porosity, 66%.

Depth in Sediment, m	Interval Pressure, kg/cm²	Cumulative Pressure, kg/cm²	Void ratio, e	Density saturated, g/cm	Submerged Weight g/cm²	Porosity, n, %
0	0	0	1.90	1.62	0.57	66
10	0.57	0.57	1.88	1.62	0.57	65
20	0.57	1.14	1.85	1.63	0.58	65
30	0.58	1.72	1.82	1.63	0.58	65
40	0.58	2.30	1.80	1.64	0.59	64
50	0.59	2.89	1.78	1.64	0.59	64
100	3.00	5.89	1.68	1.66	0.61	63
200	6.2	12.1	1.56	1.69	0.64	61
300	6.5	18.6	1.49	1.71	0.66	60
400	6.7	25.3	1.44	1.72	0.67	59
500	6.8	32.1	1.40	1.73	0.68	58
600	6.9	39.0	1.36	1.74	0.69	58
700	7.0	46.0	1.32	1.76	0.71	57
800	7.2	53.2	1.27	1.77	0.72	56
900	7.3	60.5	1.23	1.79	0.74	55
1000	7.4	67.9	1.18	1.80	0.75	54
1500	39.0	106.9	1.01	1.87	0.82	50
2000	42.5	149.4	0.88	1.92	0.87	47

trace element concentration for the most part by absorption processes rather than by biological activity. Notice that the concentrations are particularly high in manganese nodules (see below), objects formed by slow absorptive accretion.

The water content of a freshly settled sediment increases with decreasing particle size (Table 12.6), but chemical processes and the pressure of the accumulation of further material above tend to compress and consolidate the sediment so that the water content decreases with deeper penetration of the sedimentary layer. For example, the water content of a fine-grained (1.7–2.7 μ) sediment decreased from 81% at the top to 75% at a depth of 2 m, whereas a coarser sample (16–20 μ) decreased from 65% to 56%.

Hamilton (1959, 1964) has examined the gravitational consolidation and lithification of deep-sea sediments in some detail, both experimentally and theoretically. The results of the latter are given in Table 12.7 which shows the decrease in void ratio and porosity with increasing depth. Several studies of

Table 12.8 Surface Properties and Pore Structure of Marine Sediments
(From Weiler and Mills, 1965, with permission of Pergamon Press, Ltd.)

Sample No.	Sediment Type	Pretreatment of Sample	Carbonate content %	In situ Density g/cm³	Bulk Density g/cm³	True Density g/cm³	Calculated Geometric Surface Area m²/g	True Surface Area m²/g	Average Particle Diameter Å	Total Pore Volume cm³/g	Average Pore Radius Å
1	Red Clay	None		1.42	0.92	2.57	0.68	28.2	830		
		Washed with distilled water						28.5	720		
		Cleaned with H₂O₂						32.4	270		
2	Red Clay	None	33.9		0.66	2.56		86.7	86.7		
3	Red Clay	None	27.2			2.64	0.79	54.7	350		
4	Globigerina ooze	Cleaned with H₂O₂	5.4		1.00	2.51		69.2	4700	0.017	81
		None		1.42	1.48	3.08		4.19	4300	0.022	72
		Cleaned with H₂O₂				2.28		6.14			
5	Globigerina ooze	None	46.9		0.91	2.65	<0.01	13.0	1700		
6	Foraminiferal ooze	None	78.9		0.75	2.26	0.94	28.5	930	0.154	108
7	Foraminiferal ooze	None	47.7		0.70	2.07	0.60	10.0	2900		
8	Pteropod ooze	None	95.6		0.96	2.94	<0.01	2.81	7200		
		Cleaned with H₂O₂				2.59		4.77	4900	0.046	193
9	Pteropod ooze	None	96.5		0.64	2.43		5.60	4400		
10	Radiolarian ooze	None	7.0		0.79	2.62		71.6	320	0.363	101
		Washed with distilled water			0.66			96.0	260		
		Cleaned with H₂O₂			0.68	2.42		99.2	250	0.358	72
		Cleaned with HCl				1.95		149	170	0.767	103
11	Radiolarian ooze	None	3.7		0.32	1.96		12.8	2400		
		Washed with distilled water						24.9	1200		
12	Diatom ooze	None	15.8		0.41	2.05		17.9	1700	0.168	188
		Cleaned with H₂O₂			0.41	1.95		23.9	1200	0.116	97
13	Diatom ooze	None	31.7		0.42			31.7			
14	Shelf (terrigenous) sediment	None		1.33	0.91	2.52		12.4 (110°, 150°C) 14.6 (215°C) 10.9	1900 1600	0.136	186
		None, but air-dried				2.36		(−14 + 32 mesh) 11.7		0.106	195
		Washed with distilled water			1.12	2.55		(−60 mesh) 16.6 17.9	2300	0.106	181
		Cleaned with H₂O₂	2.9					(110°, 150°C) 16.7 (210°C)	1300 1400	0.123	147

the surface properties and pore structure of marine sediments have been made (Kulp and Carr, 1955; Slabaugh and Stump, 1964; Weiler and Mills, 1965), and the results of Weiler and Mills (1965) are summarized in Table 12.8. These authors have observed that the chemically available surface area in the sediments is reduced by the blockage of pores by organic material and that surface sorption is adequate for accounting for the measured quantities of organic matter in the sediment. This should remind us once again of the ubiquity of biogenic organic material in the seas. In the seas, at the surface or bottom interfaces, or in the bulk phase, inorganic chemistry is never separable from biochemical processes. Earlier we saw that absorbed organic coatings may reduce the rate of solution of $CaCO_3$ and siliceous material in seawater (Chapters 7 and 8), and in our final section we will explore briefly the very intimate and intricate relationship between marine fouling and marine corrosion.

As indicated in Table 12.2, the pelagic sediments are the predominant bottom material over vast reaches of the world's oceans, and now we turn specifically to the question of their chemical composition. The oceanic average composition of pelagic sediments is given in Table 12.9 which is an abstract or summary of a much more detailed table given in the review paper of

Table 12.9 Chemical Composition of Pelagic Sediments
(Abstracted from a more detailed table by Chester, 1965)
A. Oceanic Average of Major Elements as Weight Percent Oxide

SiO_2	44	CaO	0.6
Al_2O_3	12	MgO	2.3
Fe_2O_3	6.5	Na_2O	1.8
TiO_2	0.7	K_2O	2.4
MnO_2	0.9	P_2O_5	0.3

B. Oceanic Average of Trace Elements Given as Parts per Million

Ba	2852	Ni	236
B	3	Pb	109
Co	3	Sn	11
Cr	69	Sr	755
Cu	345	V	193
Ga	18	Zr	180
Ge	1.6		

Chester (1965). Detailed analyses of East Pacific pelagic sediments are given by Goldberg and Arrhenius (1958). The values given, it should be emphasized, are averages, and inspection of his original table shows a wide variation in any one of the major or trace elements, for example, a range of 5.3–63.9 wt % SiO_2 or 120–1110 ppm Sr. There are marked variations depending on the type of pelagic sediment (Table 12.10), and there are other variations even in a given type or, for that matter, with sediment depth in a given location. Although based on data now quite old, two other tables (Table 12.11 and 12.12), taken from Sverdrup et al. (1942), are instructive in giving some hint of the variety and relative amounts of biogenetic and terrestrial mineral substances in pelagic sediments. On the basis of the similarity of the chemical composition of iqueous rocks and Pacific pelagic sediments, the major elements Na, K, Mg, Ca, Al, Ti, and Fe appear to be neither enriched nor depleted in the sediments as a whole (Goldberg and Arrhenius, 1958; Goldberg, 1961) in contrast to the minor elements which are frequently enriched in the sediment (Table 12.5 and Figure 12.8).

The rare earth content of sediments (Table 12.13) provides an interesting footnote; there appears to be a definite systematic dependence on the ionic

Table 12.10   Chemical Analyses of the Principal Types of Pelagic Sediments (weight % oxides) (From Wakeel and Riley, 1961, with permission of Pergamon Press, Ltd.)

	Red Clay	Calcareous Ooze	Siliceous Ooze
SiO_2	53.93	24.23	67.36
TiO_2	0.96	0.25	0.59
Al_2O_3	17.46	6.60	11.33
Fe_2O_3	8.53	2.43	3.40
FeO	0.45	0.64	1.42
MnO	0.78	0.31	0.19
CaO	1.34	0.20	0.89
MgO	4.35	1.07	1.71
Na_2O	1.27	0.75	1.64
K_2O	3.65	1.40	2.15
P_2O_5	0.09	0.10	0.10
H_2O	6.30	3.31	6.33
$CaCO_2$	0.39	56.73	1.52
$MgCO_3$	0.44	1.78	1.21
Available O	0.11	0.050	N.D.
Organic C	0.13	0.30	0.26
Organic N	0.016	0.017	—
Total	100.20	100.17	100.10
Total Fe_2O_2	9.02	3.14	4.98

Table 12.11   Physical Composition of Pelagic Sediments and Texture of Mineral Particles
(From Sverdrup, Johnson, and Fleming, 1942, with permission of Prentice-Hall, Inc.)

Physical Composition	Red Clay, %	Radiolarian Ooze, %		Diatom Ooze, %		Globigerina Ooze, %		Pteropod Ooze, %
$CaCO_3$								
Maximum	28.8	29.0	20.0	36.3	24.0	96.8	97.2	98.5
Minimum	0	0	Tr	2.0	0	80.2	30.0	44.8
Average	6.7	10.4	4.0	23.0	2.7	64.5	64.7	73.9
Planktonic foraminifera								
Maximum		27.0			Pre-	80.0	95.0	75.0
Minimum		0			domi-	25.0	15.0	15.0
Average	4.77	8.8	3.1	3.1	nant part of $CaCO_3$	53.1	58.9	34.7
Benthic foraminifera								
Maximum		3.0					10.0	10.0
Minimum		0			Present		0	Tr
Average	0.6	0.6	0.1	1.6		2.1	2.1	3.6
Other calcareous remains								
Maximum		6.3				31.8	26.0	57.0
Minimum		0			Present	1.2	Tr	15.8
Average	1.3	1.0	0.8	3.2		9.2	3.7	35.5
Siliceous remains								
Maximum		5.0	80.0	60.0	90.0	10.0	15.0	20.0
Minimum		0	30.0	20.0	40.0	4.0	Tr	Tr
Average	2.4	0.7	54.4	41.0	73.1	1.6	1.7	1.9
Texture of Mineral Particles								
> .05 mm, diameter								
Maximum	20.0	60.0	5.0	25.0	40.0	50.0	50.0	20.0
Minimum	1.0	Tr	1.0	3.0	1.0	1.0	Tr	Tr
Average	5.6	2.4	1.7	15.6	8.4	5.3	5.1	4.7
< .05 mm, diameter								
Maximum		100.0	67.0	27.9	34.0	66.6	69.3	41.8
Minimum		31.0	17.0	12.5	9.0	1.2	1.2	Tr
Average	85.4	86.5	39.9	20.4	15.8	30.6	28.5	19.6
Number of samples averaged	70	126	9	5	16	118	772	40

Table 12.12   Minerals in Pelagic Deposits (diameters greater than 0.05 mm) (From Sverdrup, Johnson, and Fleming, 1942, with permission of Prentice-Hall, Inc.)

Principal Inorganic Constituents	Red Clay, %		Radiolarian Ooze, %	Diatom Ooze, %		Globigerina Ooze, %		Pteropod Ooze, %
1. Allogenic minerals								
Amphibole	44	19	19	100	60	50	36	60
Chlorite								
Epidote	3							
Feldspar, undifferentiated	76	19	90	60	90	73	38	50
Orthoclase				40		×		
Sanidine	10					18		10
Plagioclase	29			80		20		10
Garnet	1			20		×		
Magnetite	89	23		100		80	29	40
Mica, undifferentiated	27	40	10	40	6	26	64	70
White mica								
Black mica			10	20				
Olivine	7					16		10
Pyroxene, undifferentiated						×		
Enstatite						×		
Hypersthene								
Augite	61	13	30	40	12	70	12	40
Quartz	30	58		80	80	42	68	70
Tourmaline	4			20		×		
Zircon	4			20		×		
2. Authigenic minerals								
Analcite			10					
Glauconite	9	14		20	6	11	13	
Manganese grains, nodules	79	43	70			31	6	
Palagonite	31	9	40	20		8	30	40
Phillipsite	14		40			×		
Phosphate						×		
3. Rock fragments, etc.								
Crystalline, sedimentary	7			60	12	9		
Volcanic, glass	64	23	70	80	70		60	65
Lapilli, scoria	16		10	20				
Pumice	49	8	20	40	30		40	30
Rock fragments								
4. Magnetite, other (cosmic)								
Spherules	11		30			×		

Column header note: *Percent of Samples Analyzed in which Substance is Recorded*

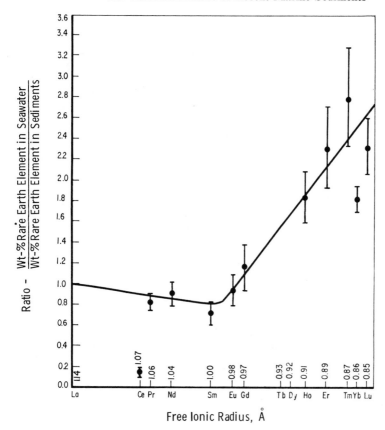

Figure 12.9   Relative Rare Earth Element Concentration in Seawater and Sediments as a Function of Ionic Radius. From Goldberg, Koide, Schmitt, and Smith (1963), with permission of American Geophysical Union and the authors.

radius (Figure 12.9) with the members of the series, ionic radii less than that of Sm being increasingly excluded, relatively speaking, from the sediment as the ionic radius decreases.

As a participant in a complicated chemical system the $CaCO_3$ content of deep-sea sediments shows some interesting behavior. Figure 12.10 shows the distribution of $CaCO_3$ in Atlantic sediments. A puzzling feature of the $CaCO_3$ distribution is the abrupt decrease in the $CaCO_3$ content of sediments below 4000 m (Figures 12.11 and 12.12), sometimes called the "compensation depth." The phenomenon is presumably due to the increased solubility of $CaCO_3$ with increasing hydrostatic pressure (Pykowicz and Conners, 1964), but the shape of the curve is unexpected. Seawater is undersaturated with respect to $CaCO_3$

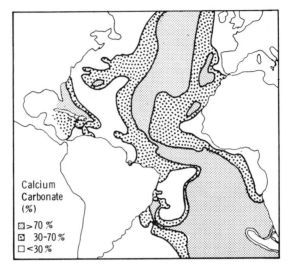

Figure 12.10   Distribution of CaCO₃ in Atlantic Sediments. From Turekian (1965), with permission of Academic Press, Inc.

at depths of only 500 m; why, then, the sudden increased solubility at much greater depths? Again the rate of solution, in particular the retardation of the dissolution process by biogenetic organic materials, may provide an answer. Turekian (1965) notes that in areas of the Atlantic Ocean such as the Cape Basin area, where CaCO₃ production is very high, high CaCO₃ contents are

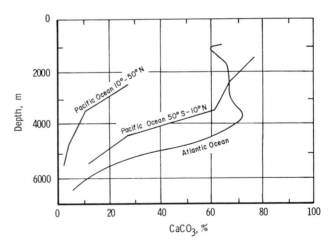

Figure 12.11   Depth Dependence of the Average CaCO₃ Content of Deep-Sea Sediments. From Dietrich (1963), with permission of John Wiley & Sons.

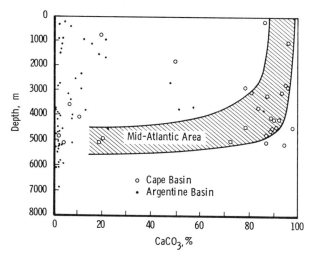

Figure 12.12   Calcium Carbonate Content of Core Tops as a Function of Depth for Three Atlantic Ocean Areas. From Turekian (1965), with permission of Academic Press, Inc.

found in the sediments at all depths, whereas in areas such as the Argentine Basin, where $CaCO_3$ solution is rapid, low $CaCO_3$ contents are found (Figure 12.12).

A second interesting distribution "anomaly" is the high concentration of Ba relative to Al along the East Pacific Rise (Figure 12.13). The high concentration of barite ($BaSO_4$) in the sediments of this region has been attributed to volcanic injection (Arrhenius, 1966).

Before turning to the subjects of the radiochemistry and biochemistry of marine sediments, I might mention that the review of Chester (1965) contains an element-by-element discussion of sediment constituents for Si, Al, Ti, Fe, Mn, Na, K, Rb, Cs, P, Cr, V, Ni, Cu, Co, Pb, Ga, Ge, Sn, Sr, Ba, B, and Zr.

The atmospherically produced radionuclides and the fission products have half-lives short compared with the oceanic residence times of the particular elements. Hence, except under very special circumstances in which the radionuclides chance to be in particulate form (including organisms) or in estuaries of rivers carrying great burdens of activity (see, for example, Gross, 1966), they rarely reach the ocean floor to be incorporated into the sediments. The elements important in the radiochemistry of marine sediments are the members of the long-lived natural radioactive series described in Chapter 10 (Picciotto, 1961).

Uranium shows little variation with depth below the sediment surface, and its concentration usually lies in the range of 0.1–6.0 ppm, nor does this

Table 12.13   Relative Rare Earth Concentration in Seawater[a] and
Sediments
(From Goldberg, Koide, and Schmitt, 1960, with permission of
American Geophysical Union and the authors)

Element REE Radius, Å	Sea-water	Phosphorite[b]		Manganese Nodule[b]		Shales	Sedi-ments	Meteor-ites
La 1.14	1	1	(1.0)	1	(1.0)	1	1	1
Ce 1.07	0.44	1.83		4.5		2.5	2.9	2.80
Pr 1.06	0.22	0.26	(1.2)	0.31	(1.4)	0.3	0.27	0.40
Nd 1.04	0.79	1.02	(1.3)	1.2	(1.5)	1.3	0.98	1.77
Pm								
Sm 1.00	0.15	0.20	(1.3)	0.56	(3.7)	0.35	0.21	0.65
Eu 0.98	0.039	0.050	(1.3)	0.09	(2.3)	0.055	0.041	0.23
Gd 0.97	0.21	0.27	(1.3)	0.4	(1.9)	0.35	0.18	0.98
Tb 0.93		0.038		0.06		0.05	0.030	0.16
Dy 0.92	0.25	0.22	(0.88)	0.26	(10)	0.25		0.95
Ho 0.91	0.076	0.059	(0.79)	0.05	(0.66)	0.065	0.041	0.23
Er 0.89	0.21	0.16	(0.76)	0.12	(0.57)	0.135	0.090	0.67
Tm 0.87	0.045	0.019	(0.42)	<0.015	(0.30)	0.011	0.016	0.10
Yb 0.86	0.18	0.12	(0.67)	0.04	(0.22)	0.14	0.098	0.56
Lu 0.85	0.040	0.021	(0.52)	<0.01	(0.25)	0.04	0.017	0.10

[a] Absolute concentration values in sequence of La in ppm units are: sea water, 2.9×10^{-6}; phosphorite, 56; manganese nodule, 11,500; shales, 7; sediments, 10.2; and meteorites, 0.32.
[b] The values in parentheses are the ratio of the REE in the mineral to that in seawater.

element appear to be particularly concentrated in manganese nodules. High U contents are found in the phosphate deposits of fish bones and shells and higher concentrations, around 30 ppm, are found in sediments deposited under anaerobic conditions. A high U content appears to correlate with a high organic content of the sediment (Burton, 1965).

Th^{232}, like U, is distributed rather uniformly with depth in the sediment, and its value tends to lie in the 1–10 ppm range but, unlike U, it does tend to concentrate in manganese nodules (10–100 ppm). In sediments the bulk of the Th^{232} is present in the lithogenous fraction. The ratio Th^{230}/Th^{232} should, in principle, be ideally suited for dating marine sediments for, since these two isotopes of the same element have identical chemistry and are thus deposited simultaneously, the ratio should be independent of any variations in the rate of sedimentation. In actual practice, however, as is ever the case, the situation is complicated by several interfering factors. Nevertheless, in general one finds the expected decrease in Th^{230}/Th^{232} as one penetrates deeper into the sediment. The Th^{230}/U^{238} ratio can also be used, provided certain conditions

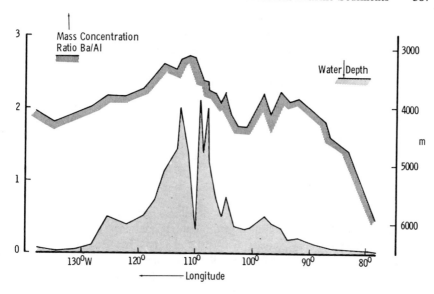

Figure 12.13 Barium Profile across the East Pacific Rise at Latitude 12°S. From K. Boström, E. Bonatti, and G. Arrhenius, *Ocean Baryte* (in preparation), with permission of Prof. Arrhenius.

obtain, to date calcareous deposits, since these materials contain U but not Th at the time of their deposition. The concentration of Th^{230} at the surface of sediments usually falls in the $1–30 \times 10^{-10}$ g/g range.

The concentration of Ra^{226} in surface pelagic sediments is even less, $\times 10^{-12}$ g/g. The Ra^{226} is formed *in situ* from Th^{230} and is expected to increase with increasing depth in the sediment until it is in equilibrium with the Th^{230} present at the level corresponding to an age of about 1.5×10^4 years. In the deeper, older sediments the Ra^{226} thus decreases until a level corresponding to about 10^6 years is reached, at which point U, Th^{230}, and Ra^{226} are all in equilibrium and the concentration of the last remains constant with greater depth (age) (Pettersson, 1937). Figure 12.14, based on data of Kröll (1955), shows the expected subsurface maximum in the Ra^{226} content but it also shows, equally clearly, the unexpected superposition of a great deal of complexity. The concentration of Ra^{226} in the sediments beneath near-shore waters tends to be slightly less than the pelagic deposits and is related to the radium content of the rocks of the adjacent land areas.

Pa^{231} is found in marine sediments in surface concentrations of $1–15 \times 10^{-11}$ g/g, and the amount decreases below the surface (Sackett, 1960). The Pa^{231}/Th^{230} ratio can be used to establish the geochronology of sediment cores.

Koczy (1956) has made quantitative estimates of the partitioning of radio-nuclides in the marine environment, including sediments (Table 12.14; see also Table 10.3) based on a steady state model, while Burton (1965) has drawn the following qualitative conclusions concerning the origin and fate of the more important radionuclides in marine sediments:

U and Th^{232} from continental sources, carried into the sea by rivers; small volcanic contribution; lost from hydrosphere by sedimentation rather than decay

Th^{230} from decay of U; deposited chiefly in pelagic sediments

Ra^{223} largely from decay of Th^{230} in sediments; small river contribution lost from hydrosphere largely by radioactive decay.

Marine sediments contain varying amounts of organic matter—carbonaceous material which in the course of geological time may eventually be converted to petroleum. The geochemistry of the formation of petroleum and the organic chemical content of these deposits, especially those fragmentary substances which are the traces of very ancient living forms, are of great interest both practically and scientifically but they are topics which, regretfully, we do not have time to treat adequately in this book. The organic content of coarse, sandy sediments is usually less ($< 1\%$) than in finer deposits, since the organic material is more readily washed away and/or oxidized in the former.

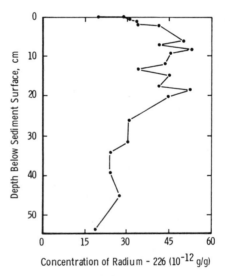

Figure 12.14   Ra^{226} Content of a Pacific Red Clay Core Taken at a Depth of 5500 m. From Burton (1965), with permission of Academic Press, Inc.

Table 12.14 Contents of, and Balance Sheet for, Radioactive Elements in Ocean Water, Calculated for 4000 m Average Depth of the Ocean and a Supply of River Water of 8 ml per cm² Ocean Floor and Year (From Koczy, 1956, with permission of Pergamon Press, Ltd.)

	g	Contents per ml		Supply to Ocean Water per cm² and per year by			Removal from Ocean Water by cm² and year by	
		Ocean	River	Decay	Rivers	Bottom	Decay	Sediment
Uranium, U_{tot}	10^{-9}	2 ± 1	1 ± 0.5	0	8	0	0.00015	8
Thorium, Th^{232}	10^{-14}	$4 {+10 \atop -2}$	1000 ± 600	0	8000	0	0.0001 (?)	8,000
Ionium, Th^{230}	10^{-17}	8 ± 4	100	15,000	800	0	72	16,000
Radium, Ra^{226}	10^{-17}	8 ± 5	4 ± 2	72	32	1,300	1,400	0
Mesothor, Ra^{228}	10^{-20}	1^a	(30?)	40	(100)	40,000	40,000	0
Radiothor, Th^{228}	10^{-21}	3^a	—	40,000	?	0	40,000	6,000

[a] Coastal centers only.

Earlier I theorized that the larger part of the organic content of deep-sea sediments is absorbed on solid surfaces and that the amount of biomaterial surviving the long settling process is relatively small. We might expect this adsorbed nutrient material to support considerable animal life. But we must remember that the water in the marine sediment is stagnant, that the rapid processes of macroscopic oceanic circulation are no longer operative, and that the renewal of oxygen on which animal life depends becomes in turn directly dependent on the relatively slow microscopic process of molecular diffusion. In the first 5–10 cm of sediment, aerobic bacterial forms may still flourish, but the supply of available oxygen soon becomes depleted and anaerobic bacteria are much in evidence at depths of 40–60 cm, below which even they begin to die off (ZoBell, 1939). The important point to note is that anaerobic conditions commonly obtain in marine sediments with the charac-teristic chemical processes of sulfur metabolism, H_2S and hydrocarbon production, and oxidation potential and metal valency changes mentioned earlier. Thus the chemistry within a marine sediment may be and often is entirely different from the chemistry of the free ocean water above the sedi-ment. The sea-bottom interface therefore can mark the boundary between two drastically different chemical environments, and this can have consequences of practical significance; for example, metal structures erected in the sea often exhibit extensive corrosion at the mud line.

4   Ion Exchange

We are inclined, I think, to look upon mineral substances as insoluble and highly inert chemically. From this opinion it is a short step to the conclusion that the terrigenous material in marine sediments should be chemically identical with the parent mineral as it is found on land. But such need not be the case, for we have overlooked one very important type of mineral chemical reaction—ion exchange.

Remember that in aqueous solution the ionic species are in constant *dynamic* equilibrium. For example, we can write the simple metathesis

$$(12.1) \qquad Na^+Cl^- + K^+ \rightleftharpoons Na^+ + K^+Cl^-$$

but this equilibrium is so rapid and the formation constants, $K_{Na^+Cl^-}$ and $K_{K^+Cl^-}$, are so small that for all practical purposes it has no chemical conse-quences and we can just as well imagine that it does not exist. All the species in system (12.1) are highly mobile. If, however, we can somehow immobilize one of the participants in a simple metathesis such as equilibrium (12.1), then the situation becomes one of great theoretical and practical importance:

$$(12.2) \qquad Na^+R + K^+ \rightleftharpoons Na^+ + K^+R, \qquad K = (Na^+)/(K^+)$$

where R is now a solid material, an enormous polymeric substance either inorganic or organic in nature and, in our particular example, interspersed throughout with negative charge sites capable of attracting cations. Substances such as R, consisting of a solution-penetratable insoluble matrix studded with charge sites, are called ion exchangers.

Although ion exchange was known in antiquity and duly mentioned by Aristotle, the modern history of ion exchange technology began in the middle of the nineteenth century with the rediscovery of the cation-exchanging capability of certain soils. These natural materials, notably the zeolites, found application as water softeners:

(12.3) $2 Na^+ R + Ca^{2+}$ (responsible for water hardness) $\rightleftharpoons Ca^{2+} R_2 + 2 Na^+$

Between World War I and World War II, synthetic, organic ion exchange resins came into being. These substances were found to be very useful in separating fission product elements and, as a by-product of the development of the bomb, a whole new area in solution physical chemistry was opened up, an area which was exploited with very great vigor in the postwar years.

But, in spite of this enormous effort, and the tremendous amount of information available about ion exchange (for a good review see Helfferich, 1962), the details of the microscopic processes which go in these materials are far from being adequately understood. Hence here we content ourselves simply with a listing of the factors, first those of the resin and second those of the exchanging ions, which we are almost certainly determining in the ion exchange process:

The Ion Exchange Material.

ACCESSIBILITY OF THE CHARGE SITES. If the interior of the material is not accessible, exchange is limited to surface sites. Accessibility depends on many factors: the pore sizes, the tortuosity, the degree of cross-linking in synthetic resins, the extent of fractures in mineral substances. The pores may be large enough to admit small ions but block large ones. Many exchangers swell when moistened. Accessibility can affect the exchange equilibria, and it is of the greatest importance in determining the rates of ion exchange.

NUMBER, DISTRIBUTION, AND NATURE OF THE CHARGE SITES. The ion exchange capacity of the material depends on the number of available charge sites. Particularly when dealing with polyvalent ions, the spacing of the charge sites and whatever limited mobility they might have are significant. The nature of the charge sites is half of the story in determining the strength of the mobile ion-charge site interaction and its ionic and/or covalent character (see below). Cation exchangers have negative sites, anion exchanger positive ones.

DIELECTRIC CONSTANT AND WATER STRUCTURE. The strength of ion-ion interactions depends on the dielectric constant, yet there has never been agreement as to what is the value of the dielectric constant within the exchanger, if indeed the microscopic dielectric constant has any meaning at all in such a system. The water structure in the exchanger is certainly very important, and virtually nothing is known about it. If the pores are small and/or the concentration of charge sites and internal mobile electrolytes is large, most of the water will be "abnormal." I venture, on the basis of our earlier discussion, that the water near the uncharged portion of the matrix is in the same form as the hydrophobic hydration associated with interfaces, and that that near charge sites is of the coulombic hydration form.

The Mobile, Exchangeable Ion.

CHARGE DENSITY. The most important characteristic of the exchangeable ion appears to be its charge density, that is, its charge divided by its *effective* surface area. From this the important conclusion can be drawn that the ion-ion exchanger-charge site interaction is, in the majority of instances, largely coulombic in nature.

CHARGE. Generally speaking, the greater its charge, the more strongly is an ion associated with the exchanger, and polyvalent ions tend to displace monovalent ions on the exchanger. Thus Na^+ is readily replaced by Ca^{2+} (see reaction 12.3), so that sodium is rarely found in clays that have been in contact with calcium-containing waters. For a given series of comparable ions such as the alkali metal cations, the affinity of the exchange sites sometimes increases with decreasing crystal radii, but more commonly the affinity increases with decreasing radii of the hydrated ion (which is just the reverse of the order of crystal radii) giving the following series of affinities: $Cs > Rb > K > Na > Li$. Exceptions to these tendencies are not rare; the order for basic sodalite, for example, is $Na > Li > K$. Li^+ sometimes seems to be held more strongly than it ought to be, and the suggestion has been made that its very small crystal radius enables it to fit very tightly into mineral lattices. In addition to its direct effect on affinities, the charge also strongly influences the mobility of ions in the solution-filled matrix. Cations diffuse 100 to 10,000 times slower in synthetic organic ion exchange resin than in free solutions, and the retardation of their movement increases precipitously with increasing charge, the self-diffusion coefficients for Na^+, Zn^{2+}, and La^{3+} being 2.4×10^{-7}, 11.2×10^{-8}, and $5.1 \times 10^{-10} \, cm^2 \, sec^{-1}$, respectively (Boyd and Soldano, 1953). But of course we must not forget that the higher-charged ions tend to be more heavily hydrated and thus bigger. The isolation of properties due specifically to ionic charge from those due to hydration is not an easy task. Typically the affinities follow the well-known Hofmeister or lyotropic series: $Sr^{2+} > Ca^{2+} > Mg^{2+} > H^+ > Cs^+ > Rb^+ > NH_4^+ > K^+ > Na^+ > Li^+$ for cations, and $Cl^- > Br^- > NO_3^- > I^- > SCN^-$ for anions. In general, the charge effects are much more pronounced for cations than for anions.

Some progress has been made in calculating the absorption constants for monovalent cations on bentonite from a consideration of the differences in hydration and polarization energies (Shainberg and Kemper, 1967).

Montmorillonite, a familiar mineral constituent of marine sediments, consists ideally of silicate sheets (Figure 12.15) with 1–4 layers of water molecules sandwiched between them. If a higher-valence element is isomorphically replaced in the crystal lattice by a lower-valence one, such as Al(III) by Mg(II) or Si(IV) by Al(III), then the sheets are left with a negative charge. This negative charge is then neutralized by exchangeable cations (K^+ in Figure 12.15). If one Si(IV) out of 400 is replaced by Al(III), an exchange capacity as large as 2 meq/100 g can result (Schofield and Samson, 1954). Kaolinite and a number of other minerals behave similarly. The surface of the mineral substance can also acquire a charge through lattic defects or by gross mechanical rupture of the crystal. In the latter case, depending on which bonds are broken, the charges can be either positive or negative and hydrolysis may occur:

(12.4)

$$\text{—Si—O—Si—O—Si—} \xrightarrow{\text{break}} \text{—Si—O—Si—O—} \overset{H_2O}{\rightleftharpoons} \text{—Si—O—Si—O—H}$$
$$\big\Updownarrow$$
$$\text{—Si—O—Si—O}^-\text{H}^+$$

and, as a final complication, the charged surface probably builds up an electrical double layer (Wayman, 1967).

Ionic strength, temperature, and pressure all affect ion exchange processes. The pressure dependence of an ion exchange equilibrium is given by

(12.5)
$$\left(\frac{d \ln K^\circ}{dP} \right)_T = -\frac{\Delta V^\circ}{RT}$$

where K° is the thermodynamic equilibrium constant and ΔV° is the volume change for the total system, including both exchanger and solution. In the case of a synthetic organic ion exchange resin, pressure exerted no discernible influence on the K^+–H^+ exchange, but for the exchange Sr^{3+}–H^+ the resin's preference for the more highly charged ion increased by about 3% at 1000 atm. This increase was attributed to the greater importance of pressure-induced dehydration of the initially more heavily hydrated ion (Horne, Courant, Myers, and George, 1964).

Ion exchange is, I repeat, the most important chemical exchange across the sea-sea bottom interface, especially if we include ionic lattice substitution as well as surface effects, and it has only begun to receive the attention it deserves. Once it enters the sea, a terrigenous material suffers chemical change at a rate proportional to its permeability by seawater as a result of adsorption

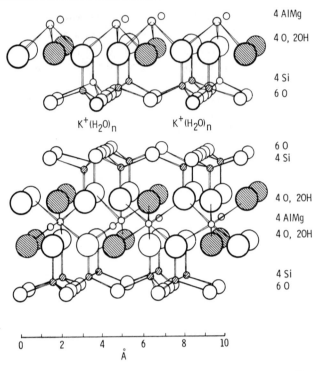

4 AlMg

4 O, 2OH

4 Si
6 O

$K^+(H_2O)_n$ $K^+(H_2O)_n$

6 O
4 Si

4 O, 2OH
4 AlMg
4 O, 2OH

4 Si
6 O

0   2   4   6   8   10
Å

Figure 12.15   Structure of Montmorillonite. Adapted from Bragg, Claringbull, and Taylor (1965).

and ion exchange. In the case of the more prevalent and exchangeable cations the changes can be quite marked in very short times, geologically speaking (Table 12.15). Even after it has settled to the ocean floor, the sedimentary particle can continue to interact with its environment, in particular to absorb trace elements from seawater. Mg and Na are abundant in marine sediments, the former because of its strong affinity for exchangers, the latter by virtue of its high concentration in seawater. Cs appears to be enriched in marine sediments (Welby, 1958).

Although Berner (1965) has reported dolomitization of Pacific atoll material by the reaction

$$(12.6) \qquad Mg^{2+} + 2\,CaCO_3 \rightleftharpoons CaMg(CO_3)_2 + Ca^{2+}$$

he found no evidence for any adsorption or exchange of Mg^{2+} or Sr^{2+} with fine-grained carbonate sediments, and he attributed the absence of any exchange to the blockage of ion transport across grain boundaries by adsorbed Mg^{2+} or coatings of organic matter (Berner, 1966). Yet, although he

Table 12.15   Changes in the Composition of Clay Minerals on Exposure to Seawater (wt % oxides) (From Chester, 1965, with permission of Academic Press, Inc.)

	Illitic Clay		Kaolinitic Clay		Montmorillonitic Clay[a]		
	Original Com- position	After 54–60 Months in Seawater	Original Com- position	After 54–60 Months in Seawater	Original Com- position	After 54–60	
						Months in Seawater	
SiO_2	52.75	52.55	45.38	45.47	58.56	50.8	59.3
Al_2O_2	24.83	24.75	38.24	38.38	18.25	16.3	18.5
Fe_2O_3	4.12	4.10	0.19	0.20	2.82	2.5	2.8
FeO	0.26	0.25	0.00	0.00	0.00	0.0	0.0
MgO	2.29	3.16	0.16	0.58	2.30	12.8	3.3
CaO	0.32	0.10	0.28	0.12	0.41	0.7	0.3
Na_2O	0.35	0.21	0.60	0.27	2.60	0.6	1.3
K_2O	5.71	6.48	0.42	0.27	0.55	3.9	0.6
TiO_2	0.62	0.61	0.76	0.77	0.08	0.08	0.09
H_2O+	7.94	7.85	13.18	13.34	7.35	8.8	7.3

[a] The montmorillonitic clay was diagenetically modified to a chlorite type (A) and an illitic type (B).

found their exchange capacity (7–30 meq/100 g) less noticeable than for the corresponding materials in soils, Zaitseva (1959) states that the organic content of Pacific sediment clays is too small to affect their exchange capacity.

5   The Composition of Interstitial Water in Marine Sediments

In view of water structuring at interfaces, ion exclusion, anaerobic bacterial action, and ion exchange and absorption processes, there is little reason for expecting that the composition of the interstitial water in marine sediments will bear any resemblance to the composition of the seawater through which the material settled or to that of the free seawater above the sediment at the time of sampling. Berner's (1966) observation that the Mg^{2+}/Cl^- and Sr^{2+}/Cl^- ratios for the interstitial water in a fine-grained carbonate sediment did not differ appreciably from the corresponding ratios in seawater may very well be the exception rather than the rule.

Recent marine muds taken off Cape Cod showed higher salinities and lower pH's than the free ocean water near the bottom in the same area, and the salt content decreased with increasing penetration of the sediment (Figure 12.16) (Siever, Garrels, Kanwisher, and Berner, 1961). The investigators attributed the lower pH values to the production of CO_2 from the bacterial decomposition of organic matter, but the high salinities are more problematical. A decrease in electrolyte concentration could be due to ion exclusion, but the fact that the salinities are higher than that of the free seawater immediately above the sediment is not readily reconciled with such an explanation. The investigators advance two hypotheses: (a) the clay acts as a semipermeable system which on compression expresses the water upwards but retains the hydrated ions; or (b) when the seawater was trapped in the accumulating sediment, there was less fresh water dilution from the Gulf of Maine than at the present time. In the latter case, the interstitial water represents a paleosalinity, that is, water trapped sometime in the geological past from ancient seas whose composition was different from that of the modern water. As I implied earlier, my own view is that structural and chemical processes are much more important and should completely erase all hint of the composition of the original seawater, except possibly in special circumstances where the compositional changes in the free seawater in a given area have been drastic and very rapid. Another puzzling feature of Figure 12.16 is the failure of the compositional ratios in the interstitial water to approach the free water values as the surface of the sedimentary deposit is approached. As the surface is approached, compaction is less and the forces of ionic diffusion should tend to reduce any concentration gradients. A much more detailed examination of a greater number of cores of different lithology from a variety of sites (Siever, Beck, and Berner, 1965) gave similar results: significant salinity differences which again are attributed to compaction

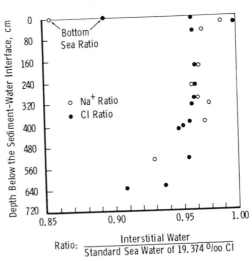

Figure 12.16 Composition of Interstitial Water in a Marine Mud Taken off Cape Cod. From Siever, Garrels, Kanwisher, and Berner (1961), with permission of AAAS and the authors.

phenomena (although the authors describe an experiment showing that squeezing a clay does not affect water composition!) and a lowering of pH. The changes in NaCl content with penetration in deep-sea cores have also been attributed to compaction (Wangersky, 1967). I might mention in passing that H_2S production from anerobic decay also would contribute to the increased acidity. Mg^{2+} tends to be depleted in the interstitial water, possibly because of uptake by chlorites and incipient dolomitization, and K^+ is enriched. Dissolved SiO_2 is also enriched, in this case by the solution of diatoms. Some typical composition and profiles of cores taken from the Gulf of California are shown in Figure 12.17. Further data on the composition of pore solution can be found in Goldberg and Arrhenius (1958). Berner (1964) has developed an idealized model for the dissolved SO_4^{2-} distribution in recent sediments. Ammonia and phosphorus are also enriched in the interstitial water, sometimes by a factor of 10 or more, as one might expect, from the decay of organic material (Brujewicz and Zaitseva, 1959). On the other hand, anaerobic utilization of SO_4^{2-} depletes its concentration from about 55 mg equiv/kg in seawater to 2–3 mg equiv/kg in the interstitial water (Shishkina, 1959). Particularly extreme conditions are found in the Black and Azov seas, where the interstitial water is characterized by a high content of organic carbon and a very low redox potential. Here the ammonia concentration is more than 4 times greater, the phosphorus concentration is less, and the silicon concentration is about the same as in the free water (Brujewicz and Zaitseva, 1959). The total salt and, especially, the carbonate CO_2 concentrations are higher (Marchenko, 1966). Interstitial waters from the Azov Sea showed a depletion of SO_4^{2-}, an increase in Br^-, considerable accumulation of ammonia, and an increase in the Mg^{2+} with increasing penetration of the sediment (Shishkina, 1961). The way in which the salts are removed from the core can give misleading results, thus casting further uncertainty about the question of the composition of the interstitial water. For example, Wangersky (1967) has noted that the suspension of air-dried sedimentary material in distilled water removes the Ca^{2+} but not the Mg^{2+} originally in the water.

Perturbation of the chemical content of seawater by solid material, as we have mentioned (Table 12.4), can be very rapid. Keuhl and Mann (1966) studied changes in the interstitial water on a beach during a tide and found that the concentration of sulfur, calcium, and magnesium decreased and that of nitrate and phosphate increased.

Friedman (1965) has found that the interstitial water from a core taken as a part of Project Mohole ". . . is depleted in deuterium by an amount that is very large in comparison with the variations in deuterium in the modern oceans," and no entirely satisfactory explanation for this very fascinating phenomenon is evident. My own suspicion is that the isotopic fractionation arises from the water structuring near interfaces, but one thing is clear, namely, here, as so

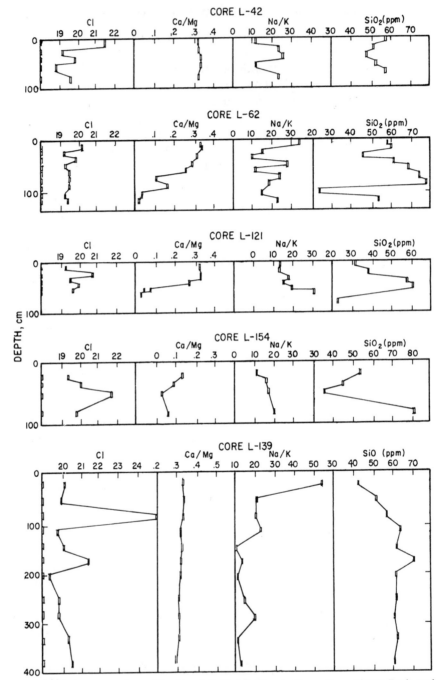

Figure 12.17 Compositional Profiles of Gulf of California Cores. From Siever, Beck, and Berner (1965), with permission of University of Chicago Press.

often elsewhere in the chemistry of the sea-sediment interface, further experimental and theoretical research is badly needed. If undertaken, it holds promise of interesting results.

Our knowledge of the chemistry of the sea-sediment interaction is obviously in its infancy, and speculation about the processes responsible for the chemical differences in interstitial water remains for the greater part exactly that— speculation. Yet in order to end this section on a note of hope I should like to mention a quantitative model which has been applied with some measure of success, for it illustrates that, complicated though these systems may be, they are nevertheless amenable to attack. The mathematical model of Anikouchine (1967), which, unlike earlier models, takes into consideration the movement of interstitial water caused by compaction, yields a relatively simple expression for the distribution of dissolved species in interstitial water:

(12.7) $$\frac{C - C_f}{C_o - C_f} = \exp\left[z(k/D)^{1/2}\right]$$

where C_f is the final or saturation concentration, z the depth in the sediment, and D the diffusivity coefficient. Although it incorporates a number of simplifying assumptions such as constant rate of sedimentation, homogeneity of the sediment, attainment of steady state, constant D, negligible temperature and pressure effects, solution (12.7) yields a very impressive fit of the data of Siever, Beck, and Berner (1965) for the concentration of dissolved silica in interstitial water (Figure 12.18). Application of this model also leads to the suggestion that the source of manganese nodules (see Section 6) is the interstitial water of the underlying sediment.

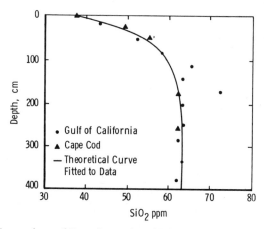

Figure 12.18 Comparison of Experimental and Theoretical Distribution of Dissolved Silicate in the Interstitial Water in Marine Sediments. From Anikouchine (1967), with permission of American Geophysical Union and the author.

6   Deposition of Nodules and other Accretions

Hitherto we have confined our attention to terrigenous and biogenous marine sediments. Now we turn to halmyrogenous deposits. These materials are composed of species once in solution which flocculated or precipitated out of the hydrosphere into the solid phase.

To begin with, $CaCO_3$, a substance we have already examined in some detail, although most of the $CaCO_3$ in marine sediments is biogenous, there is some

Figure 12.19   Continuous Manganese Pavement on the Blake Plateau. Photograph by courtesy of F. T. Manheim.

inorganic precipitation of it. But, inasmuch as most of the seas, except for the surface waters, are undersaturated with respect to $CaCO_3$, its inorganic precipitation is relatively rare and confined to special localities. If CO_2 is removed from the seawater by, for example, warming, aragonite and calcite may precipitate. In certain Bahamian sediments the $CaCO_3$ may be as much as 65–75% the result of chemical precipitation (Cloud, 1965). Aragonite and calcite can also precipitate at grain boundaries in sands and on beach rock; and the calcite and dolomite in some Pacific pelagic red clays is believed to have precipitated from hydrothermal solutions of volcanic origin (Bonatti, 1966).

Table 12.16   Manganese Nodule Distributions, Concentrations, and Sizes (From Mero, 1965, with permission of publisher and author)
A. Manganese Nodule Concentrations at the Surface of the Pacific Ocean Sea Floor Sediments as determined by Coring

Station	Location Data		Depth, m	Average Nodule Size, cm	Surface Concentration Estimate, g/cm²	Sediment Type
	Latitude	Longitude				
Eastern region						
Wig. 6	N 28°59'	W 125°41'	4000	0.3	0.2	R. C.
DWHG 4	N 24°22'	W 125°00'	4330	1 × 2 × 2	1.4	R. C.
Chub. 19	N 7°41'	W 125°37'	4416	0.6	0.4	R. C.
Chub. 9	N 10°19'	W 125°27'	4580	0.9	0.6	R. C.
Chub. 3	N 15°00'	W 125°26'	4380	1	0.7	R. C.
Chub. 17	N 8°05'	W 125°25'	4453	1	0.7	R. C.
Chub. 39	N 8°09'	W 125°02'	4360	1	0.7	R. C.
Chub. 2	N 16°02'	W 125°01'	4354	1 × 2 × 2	1.6	R. C.
Chub. 1	N 19°00'	W 121°53'	4138	0.2	0.1	R. C.
Expl. 14b	N 19°46'	W 114°44'	3438	1.5 × 3 × 2	1.8	R. C.
Expl. 14d	N 19°20'	W 114°12'	3480	3 × 3 × 1	2.3	R. C.
Acap. 114	N 10°53'	W 105°07'	3275	1	1.0	—
DWBG 147	N 1°27'	W 116°13'	4000	1 × 1.5 × 1.5	1.0	Calc. O.
DWBG 19	S 14°59'	W 136°02'	4465	0.5 × 1	0.5	R. C.
DWBG 18	S 13°37'	W 135°31'	4337	1.5	1.1	R. C.
DWBG 17	S 12°51'	W 135°13'	4318	0.5	0.4	R. C.
DWBG 16	S 6°05'	W 132°53'	4855	0.5	0.5	Rad. O.
Cap. 33 Bg.	S 12°46'	W 143°33'	4380	1	0.7	R. C.
PAS 19121	N 27°20'	W 116°10'	4030	0.5 × 2 × 2	1.2	—
Msn 148G	N 9°06'	W 145°18'	5400	1.4 × 2.5 × 2.7	1.5	R. C.
Msn 150G	N 10°59'	W 142°37'	4978	1.1 × 0.8 × 1	0.4	R. C.
Msn 153PG	N 13°07'	W 138°56'	4927	1 × 1 × 1	1.2	R. C.
Msn 157G	N 24°18'	W 126°30'	4414	0.5 × 0.8 × 0.5	0.2	R. C.
DWBG 18	S 13°37'	W 135°31'	4337	1	1.2	R. C.

A. (*continued*)

| Station | Location Data | | Depth, m | Average Nodule size, cm | Surface Concentration Estimate, g/cm² | Sediment Type |
	Latitude	Longitude				
Western region						
MP 43J	N 12°07′	E 164°52′	3290	1.5	1.1	R. C.
Msn Q	S 7°03′	E 174°12′	5378	1.5	1.2	R. C.
Cap. 30Bg	S 17°28′	W 160°59′	4710	1	0.7	R. C.
Cap. 31 Bg5	S 17°29′	W 158°40′	4890	1	0.7	R. C.
Cap. H13	S 21°25′	E 177°46′	3840	0.5	0.4	R. C.
Central region						
Cusp 8P	N 43°58′	W 140°38′	4350	1.2	0.9	R. C.
Jyn II–21	N 36°29′	E 146°43′	5720	2.5 × 1 × 1	0.7	R. C.
Cusp 15	N 37°15′	W 143°07′	5220	2	1.7	R. C.
Tet. 22	N 16°06′	W 165°45′	2400	2 × 2.5 × 3	2.1	R. C.
Tet. 24	N 15°02′	W 162°31′	5666	2 × 1.3 × 1	0.7	R. C.
Tet. 27A	N 13°05′	W 163°10′	5413	1.3 × 1 × 1	0.5	R. C.
Msn G	N 14°11′	W 161°08′	5632	0.5 × 1 × 3	1.4	R. C.
Msn K	N 6°03′	W 169°59′	5400	1.5 × 1.5	1.2	R. C.
Msn J	N 7°47′	W 168°00′	4994	2 × 2.5 × 2.5	1.9	R. C.
Msn 128G	S 13°53′	W 150°35′	3623	3 × 3 × 2	2.9	Coral S.
Msn 121G	S 29°35′	W 158°58′	5252	1.5 × 1.5 × 2.5	1.2	R. C.
Msn 125G	S 26°01′	W 153°59′	5038	3 × 3.5 × 4	3.0	R. C.
Msn 126G	S 24°41′	W 154°45′	4542	1.1 × 3 × 3	2.6	R. C.
DWBG 37	S 29°09′	W 143°01′	4120	2 × 4 × 4	3.8	R. C.
Msn 116P	S 35°50′	W 163°01′	4950	2 × 2 × 1	2.0	R. C.
DWBG 44	S 34°25′	W 138°47′	4860	0.5	0.8	R. C.
DWBG 47	S 36°33′	W 137°24′	4700	2 × 1.5 × 2	0.9	R. C.
DWBG 46	S 36°23′	W 137°15′	4680	1.5	1.1	R. C.
DWBG 48	S 37°05′	W 137°10′	4940	1.5	1.1	R. C.
DWBG 52	S 40°36′	W 132°49′	5120	3	3.0	Calc. O.
DWHG 31	S 35°11′	W 135°32′	4700	1	0.7	—
DWBG 54	S 41°24′	W 129°06′	4880	2–3	2.2	R. C.
DWHG 34	S 44°13′	W 127°20′	4600	0.5–2.5	1.4	Calc. O.
DWBG 56	S 42°16′	W 125°50′	4560	0.5–1.5	0.8	R. C.
DWBG 57A	S 42°50′	W 125°32′	4560	0.5 × 2 × 2	1.5	Calc. O.
DWBG 58	S 43°07′	W 125°23′	4640	2 × 3 × 3	2.6	R. C.
DWBG 59	S 44°23′	W 124°39′	4500	2	1.7	R. C.
DWHG 48	S 42°00′	W 102°00′	4240	3.5	3.8	R. C.
DWBG 78	S 44°08′	W 100°58′	4100	1–2	1.2	Calc. O.
Msn 85G	S 57°43′	E 169°12′	5288	2 × 2 × 3	1.5	R. C.
Msn 98P	S 54°31′	W 177°12′	5274	1 × 1 × 1.7	0.6	R. C.
Msn 90G	S 63°04′	E 178°29′	3583	2.5 × 3.5 × 1	2.5	Sil. O.
Msn 91G	S 64°11′	W 165°56′	2932	2.3 × 2.5 × 2.5	2.5	Sil. O.

Sea Floor Concentration of Manganese Nodules as Determined by
ottom Scoop Methods

Station	Location Data		Depth, m	Percent of Area Covered by Nodules, %	Surface Concen- tration Estimate, g/cm²	Surface Concentration Estimate by Photos at Nearby Station, g/cm²
	Latitude	Longitude				
Vit. 4243	N 24°56′	W 139°51′	4368	4.1	0.05	—
Vit. 4245	N 25°00′	W 137°19′	4645	20.1	0.18	—
Vit. 4273	N 19°59′	W 113°57′	3820	6.2	0.11	0.60
Vit. 4285	N 19°57′	W 126°06′	4576	25.0	0.23	0.36
Vit. 4289	N 20°00′	W 130°01′	5005	6.7	0.11	—
Vit. 4343	N 24°00′	E 179°58′	5815	22.2	0.40	—
Vit. 4347	N 24°00′	E 173°36′	5318	50.0	0.60	—
Vit. 4351	N 23°57′	E 170°58′	5817	10.0	0.17	—
Vit. 4355	N 24°02′	E 167°24′	6052	16.0	0.60	—
Vit. 4359	N 24°01′	E 163°02′	5542	36.0	1.00	1.90
			5096	20.0	0.35	

The existence of manganese-rich concretions on the floor of the world's three major oceans has been known since the *Challanger* expedition of 1873–1876 and, in recent years, perhaps as an outgrowth of the talk of the economic feasibility of mining these deposits, there has been a great deal of interest, in the form of both research and speculation, in these concretions.

The forms of ferromanganese deposits are highly varied. They range from thin stains and coatings on just about everything imaginable, including naval shell fragments which can accumulate coatings "several millimeters thick in the span of a few tens of years" (Mero, 1965), to granules, to nodules the size of a potato and a cannonball. The nodules can be so closely spaced as to form a mosaic. A spectacular formation recently reported is a continuous manganese pavement on the Blake Plateau (Figure 12.19) some 5000 km² in area grading into round manganese nodules on the south and east and into phosphate nodules in the westerly direction (Pratt and McFarlin, 1966). In this area the Gulf Stream assures a fresh supply of seawater while at the same time preventing the accumulation of sediments. Nodules are relatively scarce in the Atlantic Ocean compared to the Pacific. They are especially common in red clay regions. They are formed in the deep ocean, in shallow waters, in bays and seas, and even in lakes (Manheim, 1965). Table 12.16 gives information about distribution, concentration, and size of nodules, and Table 12.17, based on data in Table 12.16, gives some inkling of the enormous total quantities involved.

Table 12.17   Statistics on Total Manganese Nodule Quantities in the Pacific Ocean
(From Mero, 1965, with permission of publisher and author)

Statistics	Eastern Region	Central Region	Western Region	Pacific Ocean
Number of photographs	11	13	5	29
Concentration estimates				
Maximum, g/cm^2	1.2	2.5	1.5	2.5
Minimum, g/cm^2	0.36	0.9	0.46	0.36
Average, g/cm^2	0.86	1.60	0.90	0.97
Number of grab samples	5	5	0	10
Concentration estimates				
Maximum, g/cm^2	0.23	1.00	—	1.00
Minimum, g/cm^2	0.05	0.17	—	0.05
Average, g/cm^2	0.14	0.56	—	0.35
Number of cores	24	33	5	62
Concentration estimates				
Maximum, g/cm^2	2.3	3.8	1.2	3.8
Minimum, g/cm^2	0.1	0.5	0.4	0.1
Average, g/cm^2	0.89	1.71	0.82	1.32
Total of all measurements	40	51	10	101
Average concentrations of all methods, g/cm^2	0.78	1.45	0.86	1.12
Area in region, $km^2 \times 10^6$	44.9	62.1	47.2	154.2
Tonnage of nodules (billions of metric tons)	350	900	406	1,656

The details of the structure of the nodules are curious. At the center is commonly some foreign material such as a shark's tooth, and then the nodule is formed around this nucleus in a series of concentric rings, reminiscent of the growth rings of a tree (Figure 12.20). The alternating light and dark layers correspond to accretion zones of high and low goethite ($FeO \cdot OH$) content (Arrhenius, 1963).

Extensive analytic results have been obtained on the chemical composition of ferromanganese concretions. The nodules consist largely of hydrated oxides of manganese and iron (Table 12.18). Mineral substances common in the environment are also incorporated, and the deep-sea nodules have a pronounced ability to concentrate trace elements from seawater such as Co, Ni, Zn, and Pb. This is apparent in Table 12.19 which compares concretion concentrations with those in seawater, sediments, and crystal rock.. Mero

Figure 12.20 Manganese Nodule Growing around a Core of Phosphorite. Taken from the Blake Plateau at a Depth of about 800 m. From Manheim 1965 with permission of the author.

(1965) has summarized the chemical content of ferromanganese concretions in a very detailed set of tables. The composition of nodules is highly variable from place to place (Table 12.18B), and Table 12.20 lists some maximum, minimum, and mean values. These variations are sometimes systematic and may be the result of different origins of the nodules. For example, mid-Pacific nodules tend to have high cobalt contents (Mn/Co < 300), while samples taken near Japan or off the coasts of North and South America tend to be cobalt-deficient with Mn/Co atom ratios greater than 300. The latter concretions may have formed slowly from dilute Mn solutions of continental origin, whereas the former may have been precipitated rapidly from volcanogenic seawater (Arrhenius, Mero, and Korkisch, 1964).

This brings us to the question of the origin of the nodules. The idea that organisms are involved has been persistent. Indeed, the nodules do contain

Table 12.18　Composition of Manganese Nodules
(From Chester, 1965, with permission of Academic Press, Inc.)

A. Manganese Nodule Analyses (wt % oxides)

	1	2	3	4	5	6	7	8	9	10	11	12	13
Fe	22.44	14.0	12.16	15.3	11.4	17.0	14.6	9.6	12.4	—	14.0	17.5	14.58
Mn	32.04	19.0	19.18	—	—	—	—	23.2	20.5	—	24.2	16.3	22.06
Al	2.95	0.73	2.87	—	—	—	—	1.59	—	0.2, 2.0	2.9	3.1	2.36
Sb	N.D.[a]	—	—	—	—	—	—	—	—	—	—	—	—
Ba	0.59	—	—	—	—	—	—	—	—	—	0.18	0.17	0.31
Bi	N.D.[a]	—	—	—	—	—	—	—	—	—	—	—	—
Be	N.D.[a]	—	—	—	—	—	—	—	—	0.0001, 0.0003	—	—	—
B	0.005	—	—	—	—	—	—	—	—	0.01, 0.03	0.029	0.03	0.0295
Cd	0.0010	—	—	—	—	—	—	—	—	0.0003, 0.001	—	—	0.0010
Ca	0.27	—	1.81	—	—	—	—	2.21	—	present	1.9	2.7	1.78
Cr	<0.0005	—	—	—	—	—	—	—	—	<0.0005	0.001	0.002	<0.001
Co	0.23	0.28	0.36	0.30	0.39	0.17	0.29	0.50	0.59	0.002, 0.006	0.35	0.31	0.34
Cu	0.59	0.55	—	0.12	0.39	0.29	0.27	0.06	—	0.003, 0.01	0.53	0.20	0.33
Ga	0.0024	—	—	—	—	—	—	—	—	<0.0005	0.001	—	0.0017
Ge	0.0007	—	—	—	—	—	—	—	—	<0.0010	—	—	—
La	0.03	—	—	—	—	—	—	—	—	—	0.016	—	0.023
Pb	0.25	—	—	—	—	—	—	—	—	0.01, 0.03	0.09	0.10	0.15
Mg	0.27	—	1.47	1.8	1.6	1.8	1.7	1.15	—	0.1, 1.0	1.7	1.7	1.47
Mo	0.054	—	—	0.044	0.056	0.054	0.051	0.20	—	<0.0005	0.052	0.035	0.068
Ni	0.60	0.46	0.47	0.56	0.49	0.68	0.58	0.63	0.42	0.01, 0.003	0.99	0.42	0.57

	1	2	3	4	5	6	7	8	9	10	11	12	13
P	0.17	0.54	0.18	—	—	—	—	0.22	—	0.1 ⎱	—	—	0.28
K	1.51	—	—	—	—	—	—	—	—	1 ⎰	0.8	0.7	1.00
Sc	0.0010	—	—	—	—	—	—	—	—	—	0.001	0.002	0.0013
Ag	0.0019	—	—	—	—	—	—	—	—	—	0.0003	—	0.0011
Na	2.74	—	—	—	—	—	—	—	—	0.3 ⎱	2.6	2.3	2.55
Si	—	—	8.05	7.2	10.0	8.5	8.6	4.06	—	3.0 ⎰	9.4	11.0	8.35
Sr	0.13	—	—	—	—	—	—	—	—	—	0.081	0.09	0.10
Tl	0.013	—	—	0.007	0.007	0.007	0.007	—	—	0.0001 ⎱	—	—	0.007
Sn	0.03	—	—	—	—	—	—	—	—	0.0003 ⎰	—	—	—
Ti	0.56	0.81	0.60	0.38	0.81	0.37	0.52	0.55	—	present	0.67	0.80	0.61
W	0.009	—	—	—	—	—	—	—	—	0.0003 ⎱	—	—	—
V	0.064	—	—	0.047	0.053	0.060	0.053	0.073	—	0.0018 ⎰	0.054	0.07	0.059
Y	0.003	—	—	—	—	—	—	—	—	0.01 ⎱	0.016	0.018	0.012
Zn	0.66	—	—	—	—	—	—	—	—	0.03 ⎰	0.047	—	0.35
Zr	0.012	0.0064	—	—	—	—	—	—	—	—	0.063	0.054	0.034

[a] N.D. = not detected.
1. Riley and Sinhaseni (1958); acid-soluble fraction of 3 Pacific modules.
2. Goldberg (1954); average of 33 nodules from 11 East Pacific stations.
3. Skornyakova et al. (1962); average of 31 nodules from the Pacific Ocean.
4. Willis and Ahrens (1962); 8 nodules from the Atlantic Ocean.
5. Willis and Ahrens (1962); 7 nodules from the Pacific Ocean.
6. Willis and Ahrens (1962); 4 nodules from the Indian Ocean.
7. Willis and Ahrens (1962); average of the 19 nodules given above.
8. Dietz (1955); modified from the analysis of a nodule from the Gilbert Seamount, Pacific Ocean.
9. Dietz (1955); 4 manganese crusts and nodules from the Pacific Ocean.
10. Manheim (1961); semiquantitative analyses of outer layer of a nodule from the Baltic Sea.
11. Mero (quoted in Arrhenius, 1963); average bulk composition of Pacific Ocean nodules.
12. Mero (quoted in Arrhenius, 1963); average bulk composition of Atlantic Ocean nodules.
13. Average values obtained from this table (excluding column 10).

B. Chemical Composition of the Pacific Ocean Iron-Manganese Concretions, wt %
(From Skornyakova, Andruschenko, and Fumina, 1964, with permission of Pergamon Press, Ltd.)

	Station[a]																	
	3633	3802	3899	3996a	3996b	4080	4090a	4090b	4990c	4104	4191a	4191b	4191c	4281	4351	4370	HD = 4	BDB = 15
SiO_2	13.31	14.71	26.61	15.31	13.82	27.72	18.75	18.20	22.88	31.59	29.89	18.59	22.05	14.05	12.97	15.34	1.11	12.51
Al_2O_3	5.23	6.31	8.54	4.96	3.69	6.85	6.34	5.12	5.97	7.63	8.75	5.68	4.97	5.29	4.98	5.88	1.06	5.34
TiO_2	1.76	0.76	1.57	0.79	0.95	0.68	1.38	0.94	0.84	0.74	0.70	1.04	1.04	0.58	1.00	1.17	1.74	1.09
Fe_2O_3	20.53	10.32	27.16	17.16	21.17	15.47	17.34	15.11	15.94	12.75	10.42	14.99	17.93	11.85	17.03	20.86	16.41	18.34
MnO	22.20	31.43	6.53	24.56	21.13	16.84	20.37	24.14	20.59	17.10	16.25	21.79	19.66	30.36	26.08	20.75	30.72	24.63
NiO	0.42	0.95	0.06	0.83	0.54	0.37	0.52	0.81	0.78	0.55	0.75	0.75	0.47	1.20	0.67	0.52	0.39	1.03
CoO	0.67	0.31	0.10	0.25	0.18	0.28	0.42	0.40	0.46	0.40	0.15	0.31	0.45	0.56	0.59	0.46	1.93	0.49
CaO	2.50	3.00	3.07	2.81	2.19	2.17	2.19	1.93	2.30	2.70	1.81	2.24	2.60	2.46	2.46	2.20	4.01	2.82
MgO	2.26	3.96	0.40	2.17	2.18	2.47	2.24	2.57	2.86	2.43	2.89	2.22	2.45	3.23	3.01	2.72	2.74	3.35
P_2O_3	0.40	0.43	0.48	0.31	0.46	0.35	0.49	0.48	0.36	0.31	0.26	0.45	0.42	0.36	0.38	0.50	0.49	0.64
Calcination loss	27.72	21.77	20.17	25.97	27.27	20.77	23.91	21.87	22.80	20.14	19.87	25.57	24.31	24.10	27.00	26.99	33.84	24.66
Active oxygen	4.45	6.76	1.31	5.27	4.50	3.65	4.32	5.22	4.59	3.55	3.36	4.92	3.95	6.51	5.70	4.52	7.00	5.30

Station[b]

	3150	3729	4074	4217	4265a	4265b	4331	4359	4362b	4362b	CD = 47a	HD = 47b	HD = 72
SiO_2	6.09	13.53	30.87	19.63	7.18	11.40	14.20	20.30	13.10	11.63	10.91	10.06	14.99
Al_2O_3	1.04	4.09	8.21	6.50	3.52	4.97	2.84	7.42	3.65	3.22	7.93	4.90	1.24
TiO_2	1.16	1.24	0.66	0.75	—	—	1.85	1.22	1.11	1.80	0.61	0.65	0.39
Fe_2O_3	11.10	27.80	15.30	20.08	1.22	1.72	25.27	14.59	16.12	18.84	13.00	14.22	19.49
MnO	N.D.	N.D.	N.D.	N.D.	3.90	2.65	4.35	N.D.	6.31	6.70	N.D.	N.D.	2.38
MnO_2	53.60	22.42	19.00	24.05	57.30	52.00	21.25	27.27	27.89	27.84	38.30	38.20	35.14
NiO	0.53	0.50	0.28	0.55	0.16	0.23	0.34	0.68	0.62	0.51	1.57	1.16	0.35
CoO	0.51	—	0.15	0.16	Tr	Tr	0.59	0.41	0.60	0.65	0.27	0.21	0.34
	0.53		0.17	0.18								0.29	
MgO	2.74	1.79	1.86	1.85	1.74	2.20	0.70	2.08	1.33	1.20	3.90	3.35	3.23
CaO	2.65	2.21	2.61	2.68	3.19	2.56	3.25	3.30	3.92	3.09	2.45	3.34	1.64
NaO_2	2.16	2.09	2.98	2.94	5.15	4.87	2.36	2.60	2.64	2.27	2.67	2.57	1.90
K_2O	1.46	0.76	1.16	1.04	0.87	1.16	0.63	1.11	0.97	0.80	1.21	0.99	1.54
H_2O^-	7.27	8.59	10.30	12.11	8.43	8.15	13.70	11.85	13.36	13.29	8.76	10.36	8.66
H_2O^+	8.23	13.07	5.62	6.33	7.14	7.17	7.87	6.72	7.23	7.48	8.47	9.16	7.79
CO_2	0.48	1.03	N.D.	—	0.18	0.70	0.56	0.53	0.57	0.60	—	0.45	N.D.
BaO	1.09	—	—	—	—	—	—	—	—	—	—	—	—
PbO	—	—	0.05	0.11	—	—	—	—	—	—	0.08	0.08	—
CuO	—	—	—	—	—	—	—	—	—	—	0.28	0.22	—

[a] Analyses carried out in the Marine Geology Section, Institute of Oceanology, Academy of Sciences, USSR.
[b] Analyses carried out in the Exogene Section of the I.G.Ye.M., Academy of Sciences, USSR.

Table 12.19 Comparison of the Mean Chemical Composition of Pacific Ocean Concretions with the Clarke Content of Elements in Seawater and the Crust Contents, wt. % (From Skornyakova, Andruschchenko, and Fumina, 1964, with permission of Pergamon Press, Ltd.)

Element[a]	Concretions[b]	Seawater	Crust	Basalts	Granites	Recent Sediments	
						Carbonate	Red Clay
B	—	0.029	4.5×10^{-4}	5×10^{-8}	0.0005	0.008	0.0055	0.023
Na	—	2.6	1.05	2.40	1.8	2.71	2.0	4.0
Mg	1.47	1.7	0.13	2.35	4.6	5.5	0.4	2.1
Al	2.87	2.9	1×10^{-6}	7.45	7.8	7.7	2.0	8.4
Si	8.05	9.4	—	—	23.0	33.0	3.2	25.0
P	0.18	—	varies	0.12	0.11	0.076	0.035	0.15
K	—	0.8	3.8×10^{-2}	2.35	0.83	3.6	0.29	2.5
Ca	1.81	1.9	4×10^{-2}	3.25	7.6	1.29	31.24	2.9
Sc	—	0.001	4×10^{-7}	6×10^{-4}	0.003	0.001	0.0002	0.0019
Ti	0.60	0.67	1×10^{-7}	0.61	1.3	0.23	0.077	0.46
V	—	0.054	3×10^{-7}	0.02	0.0230	0.0066	0.002	0.012

Cr	—	0.001	1×10^{-2}	0.03	0.017	0.0013	0.0011	0.009
Mn	19.18	24.2	1×10^{-7}	0.10	0.15	0.0046	0.1	0.67
Fe	12.16	14.0	1×10^{-7}	4.20	8.65	2.19	0.9	6.5
Co	0.36	0.35	4×10^{-8}	2×10^{-3}	0.0048	0.0004	0.0007	0.0074
Ni	0.47	0.99	2×10^{-7}	0.02	0.013	0.001	0.003	0.022
Cu	—	0.55	2×10^{-7}	0.01	0.0087	0.002	0.003	0.025
Zn	—	0.047	5×10^{-7}	0.02	0.01	0.005	0.0035	0.016
Ga	—	0.001	3×10^{9}	1×10^{-4}	0.0017	0.0017	0.0013	0.002
Sr	—	0.081	8×10^{-4}	0.035	0.046	0.027	0.2	0.018
Y	—	0.018	—	—	0.0021	0.0038	0.0042	0.009
Zr	—	0.063	—	0.025	0.014	0.016	0.002	0.015
Mo	—	0.052	1.2×10^{-2}	1×10^{-3}	1.1×10^{-4}	1.1×10^{-4}	0.0003	0.0027
Ag	—	0.0003	1.5×10^{-8}	1×10^{-5}	1.1×10^{-5}	4×10^{-5}	1×10^{-5}	1.1×10^{-5}
Ba	—	0.18	$5 - 10^{-8}$	0.05	0.033	0.063	0.019	0.25
La	—	0.016	3×10^{-8}	—	0.0015	0.005	0.001	0.011
Yb	—	0.0031	—	—	2×10^{-4}	4×10^{-4}	0.0001	0.0015
Pb	—	0.09	5×10^{-7}	1.6×10^{-3}	6×10^{-4}	0.0017	9×10^{-4}	0.008

[a] Mean according to the data of chemical analyses at the I.O.A.N. and the I.G.Ye.M.
[b] Mean according to Mero's x-ray spectral analyses.

Table 12.20   Composition of Pacific Ocean Ferromanganese Concretions (From Skornyakova, Andruschenko, and Fumina, 1964, with permission of Pergamon Press, Ltd.)

Element	Content, wt %			Element	Content, wt %		
	Min.	Max.	Mean		Min.	Max.	Mean
B	0.007	0.06	0.029	Ni	0.16	2.0	0.99
Na	1.5	4.7	2.6	Cu	0.028	1.6	0.53
Mg	1.0	2.4	1.7	Zn	0.04	0.08	0.047
Al	0.8	6.9	2.9	Ga	0.0002	0.003	0.001
Si	1.3	20.1	9.4	Sr	0.24	0.16	0.081
K	0.3	3.1	0.8	Y	0.033	0.045	0.016
Ca	0.8	4.4	1.9	Zr	0.009	0.12	0.063
Sc	0.001	0.003	0.001	Mo	0.01	0.15	0.052
Ti	0.11	1.7	0.67	Ag	—	0.0006	0.0003
V	0.021	0.11	0.094	Ba	0.08	0.64	0.18
Cr	0.001	0.007	0.001	La	0.009	0.024	0.016
Mn	8.2	52.2	24.2	Yb	0.0013	0.0066	0.0031
Fe	2.4	26.6	14.0	Pb	0.02	0.36	0.09
Co	0.014	2.3	0.35				

an appreciable amount of organic carbon (Table 12.21), but so does just about every other solid surface in the sea; in fact, the C/N ratios are similar to those for fine-grained sediments (Manheim, 1965). So ubiquitous are organic materials, and even living organisms on solid surfaces in the sea, that their presence is certainly no evidence that they are playing a functional role. Until evidence to the contrary is forthcoming, the intermediacy of biological processes in nodule formation remains an unnecessary hypothesis. To a chemist the mystery which has been confected concerning nodule formation is itself something of a puzzle. Elements capable of existing in many valence states commonly have one in which they appear to be especially happy, and this is particularly true of manganese, whose valence ranges from $+2$ to $+7$ but which under ordinary circumstances has a very strong preference for $Mn(IV)O_2$. Manganese solutions all tend to oxidize slowly or reduce to this preferred state—the brown stains which form on flasks of old potassium permanganate solutions are an example familiar to every chemist. Barnes (1967) has given the reactions for nodule formation as

(12.8) $$Mn^{2+} + \tfrac{1}{2} O_2 + 2\,OH^- \rightleftharpoons MnO_2 + H_2O$$

(12.9) $$5\,Mn^{2+} + 2\,O_2 + 10\,OH^- \rightleftharpoons 4\,MnO_2 \cdot Mn(OH)_2 \cdot 2\,H_2O + 2\,H_2O$$

and

(12.10) $$4\,MnO_2 \cdot Mn(OH)_2 \cdot 2\,H_2O + \tfrac{1}{2} O_2 \rightleftharpoons 5\,MnO_2 + 3\,H_2O$$

Table 12.21   Organic and Carbonate Carbon Content of Manganese-Iron Concretions

(From F. T. Manheim, 1965, with permission of the author)

Vessel and Location of Sample	Description	C_{total}	C_{org}	CO_2
Challenger, Pacific Sta. 276, 4260 m	Large (25 cm), dense, gray-black nodule	0.07	0.03	0.14
Challenger, Pacific Sta. 289, 4590 m	?	—	0.37	—
Chain, Indian O., 1° 37'S, 58° 37'E, 3870 m	Dark brown-black friable mass	0.17	0.12	0.16
Atlantis II, Blake Plateau, 30° 53'N, 78° 47'W, 800 m	"Potato ore," similar to small nodule	1.13	0.11	3.8
Gosnold, Gulf of Maine, composite	1–4 mm black Mn crusts on igneous boulders	2.72	2.60	0.42
Meotid, Black Sea, 44° 35'N, 33° 17'E, 70–120 m	Concretions around *modiolus* shells, avg. 3 analyses	3.84	2.75	4.00
Ermak, Barents Sea, approx. 78° 7'N, 63° 33'W, 340 m	Avg. 2 analyses, irregular brown cakes	2.73	1.48	4.60
Skagerak and *Aranda*, Baltic Sea (proper), 50–100 m	Composite, several dozen crusts on pebbles and cobbles	2.70	2.50	0.74
Gulf of Riga (Baltic), 24–48 m	Avg. 9 analyses, concretions and crusts	2.50	1.61	3.27
Finnish lake ore	Avg. 20 samples from 12 lakes	—	1.45	—
Swedish lake ore	Composite sample, south-central Sweden	1.74	1.66	0.28

He points out that the partial pressure of dissolved oxygen is a crucial parameter and proposes that the hydrostatic pressure may be controlling in the equilibria and kinetic processes, albeit in ways not yet thoroughly understood. The problem, then, becomes one, not so much of the chemistry of the precipitation process as of the source of the Mn^{2+}. There are, in all likelihood, two sources, both important: the free seawater above the sea-sediment interface, and the interstitial water in the sediment below (Manheim, 1965). Volcanic activity contributes to the manganese content of seawater (see above), and Goldberg and Arrhenius (1958) have argued that the contribution of streams and rivers alone is adequate to account for the Mn and Fe contents of seawater and marine sediments. The probable sequence of events is as follows:

1. Slightly alkaline seawater (pH 8) rich in dissolved O_2 is saturated with Mn and Fe in solution.

2. Hydroxides of these elements precipitate and form colloidal particles, which as they aggregate and settle act as effective scavengers for trace elements.

3. Once on the sea floor, the particles are swept along until they intercept some surface, in particular conducting surfaces which can remove their charge, to which they can adhere.

4. The growing nodule can concentrate further trace elements by adsorption.

5. One hypothesis proposes that the growing nodules, by means unknown, appear to be rolled along the ocean floor until they reach a certain critical size, too heavy to be moved further; they are then covered over by sediment and preserved.

Table 12.22 gives some estimates of the rate of accretion of shallow water nodules. Growth rates as slow as 1 mm/100,000 years and as fast as 10 cm/100 years (on a conductive naval shell) have been determined (Manheim, 1965). However, Bender, Ku, and Broecker (1966), utilizing the results of

Table 12.22   Iron Manganese Crust and Concretion Accumulation Rates in Shallow Marine Areas
(From Manheim, 1965, with permission of the author)

Area	Depth, m	Salinity, ‰	Est. Rate of Accum., mm/1000 yr
Atlantic Ocean (Blake Plateau)	300–900	35–26	0.1–?
Atlantic (Gulf of Maine-N. England-Newfoundland Banks)	50–200	32–35	20–500
Baltic Sea (central, northern and Gulf of Finland)	15–220	3.5–12	20–1000
Barents Sea	170–1000	34–36	1–500
Black Sea	60–150	18–19	50–1000
Caspian Sea	20	?	100–1000
Kara Sea	30–120	26–34	10–1000
North Sea (Clyde Estuary)	25–160	32–34	50–1000
Pacific Ocean (Southern California)	120–800	35–36	?
Pacific Ocean (Japan)	114–260	35	?
Petschora Sea	Transitional, Barents-Kara		
South China Sea	Poorly documented		
White Sea	30–160	26–32	50–3000

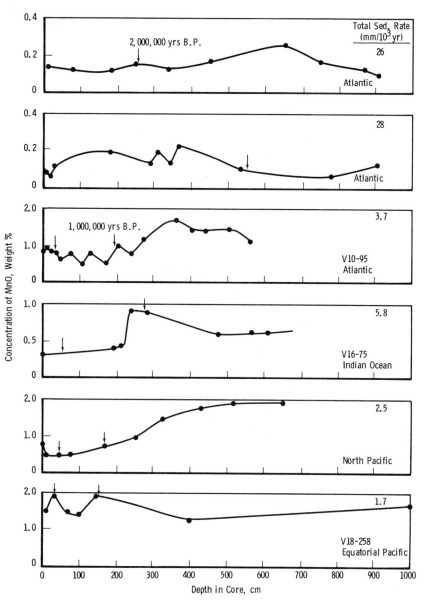

Figure 12.21   Manganese Profiles of Some Deep-Sea Cores. From Bender, Ku, and Broecker (1966), with permission of AAAS and the authors.

Th[230] dating, have concluded that, generally speaking, the rate of Mn deposition is nearly constant over the world's oceans in both nodules and sediments. They have developed a simple model which gives the rate of growth of the nodule radius by

$$(12.11) \qquad \frac{dr}{dt} = \frac{S}{4\rho f}.$$

where S is the sedimentation rate of MnO [sic], ρ the density of the nodule, and f the fraction of MnO. The same authors also give MnO profiles of some sediment cores (Figure 12.21).

In addition to its occurrence in the ferromanganese nodules, iron is also found in marine sedimentary ores deposited in bogs, marshes, and near-

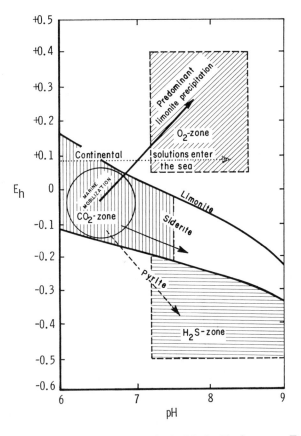

Figure 12.22 Precipitation of Iron Minerals in the Marine Environment. From Borchert (1965a), with permission of Academic Press, Inc.

Table 12.23   Phosphorite Compositions
(From Mero, 1965, with permission of the publisher and author)
A. Chemical Composition of Phosphorite Nodules from the California Borderland Area[a]

Constituent	Geographic Location					
	Forty Mile Bank	Santa Monica Canyon	Redondo Canyon	Outer Banks	Thirty Mile Bank	Patton Escarpment
CaO	47.35	45.43	45.52	46.58	37.19	47.41
R_2O_3[b]	0.43	0.30	2.03	0.70	3.93	1.40
P_2O_5	29.56	29.19	28.96	29.09	22.43	29.66
CO_2	3.91	4.01	4.30	4.54	4.63	4.87
F	3.31	3.12	3.07	3.15	2.47	3.36
Organic	0.10	1.90	2.25	0.44	0.35	1.50
Insoluble in HCl	2.59	3.57	4.45	3.57	20.99	2.12
Totals	87.25	87.52	90.58	88.07	91.99	90.32

[a] Compositions in weight percentages. Remaining portions largely $MgCO_3$, H_2O, and soluble SiO_2.
[b] R denotes metals.

B. Chemical Composition of Phosphorite from Various Locations of the World[a]

	Sea Floor		Land					
Constituent	Forty Mile Bank off California	Agulbas Bank off South Africa	Idaho	Florida	Russia	Curaco Islands	Tunisia	Morocco
CaO	47.4	37.3	48.0	36.4	27.9	50.0	44.3	51.6
R_2O_3	0.43	9.4	1.2	12.7	3.5	—	—	—
P_2O_5	29.6	22.7	32.3	31.2	17.9	37.9	29.9	32.1
CO_2	3.9	7.1	3.1	2.2	3.7	3.9	5.8	5.5
F	3.3	—	0.5	2.0	2.0	0.7	3.6	4.2
Organic	0.1	—	—	6.2	3.2	—	—	—
Totals	84.7	76.5	85.1	90.8	58.2	92.5	83.6	93.4

[a] Compositions in weight percentages.

Table 12.24 Salts Precipitated During the Concentration of Seawater (From Borchert, 1965b, with permission of Academic Press, Inc.)

Density	Volume, l	CaCO₃, g/kg	CaSO₄ 2H₂O, g/kg	NaCl, g/kg	MgCl₂, g/kg	MgSO₄, g/kg	NaBr, g/kg	KCl, g/kg
1.026	1.000							
1.050	0.533	0.0642						
1.126	0.190	0.0530	0.5600					
1.202	0.112		0.9070					
1.214	0.095		0.0508	3.2614	0.0040	0.0078		
1.221	0.064		0.1476	9.6500	0.0130	0.0356		
1.236	0.039		0.0700	7.8960	0.0262	0.0434	0.0728	
1.257	0.0302		0.0144	2.6240	0.0174	0.0150	0.0358	
1.278	0.023			2.2720	0.0254	0.0240	0.0518	
1.307	0.0162			1.4040	0.5382	0.0274	0.0620	
Total deposit		0.1172	1.7488	27.1074	0.6242	0.1532	0.2224	
Salts in last bittern (g)				2.5885	1.8545	3.1640	0.3300	0.5339
Sum, g		0.1172	1.7488	29.6959	2.4787	3.3172	0.5524	0.5339

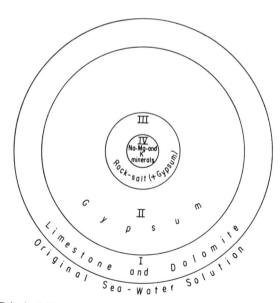

Figure 12.23 Principal Stages in the Evaporation of Seawater. From Borchert, 1965, with permission of Academic Press, Inc.

shore waters. (See the review by Borchert, 1965a.) The particular mineral obtained, whether limonite (60–70% Fe_2O_3), siderite ($FeCO_3$), or pyrite (FeS_2), depends on the redox potential (E_h) and the acidity (pH) of the environment (Figure 12.22). Then, too, we must not overlook the fact that iron oxides are responsible for the color of the red clay which represents such a large fraction of pelagic sediments.

Also we must not forget phosphorite deposits and, although they are biogenous in origin, I should like to mention them briefly here. Phosphorite or phosphate rock is a common continental shelf material in certain areas, notably off the coast of California. This mineral, like MnO_2, is found as nodules; Table 12.23 gives the results of some analyses.

Finally let us say a few words about the great deposits of salts laid down by the evaporation of shallow, isolated, ancient seas (for a review see Borchert, 1965b). As the seawater evaporates, various salts precipitate out from the brine (Table 12.24), the more insoluble first, the more soluble last, in a regular sequence beginning with limestone ($CaCO_3$) and dolomite [$CaMg(CO_3)_2$] and ending with salts such as the chlorides of Na and K (Figure 12.23).

7   The Origin and Chemical History of the Earth's Oceans

Where did the water and the chemical constituents of the Earth's oceans come from? How have their quantities varied in the geologic past? Descartes, I believe, remarks that there is no opinion so absurd that it has not been stoutly defended by some philosopher at one time or another, and I think this rather cynical remark nicely applies to theories concerning the chemical origin and evolution of the seas. For example, with respect to the question of the volume of the ocean, the theories advanced cover all possibilities (Figure 12.24). Some hold that the volume has remained pretty much as it is, others that the oceans are relatively recent, and the remaining theories run the spectrum from slow to rapid accumulation. But, lest we be too quick to criticize, we must not forget that this diversity of opinion is an immediate consequence of the extreme difficulty and complexity of these problems. Rubey (1951) warns us that the history of the Earth's atmosphere and hydrosphere "cannot be told until we have solved nearly all other problems of earth history," and he wryly suggests that his address on the "Geologic History of Sea Water" could be more aptly titled "The Problem of the Source of Sea Water and Its Bearing on Practically Everything Else." In the face of such horrendous complexity and fragmentary data the progress that has been made becomes perhaps much more impressive. Since Conway's (1942, 1943) magnificent pioneering work, Holland (1965) cites three milestones marking this advance: Rubey's (1951) concept of excess volatiles; the appreciation of the geologically

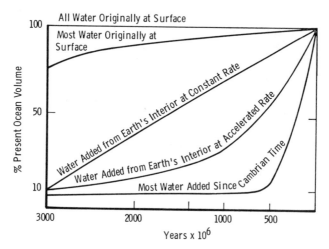

Figure 12.24 Various Theories of The Volume of the Oceans in the Geological Past. From Revelle (1955), with permission of Sears Foundation for Marine Research.

brief residence times of cations in the sea (Barth, 1952; Goldberg and Arrhenius, 1958; Goldberg, 1963); and finally Sillen's (1961) recent fresh attack on the problem of the mechanism for pH control in the seas.

Our first thought on addressing the problem of the origin of the water of the world's oceans might be that it was the consequence of the condensation of a primordial envelope of steam as the hot Earth cooled. This hypothesis appears to be untenable. The gases neon, argon, krypton, and xenon are 10^6 to 10^{10} times less abundant on Earth than in the atmospheres of stars, presumably because these inert gases, not being chemically bound, have escaped from this planet. Indeed, as one would expect, the larger discrepancies are found for the lighter gases. Similarly, gases of comparable molecular weight, H_2O, N_2, O_2, CO, and CO_2, would also have been lost from the atmosphere of the primitive earth. It follows that these substances were not free but were rather chemically bound or occluded in the condensed phase substance of the Earth (Rubey, 1951); that they were subsequently released by volcanic and subterranean hydrothermal activity; and that, after the Earth had sufficiently cooled, they accumulated to form the oceans. The striking differences between Earth and Mars are entirely compatible with this point of view. The atmosphere of Mars is very tenuous as the consequence of the escape of gases from that planet. In contrast with the great oceans of the Earth, undoubtedly the most spectacular visual feature of our world to an out-in-space observer, there is only enough moisture left on the surface of Mars to form an ever-so-light frost. Then, too, while the surfaces of our continents are ravaged with the traces of volcanism and other titanic geological processes, the bland face of Mars shows little evidence of volcanic or other activity of the type that released the waters of the oceans here (Revelle, 1955).

But the quantity of water contained in the oceans is so immense! Could all this water have conceivably been squeezed out of the interior of the planet? The answer is yes, easily. Rubey (1951) has estimated that, even at their present rate of discharge, which may be considerably less now than earlier in the Earth's geological history, hot springs on the continents and ocean floor release some 6.6×10^{16} g of water per year. Over a period of 3×10^9 years this amounts to 2.0×10^{26} g of water. Yet Rubey estimates the total quantity of water in the atmosphere, hydrosphere, biosphere, and buried in ancient sedimentary rocks to be 1.7×10^{24}, or only about 0.8% of the discharge. Thus, even if only a small fraction of the total quantity was derived from a primary magmatic source, in the long course of geologic time hot springs can readily account for the total mass of the Earth's oceans.

There now appears to be fairly general agreement that the origin of the oceans was as described above, but there is considerably less agreement about the details of the ocean's development. The fossil record provides arguments that the volumes of the oceans have remained relatively constant for at least the last sixth of geologic time. Yet, on the other hand, the existence of atolls and of flat-topped sea mounts in the Pacific, deeply buried remains of shallow water creatures, and the surprising thinness of certain sedimentary deposits can be interpreted as supporting the idea that the oceans are relatively recent, that as much as one-fourth of all the water in the seas appeared since the Mesozoic Era in the last thirtieth or fortieth of geologic time (Revelle, 1955). Still there seems to be no evidence for the progressive flooding of the continental landmasses which one would expect if the volume of the oceans has been steadily increasing (Nicolls, 1965).

There have been minor (5–10%) variations in the total volume of the oceans as a consequence of the growth and melting of ice sheets.

The answer to our second question—where did the electrolytes in seawater come from?—is fairly obvious: from the condensed phase of the Earth, but the paths by which they found their way into the oceans are much less clearly defined. Careful examination and comparison of the elemental composition of seawater and the Earth's crust is most instructive (Table 12.25). Again our intuition fails us; it is perfectly obvious from Table 12.25 that there is no direct relationship between marine and lithospheric abundances, that the electrolytes in seawater are not simply leached out of the Earth's crust. But, as we study the last column of this table more closely, we find illustrated in it many of the most fundamental principles of marine chemistry, described earlier in this book; the disorder evaporates; and we discover the elements falling into a relatively few remarkably well-defined categories. Let us take sodium as our

Table 12.25   Elements in Seawater and the Earth's Crust
(From Sverdrup, Johnson, and Fleming, 1942, with permission of
Prentice-Hall, Inc.)

Element	Seawater, $S = 35\%_0$ mg/kg	Potential "Supply" in 600 g of rock (mg/kg of seawater)	Percentage in Solution
Silicon	4	165,000	0.002
Aluminum	0.5	53,000	0.001
Iron	0.02	31,000	0.0001
Calcium	408	22,000	1.9
Sodium	10,769	17,000	65
Potassium	387	15,000	2.6
Magnesium	1,297	13,000	10
Titanium	—	3,800	?
Manganese	0.01	560	0.002
Phosphorus	0.1	470	0.02
Carbon	28	300	9
Sulphur	901	300	300
Chlorine	19,353	290	6700
Strontium	13	250	5
Barium	0.05	230	0.02
Rubidium	0.2	190	0.1
Fluorine	1.4	160	0.9
Chromium	p[a]	120	?
Zirconium	—	120	?
Copper	0.01	60	0.02
Nickel	0.0001	60	0.0002
Vanadium	0.0003	60	0.0005
Tungsten	—	41	?
Lithium	0.1	39	0.2
Cerium	0.0004	26	0.002
Cobalt	p	24	?
Tin	p	24	?
Zinc	0.005	24	0.02
Yttrium	0.0003	19	0.002
Lanthanum	0.0003	11	0.003
Lead	0.004	10	0.04
Molybdenum	0.0005	9	0.005
Thorium	<0.0005	6	0.01
Cesium	0.002	4	0.05
Arsenic	0.02	3	0.7
Scandium	0.00004	3	0.001
Bromine	66	3	2000
Boron	4.7	2	240
Uranium	0.015	2	0.8
Selenium	0.004	0.4	1
Cadmium	p	0.3	?

Table 12.25 (Continued)

Element	Seawater, $S = 35\%_0$ mg/kg	Potential "Supply" in 600 g of rock (mg/kg of seawater)	Percentage in Solution
Mercury	0.00003	0.3	0.001
Iodine	0.05	0.2	25
Silver	0.0003	0.06	0.5
Gold	0.0_56	0.003	0.3
Radium	0.0_93	0.0_66	0.05

[a] p = present.

basis, since it has so little chemistry in the seas. With respect to this element many of the remaining elements fall into three categories: those with percentages in solution comparable to Na; those with very much greater percentages; and those with very much smaller percentages. The very small percentages are associated with elements, such as Si and Al, whose crustal compounds are highly insoluble and/or with elements, such as Fe, Mn, Ni, V, which are known to be scavenged out of seawater by precipitation, strong sorption process, and/or nodule formation. The concentrations of a subclass of elements are depleted in seawater by the intervention of biological processes. Thus, in Table 12.25 P shows depletion but Ca and C unexpectedly are not conspicuous. However, if we compare the concentration of Ca^{2+} and of CO_3^{2-} in the seas and the amounts added by rivers (Table 12.26), this depletion by biological precipitation of $CaCO_3$ becomes quite evident. The depletion of the nutrients NO_3^- and SO_4^{2-} is also shown in Table 12.26; it is about 10 times greater in the case of the NO_3^-, the more important nutrient.

Table 12.26 Present Amount of Ions in Sea, and Amount of Dissolved Ions Added by Rivers in 100 Million Years, in moles/cm² of Total Earth Surface
(From Sillen, 1967, with permission of The Chemical Society and the author)

	Na^+	Mg^{2+}	Ca^{2+}	K^+	Cl^-	SO_4^{2-}	CO_3^{2-}	NO_3^-
Present in ocean	129	15	2.8	2.7	150	8	0.3	0.01
Added in 100 million years	196	122	268	42	157	84	342	11

The elements found in seawater in much greater amounts, Cl, B, Br, and S, all occur in anionic form—a first hint that anionic or acidic materials might take a different path from the lithosphere to the hydrosphere than do cationic or basic materials (Figure 12.25). The elements in these anionic electrolytes are the excess volatiles discussed by Rubey (1951) (Table 12.27). If they do not result entirely from the weathering of the crustal material, then what is the additional source of these elements? His answer is that, like the water, they come from within the Earth and are spewed into the sea, directly, via rivers or via the atmosphere from volcanic and other geothermal activity—an explanation supported by their presence in present-day emanations (Table 12.28). The Cl concentration of seawater is particularly large. Most of the elements have definite finite residence times in the ocean (Table 5.3) indicating that they are circulated in the dynamics of the atmosphere-lithosphere-hydrosphere system (Figure 12.24), but chlorine does not. Of the chlorine

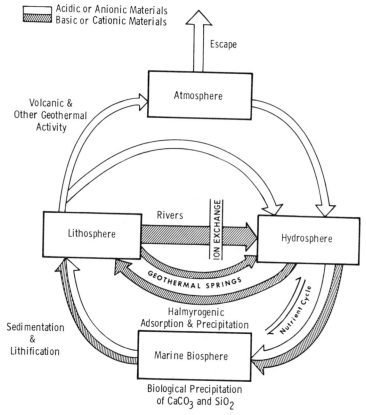

Figure 12.25 The Cycle of Cationic and Anionic Species in Seawater.

Table 12.27   Estimated Quantities (in units of 10^{20}) of Volatile Materials now at or near the Earth's Surface, Compared with Quantities of these Materials that Have Been Supplied by the Weathering of Crystalline Rocks (From Rubey, 1951, with permission of Geological Society of America and the author)

	H_2O	Total C as CO_2	Cl	N	S	H, B, Br, A, F, etc.
In present atmosphere, hydrosphere, and biosphere	14,600	1.5	276	39	13	1.7
Buried in ancient sedimentary rocks	2,100	920	30	4.0	15	15
Total	16,700	921	306	43	28	16.7
Supplied by weathering of crystalline rocks	130	11	5	0.6	6	3.5
"Excess" volatiles unaccounted for by rock weathering	16,600	910	300	42	22	13

added from the interior of the Earth, 75% is still present in the oceans; that is, the hydrosphere is a sink for Cl as well as H_2O (Holland, 1965). Rubey (1951) compares two alternative hypotheses concerning the rates at which the excess volatiles were added to seawater; the "quick-soak" (columns 1, 2, and 3 in Table 12.29) and the "slow-soak" (columns 4 and 5) hypotheses. According to the "quick-soak" view, all the excess volatiles were originally dissolved into a primitive ocean, according to the "slow-soak" view, these materials have been dissolving steadily throughout geologic time. The dissolution of the excess volatiles produces acid substances; therefore the primitive "quick-soak" oceans should have been very acid with a pH of only about 0.3 (Table 12.29) and the present ocean would have a pH of about 7.3 and a salinity roughly twice its present value, and more than half of the original CO_2 would still be in the atmosphere. Then, too, the neutralization of the highly acid primitive ocean would have required the weathering of a much more igneous rock and the subsequent deposition of much greater quantities of carbonate rocks than indicated by the existing geological record. Rubey (1951) in the light of these insurmountable difficulties is obliged to reject the "quick-soak" hypothesis and to conclude that both the quantity of water and the excess volatiles have increased gradually, so that the composition of both the hydrosphere and the atmosphere "has varied surprisingly little, at least since early in geologic time" and that "these variations have probably been within relatively narrow limits."

Table 12.28   Volume Percentages of Gases from Volcanoes, Rocks, and Hot Springs
(From Rubey, 1951, with permission of Geological Society of America and the author)

| | Volcano Gases from Kilauea and Mauna Loa | | | Gases from Rocks | | | | | | | Gases from Fumaroles of the Katma Region and from Steam Wells and Geysers of California and Wyoming | | |
| | | | | Basaltic Lava and Diabase | | | Obsidian, Andesitic lava, and Granite | | | | | | |
	Min.	Max.	Median	Min.	Max.	Median	Min.	Max.	Median		Min.	Max.	Median
CO_2	0.87	47.68	11.8	0.89	15.30	8.1	0.08	20.26	2.0	CO_2	0.03	1.24	0.2
CO	0.00	3.92	0.5	0.02	8.28	0.2	0.01	2.22	0.5	CO	—	0.01	Tr
H_2	0.00	4.22	0.4	0.38	6.18	1.2	0.08	11.60	0.4	O_2	0.00	0.08	Tr
N_2	0.68	37.84	4.7	0.27	7.21	2.0	0.03	3.90	1.2	CH_4	0.00	0.30	0.11
A	0.00	0.66	0.2	0.00	0.04	Tr	0.00	0.02	Tr	H_2	0.00	0.29	0.15
SO_2	0.00	29.83	6.4	—	—	—	—	—	—	$N_2 + A$	0.00	0.31	0.02
S_2	0.00	8.61	0.2	0.08	1.96	1.1	0.00	2.89	0.2	NH_3	—	0.02	0.01
SO_3	0.00	8.12	2.3	—	—	—	—	—	—	H_2S	0.00	0.10	0.02
Cl_2	0.00	4.08	0.05	0.06	1.33	0.5	0.01	10.59	0.5	HCl	0.01	0.57	0.06
F_2	—	—	—	0.00	14.12	3.8	0.25	7.80	2.3	HF	0.00	0.10	0.03
H_2O	17.97	97.09	73.5	71.32	92.40	83.1	69.44	98.55	92.9	H_2O	98.04	99.99	99.58
			100.0			100.0			100.0				100.00

Table 12.29 Composition of Atmosphere and Seawater under Alternative Hypotheses of Origin, Compared with Present-Day Conditions (From Rubey, 1951, with permission of Geological Society of America and the author)

	All "Excess" Volatiles in Primitive Atmosphere and Ocean (original P_{CO_2} very high)			Only a fraction of Total Volatiles in Primitive Atmosphere and Ocean (original $P_{CO_2} \leq 1.0$; life begins early)		Present-Day Conditions (6)
	Initial Stage: before Rock weathering (1)	Intermediate stage: CaCo$_2$ begins to precipitate (2)	Late Stage: life begins at $P_{CO_2} = 1.0$ (3)	Initial Stage: before Rock weathering (4)	Intermediate Stage: CaCO$_2$ begins to precipitate (5)	
Atmosphere (kg/cm^2)	14.2	13.8	2.1	<1.1	<1.1	1.0
N$_2$	9	9	50	7	7	78
CO$_2$	89	89	47	90	90	0.03
H$_2$S	2	2	3	3	3	—
O$_2$, others	—	—	—	—	Tr	22

(% by volume)

Ocean ($\times 10^{20}$ g)	16,600	16,600	16,600	<990	<990	14,250
Cl, F, Br	18.3	18.3	18.3	18.3	18.3	19.4
ΣS, ΣB, others	0.8	0.8	1.3	0.1	0.1	2.8
ΣCO_2	14.3	15.8	25.2	<1.1	<1.7	0.1
Ca (g/kg)	—	5.9	Tr	—	<5.5	0.4
Mg	—	1.3	5.2	—	<1.2	1.3
Na	—	3.1	12.5	—	<2.9	10.8
K	0.5	1.2	4.7	0.5	<1.1	0.4
H	33.9	Tr	—	<20.0	Tr	—
"Salinity" ‰	0.3	46.4	67.2		<30.8	35.2
pH		5.1	7.3	0.3	5.7	8.2
CaCO₃ pptd ($\times 10^{20}$ g)	None	None	980	None	None	1500±
Igneous rock eroded ($\times 10^{20}$ g)	None	4200±	17,000±	None	<240	11,000±

Whereas the anion content of seawater is largely plutonic in origin, the cation content is largely a product of the weathering of the Earth's crust (Figure 12.24). Notice in Tables 12.25 and 12.26 that K, although like Na it has little chemistry, is noticeably depleted in seawater. Conway (1942, 1943) believes that the cation contents in seawater are intimately interdependent and controlled primarily by the removal of K, and that furthermore the rate of this removal has not deviated widely from its present value. While the total amount of Na + K has remained relatively constant (Table 12.30), the Na/K ratio has undergone some rather drastic changes, especially around about 270×10^6 years ago.

Table 12.30   Average Contents of Sodium and Potassium and Average Values of Na/K Ratio in Argillaceous Rocks of Various Ages (From Nicolls, 1965, with permission of Academic Press, Inc.)

Time Scale, in 10^6 yr	Period	No. of Analyses	Na, %	K, %	(Na + K), %	Na/
1	Quaternary	5	0.55	2.6	3.15	0.2
70	Tertiary	12	0.88	1.9	2.78	0.4
	Cretaceous, upper	4	0.85	1.6	2.45	0.5
		17	0.81	2.07	2.88	0.3
135	Cretaceous, lower	8	0.99	2.2	3.19	0.4
180	Jurassic	16	1.04	2.6	3.64	0.4
225	Triassic	4	0.76	1.9	2.66	0.4
	Permian	28	0.96	2.2	3.16	0.4
270	Carboniferous, upper	9	0.36	2.51	2.87	0.1
	Carboniferous, middle	19	0.54	4.2	4.74	0.1
	Carboniferous, lower	17	0.39	2.2	2.59	0.1
350	Devonian	18	0.86	3.07	3.93	0.2
	Devonian, top upper	9	0.47	3.4	3.87	0.1
	Devonian, bottom upper	41	0.43	2.9	3.33	0.1
	Devonian, middle	41	0.45	2.9	3.35	0.1
400	Silurian	6	0.51	3.7	4.21	0.1
440	Ordovician	8	0.59	3.3	3.89	0.1
500	Cambrian	14	0.36	4.0	4.36	0.0
600		7	0.84	2.86	3.70	0.2
	Keweenawen	4	1.68	2.38	4.06	0.7
?900	Upper Huronian	10	1.10	3.38	4.48	0.3
	Lower Huronian	15	0.26	3.54	3.80	0.0
	Pre-Huronian	4	2.07	2.45	4.52	0.8

Another important point which must be borne in mind is that the sedimentary material weathered from the lithosphere and carried to the sea by rivers probably undergoes profound chemical changes at rates rapid on an oceanographic or geological scale when entering the marine environment as a result of ion exchange (Figure 12.24). In particular Mg^{2+}, Na^+, and K^+ are exchanged for Ca^{2+}.

The electrolyte concentration of the Earth's oceans is the result of a massive neutralization of plutonic acids by crustal bases, and we are brought back again to a question we have examined before, namely, what are the chemical processes which have managed to attain and maintain the pH of the oceans. The traditional answer—the carbonate system—has by no means been invalidated, but Sillen's (1961b) suggestion that silicate minerals and equilibria of the sort

$$(12.12) \quad 1.5 \underset{\text{Kaolinite}}{Al_2Si_2O_5(OH)_4} \text{ (s)} + K^+$$

$$\rightleftharpoons \underset{\text{K-mica}}{KAl_3Si_3O_{10}(OH)_2} \text{ (s)} + 1.5 H_2O + H^+$$

may also be playing an important role is gaining ground (Holland, 1965; Garrels, 1965; Sillen, 1967b).

8 Summary

An aqueous solution near a solid interface is organized into a complex structure. If the solid surface is charged, an electrical double layer is formed by the ions in solution. Even if the surface is not charged, the water structure adjacent to it is profoundly altered. The physical-chemical properties of this interfacial water structure are very different from those of the "normal" water in the bulk phase and, again, as in the case of the gas-liquid water interface, recent evidence, especially from studies on porous media and thin capillaries, indicates that the layer of structured water near a surface is quite thick. In addition, there may also be specific sorption on the solid surface.

Marine sediments fall into two general compositional categories: calcareous ($CaCO_3$) and siliceous (SiO_2). Their water content depends on compaction and other factors, and the ionic composition of this interfacial water, like that of the marine aerosols, may be appreciably different from that of the free parent seawater. The details of their particle size, porosity, distribution, and chemistry, including their trace element and radionuclide contents, give us many insights into physical and chemical processes both past and present occurring in the oceans and in the Earth's lithosphere as well. One of the most important types of chemical interaction of the sediments and other lithosphere materials with seawater is ion exchange.

The sediments are largely terrigenous or biogenous in their origin, while accretions such as manganese nodules are halmyrogenous or precipitate directly out of species in solution in seawater. The manganese nodules exhibit great variety both in their physical form and in their chemical composition, and they are widely distributed on the floors of the oceans.

Other marine deposits worthy of mention are iron deposits, phosphorite deposits, and, of course, the precipitation of salts from the solar evaporation of shallow seas.

The water of the seas comes not from the condensation of the Earth's primitive atmosphere but from the lithosphere via thermal springs. As for the electrolytes in seawater, the anionic substances, or excess volatiles, are plutonic in origin, while the cations are added to the oceans from the leaching of the superficial lithosphere. In recent years the importance of chemical equilibria with mineral silicates in controlling the pH of seawater has been given new emphasis.

REFERENCES

Anikouchine, W. A., *J. Geophys. Res.*, **72**, 505 (1967).

Antoniou, A. A., *J. Phys. Chem.*, **68**, 2754 (1964).

Arrhenius, G. O. S., in M. N. Hill, ed., *The Sea*, Interscience, New York, 1963, Vol. 3.

Arrhenius, G., "Sedimentary Record of Long Range Phenomena," in P. M. Hurley, ed., *Advances in Earth Science*, M.I.T. Press, Cambridge, Mass., 1966.

Arrhenius, G., J. Mero, and J. Korkisch, *Science*, **144**, 172 (1964).

Barnes, S. S., *Science*, **157**, 63 (1967).

Barth, T. F. W., *Theoretical Petrology*, John Wiley & Sons, New York, 1952.

Bender, M. L., T-L. Ku, and W. S. Broecker, *Science*, **151**, 325 (1966).

Berner, R. A., *Geochim. Cosmochim. Acta*, **28**, 1497 (1964).

Berner, R. A., *Science*, **147**, 1297 (1965).

Berner, R. A., *Am. J. Sci.*, **264**, 1 (1966).

Bijl, A. J., *Rec. Trav. Chim.*, **46**, 767 (1927).

Bockris, J. O'M., M. A. V. Devanathan, and K. Müller, *Proc. Roy. Soc. (London)*, **274A**, 55 (1963).

Bonatti, E., *Science*, **153**, 534 (1966).

Borchert, H., "Formation of Marine Sedimentary Iron Ores," in J. P. Riley and G. Skirrow, eds., *Chemical Oceanography*, Academic Press, London, 1965a, Vol. 2, Chap. 18.

Borchert, H., "Principles of Oceanic Salt Deposition and Metamorphism," in J. P. Riley and G. Skirrow, eds., *Chemical Oceanography*, Academic Press, London, 1965b, Vol. 2, Chap. 19.

Boyd, G. E., and B. A. Soldano, *J. Am. Chem. Soc.*, **75**, 6105 (1953).

Bragg, L., G. F. Claringbull, and W. H. Taylor, *Crystal Structure of Minerals*, Cornell Univ. Press, Ithaca, N.Y., 1965.

Brindley, G. W., and G. L. Millhollen, *Science*, **152**, 1385 (1966).

Brujewicz, S. W., and E. D. Zaitseva, "Chemical Features of Marine Interstitial Solutions," Preprint, Intern. Congr. Oceanog., Am. Assoc., Advan. Sci., Washington, D.C., 1959.

Burton, J. D., "Radioactive Nuclides in Sea Water, Marine Sediments, and Marine Organisms," in J. P. Riley and G. Skirrow, eds., *Chemical Oceanography*, Academic Press, London, 1965, Vol. 2, Chap. 22.

Chahal, R. S., and R. D. Miller, *Brit. J. Appl. Phys.*, **16**, 231 (1965).

Chester, R., in J. P. Riley, and G. Skirrow, eds., *Chemical Oceanography*, Academic Press, London, 1965.

Cloud, P. E., Jr., "Carbonate Precipitation and Dissolution in the Marine Environment," in J. P. Riley and G. Skirrow, eds., *Chemical Oceanography*, Academic Press, London, 1965, Vol. 2, Chap. 17.

Conway, E. J., *Proc. Roy. Irish Acad.*, **B48**, 119 (1942); **B48**, 161 (1943).

Conway, B. E., *Theory and Principle of Electrode Processes*, Ronald Press, Co., New York, 1965.

Delahay, P., *Double Layer and Electrode Kinetics*, Interscience, New York, 1965.

Derjaguin, B. V., I. G. Ershova, B. V. Telesnyi, and N. V. Churajev, *Dokl. Akad. Nauk SSSR*, **170**, 876 (1966).

Deryagin, B. V., and N. N. Fedyakin, *Dokl. Akad. Nauk SSSR*, **147**, 403 (1962).

Devanthan, M. A. V., and B. V. K. S. R. A. Tilak, *Chem. Rev.*, **65**, 635 (1965).

Dietrich, G., *General Oceanography*, Interscience, New York, 1963.

Drost-Hansen, W., "The Resistivity of Brines in Capillaries," *Pan-Am. Petrol Corp.*, *Res. Dept. Rept.*, No. F59-G-2 (July 10, 1959).

Duff, A. W., *Phil. Mag.*, **9**, 685 (1905).

Friedman, I., *J. Geophys. Res.*, **70**, 4066 (1965).

Garrels, R. M., *Science*, **148**, 69 (1965).

George, J. H. B., and R. A. Courant, *J. Phys. Chem.*, **71**, 246 (1967).

Goldberg, E. D., "The Oceans as a Chemical System," in M. N. Hill, ed., *The Sea*, Interscience, New York, 1963, Vol. 2, Chap. 1.

Goldberg, E. D., in L. H. Ahrens, F. Press, K. Rankama, and S. K. Runcorn, eds., *Physics and Chemistry of the Earth*, Pergamon Press, London, 1961, Vol. 4.

Goldberg, E. D., and G. O. S. Arrhenius, *Geochim. Cosmochim. Acta*, **13**, 153 (1958).

Goldberg, E. D., M. Koide, R. A. Schmitt, and R. H. Smith, *J. Geophys. Res.* **67**, 4209 (1963).

Gross, M. G., *J. Geophys. Res.*, **71**, 2017 (1966).

Hambleton, F. H., J. A. Hockey, and J. A. G. Taylor, *Trans. Faraday Soc.*, **62**, 795 (1966).

Hamilton, E. L., *Bull. Geol. Soc. Am.*, **70**, 1399 (1959).

Hamilton, E. L., *J. Geophys. Res.*, **69**, 4257 (1964).

Helfferich, F., *Ion Exchanges*, McGraw-Hill Book Co., New York, 1962.

Henniker, J. C., *Rev. Mod. Phys.*, **21**, 322 (1949).

Holland, H. D., *Proc. Natl. Acad. Sci. U.S.*, **53**, 1173 (1965).

Horne, R. A., *Surv. Progr. Chem.*, **4**, 1 (1968).

Horne, R. A., *J. Phys. Chem.*, **70**, 1335 (1966).

Horne, R. A., R. A. Courant, B. R. Myers, and J. H. B. George, *J. Phys. Chem.*, **68**, 2578 (1964).

Horne, R. A., A. F. Day, R. P. Young, and N-T. Yu, *Electrochim. Acta*, **13**, 397 (1968).

Koczy, F. F., *Deep-Sea Res.*, **3**, 93 (1956).

Korringa, J., *Bull. Am. Phys. Soc.*, **1** ser. II, 216 (1956).

Krauskopf, K. B., *Geochim. Cosmochim. Acta*, **9**, 1 (1956).

Kroll, U. S., *Rept. Swedish Deep-Sea Expedition, 1947–48*, 10, Fasc. 1.

Kuehl, H., and H. Mann, *Helgolaender Wiss. Meeresuntersuch.*, **13**, 238 (1966).

Kuenen, P. H., "Geological Conditions of Sedimentation," in J. P. Riley and G. Skirrow, eds., *Chemical Oceanography*, Academic Press, London, 1965, Vol. 2, Chap. 14.

Kulp, J. L., and D. R. Carr, *J. Geol.*, **60**, 148 (1952).

Maatman, R. W., *J. Phys. Chem.*, **69**, 3196 (1965).

McConnell, B. L., K. C. Williams, J. L. Daniel, J. H. Stanton, B. N. Irby, D. L. Dugger, and R. W. Maatman, *J. Phys. Chem.*, **68**, 2941 (1964).

Macey, H. H., *Trans. Brit. Ceram. Soc.*, **41**, 73 (1942).

Manheim, F. T., " Manganese-Iron Accumulations in the Shallow Marine Environment," in D. K. Schink and J. T. Corless, eds., Symp. Marine Geol., Univ. Rhode Island Occas. Publ., No. 3 (1965).

Marchenko, A. S., *Soviet Geol.*, **9**, 155 (1966).

Mero, J. L., *The Mineral Resources of the Sea*, Elsevier Pub. Co., Amsterdam, 1965.

Nicolls, G. D., "The Geochemical History of the Oceans," in J. P. Riley and G. Skirrow, eds., *Chemical Oceanography*, Academic Press, London, 1965, Vol. 2, Chap. 20.

Parravano, C., J. D. Baldeschwieler, and M. Boudart, *Science*, **155**, 1535 (1967).

Pettersson, H., *Anz. Akad. Wiss. Wien.*, **127** (1937).

Picciotto, E. E., in M. Sears, ed., *Oceanography*, Am. Assoc., Advan. Sci. Publ., No. 67, Washington, D.C., 1961.

Pratt, R. M., and P. F. McFarlin, *Science*, **151**, 1080 (1966).

Pytkowicz, R. M., and D. N. Connors, *Science*, **144**, 840 (1964).

Revelle, R., *J. Marine Res. (Sears Found. Marine Res.)*, **14**, 446 (1955).

Revelle, R., and F. P. Shepard, "Sediments off the California Coast," in P. D. Trask (ed)., *Recent Marine Sediments*, Amer. Assoc. Petroleum Geologists, Tulsa, 1939.

Rubey, W. W., *Bull. Geol. Soc. Am.*, **62**, 1111 (1951).

Sackett, W. M., *Science*, **132**, 1761 (1960).

Schofield, R. K., and H. R. Samson, *Discussion Faraday Soc.*, **18**, 135 (1954).

Schufle, J. A., *Chem. Ind.*, 690 (1965).

Schufle, J. A., and M. Venugopalan, *J. Geophys. Res.*, **72**, 3271 (1967).

Shainberg, I., and W. D. Kemper, *Soil Sic.*, 1034 (1967).

Shishkina, O. V., "On the Salt Composition of the Marine Interstitial Waters," Preprint, Intern. Congr. Oceanog., Am. Assoc. Advan. Sci., Washington, D.C., 1959.

Shishkina, O. V., *Okeanol.*, **1**, 646 (1961).

Siever, R., K. C. Beck, and R. A. Berner, *J. Geol.*, **73**, 39 (1965).

Siever, R., R. M. Garrels, J. Kanwisher, and R. A. Berner, *Science*, **134**, 1071 (1961).

Sillen, L. G., *Chem. Britain*, **3**, 291 (1967a).

Sillen, L. G., "The Physical Chemistry of Sea Water," in M. Sears, ed., *Oceanography*, Am. Assoc. Advan. Sci. Publ., No. 67, Washington, D.C., 1961.

Sillen, L. G., *Science*, **156**, 1189 (1967b).

Slabaugh, W. H., and A. D. Stump, *J. Phys. Chem.*, **68**, 1251 (1964); *J. Geophys. Res.*, **69**, 4773 (1964).

Skornyakova, N. S., P. F. Andrushchenko and L. S. Fumina, *Deep-Sea Res.*, **11**, 93 (1964).

Slonimskaya, M. V., and T. M. Raitburd, *Dokl. Akad. Nauk SSSR*, **162**, 176 (1965).

Sverdrup, H. U., M. W. Johnson, and R. H. Fleming, *The Oceans*, Prentice-Hall, Englewood Cliffs, N.J., 1942.

Tien, H. T., *J. Phys. Chem.*, **69**, 350 (1965).

Turekian, K. K., "Some Aspects of the Geochemistry of Marine Sediments," in J. P. Riley and G. Skirrow, eds., *Chemical Oceanography*, Academic Press, London, 1965, Vol. 2, Chap. 16.

Wakeel, S. K. El., and J. R. Riley, *Geochim. Cosmochim. Acta*, **25**, 110 (1961).

Wangersky, P. J., *J. Geol.*, **75**, 733 (1967).

Wayman, C. H., "Adsorption on Clay Mineral Surfaces," in S. D. Faust and J. V. Hinter, eds., *Principles and Applications of Water Chemistry*, John Wiley & Sons, New York, 1967.

Weiler, R. R., and A. A. Mills, *Deep-Sea Res.*, **12**, 511 (1965).

Weiss, A., *Kolloid-Z., Z. Polymer.*, **211**, 94 (1966).

Welby, G. W., *J. Sediment. Petrol.*, **28**, 431 (1958).

Wolkowa, Z. W., *Kolloid-Z.*, **67**, 280 (1934).

Yu, N-T., *M.S. Thesis*, New Mexico Highlands Univ., 1966.

Zaitseva, E. D., "Exchange Capacity and Exchangeable Cations in Sea Sediments," Preprint, Intern. Oceanog. Congr., Am. Assoc. Advan. Sci., Washington, D.C., 1959.

Zimmerman, J. R., and W. G. Brittin, *J. Phys. Chem.*, **61**, 1328 (1957).

ZoBell, C. E., "Occurrence and Activity of Bacteria in Marine Sediments," in P. D. Trask, ed., *Recent Marine Sediments*, Am. Assoc. Petroleum Geologists, Tulsa, Okl., 1939.

Part IV
Selected Topics

13 Selected Topics in Chemical Oceanography

1 Marine Corrosion

The sea is a harsh environment. As man tries to trespass in this forbidding world, the sea will strive relentlessly to undo his handiwork. Waves will crush his ships and undermine his ramparts, and any object which he places in or even near the sea the sea will attempt to devour.

The deterioration of materials and structures in the sea may be brought about by the single or combined agencies of physical abrasion and chemical corrosion. Physical abrasion in the sea is highly localized, occurring only in near-shore waters or on the bottom, where swiftly moving water impinges solid material upon a surface. The roughening and freshening of a surface by abrasion can greatly accelerate corrosion. Marine corrosion, in the absence of abrasion, of common metals and alloys, on the other hand, shows surprisingly small variations from place to place throughout the oceans of the world, despite differences in the temperature, salinity, and organic content of sea-water (LaQue, 1948).

Marine corrosion has been an area in which communication between the man engaged in basic research and the man confronted with practical problems has been particularly feeble. Effort has tended toward extremes—fundamental electrochemical studies in the laboratory, and the empirical approach of simply exposing samples to the sea—with very little mediating them. However, the pressure of problems connected with nuclear submarines, anti-submarine warfare surveillance systems, desalination plants, and increased offshore oil drilling seems to be in the process of remedying this state of affairs. A fair insight into the status of research on marine corrosion may be obtained from the published proceedings of the Congress International de la Corrosion Marine et des Salissures held in Cannes, France, in 1964. One particularly serious applied problem upon which the full weight of scientific knowledge has been brought to bear is stress-corrosion fracture, or the yielding of alloys in some chemical environments under stresses which they should be able to endure (Rhodin, 1959; Hoar, 1963; Parkins, 1964).

In order for a metal to go into aqueous solution, oxidation or a change in valence state must occur (here we consider only the corrosion of metals and their alloys). Metals above hydrogen in the electromotive series (Table A.14 in the Appendix) displace protons, but seawater is too nearly neutral for such acidic attack to be present except in the case of the most active metals (Na, Ca, etc.). Elze and Oelsner (1958) published a practical electromotive series

$$(13.1) \qquad M + n\,H^+ \rightarrow M^{n+} + \frac{n}{2}\,H_2\,(g)$$

for a number of metals in synthetic seawater. Seawater is an electrolyte, a fairly good conductor of electrical current, and, with seawater forming one leg of the circuit, electrically connected metallic heterogeneities in the seas will immediately establish a voltaic cell at the expense of the more active (or less "noble") heterogeneity. Two distinctly different metals need not be in proximity for a voltaic cell to be established and corrosion to occur. Differences in chemical potential can also arise from slight local irregularities in the composition of an alloy, concentration differences in the seawater (concentration cells), especially differences in the concentration of dissolved oxygen (Riggs, Sudbury, and Hutchinson, 1960), pressure, and stress. In fact it is even possible to form a cell with two identical metal electrodes if they are at different temperatures (thermocells) (Agar, 1958).

Speller (1951) lists the following as among the more important factors influencing corrosion:

FACTORS ASSOCIATED MAINLY WITH THE METAL
Effective electrode potential of the metal in solution
Overvoltage of hydrogen on the metal
Chemical and physical homogeneity of the metal
The metal's ability to form an insoluble protective film

FACTORS ASSOCIATED MAINLY WITH THE ENVIRONMENT
Acidity
Nature and concentration of dissolved gases, especially oxygen
Nature, concentration, and distribution of ions in solution
Flow rate of solution past the metal
The environments' ability to form a protective coating on the metal
Temperature
Mechanical stresses
Proximity of dissimilar materials
The nature and concentration of organisms

If the metallic cation complexes strongly with an anion, this will of course tend to bring the metal into solution more readily. The highly corrosive nature of seawater is in large part due to the effectiveness of Cl^- in this respect. Concentration gradients in the amount of dissolved O_2 or H_2S in seawater can give rise to accelerated corrosion notably at the sea-air and sea-mud interfaces (but see below). O_2 concentration cells appear to be, at least in part, responsible for the deterioration of steel-reinforced concrete structures in the marine atmosphere (Finley, 1961). Working with fresh water, Watkins, and Kincheloe (1958) observed the highest corrosion rates of steel when the water was saturated with O_2 and contained a lot of CO_2 as well. The same authors obtained erratic results with H_2S. This gas seems sometimes to inhibit corrosion by the formation of a protective sulfide coating. Similarly Starkey (1958) was unable to demonstrate rapid corrosion under anaerobic conditions, even though FeS was formed. Communities of fouling organisms (see below) tend to form both O_2 and, on their death and subsequent decay, H_2S concentration gradients; and through the biochemistry of CO_2, H_2S, and NH_3, they also produce local perturbations of the pH.

Like most diffusion-controlled reactions in aqueous solution, agitation by supplying fresh reactants and removing products accelerates corrosion. Uusitalo (1961), for example, found that a low carbon steel corrodes up to 15 times faster at a flow rate of 5 m/sec than in standing seawater, and Juchniewicz (1958) has noted that the current density necessary to cathodically protect a moving ship is about twice as great as for an anchored vessel.

Table 13.1 compares the marine corrosion and fouling properties of some common metals and alloys.

Only in very recent years have studies been undertaken on corrosion at great depth in the seas under high hydrostatic pressure. The effect of pressure on the electrical potential E of a cell at constant temperature is given by

$$(13.2) \qquad \left(\frac{dE}{dP}\right)_{C,T} = \frac{\Delta V}{n\mathscr{F}}$$

where ΔV is the volume change upon the passage of n faradays. As long as only condensed phases are involved, the effect is small, but any electrochemical process involving the evolution of a gas, such as H_2, should be markedly repressed by the application of pressure. Disteche (1962) has measured some cell potentials under high pressure, and Horne, Myers, and Frysinger (1964) have examined the dissociation of the $FeCl^{2+}$ complex under pressure, since this dissociation affects the value of the redox potentials of the iron system. The electrochemical effects suggested by these studies are small, an expectation confirmed by *in situ* immersion testing (Gray, 1964; DeLuccia, 1966). For all the great difficulty involved in the performance of these tests the results have not been very interesting. Generally speaking a recent survey (Newton, 1967), indicated that, except at the mud line, corrosion at great depth tends to be less than at the surface. This can probably be attributed to the lower temperature, the slower ocean currents, and lower concentration of

Table 13.1   Corrosion and Fouling Properties of Some Common Metals and Alloys

Material	Resistance to Saltwater Corrosion (corrosion rate in inches/yr)	Fouling Tendency
Aluminum	Fair to good	Fouls; antifouling if rapid corrosion
Brass	Variable	Antifouling if corrosion
Bronze	Fair to good	Antifouling if corrosion
Copper	Variable	Antifouling if corrosion
Hastelloy	Good to excellent, 0.00005–0.001	Fouls
Iron, cast	Fair to good	Fouls
Iron, wrought	Fair	Fouls
Inconel	Fair to excellent, 0.0008	Fouls
Invar	0.001	Fouls
Iron, high-Si	Excellent	Fouls
Lead	Good, 0.0006	Fouls
Magnesium	Poor to fair	Fouls; antifouling if rapid corrosion
Monel	Good to excellent, 0.00001–0.0001	Fouls
Nickel	Good, 0.0003–0.001	Fouls
Steel	Variable to good	Fouls
Steel, low C	Fair	Fouls
Steel, Cr	Poor to good	Fouls
Steel, galvanized	0.001–0.002	Variable fouling
Stellite	Excellent	Fouls
Tantalum	Variable	Fouls
Tin	0.00003	Fouls
Zinc	0.001–0.004	Variable fouling; Antifouling if corrosion

dissolved oxygen. However, crevice corrosion is sometimes greater at great depths (Forgeson, Waldron, Peterson, and Brown, 1962).

Short-period (1–3 years) protection against marine corrosion can be realized by several means, singly or in combination:

Avoidance of dissimilar materials and all conditions which can give rise to differences in chemical potential such as temperature and concentration gradients, differences in crystalline state, and mechanical stress.

Metallic protective coatings such as galvanizing and cladding, in which the metal is protected by a second metal less subject to corrosion (more "noble").

Nonmetallic protective coatings including paints, glasses, and plastic materials (Bobalek, 1965).

Cathodic Protection. A passive cathodic protection system utilizes a difference in chemical potential in order to prevent the corrosion of, for example, a ship's steel hull, by the controlled corrosion of a more active sacrificial metal such as Zn, Mg, or Al (Christie, 1958; Mussad and Holmes, 1957; Scrieber, 1958), whereas an active system uses a noble metal anode and an impressed electrical current (Nelson, 1957; Preiser and Cook, 1957; Schwerdtfeger, 1958).

Fouling. Another phenomenon which is responsible for the deterioration of the functionality of man-made objects in the sea is fouling by the growth of colonies of marine organisms (for a now old but excellent review see Woods Hole Oceanographic Institute, 1952). The relationships between marine corrosion and fouling are extremely intimate, complex, and imperfectly understood. In some instances the growth of a colony of marine organisms appears to act as a protective layer which reduces the rate of corrosion, whereas in other instances the biological production of corrosive agents and the formation of oxygen concentration gradients (Clapp, 1948) accelerate corrosion. Sometimes, for example, pits form only under barnacles, and at other times they form only in the adjacent surface uncovered by barnacles [cf. Figures 16R (p. 15) and 17 (p. 16) in Woods Hole Oceanographic Institute, 1952]. Nor is it clear whether bioproduction of H_2S accelerates or retards corrosion. The initial phase in the growth of a fouling community is the formation of a bacterial slime which appears to have a particularly important, but as yet largely unknown, relationship to metal corrosion.

In some respects, fouling seems to be the converse of corrosion. We saw that corrosion is greater at higher water velocities; fouling, on the other hand, is a problem, especially on stationary structures and objects and on ships in port, and relatively little fouling accretion is made on ships under way in the open sea. If the metal surface corrodes so rapidly that material sloughs off, organisms may encounter difficulty in finding and maintaining a toehold (see Al and Mg in Table 13.1). Also, if when the metal corrodes it releases a toxic cation such as Cu^{2+}, corrosion will tend to discourage the fouling process (see brass, bronze, and Cu in Table 13.1).

Fouling is much more variable from place to place in the seas than is corrosion. Fouling is particularly troublesome in warm, fertile, coastal waters, and it drops off very abruptly with depth in the sea, although it does not disappear even in the abyssal depths.

Since ancient times, marine fouling has been a more vexsome problem than marine corrosion, and the tenacity of life assures that it will continue to be an intransigent obstacle to man's conquest of the seas. Electrical shocks and even radioactivity have proved to be relatively ineffective in discouraging the growth of fouling communities. Only corroding or slowly dissolving alloys and coatings which release toxic chemicals have proved to be useful, and their lifetimes are, by their nature, limited.

2   Desalination and the Recovery of Chemicals from the Sea

The seas and their immediate environment, both existing and ancient, represent an enormous, but not inexhaustible, reservoir of chemical substances. Here we will be briefly concerned only with those materials which come from the waters of existing seas, and we will make no further mention of the vast deposits of salts, petroleum, sulfur, limestone, and other mineral substances laid down by ancient seas, nor will we discuss the problems of recovery of the great treasure of manganese nodules (Chapter 12) at the bottom of the sea—a subject treated with thoroughness and enthusiasm in Mero's book, *The Mineral Resources of the Sea* (1965).

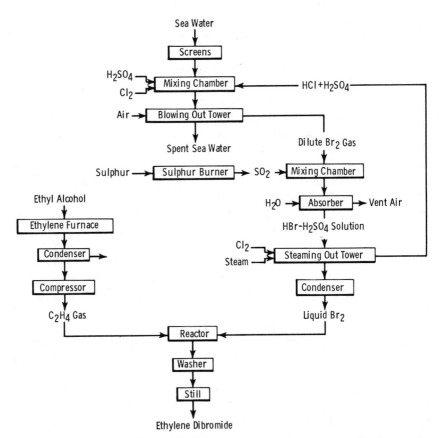

Figure 13.1 Schematic Flow Diagram of the Ethyl-Dow Process for Recovering Bromine from Seawater. From Shigley (1951), with permission of Metallurgical Society of AIME.

Salt. The most obvious chemical recovered in large quantities from the sea is NaCl. About 6,000,000 tons of salt are produced annually from seawater by evaporation in shallow basins. About 5% of the salt consumed in the United States is produced in this manner, largely in the San Francisco Bay area. As the seawater evaporates, various dissolved salts begin to sequentially crystallize out (Table 12.24 and Figure 12.23). First the $CaSO_4$, then the brine is concentrated until NaCl begins to precipitate; finally, when the specific gravity exceeds 1.28, magnesium salts begin to precipitate. The remaining concentrated brine solution is called "bitterns" and may be further processed to recover magnesium compounds, halides, and other salts.

Halides. The element iodine was first discovered by Courtois in 1811 (isolated from the ash of seaweed), and in 1825 another Frenchman, Balard, discovered the sister element bromine in bitterns prepared from salt marsh waters (Weeks, 1945). The production of iodine from seaweed ash has long

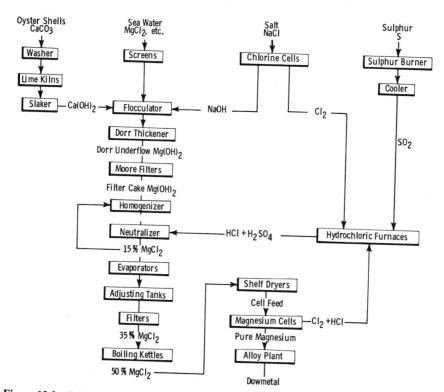

Figure 13.2 Schematic Flow Diagram of the Dow Process for Recovering Magnesium from Seawater by the Electrolysis of $MgCl_2$ Solutions. From Shigley (1951), with permission of Metallurgical Society of AIME.

since been replaced by other sources and processes, but bromine is still obtained from seawater both as a by-product of salt production and by direct precipitation of the insoluble tribromoaniline formed on treating unconcentrated but acidified seawater with aniline and chlorine. The latter process has been refined and modified, notably by the use of SO_2 and air in the stripping steps (Figure 13.1), and it now accounts for the majority of the bromine production in the United States. Chlorine is prepared by the electrolysis of fused NaCl (Downs process)

$$2\,Na^+ + 2\,e^- \rightarrow 2\,Na\;(cathode)$$
(13.3)
$$2\,Cl^- \rightarrow Cl_2 \uparrow + 2\,e^-\;(anode)$$

or of concentrated brine (Vorce process)

$$2\,H^+ + 2\,e^- \rightarrow H_2 \uparrow (cathode)$$
(13.4)
$$2\,Cl^- \rightarrow Cl_2 \uparrow + 2\,e^-\;(anode)$$

but mined salt from ancient seas rather than solar-evaporated sea salt is largely used. This chlorine production forms the nucleus of a cluster of heavy industrial chemicals.

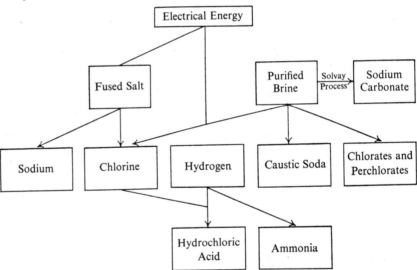

Magnesium. Next to Na, Mg is the most abundant metallic cation in seawater. Since World War II the Mg metal consumed in the United States has come largely from the sea. In 1964 the annual world production of this strong, lightweight metal was about 150,000 tons. The Dow Chemical Company Mg production plant at Freeport, Texas, processes 1,000,000 gal/hr of seawater from the Gulf of Mexico (Figure 13.2). In addition to the metal, MgO, $Mg(OH)_2$, and $MgCl_2$ are also produced from seawater and brines.

Gold. Human nature is tempted by the lure of something for nothing and, ever since its presence was detected there, some men have schemed to devise ways to recover gold from seawater. The oceans of the world are said to contain enough gold to make everyone on Earth a millionaire. The recovery of gold from seawater, curiously enough, appears particularly to appeal to the Germans who, we recall, were also avid alchemists. No less a German chemist than Fritz Haber dreamed of paying his nation's World War I debt with gold extracted from the sea. But after the expenditure of ten years of effort he was forced reluctantly to conclude that the gold concentration is far too small (Table 5.3) and the quantity of seawater that would have to be processed far too large to make the extraction profitable. Some years ago a colleague of mine at M.I.T. (incidentally also a German) dragged a sack containing a very strong anion exchange resin through the waters off Plumb Island in order to recover $AuCl_4^-$, but all he ever got for his pains was a bad chill. After processing 15 tons of seawater in their North Carolina bromine extraction facility, the Dow Company succeeded in recovering 0.09 mg of gold worth about $0.0001.

Biomaterial. The marine biosphere yields many materials useful to man, the most obvious being, of course, the food materials, but also seaweed as a source of sodium alginate and fertilizer as well as food. Not so well known but attracting increasing scientific attention is the marine biosphere as a source of physiologically active substances ranging from toxins (see the magnificent volume on toxic marine animals by Halstead and Courville, 1965) to antibiotics (Olesen, et al., Maretzki, 1964; Aubert, Aubert, et al., 1966; and Nigrelli, Stempien, et al. 1967). In August of 1967 the Marine Technology Society sponsored a conference on "Drugs from the Sea" at the University of Rhode Island, readers interested in this fascinating topic are referred to the detailed bibliography prepared by A. D. Marderosian for that conference.

Desalination. In the not too distant future by far the most important chemical recovered from the sea may very well be water. The unchecked spread of the human infestation of this smallish planet not only is making an ever-increasing demand on water resources but is at the same time diminishing the usability and debasing the quality of those resources by pollution. Just as he has turned to the oceans as a panacea for his food problem, man is turning to the seas as a panacea for the water shortages he has created.

Earlier in this book we examined the powerful solvent properties of water, the ease with which substances dissolve, the large energies involved, and the great strength of their solvation envelopes. Clearly, to reverse this process, to separate these affianced substances, the salts from the water, is not an easy task. But, in spite of all this difficulty, so frightening is the urgency that processes designed to accomplish this end have already been pushed from research and development into engineering and production phases (for a good review see Spiegler, 1966).

Glueckauf (1966) classifies processes for desalting water into three main categories:

1. Processes involving a phase change
   Evaporation
   Distillation
   Freezing
2. Processes involving semipermeable membranes
   Electrodialysis
   Reverse osmosis (hyperfiltration)
3. Processes involving chemical equilibria
   Ion exchange
   Hydration

The method of obtaining fresh water from salt water with a minimum of equipment that first comes to mind is probably solar distillation. In a simple solar still such as shown in Figure 13.3 solar energy is utilized to evaporate the water which is then condensed on a cooler surface and collected. More complicated but closely related in principle is the diffusion humidification process (Figure 13.4) in which the water evaporates from a rotating disk, diffuses across a narrow airspace, and collects on a stationary, cooled condenser plate. The equipment requires an energy source both to warm the salt

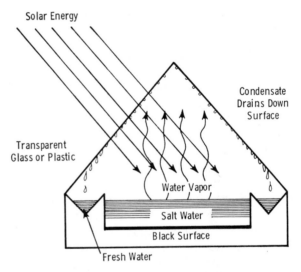

Figure 13.3   A Simple Solar Still.

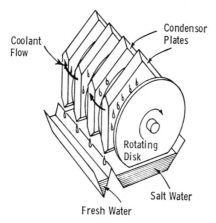

Figure 13.4   A Diffusion Humidification Unit.

water and to rotate the apparatus, but nevertheless it appears to represent a feasible type of installation for small-scale applications.

The classical method of purifying water is, of course, distillation. Distillation is one of the most important large-scale desalination techniques, and the method admits of many engineering ramifications such as multistage flash distillation (Figure 13.5), long-tube vertical distillation, multistage-multi-effect distillation, and vapor compression distillation. But, as we saw in Chapter 1, water molecules are very disinclined to be separated and, compared to other substances, enormous quantities of energy are required to convert

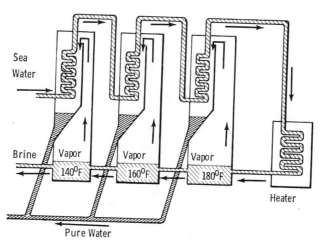

Figure 13.5   Multistage Flash Distillation Process.

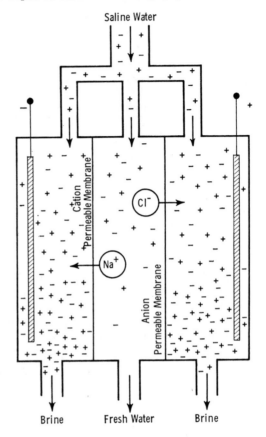

Figure 13.6   Electrodialysis.

water from the liquid to the gaseous state. The conversion of the liquid to the solid, however, requires less than one-sixteenth as much energy as evaporization. As an additional dividend, the lower temperatures tend to minimize scale and corrosion problems; hence freezing appears to offer an attractive phase transition method of desalinating seawater. If an aqueous electrolytic solution is not frozen too rapidly, crystals of pure ice are formed and the remaining liquid becomes more concentrated in salt (see also Section 4, on sea ice). A major problem in the freezing process which, like distillation, also admits of several engineering variations, is the mechanical separation of the solid and liquid phases.

In electrodialysis, under the influence of an imposed electric field, ions are transported out of the saline solution through ion-permeable membranes (Figure 13.6). Inasmuch as the current requirements depend on the electrolyte concentration, at the present time the process appears to be better suited for the desalination of less concentrated brackish waters than of seawater. The anion-permeable membrane tends to deteriorate under operating conditions, and in a variant on the process—transport depletion—it is replaced by a non-selective membrane.

If pure water and a saline solution are separated by a membrane permeable only to water molecules and not electrolytes, water molecules diffuse through the membrane into the saline solution, diluting it and building up an osmotic pressure (Figure 13.7). Conversely, if pressure in excess of the osmotic pressure is applied to the solution, water molecules can be forced out of the solution through the membrane; and this forms the basis of desalination by reverse osmosis. This process also is presently most applicable to brackish waters, although it is hoped to be able to extend its use to seawater.

Ion exchange (see Section 4, Chapter 12) can also be used to purify waters with relatively low salt contents. A cation exchange resin replaces the Na^+ with H^+

(13.5) $$R_cH + Na^+ + Cl^- \rightleftharpoons R_cNa + H^+ + Cl^-$$

and an anion exchange replaces Cl^- with OH^-

(13.6) $$R_aOH + H^+ + Cl^- \rightleftharpoons R_aCl + H^+ + OH^-$$
$$\Updownarrow$$
$$H_2O$$

but regeneration of the exchangers tends to make the process costly.

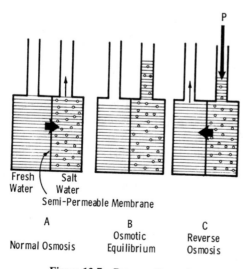

Figure 13.7   Reverse Osmosis.

Finally, desalination processes based on the phenomenon of hydration have been suggested (see Chapter 2). Crystalline hydrate forms of substances, such as propane (C_3H_8) or carbon dioxide, are formed, separated from the concentrated brine, washed, and then decomposed by a temperature and/or pressure change to give separated hydrating agent, which is then reused, and pure water.

The cost of pure water from conventional sources is slowly rising, whereas that from desalination, thanks to research, development, and engineering, is rapidly falling, especially for large-capacity installations. At present the costs of both sources are falling in the 50¢/1000 gal range. At this price the water in a cubic mile of seawater is worth some 10^{10}, whereas the gold is worth only a paltry 10^5. Gold is a luxury, but without water we shall perish. Table 13.2 summarizes the location, capacity, and method of the world's major desalination installations. I apologize for the table for, although it is only four

Table 13.2   World's Major Saline Water Conversion Plants (as of 1965)[a] (From *Business Week*, No. 1883, 120, Oct. 16, 1965, with permission of McGraw-Hill Pub. Co.)

In operation or under construction

Middle East		
Quatar, Doha	360,000 gal/day	Multistage flash distillation
Quatar, Doha	1,800,000	
Saudi Arabia, Dhahran	105,000	Electrodialysis
Saudi Arabia, Dhahran	200,000	Vapor compression
Neutral Zone, Arabia	500,000	
Kuwait	5,520,000	
Kuwait	2,400,000	Multistage flash distillation
Ahmadi	360,000	
Ahmadi	720,000	
Al Shaiba	6,000,000	
Kuwait	240,000	Electrodialysis

Atlantic Ocean–Mediterranean		
United Arab Republic, Sinai	600,000	Multistage flash distillation
Italy, Taranto	1,200,000	
Malta, Valletta	1,200,000	
Israel, Eilat	1,000,000	
Israel, Eilat	240,000	Freezing
Morocco, Centa	1,060,000	Multistage flash distillation
Guernsey	600,000	
Canary Islands	650,000	
Libya	200,000	

Table 13.2 (cont.)

Caribbean–South America		
Dutch West Indies		
Curacao	3,400,000	
Curacao	1,600,000	
Aruba	2,650,000	
Aruba	795,000	
Virgin Islands		Multistage flash
St. Thomas	1,000,000	distillation
St. Thomas	275,000	
St. Croix	1,500,000	
Venezuela		
Port Cardon	1,270,000	
Bermuda Air Base	200,000	Vapor compression
Cuba, Quantanamo	2,250,000	Multistage flash
Chile, Chanaral	240,000	distillation
Bahama, Nassau	1,440,000	
United States		
California, Catalina	150,000	Multistage flash distillation
Arizona, Buckeye	650,000	Electrodialysis
S. Dakota, Webster	250,000	
N. Carolina, Wrightsville	200,000	Freezing
New Mexico, Roswell	1,000,000	Vapor compression
Texas, Freeport	1,000,000	Long-tube vertical
Texas, Chocolate Bayou	900,000	
Virginia, Possam Pt.	188,000	Multistage flash
California, Moss Landing	376,000	distillation
Tennessee, Paradise	418,000	
Florida and Key West	2,620,000	
U.S. Mexican Border	100,000,000	Undecided
Soviet Union		
20 plants (location unknown)	1,000,000	Unknown
Kazakhstan, Shevchenko	1,300,000	Long-tube vertical
Being Planned		
Egypt	50,000,000	
Hong Kong	3,000,000	
New York, Riverhead	1,000,000	
Southern California	150,000,000	
Israel	100,000,000	Unknown
Saudi Arabia	2,000,000	
Spain, Southeast Coast	2,000,000	
Soviet Union, Don Basin	150,000,000	

[a] A more recent and detailed listing can be found at the end of *Saline Water Conversion Report for* 1966, Office of Saline Water, U.S. Dept. of Interior, Washington, D.C., 1966.

years old, so rapidly are developments occurring in the field that it is already out of date.

3   Physical Chemistry of Underwater Sound Transmission

Man is determined to explore, exploit, and fight in the seas, and for these ends, good and bad, it is necessary that he be able to "see" what he and the enemy are doing. Of his several senses, man relies most heavily on sight, but the penetration of light in the seas is feeble and the oceans depths lie in darkness. During World War II the M.I.T. Lincoln Laboratory developed a defensive radar system to protect the North American continent from attack by hostile aircraft. Now it has become particularly urgent, especially with the development of Polaris-type missiles, to develop a comparable defensive network to protect our shores from approach by enemy surface and submarine vessels. A number of means of detecting submerged objects has been explored, but hitherto there appears to be only one phenomenon which can provide a satisfactory "window" in the sea, and that is the transmission of acoustic energy. Let me hasten to add, however, that sonar, or the technology of underwater sound transmission (for more detailed discussions of the principles of underwater acoustics see Horton, 1959; Albers, 1960, 1961; and Frosch, 1964, is by no means restricted in its application to naval operations. On the contrary, it provides an invaluable research tool for the exploration of the topology and geology of the ocean floor and even subfloor, while the closely related sister technique of ultrasonics is being increasingly used in fundamental research on the properties and structure of liquids, including water and aqueous electrolytic solutions.

Sonar systems fall conveniently into two types: *Active systems*, in which a sound signal is bounced off from an object and the reflected signal detected; and *passive systems*, which simply listen to detect any sound signal originated by the object. In traveling through the seawater the sound signal is subject to degradation and the degradation can also be divided into two types (a) dispersion, scattering, and reflection, which represent deflections or changes in the direction of the sound ray; and (b) absorption which, together with the unavoidable diminution due simply to spherical spreading, represents a reduction of the energy level or intensity of the sound field.

As we have pointed out elsewhere (Horne and Eden, 1966), the physical chemistry of seawater and the structural changes they reflect have remarkably little effect on the velocity of sound, but they are directly responsible for the absorption of sound in seawater. The velocity of sound in a medium depends on the density and the compressibility. Hence those parameters which affect the density and the compressibility, and in the same order of relative importance, namely, temperature, pressure, and salinity, will determine the velocity of sound in seawater (Table A-28 of the Appendix). The relative importance of T, P, and S is clearly shown in the first-order terms in the emprical equation given by Wilson (1960a).*

(13.6) $$V = 1449.14 + V_T + V_P + V_S + {}_{STP}$$

where

$$V_T = 4.5721T - 4.4532 \times 10^{-2}T^2 - 2.6045 \times 10^{-4}T^3 + 7.9851 \times 10^{-6}T^4$$
$$V_P = 1.60272 \times 10^{-1}P + 1.0268 \times 10^{-5}P^2 + 3.526 \times 10^{-9}P^3 - 3.3603 \times 10^{-12}P^4$$
$$V_S = 1.39799(S - 35) + 1.69202 \times 10^{-3}(S - 35)^2$$
$$V_{STP} = (S - 35)(-1.1244 \times 10^{-2}T + 7.771 \times 10^{-7}T^2 + 7.016 \times 10^{-5}P$$
$$- 1.2943 \times 10^{-7}P^2 + 3.15 \times 10^{-8}PT + 1.5790 \times 10^{-9}PT^2)$$
$$+ P(-1.8607 \times 10^{-4}T + 7.4812 \times 10^{-6}T^2 + 4.5283 \times 10^{-8}T^3)$$
$$+ P^2(-2.5294 \times 10^{-7}T + 1.8563 \times 10^{-9}T^2)$$
$$+ P^3(-1.9646 \times 10^{-10}T)$$

for the ranges $T = -4$ to $+30°C$, $P = 1–1000$ kg/cm², and $S = 0–37‰$.

Dissolved gases have a negligible effect on the velocity of sound in seawater (Greenspan and Tschiegg, 1956), but large concentrations of bubbles, small compared with the dimensions of the sound pulse, can alter the density and compressibility of the medium and thus the sound velocity.

If there is a density and/or compressibility gradient in the sea, the sound rays tend to bend toward the region of lower velocity. Thus, inasmuch as the pressure increases with depth and temperature and salinity usually exhibit complex profiles (Figure I.3 of the Introduction), the path of the sound rays in real situations is commonly complex (Figure 13.8), a complexity which can be utilized to advantage. In particular, a vertical thermal profile such as shown in Figure I.3 can result in a surface duct (Figure 13.8B) which serves as a sort of wave guide for the acoustic energy. Another sort of wave guide, the deep sound channel (Figure 13.8D), results from the entrapment of the sound ray by alternate upward deflection from the higher-pressure regions and downward deflection from the higher-temperature regions.

More violent deflections, that is, scattering, result from heterogeneities in the sea, such as the large-scale deflections from the sea's two interfaces, the surface and the bottom, which if reasonably coherent, can be exploited in the bottom bounce mode of sonar transmission (Figure 13.8C), and small scale deflection from solid particles, organisms, and gaseous bubbles suspended in the seawater. In some areas of the ocean there is a deep scattering layer whose depth shows a definite diurnal cycle. The cause of this mysterious

* This is Wilson's (1960a) second equation, which differs slightly from his first equation as given in Table A-28.

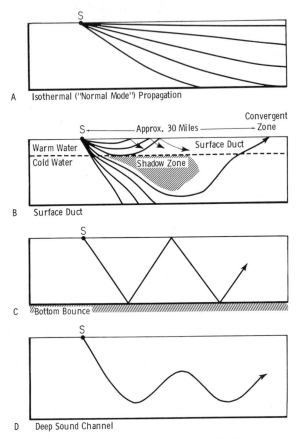

A   Isothermal ("Normal Mode") Propagation

B   Surface Duct

C   Bottom Bounce

D   Deep Sound Channel

Figure 13.8   Some Sonar Paths in the Sea.

phenomenon is now known to be great populations of organisms which rise to the surface at night to feed and then during the daylight hours sink back into the darkness of the depths to escape predators.

Not only does the velocity of sound vary from place to place, both horizontally and vertically in the sea, but at a given location the sound velocity can vary quite widely in time, often chaotically, owing to fluctuations in the thermal microstructure of the waters (Whitmarsh, Skudrazyk, and R. J. Urik, 1957; Piip, 1964, but sometimes systematically as a consequence of periodic processes such as internal waves (Crease, 1961; Lee, 1961).

Most of the loss of acoustic energy is attributable simply to geometric spreading, but the sound field is also attenuated very appreciably by absorption processes. In addition to scattering, heterogeneities in the sea can give rise to absorption, and bubbles, including air spaces in organisms, are particularly effective in this respect, far more effective in fact than suspended solid particles of comparable size. But, even in the absence of heterogeneities, seawater absorbs strongly, especially at high frequencies (Figure 13.9). Two types of absorption processes are evident: the "normal" absorption which occurs in pure water as well as seawater, and an additional "abnormal" absorption which is found in seawater but not in pure water. Hydrodynamic theory predicts that the attenuation α in a fluid should be given by

$$(13.7) \qquad \frac{\alpha}{f^2} = 8\pi^2 \frac{\mu V_s}{3V^3}$$

where f, μ, V_s, and V are the frequency, shear viscosity, specific volume, and velocity of sound, respectively. But in the case of pure water the observed value of α/f^2 is about three times greater than the expected calculated value. Hall (1948) has attributed this excess component of the "normal" absorption to structural relaxation in water, that is, to the displacement of an equilibrium between two different structural states in liquid water. The theory was subsequently modified somewhat by Litovitz and Carnevale (1955), and more recently Davis and Litovitz (1965) have proposed a two-structure model for liquid water (see Chapter 1, especially Figure 1.12) based directly on acoustic considerations. Yet, while the structural relaxation theory of the excess "normal" absorption of acoustic energy in pure water is universally accepted,

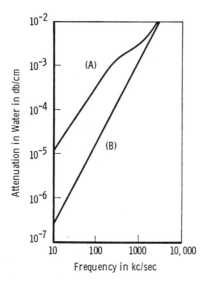

Figure 13.9 Ultrasonic Attenuation in Pure Water (B) and in Seawater (A). From Kinsler and Frey (1950), with permission of John Wiley & Sons.

at least in outline, the details of exactly what is happening, specifically which of the two structural forms represents the higher-energy state, are still not entirely clear.

To turn now to the "abnormal" absorption in seawater: studies have uncovered $MgSO_4$ as the guilty constituent. $MgSO_4$ is not unique in this respect; other 2:2 electrolytes also show strong absorption at certain frequencies (Verma, 1959), but they are relatively minor constituents of seawater. We had earlier clues for suspecting $MgSO_4$: in aqueous solutions this electrolyte is heavily hydrated (Table A-6), and its effect on the viscosity of water is very pronounced (Figure 1.25). It has a very high viscosity B coefficient, that is, it profoundly alters the water structure in its immediate vicinity. In fact, so bulky is this hydrated aggregate that it appears to be too large to fit into the available distribution of hole sizes in liquid water so that the energetics of its transport resembles that of viscous flow of water rather than that of other "normal" ionic species (Figure 3.12). Little wonder that so complex an aggregate with its many bonds and vibrational modes should be capable of absorbing appreciable energy. Acoustic energy is a pressure pulse, and we saw earlier that the dissociation of weak electrolytes, such as $MgSO_4$, increases with increasing pressure (Chapter 1). Eigen and Tamm (1962) have proposed a multistep dissociation process:

$$MgSO_4 \rightleftharpoons MgO\!\!\begin{array}{c} {}^{\diagup H} \\ {}_{\diagdown H} \end{array}\!\! SO_4 \quad\rightleftharpoons\quad MgO\!\!\begin{array}{c} {}^{\diagup H} \\ {}_{\diagdown H} \end{array}\!\! O\!\!\begin{array}{c} {}^{\diagup H} \\ {}_{\diagdown H} \end{array}\!\! SO_4 \quad\rightleftharpoons\quad Mg^{2+} + SO_4^{2-}$$

I   II   III   IV

(13.8)

Fisher (1962, 1965a, 1965b) has published a series of papers on the ultrasonic absorption and effect of pressure on the dissociation of $MgSO_4$ which apply this model with quantitative success.

Thorp (1965) has reported an apparent systematic anomaly in low-frequency sound attenuation in the deep ocean, but it is not yet clear whether this phenomenon is a result of boundary losses or of structural relaxation processes involving water or electrolyte such as those discussed above.

4 The Freezing Process and Sea-Ice

Vast expanses of the surface of the polar seas are covered with ice (Figure 13.10). The chemical composition and physical properties of this ice are highly variable, reflecting the highly variable conditions of its formation, aging, and melting. As a result there are a number of ways of conveniently

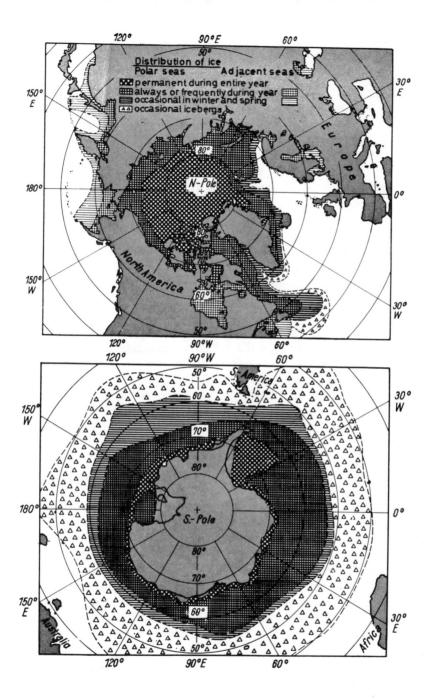

Figure 13.10 Boundaries of the Ice Cover of Polar Regions. From Dietrich (1963), with permission of John Wiley & Sons.

classifying this ice. For example, I believe the distinction between sea ice, that is, ice frozen from seawater, and land ice, consisting of glacial ice and river ice, is a useful one although the distinction is blurred by intermediate or rather composite forms such as the shelf ice so important in the Antarctic. In this section we restrict our attention to sea ice. Other useful classifications distinguish between solid ice and drift ice and between younger winter ice (less than 1 year old) and older (2 or more years) polar ice.

Let me first say a word about pure ice. (For a review of the water-ice interface see Drost-Hansen, 1966.) The structure of the familiar solid form stable at 1 atm—Ice-I_h—unlike that of liquid water, is, as we saw in Chapter 1 (Figures 1.5 and 1.7), well known (Owston, 1951, 1958). But such knowledge establishes only the average spacial locations of, primarily, the centers of the oxygen atoms. The translational, vibrational, and rotational movements of the atoms, especially the hydrogens, is quite another matter and still lies largely hidden in ignorance's domain. And, although elaborate theories have been concocted to account for the dielectric and electrically conductive properties of ice (Onsager, 1962), including the protonic conduction (Eigen and DeMaeyer (1956, 1958), my own feeling is that relaxation and transport processes in this substance are still very inadequately understood. Some of the physical properties of pure ice are summarized in Table A.5.

Even deeper mystery shrouds the nature of the freezing process; it is phenomenologically reflected, as we saw previously, in the marked tendency of water to supercool. If I may again indulge in speculation, I suspect that many of the anomalies of the solidification process in the case of water arise from the necessity of disrupting the strong and peculiar interfacial water structure near the water-ice boundary. Research on the freezing process is receiving new impetus as a consequence of the interest in desalination (see above). Let us hope that some of the mystery will be resolved as a fruit of this effort.

Table 13.3   Freezing (Melting) Point of Seawater

S	T	S	T	S	T	S	T
1	0.055	11	0.587	21	1.129	31	1.683
2	0.108	12	0.640	22	1.184	32	1.740
3	0.161	13	0.694	23	1.239	33	1.797
4	0.214	14	0.748	24	1.294	34	1.853
5	0.267	15	0.802	25	1.349	35	1.910
6	0.320	16	0.856	26	1.405	36	1.967
7	0.373	17	0.910	27	1.460	37	2.024
8	0.427	18	0.965	28	1.516	38	2.081
9	0.480	19	1.019	29	1.572	39	2.138
10	0.534	20	1.074	30	1.627	40	2.196

If an aqueous salt solution such as seawater is frozen very slowly in the laboratory, pure ice commences to crystallize out at a temperature dependent on the solute concentration (Figure 1.23 and Table 13.3), and the remaining liquid solution becomes more concentrated in electrolyte. If the temperature is further lowered at about $-8.2°C$, $Na_2SO_4 \cdot 10 H_2O$ begins to crystallize out of the concentrated brine, then the NaCl at $-23°C$; there is also some precipitation of $CaCO_3$, and finally below $-55°C$ the entire system is frozen. Analysis of 1 kg of 35.05‰ S seawater frozen at $-30°C$ gave (Sverdrup, Johnson, and Fleming, 1942):

Solid
931.9 g of ice crystals
20.23 g of crystalline NaCl
3.95 g of crystalline Na_2SO_4
Trace of crystalline $CaCO_3$

Brine (43.95 g)
23.31 g of H_2O
1.42 g of Na^+
1.31 g of Mg^{2+}
0.38 g of K^+
0.39 g of Ca^{2+}
7.03 g of Cl^-
0.08 g of Br^-
0.03 g of SO_4^{2-}

On the other hand, if the solution is rapidly frozen the electrolyte can be trapped in the ice either by ion entrapment or by entrapment of small pockets of brine. The ion entrapment process is selective; it takes up some ions more readily than others. This can result in a charge separation and the development of very appreciable electrical potentials. This curious phenomenon is known as the Workman-Reynolds (1950) effect and was originally proposed as a cause of electrification in thunderstorms.

The sequence of events described above might be expected to give rise to ionic fractionation, notably in the case of Cl^- and SO_4^{2-}, but, although some results have been reported to the contrary (Wiese, 1930), generally speaking the SO_4^{2-}/Cl^- ratio remains the same in sea ice as in seawater (Malmgren, 1927). In the sea, in contrast to in the laboratory, the more dense, concentrated brine solution sinks away from the advancing ice front so that the ice largely grows in fresh seawater rather than in concentrated brine, except, of course, for the pockets of already trapped seawater, which again cause no perturbation since they are incorporated entire into the analysis.

In fresh water the crystal plates are parallel to the freezing surface, whereas in seawater, owing to electrolyte transport processes, they tend to be vertical to the freezing surface, thus giving rise to the observed fibrous structure and vertical fracture planes of sea ice (Defant, 1961). The directionality of the crystals apparently has an important effect on the salt content; in the case of natural lake ices, milky ice with a vertical c axis has a much greater content of incorporated salts than clear ice with a horizontal c axis (Cobb, 1963).

Some properties of sea ice are given in Table A-29. The two most important factors in determining the physical-chemical properties are the salinity and the porosity or air content; they in turn depend on the particular conditions obtaining during the freezing and aging processes. Ice frozen from highly saline water tends to be more saline than ice frozen from less saline seas. Because of greater entrapment, rapidly frozen ice is more saline than more slowly frozen ice, and inasmuch as the rate of freezing depends on the air temperature, the salinity of the ice can increase with decreasing air temperature ranging from, for example, 5.64‰ S at $-16°C$ to 10.16‰ S at $-40°C$. Differences in freezing rates can also result in a dependence of salinity on the depth within the sample, but here the situation is complicated by the fact that the brine pockets are mobile and slowly migrate in the ice (Kingery and Goodnow, 1963; Hoekstra, Osterkamp, and Weeks, 1965; Harrison, 1965). Reported migrational velocities range from millimeters per day to millimeters per hour and appear to be independent of droplet size and largely dependent on the temperature gradient. Harrison (1965) also noticed that the droplets increase in size and accelerate as they approach the ice-liquid interface. Weeks and Lee (1962) found a systematic decrease in salinity with the age of the sea ice.

This brings us to the topic of the aging of sea ice. When a sample of pure ice is subjected to a slow temperature cycle it may be annealed or strengthened but, when sea ice ages quite the reverse happens and the structure tends to be weakened to such an extent that in the final stages of the aging process the ice is said to be "rotten." As the ice ages, the entrapped brine tends to escape and is replaced by air pockets. Ice from less saline seas tends, thereby, to contain less air than ice from saltier seas: 4% of the volume of the ice from the relatively fresh waters of the Gulf of Finland is air, in contrast to 8% in ice from the highly saline Barents Sea. But only a part of the gas content of sea ice can be attributed to brine replacement; the rest, like the brine itself, is entrapped in the freezing process, the quantities involved being greater the more rapid the freezing. Some of the gas may also come from changes in solubility of atmospheric gases in the brine with temperature, pressure, and electrolyte concentration. The analyses of Matsuo and Miyake (1966) (Table

13.4) appear to point to a somewhat different sequence of events. They found a higher gas content in young rather than old ice, a definite correlation with the chloride content, and they concluded that the dissolved gases tend to escape along with the brine. Older tests indicated that the relative O_2 content of the gas in sea ice was greater than that of air but less than that of the dissolved gas in seawater, whereas the results of Matsuo and Miyake (1966) indicate that N_2 is more strongly retained than the other elemental atmospheric gases, and they attribute this preference to the formation of a N_2-clathrate in ice. Table 13.4B shows that CO_2 is also strongly retained in ice. The gas content affects the density of ice much more strongly than the salt content (Table A-29A), and the high degree of variability of the thermal conductivity of sea ice is attributable to variations in air content or porosity. Snow and ice are good insulators, and the thermal conductivity of the sea ice cover must be taken into consideration in analyzing the heat budget of polar seas (Untersteiner, 1964; Donn and Shaw, 1966).

5   The Origin and Evolution of Life in the Seas

Ever since Socrates' adoption of the Delphic maxim "Know thyself" as his goal, the central question of natural philosophy has been that of the nature of man. Even the physical sciences fascinate us ultimately not for what they tell us about the external cosmos, but rather for what they tell us about ourselves. All the disciplines, not only the biological and behavioral sciences, converge upon the same central questions. "What are we? "Where did we come from?" and, if we dare, "Where are we going?"

The history of science has described a great cycle. Many of the world's major religions, past and present, have in their cosmological mythologies fashioned the world out of water. Science, that is, logical empiricism, began with Thales of Miletus who held that "Water is best," that water is the source of all things, that all things are forms of water. His successors, such as Empedocles, explicitly envisioned a process of biological evolution with man himself emerging from fish-like forms. Today the scientist again faces the sea. He has explored the continents, the stars, the moon, the galaxies, the atom, but now he is brought to the shore of the sea once more. He has come to appreciate that it was here that life began (Oparin, 1953; Fox, 1953; see also the *Proceedings of the First International Symposium on the Origin of Life in Earth, Moscow, 1957*, Pergamon Press, London, 1959). He realizes that, if careless time has left any clues to the origin of life on this planet and to the course of the long evolutionary process, those clues are probably somewhere in the oceans of Earth. The keys to our beginning, to our being, perhaps even to our future are hidden in the sea. Here are the innermost secrets of life, in

Table 13.4   Gas Content of Sea and Other Ices
(From Matsuo and Miyake, 1966, with permission of American
Geophysical Union and the authors)
A. Analysis of the Air in Ice Samples

Ice Sample	Density of Ice, g/cm³	Amount of Gas, cm³/kg (STP)	Pressure of Gas bars	Gas Composition			
				N_2	O_2	Ar	CO_2
Iceberg ice							
1	0.906	25.2	1.9	77.7	21.3	0.94	0.084
2	0.880	51.9	1.1	77.9	21.2	0.93	0.031
3	0.894	29.2	1.0	78.0	21.0	0.93	0.098
4	0.877	34.6	0.7	78.3	20.8	0.93	0.04
5	0.885	75.4	1.8	78.1	20.9	0.93	0.10
6	0.891	67.1	2.2	78.3	20.7	0.93	0.08
7		40.8		77.9	21.2	0.93	0.030
Glacier ice							
1	0.886	36.0	0.9	78.3	20.7	0.93	0.028
2	0.913	15.8	3.3	73.2	25.3	1.22	0.29
Sea ice							
1	0.909	8.9	0.9	76.8	21.3	0.93	0.98
2	0.915	2.2	0.9	54.2	20.6	0.99	24.3
3	—	21.2	—	69.3	29.0	1.1	0.53
4	—	19.8	—	71.4	26.5	1.1	1.0
5	—	10.7	—	69.5	29.0	0.97	0.60
6	—	16.4	—	67.7	31.0	0.94	0.41
7	—	15.7	—	74.3	23.8	1.1	0.84
8	—	19.9	—	72.5	25.4	1.1	0.93
Pond ice							
1	0.916	1.0	1.0	52.2	22.0	1.0	24.8

B. Ratios of the Content of Gas Components in Various Samples

Sample	Total Gas Content (cm³/kg) (STP)	Retention Ratio of Gas (%)	N_2/Ar	N_2/O_2	N_2/CO_2	Cl^- (ppm)
Sea ice						
1	8.9	38.5	82.6	3.61	78.4	8.3
2	2.2	9.5	54.7	2.64	2.23	16.4
3	21.2	91.8	62.4	2.39	131.	1,865
4	19.8	85.7	63.2	2.69	71.4	328
5	10.7	46.3	71.7	2.40	116.	
6	16.4	71.0	72.0	2.18	165.	1,809
7	15.7	68.0	67.5	3.12	88.5	1,880
8	19.9	86.1	65.9	2.85	78.0	2,800
Pond ice	1.0	3.3	51.2	2.37	2.1	
Distilled water at 0°C	29.9	100	37.9	1.80	(42)	
Seawater with 34.5‰ salinity at 0°C	23.1	100	37.5	1.81	(30)	19,100
Atmospheric air			84.0	3.73	2400	

the myriad living forms, in the geological record of ancient seas, in the sea-water itself.

A living cell is an enormously complex dynamic structure. Even the most elaborate chemical processing plants built by man are pitifully simple by comparison. Yet, when I first turned my attention to the origin and evolution of life, I was amazed, not because we know so little, but rather because we know so much. In the words of Professor Bernal* (1961), "we now have the skeleton of a story of the origin of life . . ." While the details remain scarce and while many questions will remain unanswered for many years to come, perhaps forever, the answers to the principal questions now seem to be all at least foreshadowed; the principal conceptual barriers have already been breached.

The first of these barriers was an emotional and metaphysical one—the conviction that "organic" compounds could be formed only through the agency of living cells, that their genesis was in a sense supernatural and necessitated a vital or "life force." This belief was shattered in 1824 when Wöhler demonstrated that the familiar organic compound urea, an end product of protein metabolism, can be prepared simply by the evaporation of a solution of a typical inorganic salt, ammonium cyanate. Just as the Copernican theory of heliocentricity cleansed cosmology of its mystical and anthropormorphic elements and placed the solar system in true perspective in relation to the rest of the natural order, so Wöhler's simple experiment revolutionized biological thought, set organic chemistry in its proper perspective in relation to chemistry in general, and revealed that life is a natural process subject to elucidation by scientific method, thereby establishing the basis of a true science of biochemistry.

The inorganic synthesis of urea opened the possibility that organic compounds, and subsequently life, could have resulted from purely "nonliving" or inorganic substances and processes at some time in the Earth's distant geological past. Yet subsequently the long-flourishing theory of the spontaneous generation of living from nonliving forms was finally laid to rest by the work of Pasteur. Does this mean that life could not have originated on Earth from purely natural causes? No. We now know that it means simply that the process of spontaneous generation requires a span of time far in excess of the total lifetime of the human species and/or that the environmental conditions necessary for spontaneous generation no longer obtain on this planet.

Let me note parenthetically here that theories of the extraterrestrial inoculation of Earth with life (see Shklovskii and Sayan, 1966, for a good review)

---

* I hope that it does not escape the reader's attention, or seem to be a coincidence, that one of the most prominent scientists concerned with the problem of biogenesis is the father of modern theories of the structure of water.

do not answer the question of the origin of life they simply remove the problem to a different and nonexaminable frame of reference.

More than one hundred years separate Wöhler's demonstration that life *in principle* could have arisen from nonliving substances by natural processes from the first *actual* synthesis of proto-biomaterials from inorganic substances under simulated primitive Earth conditions. At the suggestion of Professor Urey, Miller (1953, 1955) passed an electrical discharge through a gaseous mixture of CH_4, NH_3, H_2O and H_2 gases believed to be constituents of the primitive atmosphere of the earth—for time intervals of roughly a week—and obtained polyhydroxy compounds and amino acids, the fundamental building blocks of the life substance (Chapter 9). Not only were yields surprisingly abundant, but also the number of compounds identifiable was considerable (Table 13.5).

In recent years this work has been widely repeated, confirmed, and varied. In addition, other processes for the production of simple organic from inorganic materials have been discussed, such as the formation of hydrocarbons from the reaction of certain carbides with water, and Eck, Lippincott, et al. (1966) have calculated the concentration of a number of organic substances such as alcohols, carboxylic acids, hydrocarbons, ketones, amines, ethers, and heterocyclic compounds that might result from simple equilibria with inorganic reactants.

But the stones are not a temple; once in hand, the building blocks must be put together. Now we must confront the third barrier. How are the pieces brought together? The putting together of the pieces was a long, tedious, and delicate sequence and each step in the sequence was highly improbable. Fortunately the time span allotted to the beginnings of life was exceedingly long, perhaps several billion years (Bernal, 1961), so that the improbable was not necessarily the impossible. Biogenesis is pushed further into the realm of possibility if there were mechanisms operative for the concentration of the pieces and, in order to outrace the forces of dissolution, for the stabilization of the pieces and their combinations.

In terms of proto-biological reactants the ancient seas were a very dilute broth, and various mechanisms have been suggested that could have resulted in large enough concentrations of these substances for the improbable life-building steps to become probable. Darwin suggested that the proto-biological broth may have become concentrated by the evaporation of shallow pools. However, I think the alternative mechanism suggested by Professor Bernal (1961) is far more attractive, namely, concentration by interfacial absorption processes, which we have encountered repeatedly in this book, on bubbles (Wangersky, 1965; Barber, 1966), on the sea's surface, or on solid particles. I am taken by his analogy that life, like Aphrodite, the goddess of love, was born of the sea foam. Let us imagine, then, the proto-biological substances

Table 13.5   Yield of Organic Compounds Resulting from Electrical Discharge in a Simulated Primitive Earth Atmosphere (From Miller, 1955, with permission of ACS)

A. Yield of Amino Acids

Amino Acid	Moles $\times 10^5$	mg	Mole Ratio (glycine = 1)	Yield (%)
Run 1				
Glycine	63	47	1.00	2.1(4.0)
Alanine	34	30	0.54	1.7(3.2)
Sarcosine	5	4	0.08	0.3(0.5)
β-Alanine	15	13	0.24	0.8(1.5)
B_1	1	1	.02	0.07(0.13)
α-Aminobutyric acid	5	5	0.08	0.3(0.6)
B_2	3	3	0.05	0.2(0.4)
Run 2				
Glycine	55	41	1.00	1.8(3.2)
Alanine	36	32	0.65	1.8(3.2)
Sarcosine	2	2	0.04	0.1(0.2)
β-Alanine	18	16	0.33	1.0(1.8)
B_2	0.1	0.1	0.002	0.01(0.02)
α-Aminobutyric acid	3.0	3.0	0.054	0.2(0.34)
B_1	0.4	0.4	0.007	0.03(0.05)
Run 3				
Glycine	80	60	1.0	0.46(2.1)
Alanine	9	8	0.11	0.08(0.3)
Sarcosine	86	77	1.07	0.74(3.4)
β-Alanine	4	3	0.05	0.03(0.1)
B_1	12.5	12.9	0.16	0.14(0.63)
α-Aminobutyric acid	1	1	0.01	0.01(0.05)
B_2	14.7	15.2	0.18	0.17(0.76)

B. Yields of Acids

Acid	Moles $\times 10^5$	Mg	Yield (%)
Run 1			
Formic	233	107	3.9(7.4)
Acetic	15.2	9.1	0.5(1.0)
Propionic	12.6	9.1	0.6(1.2)
Glycolic	56	42	1.9(3.5)
Lactic	39	35	1.8(3.4)
Run 3			
Formic	149	69	0.4(2.0)
Acetic	135	81	0.7(3.5)
Propionic	19	14	0.2(0.7)
Glycolic	28	21	0.2(0.7)
Lactic	4.3	3.9	0.03(0.2)

being absorbed on bubble surfaces, transported upward to the sea's surface, and joined with other material absorbed there, then tossed by the waves and carried by the sea spray (Chapter 11) up onto the beaches and the estuarine mud where, in the richer, warmer waters the pieces begin to react and the aggregates to grow. Earlier life was described as a struggle against the forces of destruction. These forces were operating long before anything identifiable as life came into being, constantly endangering the existence of not only the aggregates but even the initial organic building blocks as well. In the present world the forces of degradation of organic materials are bacteriological decomposition, oxidation, and the reversal of chemical equilibria.

Bacterial decomposition obviously was not a problem in the prebiotic world. Neither was oxidation. The bulk of evidence indicates that the primitive atmosphere of the Earth was a reducing one, and that the present oxidative atmosphere and hydrosphere are relatively recent and, in fact, the product of biological activity. But, what stayed the hand of dissolution by the reversal of equilibria? The apparent answer is a fascinating one and one which returns us to the central theme of this book—structural stabilization. Once polymerization of the amino acid fragments has occurred the danger of dissolution diminishes, for the polymer may organize itself and its fellows into a stable configuration. Hydrophobic bonding comes into play, and the polymer is surrounded by a hydration sheath which in effect protects it from the solvent powers of the bulk water (Chapter 1). These structural configurations may be further stabilized by the electrolytes in seawater. As an example, Wald (1954) calls attention to the astonishing transformation of an aqueous solution of highly random collagen filaments into highly organized collagen filaments on the addition of 1% NaCl.

At this point the aggregates may assume the form of coacervate droplets. Smith and Bellware (1966) have demonstrated that suspended proteinoid microspheres assume a coacervate structure upon repeated dehydration and rehydration such as might have occurred in the rocky crevices along the edge of some ancient sea (Hinton and Blum, 1965). The coacervate droplets are, according to Oparin (1961), "the original precursors of primitive organisms." In structure, protoplasm is a complex coacervate. Oparin (1961) also theorizes that it was at this level that the mechanism of evolution, natural selection, came into operation. Coacervates more stable and more successful than their fellows in the competition for "nutrient" material survived, while their weaker contemporaries perished and their substance was devoured.

One important structural feature separates the coacervate droplet from the cell; the cell has a wall. Lipids have one hydrophobic and one hydrophilic end on their long-chain molecules; they tend to pack in sheets, and such sheets or membranes made possible the isolation of the interior of the coacervate from its surroundings.

To repeat, Bernal (1961) claims that, with respect to the first beginnings of life, "there are, roughly, only about six [stages of molecular complexity] between simple atoms and something you can see in the microscope." He represents this hierarchy of polymer complexity first by the simple polypeptide chain (Chapter 9, Section 3), then the α-helix (Figure 9.2), next the folded coil, then linked folded coils as in myoglobin (Figure 9.6), and finally the linking of different folded coils together with a ribonucleic acid chain as in the tobacco mosaic virus—one of the simplest living things.

Although the beginnings were painfully slow, once certain crucial milestones were passed, the evolutionary life process accelerated at an ever-increasing rate as much more effective measures were realized to accomplish necessary ends, such as the evolution of vastly more effective biocatalysts (Chapter 9, Section 6) and the development of superior techniques of supplying the energy necessary to hold at bay the forces of disintegration. The earliest energy source may have been analogous to fermentation, the conversion of organic material to waste products, CO_2, and utilizable energy. But, if no alternative energy source were forthcoming, life would have been doomed as the organic content of seawater was depleted. This catastrophe was averted by the evolution of an improved energy source, the utilization of CO_2, H_2O, and solar energy—photosynthesis [Eqs. (9.2) and (9.3)]. The photosynthetic process in turn released O_2, a powerful oxidant, and thereby laid the basis for a still more effective energy source—respiration [Eqs. (9.4)]. "To use an economic analogy, photosynthesis brought organisms to the subsistence level; respiration provided them with capital. It is mainly this capital that they invested in the great enterprise of organic evolution" (Wald, 1954).

At the same time, the release of O_2 by photosynthesis resulted in the formation of a layer of ozone high in the Earth's atmosphere that shielded the surface from lethal ultraviolet radiation so that life forms could finally leave the shelter of the seas and venture upon the dry land. The rest was almost an anticlimax. The stage was set, the actors were in the wings, for the appearance of man.

No creature reigns forever. But, when our kind has gone, possibly as the result of our own folly, if not on the ruined Earth then on other planets in this and distant galaxies, wherever there is adequate water, life will continue to appear, flourish, and evolve.

> Nor can those motions that bring death prevail
> Forever, nor eternally entomb
> The welfare of the world; nor, further, can
> Those motions that give birth to things and growth
> Keep them forever when created there.

Thus the long war, from everlasting waged,
With equal strife among the elements
Goes on and on. Now here, now there, prevail
The vital forces of the world—or fall.
Mixed with the funeral is the wildered wail
Of infants coming to the shores of light.

Lucretius, *De Rerum Natura*

REFERENCES

Agar, J. N., *Rev. Pure Appl. Chem.*, **8**, 1 (1958).

Albers, V. M., *Underwater Acoustics Handbook*, Pennsylvania State Univ. Press, University Park, Pa., 1960.

Albers, V. M., ed., *Underwater Acoustics*, Plenum Press, New York, 1961.

Aubert, M., J. Aubert, M. Gauthier, and S. Daniel, *Rev. Intern. Oceanogr. Med.*, **1**, 10 (1966).

Barber, R. T., *Nature*, **211**, 257 (1966).

Bernal, J. D., in M. Sears, ed., *Oceanography*, Am. Assoc. Advan. Sci. Publ., No. 67. Washington, D.C., 1961.

Bobalek, E. G., *Corrosion*, **10**, 73 (1954).

Christie, G. L., *Corrosion*, **14**, 337t (1958).

Clapp, W. F., "Macro-Organisms in Sea Water and Their Effect on Corrosion," in H. H. Uhlig, ed., *Corrosion Handbook*, John Wiley & Sons, New York, 1948.

Cobb, A. E., *Science*, **141**, 733 (1963).

Crease, J., "Internal Waves," in U. N. Albers, ed., Plenum Press, New York, 1961.

Davis, C. M., Jr., and T. A. Litovitz, *J. Chem. Phys.*, **42**, 2563 (1965).

Defant, A., *Physical Oceanography*, Pergamon Press, New York, 1961, Vol. 1, Chap. 8.

DeLuccia, J. J., *Materials Protect.*, **5**, (8), 49 (1966).

Dietrich, G., *General Oceanography*, Interscience, New York, 1963.

Disteche, A., *J. Electrochem. Soc.*, **109**, 1084 (1962).

Donn, W. L., and D. M. Shaw, *J. Geophys. Res.*, **71**, 1087 (1966).

Dorsey, E. N., *Properties of Ordinary Water-Substance*, Reinhold Pub. Corp., New York, 1940.

Drost-Hansen, W., "The Water-Ice Interface as Seen from the Liquid State," Paper presented at the Am. Chem. Soc. Symposium on Interfacial Surface Properties of Ice, Pittsburgh, 1966. Inst. Marine Sci., Univ. of Miami, 1966.

Eck, R. V., E. R. Lippincott, M. O. Dayhoff, and Y. T. Pratt, *Science*, **153**, 628 (1966).

Eigen, M., and L. DeMaeyer, *Z. Elektrochem.*, **60**, 1037 (1956).

Eigen, M., and L. DeMaeyer, *Proc. Roy. Soc. (London)*, **247A**, 505 (1958).

Eigen, M., and K. Tamm, *Z. Elektrochem.*, **66**, 93, 107 (1962).

Elze, J., and G. Oelsner, *Metalloberflache*, **12**, 129 (1958).

Finley, H. F., *Corrosion*, **17**, 104t (1961).

Fisher, F. H., *J. Phys. Chem.*, **66**, 1607 (1962).

Fisher, F. H., *J. Phys. Chem.*, **69**, 695 (1965a).

Fisher, F. H., *J. Acoust. Soc. Am.*, **28**, 805 (1965b).

Forgeson, B. W., L. J. Waldron, M. H. Peterson, and B. F. Brown, "Abyssal Corrosion and Its Mitigation," U.S.N. Res. Lab. NRL Memo Rept. 1383 (Dec. 1962).

Fox, S. W., ed., *The Origins of Prebiological Systems*, Academic Press, New York, 1964.

Frosch, R. A., *Science*, **146**, 889 (1964).

Glueckauf, E., *Nature*, **211**, 1227 (1966).

Gray, K. O., *Mat. Protection*, **3** (7), 46 (1964).

Greenspan, M., and C. E. Tschiegg, *J. Acoust. Soc. Am.*, **28**, 501 (1956).

Hall, L., *Phys. Rev.*, **73**, 775 (1948).

Halstead, B. W., and D. A. Courville, *Poisonous and Venomous Marine Animals of the World*, U.S. Government Printing Office, Washington, D.C., 1965.

Harrison, J. D., *J. Appl. Phys.*, **36**, 3811 (1965).

Hinton, H. E., and M. S. Blum, *New Sci.*, **28**, 270 (1965).

Hoar, T. P., *Corrosion*, **19**, 331t (1963).

Hoekstra, P., T. E. Osterkamp, and W. F. Weeks, *J. Geophys. Res.*, **70**, 5035 (1965).

Horne, R. A., and H. F. Eden, "Sound Propagation and Pressure-Induced Structural Changes in Sea-Water," (Unclass.), in Unpublished Report, Office of Naval Research, Code 468 (Oct. 1966).

Horne, R. A., B. R. Myers, and G. R. Frysinger, *Inorg. Chem.*, **3**, 452 (1964).

Horton, J. W., *Fundamentals of Sonar*, U.S. Naval Institute, Annapolis, Ind., 1959.

Juchniewicz, R., *Werkstoffe Korrosion*, **9**, 360 (1958).

Kingery, W. D., and W. H. Goodnow, in W. D. Kingery, ed., *Ice and Snow Properties*, Technology Press, Cambridge, Mass., 1963.

Kinsler, L. E., and A. R. Frey, *Fundamentals of Acoustics*, John Wiley & Sons, New York, 1950.

LaQue, F. L., "Behavior of Metals and Alloys in Sea-Water," in H. H. Uhlig, ed., *The Corrosion Handbook*, John Wiley & Sons, New York, 1948.

Lee, O. S., *J. Acoust. Soc. Am.*, **33**, 5 (1961).

Litovitz, T. A., and E. M. Carnevale, *J. Appl. Phys.*, **26**, 816 (1955).

Malmgren, F., *Sci. Results of the "Maud" Expedition*, *(1918–1925)*, **1**, No. 5 (1927).

Matsuo, S., and Y. Miyake, *J. Geophys. Res.*, **71**, 5235 (1966).

Mero, J. L., *The Mineral Resources of the Sea*, Elsevier Pub. Co., Amsterdam, 1965.

Miller, S. L., *Science*, **117**, 528 (1953).

Miller, S. L., *J. Am. Chem. Soc.*, **77**, 2351 (1955).

Mussad, A. H., and B. G. Holmes, *Petrol. Engr.*, **29**, No. 5, B73 (1957).

Nelson, E. E., *Corrosion*, **13**, 122t (1957).

Newton, E., private communication, 1967.

Nigrelli, R. F., M. F. Stempien, Jr., G. D. Ruggieri, V. R. Ligouri, and J. T. Cecil, *Fed. Proc.*, **26**, 1197 (1967).

Olesen, P. E., A. Maretzki, and L. A. Almodovar, *Botan. Marina*, **6**, 224 (1964).

Onsager, L., *Vortex*, **23**, 138 (1962).

Oparin, A. I., *Origin of Life*, Dover Pub. New York, 1953.

Oparin, A., in M. Sears, ed., *Oceanography*, Am. Assoc. Advan. Sci. Publ., No. 67, Washington, D.C., 1961.

Owston, P. G., *Quart. Rev.*, **5**, 344 (1951).

Owston, P. G., *Advan. Phys.*, **7**, 171 (1958).

Parkins, R. N., *Met. Rev.*, **9**, 201 (1964).

Piip, A. T., *J. Acoust. Soc. Am.*, **36**, 1948 (1964).

Preiser, H. S., and F. E. Cook, *Corrosion*, **13**, 125t (1957).

Rhodin, T. N., ed., *Physical Metallurgy of Stress Corrosion Fracture*, Interscience, New York, 1959.

Riggs, O. L., Jr., J. D. Sudbury, and M. Hutchinson, *Corrosion*, **16**, 260t (1960).

Schrieber, C. F., *Corrosion*, **14**, 126t (1958).

Schwerdtfeger, W. J., *J. Res. Natl. Bur. Std. U.S.*, **60**, 153 (1958).

Shigley, C. M., *J. Metals*, **3**, 25 (1951).

Shklovskii, I. S., and C. Sagan, *Intelligent Life in the Universe*, Holden-Day, Inc., San Francisco, 1966.

Smith, A. E., and F. T. Bellware, *Science*, **152**, 362 (1966).

Speller, F. N., *Corrosion: Causes and Prevention*, McGraw-Hill Book Co., New York, 1951.

Spiegler, K. S., ed., *Principles of Desalination*, Academic Press, London, 1966.

Starkey, R. L., *Producers Monthly*, **22**, 12 (1958).

Sverdrup, H. U., M. W. Johnson, and R. H. Fleming, *The Oceans*, Prentice Hall, Inc. Englewood Cliffs, N.J., 1942.

Thorp, W. H., *J. Acoust. Soc. Am.*, **38**, 648 (1965).

Untersteiner, N., *J. Geophys. Res.*, **69**, 4755 (1964).

Uusitalo, E., *Corrosion*, **17**, 67t (1961).

Verma, G. S., *Rev. Mod. Phys.*, **31**, 1052 (1959).

Wald, G., *Sci. Am.*, **191**, 44 (Aug. 1954).

Wangersky, P. J., *Am. Sci.*, **53**, 358 (1965).

Watkins, J. W., and G. W. Kincheloe, *Corrosion*, **14**, 341t (1958).

Weeks, M. E., *Discovery of the Elements*, J. Chem. Educ., Easton, Pa., 5th ed., 1945, Chap. 24.

Weeks, W. S., and O. S. Lee, *Arctic*, **15**, 93 (1962).

Whitmarsh, D. C., E. Skudrzyk, and R. J. Urik, *J. Acoust. Soc. Am.*, **29**, 1124 (1957).

Wiese, W., *Ann. Hydrogen Marine Meteorol.*, **58**, 282 (1930).

Wilson, W. D., *J. Acoust. Soc. Am.*, **32**, 1357 (1960a).

Woods Hole Oceanographic Institute, *Marine Fouling and Its Prevention*, U.S. Naval Institute, Annapolis, Md., 1952.

Workman, E. J., and S. E. Reynolds, *Phys. Rev.*, **78**, 254 (1950).

Appendix

Table A-1 Some Signs, Symbols, Definitions, Conventions, and Conversion Units Used in This Book

$Å$ = Angstrom $\equiv 10^{-8}$ cm
a_i = activity of the species i
$\mathring{a}$ = Debye-Hückel distance of closest approach parameter
atm = atmosphere ≈ 1.033227 kg/cm$^2 \approx 1.013250$ bar ≈ 14.6974 lb/in^2
α_i = activity coefficient of the species i
α = acoustic attenuation
α = coefficient of thermal expansion $\equiv (1/V)(\Delta V/\Delta T)$
B, B^* = Tait equation parameters
bar = pressure unit (see atm)
β = compressibility $\equiv (1/V)(\Delta V/\Delta P)$
C = various constants
$°C$ = temperatures in centigrade degrees
C_p = heat capacity at constant pressure
C_v = heat capacity at constant volume
Cl = chlorinity of seawater
c = electrolyte concentration, M or m
c = Curie $\equiv 6.56 \times 10^{-6}$ g Rn
cal = calorie ≈ 4.1840 abs joules
γ_i = activity coefficient of the species i
γ = surface tension
D = diffusion coefficient, cm^2/sec
d = infinitesimal increment
Δ = finite increment
δ = partial differential
δ_i = ionic contribution to the dielectric depression
E = electromotive force or electrical potential
$\bar{e}$ = unit of electronic charge
ϵ = dielectric constant
F = (Gibbs) free energy
$\mathscr{F}$ = Faraday's number $\approx 96,485$ abs coulombs/g equiv
f = frequency
g = gram; 453.59 g $\approx$ 1lb
g = acceleration of gravity
H = enthalpy or heat content
h = height
h = geometric depth
η = viscosity
I = ionic strength $\equiv \frac{1}{2}\sum c_i z_i^2$ (often designated μ in other texts)
I = light intensity

Table A-1 (cont.)

i = the ionic species i (as a subscript)

J = mechanical equivalent of heat, 1 mean cal $\backsimeq$ 4.1816 × 10^7 erg

K = practical equilibrium constant (in terms of concentrations)

$K°$ = thermodynamic equilibrium constant (in terms of activities)

°K = temperature, in degrees Kelvin

k = Boltzmann's constant = 1.380257 × 10^{-16} erg deg^{-1} $molecule^{-1}$

k = kilo = 1000

kcal = kilocalorie ≡ 1000 cal

kg = kilogram ≡ 1000 g $\backsimeq$ 2.2046 lb

κ = specific electrical conductance, ohm^{-1} cm^{-1}

κ = light extinction coefficient or attenuation

l = liter ≡ 1000 cm^3 = 1000 ml

Λ = equivalent electrical conductance, ohm^{-1} $equiv.^{-1}$ cm^2

$\Lambda°$ = limiting equivalent electrical conductance ≡ lim Λ versus $\sqrt{c}$

λ_i = equivalent electrical conductance of the ionic species i

λ = radioactive decay constant

M = molarity ≡ moles of solute per liter of solution ≡ g mol wt/l of solution

m = molality ≡ moles of solute per 1000 g of solution ≡ g mol wt/1000 g of solution

for dilute solutions, $m \approx M$

m = meter, 1 ft $\backsimeq$ 0.3048 m, 1 m ≡ 100 cm

mole = gram molecular weight

μ = Joule-Thomson coefficient

μ = micro = 10^{-6} or micron ≡ 10^{-6} m

μ_i = chemical potential of the species i

N = Avogadro's number $\backsimeq$ 6.02380 × 10^{23} molecules/mole

N = mole fraction; in the system A and B, N_A ≡ number of moles of A/number of moles A and B; $N_A + N_B = 1$

N = normality ≡ 1 equiv wt of solute/l of solution

n = hydration or ligand number

n = number of species

n = index of refraction

o = as superscript or subscript, refers to pure soution, infinite dilution, or standard state

P = pressure, usually in kgm/cm^2 (also atm and bar)

P = vapor pressure, usually in atm or mm Hg

Π = product

R = gas constant ≡ 8.31439 joule deg^{-1} $mole^{-1}$ ≡ 1.98719 cal deg^{-1} $mole^{-1}$

r = distance, usually interatomic, intermolecular, or interionic

ρ = density, g/cm^3

S = salinity of seawater

S = entropy

Σ = sum

T = temperature, usually in °C

$t_{1/2}$ = half-life

τ = dielectric relaxation time, sec

V = volume

Φ = osmotic pressure

x = distance

Ψ_i = electrical potential of the species i in solution

z_i = charge of the ionic species i

(From R. C. West (ed.), *Handbook of Chemistry and Physics*, Chemical Rubber Co., Cleveland, Ohio, 1964 (45th ed.), with permission of the publisher)

KEY TO CHART

Atomic Number → **50** | +2 +4 ← Oxidation States
Symbol → **Sn**
Atomic Weight → 118.69 | -18-18-4 ← Electron Configuration

1a	2a	3b	4b	5b	6b	7b	8	8	8	1b	2b	3a	4a	5a	6a	7a	0	Orbit
1 H +1 -1 1.00797 (1)																	**2** He 4.0026 (2) 0	K
3 Li +1 6.939 (2-1)	**4** Be +2 9.0122 (2-2)											**5** B +3 10.811 (2-3)	**6** C +2 +4 -4 12.01115 (2-4)	**7** N +1 +2 +3 +4 +5 -1 -2 -3 14.0067 (2-5)	**8** O -2 15.9994 (2-6)	**9** F -1 18.9984 (2-7)	**10** Ne 20.183 (2-8) 0	K-L
11 Na +1 22.9898 (2-8-1)	**12** Mg +2 24.312 (2-8-2)											**13** Al +3 26.9815 (2-8-3)	**14** Si +2 +4 -4 28.086 (2-8-4)	**15** P +3 +4 +5 -2 -3 30.9738 (2-8-5)	**16** S +4 +6 -2 32.064 (2-8-6)	**17** Cl +1 +5 +7 -1 35.453 (2-8-7)	**18** Ar 39.948 (2-8-8) 0	K-L-M
19 K +1 39.102 (-8-8-1)	**20** Ca +2 40.08 (-8-8-2)	**21** Sc +3 44.956 (-8-9-2)	**22** Ti +2 +3 +4 47.90 (-8-10-2)	**23** V +2 +3 +4 +5 50.942 (-8-11-2)	**24** Cr +2 +3 +6 51.996 (-8-13-1)	**25** Mn +2 +3 +4 +6 +7 54.9380 (-8-13-2)	**26** Fe +2 +3 55.847 (-8-14-2)	**27** Co +2 +3 58.9332 (-8-15-2)	**28** Ni +2 +3 58.71 (-8-16-2)	**29** Cu +1 +2 63.54 (-8-18-1)	**30** Zn +2 65.37 (-8-18-2)	**31** Ga +3 69.72 (-8-18-3)	**32** Ge +4 72.59 (-8-18-4)	**33** As +3 +5 -3 74.9216 (-8-18-5)	**34** Se +3 +5 -3 78.96 (-8-18-6)	**35** Br +1 +5 -1 79.909 (-8-18-7)	**36** Kr 83.80 (-8-18-8) 0	-L-M-N
37 Rb +1 85.47 (-18-8-1)	**38** Sr +2 87.62 (-18-8-2)	**39** Y +3 88.905 (-18-9-2)	**40** Zr +4 91.22 (-18-10-2)	**41** Nb +3 +5 92.906 (-18-12-1)	**42** Mo +6 95.94 (-18-13-1)	**43** Tc +4 +6 +7 (99) (-18-13-2)	**44** Ru +4 +6 +7 101.07 (-18-15-1)	**45** Rh +3 102.905 (-18-16-1)	**46** Pd +2 +4 106.4 (-18-18-0)	**47** Ag +1 107.870 (-18-18-1)	**48** Cd +2 112.40 (-18-18-2)	**49** In +3 114.82 (-18-18-3)	**50** Sn +2 +4 118.69 (-18-18-4)	**51** Sb +3 +5 -3 121.75 (-18-18-5)	**52** Te +4 +6 -2 127.60 (-18-18-6)	**53** I +1 +5 +7 -1 126.9044 (-18-18-7)	**54** Xe 131.30 (-18-18-8) 0	-M-N-O
55 Cs +1 132.905 (-18-8-1)	**56** Ba +2 137.34 (-18-8-2)	**57*** La +3 138.91 (-18-9-2)	**72** Hf +4 178.49 (-32-10-2)	**73** Ta +5 180.948 (-32-11-2)	**74** W +6 183.85 (-32-12-2)	**75** Re +4 +6 +7 186.2 (-32-13-2)	**76** Os +4 +6 +7 190.2 (-32-14-2)	**77** Ir +3 +4 192.2 (-32-15-2)	**78** Pt +2 +4 195.09 (-32-16-2)	**79** Au +1 +3 196.967 (-32-18-1)	**80** Hg +1 +2 200.59 (-32-18-2)	**81** Tl +1 +3 204.37 (-32-18-3)	**82** Pb +2 +4 207.19 (-32-18-4)	**83** Bi +3 +5 208.980 (-32-18-5)	**84** Po +2 +4 (210) (-32-18-6)	**85** At (210) (-32-18-7)	**86** Rn (222) (-32-18-8) 0	-N-O-P
87 Fr +1 (223) (-18-8-1)	**88** Ra +2 (227) (-18-8-2)	**89**** Ac +3 (227) (-18-9-2)																-O-P-Q

Transition Elements

Group 8

Transition Elements

(Continued)

Table A-2 (cont.)

	-N-O-P
	-O-P-Q

*Lanthanides													
58 +3 +4 Ce 140.12 -19-9-2	59 +3 +4 Pr 140.907 -20-9-2	60 +3 Nd 144.24 -22-8-2	61 +3 Pm (145) -23-8-2	62 +3 Sm 150.35 -24-8-2	63 +2 +3 Eu 151.96 -25-8-2	64 +3 Gd 157.25 -25-9-2	65 +3 Tb 158.924 -26-9-2	66 +3 Dy 162.50 -28-8-2	67 +3 Ho 164.930 -29-8-2	68 +3 Er 167.26 -30-8-2	69 +3 Tm 168.934 -31-8-2	70 +2 +3 Yb 173.04 -32-8-2	71 +3 Lu 174.97 -32-9-2

**Actinides													
90 +4 Th 232.038 -19-9-2	91 +4 +5 Pa (231) -20-9-2	92 +5 +4 U 238.03 -21-9-2	93 +3 +4 +5 +6 Np (237) -22-9-2	94 +3 +4 +5 +6 Pu (242) -23-9-2	95 +3 +4 +5 +6 Am (243) -24-9-2	96 +3 +4 Cm (245) -25-9-2	97 +3 +4 Bk (249) -26-9-2	98 +3 +4 Cf (249) -28-9-2	99 +3 Es (254) -29-8-2	100 Fm (252) -30-8-2	101 Md (256) -31-8-2	102 No (254) -32-8-2	103 Lw

Numbers in parentheses are mass numbers of most stable isotope of that element.

476

Table A-3 Selected Properties of Water Vapor
(Values abstracted from N. E. Dorsey, *Properties of Ordinary Water-Substance*, Reinhold Pub. Corp., New York, 1940)

Molecular weight	18.0154 g
Heat of formation	242.49 (at 100°C and 1 atm)
Viscosity	96×10^{-6} g/cm sec (at 20°C and 1 atm)
Velocity of sound	405 m/sec (at 100°C and 1 atm)
Diffusion coefficient	0.380 cm²/sec (in air at 100°C and 1 atm)
Specific volume	1729.6 cm³/g (at 100°C and 1 atm)
Specific heat	2.078 joules/g°C (at 100°C and 1 atm)
Thermal conductivity	2.44×10^{-4} watt/cm °C (at 110°C and 1 atm)

Table A-4 Selected Properties of Liquid Water[a]
(Selected values from N. E. Dorsey, *Properties of Ordinary Water-Structure*, Reinhold Pub. Corp., New York, 1940)

Molecular weight	18.0154 g
Heat of formation	285.890 kj/mole (at 25°C and 1 atm)
Ionic dissociation constant	$10^{-14}\ M^{-1}$ (at 25°C and 1 atm)
Heat of ionization	55.71 kj/mole (at 25°C and 1 atm)
Apparent dipole moment	5.59×10^{-19} cgse units/mole
Viscosity	8.949 millipoise (at 25°C and 1 atm)
Velocity of sound	1496.3 m/sec (at 25°C and 1 atm)
Density	0.9979751 g/cm³ (at 25°C and 1 atm)
Freezing point	0°C (at 1 atm)
Boiling point	100°C (at 1 atm)
Isothermal compressibility	45.6×10^{-6} atm^{-1} (at 25°C over the range 1–10 atm)
Specific heat at constant volume	4.17856 int. joule/g °C (at 25°C and 1 atm)
Thermal conductivity	0.00598 watt/cm °C (at 20°C and 1 atm)
Temperature of maximum density	3.98°C (at 1 atm)
Dielectric constant	81.0 (at 1 atm, 17°C, and 60 megacycles/sec)
Electrical conductivity	Less than 10^{-8} ohm^{-1} cm^{-1} (at 25°C and 1 atm)

[a] See also Table A-10.

Table A-5 Selected Properties of Ice (Ice-I$_h$)
(Selected Values from E. N. Dorsey, *Properties of Ordinary Water-Substance*, Reinhold Pub. Corp., New York, 1940)

Molecular weight	18.0154 g
Heat of formation	292.72 kg/mole (at 0°C and 1 atm)
Lattice parameters	$a = 4.535$ Å, $C = 7.14$ Å (hexagonal)
Young's modulus of elasticity	967 kg/mm² (at -10°C and 1 atm)
Density	0.9168 g/ml (at 0°C and 1 atm)
Coefficient of cubical thermal expansion	120×10^{-6} cm³/g °C (at 0°C and 1 atm)
Coefficient of linear thermal expansion	52.7×10^{-6} °C^{-1} (at 0°C and 1 atm)
Isothermal compressibility	12×10^{-6} bar^{-1} (at 0°C and 300 bars)
Specific heat	2.06 joule/g °C (at 0°C and 1 atm)
Thermal conductivity	21 mwatt/cm °C
Dielectric constant	79 (at -1°C, 1 atm, and 3000 cycles/sec)

Table A-6   Hydration Numbers of Ten of the Most Common Simple Ions in Seawater (at 1 atm)
(The superscripts give the number of the reference at the end of the Table).

Cr	Na$^+$	Mg^{2+}	SO$_4^{2-}$	Ca^{2+}	K$^+$	Br$^-$	CO$_3$	Sr^{2+}	F$^-$	HCO$_3^-$
23[4]	4.8[4]	3.8[1]	12[6]	4.3[1]	4.1[4]	1.7[4]	15[6]	5.0[1]	4[5]	2[6]
4[5]	5[2]	10[2]	1.4[13]	8[6]	4[5]	4[5]		3.7[16]	2.2[9]	
8–9[8]	2[2]	12[2]	12[18]	4.3[16]	4[7]	2[6]		10[11]	7[17]	
1.4[9]	4[5]	6[3]		10[11]	1.1[9]	5[17]		20[17]		
3[6]	2.1[9]	8[6]		29[17]	2[6]					
20[15]	3[6]	5.1[16]			6[10]					
4–27[13]	5[11]	12[11]			20[15]					
0.9[16]	10.8[12]	36[17]			1–5[14]					
1[11]	8–45[13]				5–30[13]					
3[18]	2–8[14]				6.2[12]					
5[17]	70[15]				4[38]					
	13[17]				0.6[16]					
	8[18]				2[11]					
					4[18]					
					7[17]					

	Method

1. Swift and Sayre, *J. Chem. Phys.*, **44**, 3567 (1966).   Proton NMR
2. Robinson and Stokes, *Electrolyte Solutions*, Butterworths,   Hydrodynamic
   London 1959.
3. de Silveira, Marques, and Marques, *Mol. Phys.*, **9**, 271   Raman
   (1965).
4. Allam and Lee, *J. Chem. Soc.*, **1966**, A5, 426   Compressibility
5. Swain and Evans, *J. Am. Chem. Soc.*, **88**, 383 (1966).   Theory
6. Padova, *Bull. Res. Council Israel*, **10A**, 63 (1961)   Membrane
7. Brady and Krause, *J. Chem. Phys.*, **27**, 304 (1957).   X-ray
8. Brady, *J. Chem. Phys.*, **28**, 464 (1958).   X-ray
9. Glueckauf, *Trans. Faraday Soc.*, **60**, 1637 (1964).   Elect. water trans.
10. van Panthaleon, van Eck, Mendel, and Boog,   X-ray
    Discussions Faraday Soc., **24**, 200 (1957).
11. Haase, *Z. Elektrochem.*, **62**, 279 (1958).   Ionic mobility
12. Mukerjee, *Indian J. Phys.*, **24**, 137 (1950).   Ionic mobility
13. Baborovsky, *Z. Physik. Chem.*, **129**, 129 (1927).   Elect. water trans.
14. Washburn, *J. Am. Chem. Soc.*, **37**, 694 (1915).   Elect. water trans.
15. Nernst, *Gott. Nachr.*, **1** (1900).   Elect. water trans.
16. Glueckauf, *Trans. Faraday Soc.*, **51**, 1235 (1955).   Activity coefficient
17. Rutgers and Hendrikx, *Trans. Faraday Soc.*, **58**, 2184   Membrane
    (1962).
18. van Ruyven, *Chem. Weekblad*, **53**, 688 (1957).   Raoult's law

Table A.7   Specific Volume
(From W. Wilson and D. Bradley, "Specific Volume,
Thermal Expansion, and Isothermal Compressibility of
Sea Water," U.S.N. Ord. Lab. Rept. No. NOL TR-66-103,
June 2, 1966, Unclass, AD-635-120, with permission of
the authors)

Specific Volume of Distilled water, in cm^3/g

P, bars[a]	T, °C				
	0	10	20	30	40
1	1.0000	1.0003	1.0017	1.0043	1.0078
100	0.9950	0.9956	0.9972	0.9999	1.0035
200	0.9902	0.9910	0.9928	0.9956	0.9992
300	0.9856	0.9866	0.9886	0.9914	0.9950
400	0.9810	0.9823	0.9844	0.9873	0.9910
500	0.9767	0.9781	0.9804	0.9833	0.9870
600	0.9724	0.9741	0.9764	0.9795	0.9832
700	0.9683	0.9702	0.9726	0.9757	0.9795
800	0.9643	0.9663	0.9689	0.9721	0.9758
900	0.9605	0.9626	0.9653	0.9685	0.9723
1000	0.9567	0.9590	0.9618	0.9650	0.9688

Specific Volume of Seawater

10‰ S

P, bars[a]	T, °C				
	0	10	20	30	40
1	0.9921	0.9926	0.9942	0.9968	1.0004
100	0.9873	0.9881	0.9899	0.9926	0.9961
200	0.9826	0.9837	0.9856	0.9884	0.9920
300	0.9781	0.9794	0.9814	0.9843	0.9879
400	0.9737	0.9752	0.9774	0.9803	0.9840
500	0.9695	0.9711	0.9735	0.9764	0.9801
600	0.9654	0.9672	0.9696	0.9727	0.9764
700	0.9614	0.9634	0.9659	0.9691	0.9728
800	0.9575	0.9597	0.9623	0.9655	0.9692
900	0.9537	0.9561	0.9588	0.9620	0.9658
1000	0.9501	0.9526	0.9554	0.9587	0.9624

Table A-7 (cont.)

20‰ S

			T, °C		
P, bars[a]	0	10	20	30	40
1	0.9842	0.9850	0.9868	0.9894	0.9930
100	0.9796	0.9806	0.9826	0.9853	0.9888
200	0.9751	0.9763	0.9784	0.9812	0.9848
300	0.9707	0.9722	0.9743	0.9772	0.9808
400	0.9664	0.9681	0.9704	0.9734	0.9770
500	0.9623	0.9642	0.9666	0.9696	0.9733
600	0.9583	0.9604	0.9629	0.9660	0.9697
700	0.9544	0.9566	0.9593	0.9624	0.9661
800	0.9507	0.9530	0.9558	0.9590	0.9627
900	0.9470	0.9495	0.9524	0.9556	0.9593
1000	0.9435	0.9461	0.9491	0.9524	0.9561

30‰ S

			T, °C		
P, bars[a]	0	10	20	30	40
1	0.9764	0.9775	0.9794	0.9821	0.9856
100	0.9719	0.9732	0.9753	0.9781	0.9816
200	0.9676	0.9690	0.9712	0.9741	0.9776
300	0.9633	0.9650	0.9673	0.9702	0.9738
400	0.9592	0.9611	0.9635	0.9665	0.9701
500	0.9552	0.9572	0.9598	0.9629	0.9665
600	0.9513	0.9535	0.9562	0.9593	0.9630
700	0.9476	0.9500	0.9527	0.9559	0.9595
800	0.9439	0.9464	0.9493	0.9525	0.9562
900	0.9404	0.9430	0.9460	0.9492	0.9529
1000	0.9370	0.9397	0.9428	0.9461	0.9498

35‰ S

			T, °C		
P, bars[g]	0	10	20	30	40
1	0.9726	0.9738	0.9757	0.9784	0.9819
100	0.9681	0.9695	0.9717	0.9745	0.9780
200	0.9638	0.9654	0.9677	0.9706	0.9741
300	0.9596	0.9614	0.9638	0.9668	0.9703
400	0.9556	0.9576	0.9601	0.9631	0.9667
500	0.9517	0.9538	0.9564	0.9595	0.9631
600	0.9479	0.9502	0.9529	0.9560	0.9596
700	0.9442	0.9466	0.9494	0.9526	0.9562
800	0.9406	0.9432	0.9461	0.9493	0.9529
900	0.9371	0.9398	0.9428	0.9461	0.9497
1000	0.9337	0.9366	0.9396	0.9429	0.9466

Table A-7 (cont.)

40‰ *S*

P, bars[a]	T, °C				
	0	10	20	30	40
1	0.9687	0.9700	0.9720	0.9748	0.9783
100	0.9644	0.9659	0.9680	0.9709	0.9744
200	0.9601	0.9618	0.9641	0.9670	0.9706
300	0.9560	0.9579	0.9603	0.9633	0.9668
400	0.9520	0.9541	0.9566	0.9597	0.9632
500	0.9481	0.9504	0.9530	0.9561	0.9597
600	0.9444	0.9468	0.9495	0.9527	0.9563
700	0.9408	0.9433	0.9461	0.9493	0.9530
800	0.9372	0.9399	0.9428	0.9461	0.9497
900	0.9338	0.9366	0.9396	0.9429	0.9465
1000	0.9305	0.9334	0.9365	0.9398	0.9434

[a] 1 bar = 10^6 dynes/cm^2. Example: For a pressure of 100 bars and a temperature of 10°, the thermal expansion is 11.2×10^{-5} (°C)$^{-1}$.

Table A-8   Compressibility Data
(From W. Wilson and D. Bradley, "Specific Volume, Thermal Expansion, and Isothermal Compressibility of Sea Water," U.S.N. Ord. Lab. Rept. No. NOLTR-66-103, June 2, 1966, Unclass, AD-635-120, with permission of the authors)

Compressibility of Distilled Water ($\times 10^6$ bars)

P, bars[a]	T, °C				
	0	10	20	30	40
1	50.7	47.9	46.1	44.9	44.4
100	49.2	46.6	44.9	43.8	43.3
200	47.9	45.4	43.8	42.7	42.3
300	46.5	44.2	42.6	41.7	41.3
400	45.3	43.1	41.6	40.7	40.3
500	44.1	42.0	40.6	39.7	39.4
600	42.9	40.9	39.6	38.8	38.5
700	41.8	39.9	38.7	37.9	37.6
800	40.7	39.0	37.8	37.0	36.8
900	39.7	38.0	36.9	36.2	35.9
1000	38.7	37.1	36.0	35.4	35.2

Table A-8 (cont.)

Compressibility of Seawater
10‰ S

P, bars[a]	T, °C				
	0	10	20	30	40
1	49.5	46.9	45.1	44.0	43.5
100	48.1	45.6	43.9	42.9	42.4
200	46.8	44.4	42.8	41.8	41.4
300	45.5	43.3	41.8	40.8	40.4
400	44.3	42.2	40.7	39.9	39.5
500	43.1	41.1	39.7	38.9	38.6
600	41.9	40.0	38.8	38.0	37.7
700	40.8	39.1	37.8	37.1	36.8
800	39.8	38.1	37.0	36.3	36.0
900	38.8	37.2	36.1	35.5	35.2
1000	37.8	36.3	35.3	34.7	34.4

20‰ S

P, bars[a]	T, °C				
	0	10	20	30	40
1	48.4	45.8	44.1	43.1	42.6
100	47.0	44.6	43.0	42.0	41.6
200	45.7	43.4	41.9	41.0	40.6
300	44.4	42.3	40.9	40.0	39.6
400	43.2	41.2	39.8	39.0	38.7
500	42.1	40.2	38.9	38.1	37.8
600	41.0	39.2	37.9	37.2	36.9
700	39.9	38.2	37.0	36.4	36.1
800	38.9	37.3	36.2	35.5	35.3
900	37.9	36.4	35.3	34.7	34.5
1000	37.0	35.5	34.5	33.9	33.7

30‰ S

P, bars[a]	T, °C				
	0	10	20	30	40
1	47.2	44.8	43.2	42.1	41.7
100	45.9	43.6	42.0	41.1	40.7
200	44.6	42.5	41.0	40.1	39.7
300	43.4	41.4	40.0	39.1	38.8
400	42.2	40.3	39.0	38.2	37.8
500	41.1	39.3	38.0	37.3	37.0
600	40.0	38.3	37.1	36.4	36.1
700	39.0	37.3	36.2	35.6	35.3
800	38.0	36.4	35.4	34.8	34.5
900	37.0	35.6	34.6	34.0	33.8
1000	36.1	34.7	33.8	33.2	33.0

Table A-8 (cont.)

35‰ S

			T, °C		
P, bars[a]	0	10	20	30	40
1	46.7	44.3	42.7	41.7	41.2
100	45.4	43.1	41.6	40.6	40.2
200	44.1	42.0	40.5	39.6	39.3
300	42.9	40.9	39.5	38.7	38.3
400	41.7	39.8	38.5	37.8	37.4
500	40.6	38.8	37.6	36.9	36.6
600	39.5	37.8	36.7	36.0	35.7
700	38.5	36.9	35.8	35.2	34.9
800	37.5	36.0	35.0	34.4	34.1
900	36.6	35.1	34.2	33.6	33.4
1000	35.6	34.3	33.4	32.8	32.6

40‰ S

			T, °C		
P, bars[a]	0	10	20	30	40
1	46.1	43.8	42.2	41.2	40.8
100	44.8	42.6	41.1	40.2	39.8
200	43.6	41.5	40.1	39.2	38.8
300	42.4	40.4	39.1	38.3	37.9
400	41.2	39.4	38.1	37.3	37.0
500	40.1	38.4	37.2	36.5	36.2
600	39.1	37.4	36.3	35.6	35.3
700	38.0	36.5	35.4	34.8	34.5
800	37.1	35.6	34.6	34.0	33.8
900	36.1	34.7	33.8	33.2	33.0
1000	35.2	33.9	33.0	32.5	32.3

[a] 1 bar = 10^6 dynes/cm^2.

Table A-9   Thermal Expansion Data
(From W. Wilson and D. Bradley, "Specific Volume,
Thermal Expansion and Isothermal Compressibility of
Sea Water," U.S.N. Ord. Lab. Rept. No. NOL TR-66-103,
June 2, 1966, Unclass., AD-635-120, with permission of
the authors)

Coefficient of Thermal Expansion of Distilled Water ($\times 10^5$ °C)

			T, °C		
P, bars	0	10	20	30	40
1	−3.4	9.0	20.0	30.3	40.7
100	−0.2	11.2	21.4	31.1	41.0
200	2.8	13.3	22.8	31.9	41.2
300	5.6	15.2	24.0	32.6	41.4
400	8.3	17.1	25.2	33.2	41.5
500	10.8	18.9	26.4	33.8	41.7
600	13.2	20.6	27.5	34.4	41.8
700	15.5	22.1	28.5	35.0	41.9
800	17.7	23.6	29.5	35.5	42.0
900	19.7	25.0	30.4	36.0	42.1
1000	21.6	26.4	31.2	36.4	42.2

Coefficient of Thermal Expansion of Seawater

10‰ S

			T, °C		
P, bars[a]	0	10	20	30	40
1	−0.2	11.1	21.2	30.7	40.4
100	2.8	13.2	22.5	31.5	40.6
200	5.7	15.2	23.8	32.2	40.8
300	8.5	17.1	25.1	32.9	41.0
400	11.1	18.9	26.2	33.5	41.2
500	13.5	20.6	27.3	34.1	41.3
600	15.8	22.2	28.4	34.7	41.4
700	18.0	23.7	29.4	35.2	41.5
800	20.1	25.2	30.3	35.7	41.6
900	22.0	26.5	31.2	36.1	41.7
1000	23.9	27.8	32.0	36.6	41.7

Table A-9 (cont.)

20‰ S

P, bars[a]	T, °C				
	0	10	20	30	40
1	3.0	13.2	22.3	31.1	40.1
100	6.0	15.2	23.6	31.9	40.3
200	8.7	17.1	24.9	32.6	40.5
300	11.4	19.0	26.1	33.2	40.7
400	13.8	20.7	27.2	33.8	40.8
500	16.2	22.3	28.3	34.4	40.9
600	18.4	23.8	29.2	34.9	41.0
700	20.5	25.3	30.2	35.4	41.1
800	22.5	26.7	31.1	35.9	41.2
900	24.4	28.0	31.9	36.3	41.3
1000	26.1	29.2	32.7	36.7	41.3

30‰ S

P, bars[a]	T, °C				
	0	10	20	30	40
1	6.2	15.3	23.5	31.5	39.8
100	9.0	17.2	24.8	32.2	40.0
200	11.7	19.1	26.0	32.9	40.2
300	14.2	20.8	27.1	33.5	40.3
400	16.6	22.5	28.2	34.1	40.4
500	18.8	24.0	29.2	34.6	40.5
600	21.0	25.5	30.1	35.1	40.6
700	23.0	26.9	31.0	35.6	40.7
800	24.9	28.2	31.9	36.0	40.8
900	26.6	29.4	32.7	36.4	40.9
1000	28.3	30.6	33.4	36.8	40.9

35‰ S

P, bars[a]	T, °C				
	0	10	20	30	40
1	7.8	16.3	24.1	31.7	39.6
100	10.6	18.2	25.3	32.4	39.8
200	13.2	20.0	26.5	33.0	40.0
300	15.6	21.7	27.6	33.6	40.1
400	18.0	23.3	28.6	34.2	40.2
500	20.2	24.9	29.6	34.7	40.3
600	22.2	26.3	30.6	35.2	40.4
700	24.2	27.7	31.4	35.7	40.5
800	26.0	29.0	32.3	36.1	40.6
900	27.8	30.2	33.1	36.5	40.6
1000	29.4	31.3	33.8	36.9	40.7

Table A-9 (cont.)

40‰ S

			T, °C		
P, bars[a]	0	10	20	30	40
1	9.4	17.3	24.7	31.9	39.5
100	12.1	19.2	25.9	32.6	39.6
200	14.7	21.0	27.0	33.2	39.8
300	17.1	22.6	28.1	33.8	39.9
400	19.4	24.2	29.1	34.3	40.1
500	21.5	25.7	30.1	34.8	40.2
600	23.5	27.1	31.0	35.3	40.2
700	25.4	28.5	31.9	35.8	40.3
800	27.2	29.7	32.7	36.2	40.4
900	29.0	30.9	33.4	36.6	40.4
1000	30.6	32.0	34.2	37.0	40.5

[a] 1 bar = 10^6 dynes/cm².

Table A-10   Selected Physical Properties of Pure Water[a]
(From R. A. Robinson and R. H. Stokes *Electrolyte Solutions*, 2nd ed.,
Butterworths Sci. Pub., London, 1959, with permission of the publisher)

Temp., °C	Density, g/ml	Specific Volume, ml/g	Vapor Pressure, mm Hg	Dielectric Constant	Viscosity, Centipoise
0	0.99987	1.00013	4.580	87.740	1.787
5	0.99999	1.00001	6.538	85.763	1.516
10	0.99973	1.00027	9.203	83.832	1.306
15	0.99913	1.00087	12.782	81.945	1.138
18	0.99862	1.00138	15.471	80.835	1.053
20	0.99823	1.00177	17.529	80.103	1.002
25	0.99707	1.00293	23.753	78.303	0.8903
30	0.99568	1.00434	31.824	76.546	0.7975
35	0.99406	1.00598	42.180	74.823	0.7194
38	0.99299	1.00706	49.702	73.817	0.6783
40	0.99224	1.00782	55.338	73.151	0.6531
45	0.99024	1.00985	71.90	71.511	0.5963
50	0.98807	1.01207	92.56	69.910	0.5467
55	0.98573	1.01448	118.11	68.344	0.5044
60	0.98324	1.01705	149.57	66.813	0.4666
65	0.98059	1.01979	187.65	65.319	0.4342
70	0.97781	1.02270	233.81	63.855	0.4049
75	0.97489	1.02576	289.22	62.425	0.3788
80	0.97183	1.02899	355.31	61.027	0.3554
85	0.96865	1.03237	433.64	59.657	0.3345
90	0.96534	1.03590	525.92	58.317	0.3156
95	0.96192	1.03959	634.04	57.005	0.2985
100	0.95838	1.04343	760.00	55.720	0.2829

[a] See also Table A-4.

Table A-11   Specific Conductivity of Seawater[a]
[Values calculated from the equation of P. Weyl, *Limnol. Oceanog.*, **9**, 75
(1964), based on data of B. D. Thomas, T. G. Thompson, and C. L.
Utterback, *J. Conseil Perm. Intern. Exploration Mer.*, **9**, 28 (1934)]

S, ‰	Temperature, °C						
	30	25	20	15	10	5	0
10	(19.127)	17.345	15.628	13.967	12.361	10.816	9.341
20	(35.458)	32.188	29.027	25.967	23.010	20.166	17.456
30	(50.856)	46.213	41.713	37.351	33.137	29.090	25.238
31	(52.360)	47.584	42.954	38.467	34.131	29.968	26.005
32	(53.859)	48.951	44.192	39.579	35.122	30.843	26.771
33	(55.352)	50.314	45.426	40.688	36.110	31.716	27.535
34	(56.840)	51.671	46.656	41.794	37.096	32.588	28.298
35	(58.323)	53.025	47.882	42.896	38.080	33.457	29.060
36	(59.801)	54.374	(49.105)	(43.996)	(39.061)	(34.325)	(29.820)
37	(61.274)	55.719	(50.325)	(45.093)	(40.039)	(35.190)	(30.579)
38	(62.743)	57.061	(51.541)	(46.187)	(41.016)	(36.055)	(31.337)
39	(64.207)	58.398	(52.754)	(47.278)	(41.990)	(36.917)	(32.094)
40	(65.667)	(59.732)	(53.963)	(48.367)	(42.962)	(37.778)	(32.851)

[a]Conductivity in $ohm^{-1} cm^{-1} \times 1000$.

Table A-12   Effect of Pressure on the Conductivity of Seawater[a]
(From A. Bradshaw and K. E. Schleicher, *Deep-Sea Res.*, **12**, 151 (1965),
with permission of Pergamon Press, Ltd.)

Pressure, db	Temp.	S, ‰			Temp.	S, ‰		
		31	35	39		31	35	39
1,000	0°C	1.599	1.556	1.512	15°C	1.032	1.008	0.985
2,000		3.089	3.006	2.922		1.996	1.951	1.906
3,000		4.475	4.354	4.233		2.895	2.830	2.764
4,000		5.759	5.603	5.448		3.731	3.646	3.562
5,000		6.944	6.757	6.569		4.506	4.403	4.301
6,000		8.034	7.817	7.599		5.221	5.102	4.984
7,000		9.031	8.787	8.543		5.879	5.745	5.612
8,000		9.939	9.670	9.401		6.481	6.334	6.187
9,000		10.761	10.469	10.178		7.031	6.871	6.711
10,000		11.499	11.188	10.877		7.529	7.358	7.187
1,000	5°C	1.368	1.333	1.298	20°C	0.907	0.888	0.868
2,000		2.646	2.578	2.510		1.755	1.718	1.680
3,000		3.835	3.737	3.639		2.546	2.492	2.438

[a]Percentage increase compared with the conductivity at 1 atm.

Table A-12 (cont.)

Pressure, db	Temp.	S, ‰			Temp.	S, ‰		
		31	35	39		31	35	39
4,000		4.939	4.813	4.686		3.282	3.212	3.142
5,000		5.960	5.807	5.655		3.964	3.879	3.795
6,000		6.901	6.724	6.547		4.594	4.496	4.399
7,000		7.764	7.565	7.366		5.174	5.064	4.954
8,000		8.552	8.333	8.114		5.706	5.585	5.464
9,000		9.269	9.031	8.794		6.192	6.060	5.929
10,000		9.915	9.661	9.408		6.633	6.492	6.351
1,000	10°C	1.183	1.154	1.125	25°C	0.799	0.783	0.767
2,000		2.287	2.232	2.177		1.547	1.516	1.485
3,000		3.317	3.237	3.157		2.245	2.200	2.156
4,000		4.273	4.170	4.067		2.895	2.837	2.780
5,000		5.159	5.034	4.910		3.498	3.429	3.359
6,000		5.976	5.832	5.688		4.056	3.976	3.896
7,000		6.728	6.565	6.402		4.571	4.481	4.390
8,000		7.415	7.236	7.057		5.045	4.945	4.845
9,000		8.041	7.847	7.652		5.478	5.369	5.261
10,000		8.608	8.400	8.192		5.872	5.756	5.640

Table A-13   Selected Stability Constants of Cations with Some
Inorganic Anionic Ligands Present in Seawater
(For a much more complete table of values and an explanation of the
notation used, see L. G. Sillen and A. E. Martell, *Stability Constants of
Metal-Ion Complexes*, Spec. Publ. No. 17, The Chemical Society, London,
1964; selected values reproduced with permission of The Chemical Society
and the authors)

Metal	Temp., °C	Medium	Log of Equilibrium Constant
		Chloride, Cl^-	
H^+	$0 \rightarrow 25$	O (corr) $\rightarrow$ conc	K_1, -7
Li^+	25	var conc	no ev cpx
Na^+	25	var conc	no ev cpx
K^+	$0 \rightarrow 25$	var conc	no ev cpx
Rb^+	18	0 (corr)	K_1, -0.77 (?)
Cs^+	25	var conc	K_1, -0.45 (?)
Be^{2+}	20	var conc HCl	K_1, -0.66 (?)
Mg^{2+}	25	var conc	weak cpx (?)

Table A-13 (cont.)

Metal	Temp., °C	Medium	Log of Equilibrium Constant
		Chloride, Cl^- (continued)	
Ca^{2+}	25	var conc	weak cpx (?)
Sr^{2+}	25	var conc	weak cpx (?)
Ba^{2+}	18	0 (corr)	K_1, -0.13
Sc^{3+}	15	0.5 ($NaClO_4$)	K_1, 1.14; K_2, 1.11
	25	0.5 ($NaClO_4$)	K_1, 1.07; K_2, 1.04
	35	0.5 ($NaClO_4$)	K_1, 1.00; K_2, 0.91
Y^{3+}	15	0.5 ($NaClO_4$)	K_1, 0.38
	25	0.5 ($NaClO_4$)	K_1, 0.36
	35	0.5 ($NaClO_4$)	K_1, 0.32
	25	O (corr)	K_1, 1.26
La^{3+}	19	1 ($NaClO_4$)	K_1, -0.11
	25	1 ($NaClO_4$)	K_1, -0.12
	40	1 ($NaClO_4$)	K_1, -0.05
Ce^{3+}	25	1 ($HClO_4$)	K_1, -0.1; K_2, -0.6
Eu^{3+}	25	1 ($HClO_4$)	K_1, -0.1; K_2, -0.6
			$Eu(OH)_2Cl$; $Eu(OH)_{2.5}Cl_{0.5}$
Dy^{3+}	25		$Dy(OH)_2Cl$; $Dy(OH)_{2.5}Cl_{0.5}$
Gd^{3+}	25		$Gd(OH)_{2.5}Cl_{0.5}$
Tb^{3+}	25		$Tb(OH)_{2.5}Cl_{0.5}$
Ho^{3+}	25		$Ho(OH)_{2.5}Cl_{0.5}$
Er^{3+}	25		$Er(OH)_{2.5}Cl_{0.5}$
Yb^{3+}	25		$Yb(OH)_{2.5}Cl_{0.5}$
Lu^{3+}	20	0.1 (KNO_3)	K_1, 1.45
Ti^{3+}	40	0.5 ($HClO_4$)	K_1, 0.34
Ti^{4+}		var conc HCl	$Ti(OH)_2Cl$; $TiOCl$; $TiCl_6^{2-}$
Zr^{4+}	20	6.54 ($HClO_4$?)	K_1, 0.92; K_2, 0.40; K_3, 0.19; K_4, -0.33
Hf^{4+}	25	2 ($HClO_4$)	K_1, 0.38; K_2, -0.31; K_3, -0.68; K_4, -0.7
Th^{4+}	25	4 ($NaClO_4$)	K_1, 0.11; K_2, -1.03; K_3, -0.51; K_4, -0.42
V^{3+}	25	var conc HCl	weak cpx
VO^{2+}	20	1 ($NaClO_4$)	K_1, 0.04
Vo_2^+	25	conc HCl	ev ani cpx
$Nb(V)$	25	conc HCl	ev ani cpx
$Ta(V)$	25	var conc HCl	ev ani cpx
$Pa(V)$	25	var conc HCl	ev ani cpx
Cr^{3+}	25	5 ($HClO_4$)	K_1, -0.65, K_2, -1.54
$Cr(V)$			$CrOCl_4^-$
$Cr(VI)$			CrO_3Cl^-
$Mo(II)$			$Mo_6Cl_8(OH)_6^{2-}$
$Mo(IV)$			$MoCl_6^{2-}$
$Mo(V)$	25	0 (corr)	K_1, 6.4 (?)

Table A-13 (cont.)

Metal	Temp., °C	Medium	Log of Equilibrium Constant
		Chloride, Cl^- (continued)	
Mo(VI)	25	0 (corr)	K_1, -0.3; K_2, -0.5; K_3, -1.89
W^{3+}	0	dil	$W_2Cl_9^{3-}$
U^{3+}	25	conc HCl	ev cpx
U^{4+}	25	2 (NaClO$_4$)	K_1, 0.26; K_2, -0.02
UO_2^{2+}	10	2 (NaClO$_4$)	K_1, -0.24
	25	2 (NaClO$_4$)	K_1, -0.06
	40	2 (NaClO$_4$)	K_1, 0.06
Np^{4+}	25	1 (HClO$_4$)	K_1, -0.3
NpO_2^{2+}	5	3 (HClO$_4$)	K_1, 0.06; K_2, -0.74
Pu^{3+}	25	0.1 (HClO$_4$)	K_1, 0.57
Pu^{4+}	25	1 (HClO$_4$)	K_1, -0.24
PuO_2^+			K_1, -0.17
PuO_2^{2+}	2	2 (HClO$_4$)	K_1, -0.41
	10	2 (HClO$_4$)	K_1, -0.34
	20	2 (HClO$_4$)	K_1, -0.25
Am^{3+}	22	1 (HClO$_4$)	K_1, -0.05
AmO_2^+			ev cpx
Cm^{3+}		0 (corr)	K_1, 1.17
Mn^{2+}	20	0.7 (HClO$_4$)	K_1, 0.59
Mn^{3+}	25	2 (HClO$_4$)	K_1, 0.95
Tc(VII)		var conc HCl	ev cpx
Re^{2+}			$ReCl_4^{2-}$
Re^{3+}		var conc HCl	$ReCl_4^-$; $Re_2Cl_6^{3-}$
Re(IV)			$ReCl_6^{2-}$; $Re_2OCl_{10}^{4-}$
Re(V)			$ReCl_5(OH)_2^{2-}$; $ReCl_5O^{2-}$
Fe^{2+}	20	2 (HClO$_4$)	K_1, 0.36; K_2, 0.04
Fe^{3+}	27	1 (NaClO$_4$)	K_1, 0.62; K_2, 0.11; K_3, -1.40
	25	1 (HClO$_4$)	K_1, 0.47
	35	1 (HClO$_4$)	K_1, 0.56
	45	1 (HClO$_4$)	K_1, 0.66
Co^{2+}	20	0.7 (HClO$_4$)	K_1, 0.69
Ni^{2+}	12		K_1, -0.29 (?)
	25		K_1, -0.24 (?)
	40		K_1, -0.21 (?)
Ru^{3+}	25	0.1 (HCl)	K_2, 1.4; K_3, 0.4
Ru^{4+}			$RuCl_6^{2-}$; $Ru(OH)_2Cl_6^{4-}$ (?)
Rh^{3+}	25	1 (HClO$_4$)	K_1, 2.45; K_2, 2.09; K_3, 1.38; K_4, 1.16
Rh^{4+}		var conc HCl	ev ani cpx
Pd^{2+}	30	0 (corr)	K_1, 6.0; K_2, 4.9; K_3, 2.4; K_4, 2.6

Table A-13 (cont.)

Metal	Temp., °C	Medium	Log of Equilibrium Constant
		Chloride, Cl^- (continued)	
Os^{3+}		var conc HCl	ev ani cpx
Ir^{3+}	25	var conc HCl	ev ani cpx
Ir^{4+}	25	var conc HCl	ev ani cpx
Pt^{2+}	25	0.3 (Na_2SO_4)	K_3, 3.3; K_4, 1.82
	25	1 (NaCl)	K_1, 10.5; K_2, 10.0; K_3, 9.5; K_4, 8.7
Pt(IV)	20	dil	$PtCl_6^{2-}$
Cu^+	20	0 (corr)	K_3, 0.2
Cu^{2+}	12		K_1, 0.59
	25		K_1, 0.74
	40		K_1, 0.84
	20	0.7 ($HClO_4$)	K_1, 0.98; K_2, -0.29; K_3, 0.14; K_4, -0.55
Ag^+	25	5 ($NaClO_4$)	K_1, 3.08; K_2, 2.32; K_3, 0.75; K_4, -0.85
	25	0 (corr)	K_1, 3.04; K_2, 2.00; K_3, 0.00; K_4, 0.26
Au^+			$AuCl_2^-$
Au^{3+}	$0 \rightarrow 25$	var conc HCl	$AuCl_4^-$
Zn^{2+}	20	0.7 ($HClO_4$)	K_1, 0.72; K_2, -0.23; K_3, -0.68; K_4, 0.37
Cd^{2+}	25	var conc NaCl	K_1, 1.76; K_2, 1.06; K_3, -0.33; K_4, -0.66
Hg_2^{2+}	25		weak cpx (?)
Hg^{2+}	25	0.5 ($NaClO_4$)	K_1, 6.74; K_2, 6.48; K_3, 0.85; K_4, 1.00
Al^{3+}	25	conc HCl	no ani cpx
Ga^{3+}	25	0 (corr)	K_1, -0.6; K_2, -1.7; K_3, -2.2
In^{3+}	20	0.7 ($HClO_4$)	K_1, 2.36; K_2, 1.27; K_3, 0.32;
Tl^+	0	0 (corr)	K_1, 0.78
	25	0 (corr)	K_1, 0.68
	40	0 (corr)	K_1, 0.64
	25	0 (corr)	K_1, 0.60; K_2, -0.43
Tl^{2+}	25		$TlCl^+$
Tl^{3+}	20	0.4 ($NaClO_4$)	K_1, 7.50; K_2, 4.50; K_3, 2.75; K_4, 2.25
Ge(IV)			$GeCl_6^{2-}$ (?)
Sn^{2+}	25	0 (corr)	K_1, 1.51; K_2, 0.73; K_3, -0.21; K_4, -0.55
$SnOH^+$	0	3 ($NaClO_4$)	K_1, 0.90
	25	3 ($NaClO_4$)	K_1, 1.04
	35	3 ($NaClO_4$)	K_1, 0.85

Table A-13 (cont.)

Metal	Temp., °C	Medium	Log of Equilibrium Constant
		Chloride, Cl$^-$ (continued)	
Sn(IV)			SnCl$_5^-$; SnCl$_6^{2-}$
Pb^{2+}	25	2 (LiNO$_3$)	K_1, 1.46; K_2, -0.26; K_3, -0.31
Pb^{4+}		var conc HCl	PbCl$_5^-$; PbCl$_6^{2-}$
As(III)			As(OH)$_2$Cl
Sb(III)		var conc HCl	SbCl$_4^-$
Sb(V)			SbCl$_6^-$; SbOHCl$_5^-$; SbCl$_4^+$
Bi^{3+}	25	1 H$^+$; 2 (NaClO$_4$)	K_1, 2.08; K_2, 2.14; K_3, 1.49; K_4, 1.47
Se(IV)			Se(OH)$_2$Cl$_2$
Te(IV)		conc HCl	ev ani cpx
Po^{2+}	22		PoCl$_3^-$; PoCl$_4^{2-}$
Po^{4+}			PoCl$_6^{2-}$; Po(OH)$_2$Cl$_2^{2-}$
		Sulfate, SO$_4^{2-}$	
H$^+$	0	$\rightarrow 0$	K_1, 1.83
	10	$\rightarrow 0$	K_1, 1.86
	20	$\rightarrow 0$	K_1, 1.90
D$^+$	18	D$_2$O	K_1, 2.00
Li$^+$	18	0 (corr)	K_1, 0.64 (?)
Na$^+$	18	0 (corr)	K_1, 0.70 (?)
K$^+$	18	0 (corr)	K_1, 0.82 (?)
NH$_4^+$	18	0 (corr)	K_1, 1.11
Be^{2+}	18	0.5 (NaClO$_4$)	K_1, 0.72
Mg^{2+}	0	0 (corr)	K_1, 1.96
	20	0 (corr)	K_1, 2.20
	30	0 (corr)	K_1, 2.35
Ca^{2+}	18	$\rightarrow 0$	K_1, 2.28
Sr^{2+}	25	0 (corr)	K_{SP}, -6.6
Ba^{2+}	25	0 (corr)	K_{SP}, -10
Ra^{2+}	25	dil	K_{SP}, -14
Y^{3+}	25	$\rightarrow 0$	K_1, 3.47
La^{3+}	25	0 (corr)	K_1, 3.66
Ce^{3+}	25	0 (corr)	K_1, 3.37 (?)
Ce^{4+}	25	2 (HClO$_4$)	K_1, 3.3 (?)
Pr^{3+}, Nd^{3+}, Sm^{3+}, Eu^{3+}, Gd^{3+}, Ho^{3+}, Er^{3+}, Yb^{3+}	25	$\rightarrow 0$	K_1, 3.5
TiO^{2+}	18		K_1, 2.40
Zr^{4+}	25	2 (HClO$_4$)	K_1, 3.79; K_2, 2.85
Hf^{4+}	25	2 (HClO$_4$)	K_1, 3.11; K_2, 2.47
Th^{4+}	25	2 (HClO$_4$)	K_1, 3.32; K_2, 2.38

Table A-13 (cont.)

Metal	Temp., °C	Medium	Log of Equilibrium Constant
		Sulfate, SO_4^{2-} (continued)	
VO^{2+}	25	0 (corr)	K_1, 2.48
Nb(V), Pa(V)			ev cpx
Cr^{3+}	25	1 (NaClO$_4$)	K_1, 1.34
Cr(VI)		H_2SO_4	ev cpx
Mo(VI)		H_2SO_4 +	ev ani cpx
U^{4+}		2 (HClO$_4$)	K_1, 1.70 (?)
UO^{2+}	25		K_1, -0.9 (?)
UO_2^{2+}	20	1 (NaClO$_4$)	K_1, 1.70; K_2, 0.84; K_3, 0.86
Np^{4+}	25	2 (NaClO$_4$)	K_1, 2.4
NpO_2^+			no ev cpx
NpO_2^{2+}	25	1 (HClO$_4$)	K_1, 1.11
Pu^{3+}	25	2 (NaClO$_4$)	K_1, 1.0 (?); K_2, 0.62 (?)
Pu^{4+}	25	1 (HClO$_4$)	K_1, 1.11
Am^{3+}			ev ani cpx
AmO_2^{2+}			ev cpx
Cm^{3+}	20 → 25	0 (corr)	K_1, 3.66; K_2, 0.81
Mn^{2+}	0	0 (corr)	K_1, 2.01
	10	0 (corr)	K_1, 2.11
	20	0 (corr)	K_1, 2.20
Fe^{2+}	25	0	K_1, 2.3 (?)
Fe^{3+}	1	0.5 (NaClO$_4$)	K_1, 1.98
	15	0.5 (NaClO$_4$)	K_1, 2.17
	25	0.5 (NaClO$_4$)	K_1, 2.31
Co^{2+}	0	0 (corr)	K_1, 2.24
	15	0 (corr)	K_1, 2.30
	25	0 (corr)	K_1, 2.36
Co^{3+}	15	2.7 (NaClO$_4$)	K_1, 1.34
Ni^{2+}	0	0 (corr)	K_1, 2.08
	10	0 (corr)	K_1, 2.18
	25	0 (corr)	K_1, 2.32
Rh^{3+}			ev cpx
Ir^{4+}			ev ani cpx
Cu^{2+}	25	0 (corr)	K_1, 2.33
Ag^+	25	3 (NaClO$_4$)	K_1, 0.23, K_2, 0.05 (?)
Au^{3+}	22	H_2SO_4	ev cpx
Zn^{2+}	0	0 (corr)	K_1, 2.08
	15	0 (corr)	K_1, 2.27
	25	0 (corr)	K_1, 2.38
Cd^{2+}	35	0 (corr)	K_1, 2.17; K_2, 1.37
Hg_2^{2+}	25	0.5 (NaClO$_4$)	K_1, 1.30; K_2, 1.10
Hg^{2+}	25	0.5 (NaClO$_4$)	K_1, 1.34; K_2, 1.1
Al^{3+}	25	0.6 (NaClO$_4$)	K_1, 1.30; K_2, 1.0
Ga^{3+}	30	0 (corr)	K_1, 2.99

Table A-13 (cont.)

Metal	Temp., °C	Medium	Log of Equilibrium Constant
		Sulfate, SO_4^{2-} (continued)	
In^{3+}	20	1 ($NaClO_4$)	K_1, 1.85; K_2, 0.75; K_3, 0.40
Tl^+	0	0 (corr)	K_1, 1.38
	25	0 (corr)	K_1, 1.37
	40	0 (corr)	K_1, 1.36
Tl^{3+}	25	4 ($LiClO_4$)	K_1, 0.3
Sn^{2+}			ev cpx
Sn^{4+}	25	0 (corr)	K_2, 2.3
Pb^{2+}	25	0 (corr)	K_{SP}, -7.8
		Carbonate, CO_3^{2-}	
H^+	25	0 (corr)	K_1, 10.4; K_2, 6.5
D^+	25	0 (corr)	K_1, 10.7
Li^+	20	var conc LiCl	K_{SP}, -1.7
Na^+	25	0 (corr)	K_1, 1.27
Mg^{2+}	4	0 (corr)	K_{SP}, -3.51
	12	0 (corr)	K_{SP}, -3.73
	22	0 (corr)	K_{SP}, -4.01
Ca^{2+}	25	0 (corr)	K_{SP}, -8.0
	25	seawater	K_{SP}, -6.2
Sr^{2+}	25	0 (corr)	K_{SP}, -9.0
Ba^{2+}	25	0 (corr)	K_{SP}, -8.3
Sc^{3+}	$0 \to 50$		ev cpx
Y^{3+}	25		ev ani cpx
Ce^{3+}			ev cpx
Ce^{4+}			ev cpx
Nd^{3+}, Eu^{3+}			ev cpx
Zr^{4+}			ev ani cpx
Th^{4+}			ev cpx
VO^{2+}			ev ani cpx
Mo^{4+}			ev cpx
U^{4+}			$U(CO_3)_4^{4-}$; $U(CO_3)_5^{6-}$
UO_2^{2+}		0.5 ($NaCO_3$)	K_3, 7.0
Pu^{3+}			ev cpx
Pu^{4+}	20	10 (KCl)	K_1, 46.96
PuO_2^{2+}	20	0 (corr)	K_1, 12 (?)
Am^{3+}			ev ani cpx
AmO_2^{2+}	0		ev cpx
Cm^{3+}			ev cpx
Mn^{2+}	25	dil	K_{SP}, -9.4
Fe^{2+}	25	0	K_{SP}, -10.68
Fe^{3+}			$Fe_4O_3(CO_3)_6^{6-}$
Co^{2+}	25	0	K_{SP}, -12.84

Table A-13 (cont.)

Metal	Temp., °C	Medium	Log of Equilibrium Constant
		Carbonate, CO_3^{2-} (continued)	
Co^{3+}			$Co(CO_3)_3^{3-}$
Ni^{2+}	25	0 (corr)	K_{SP}, -6.87
Cu^{2+}	25	0 (corr)	K_1, 6.77
AG^+	25	0 (corr)	K_{SP}, -11.09
Zn^{2+}	25	0 (corr)	K_{SP}, -10.78
Cd^{2+}	25	0	K_{SP}, -13.7
Hg_2^{2+}	25	0	K_{SP}, -16.05
Ga^{3+}			ev ani cpx
In^{3+}			ev ani cpx
Pb^{2+}	25	0 (corr)	K_{SP}, -13.24
	30	0 (corr)	K_{SP}, -23.10
		Silicate, $SiO_2(OH)_2^{2-}$ or SiO_3^{2-}	
H^+	25	0.5 (NaCl)	K_1, 12.71
Ca^{2+}	30	0 (corr)	K_{SP}, -7.60
Cr^{3+}			ev ani cpx
		Phosphate, PO_4^{3-}	
H^+	25	$\rightarrow 0$	K_1, 12.33; K_{12}, 7.21; K_{13}, 2.16
D^+	20	0 (corr)	K_{12}, 7.67; K_{13}, 2.19
Li^+	0	0.2 (Pr_4NCl)	K_1, 0.32
	25	0.2 (Pr_4NCl)	K_1, 0.72
Na^+	0	0.2 (Pr_4NCl)	K_1, 0.08
	25	0.2 (Pr_4NCl)	K_1, 0.60
K^+	0	0.2 (Pr_4NCl)	K_1, 0.08
	25	0.2 (Pr_4NCl)	K_1, 0.49
Be^{2+}	20	dil	ev cpx
Mg^{2+}	25	0 (corr)	K_1, 2.5 (?)
Ca^{2+}	37	0.15 (NaCl)	K_1, 1.86
Sr^{2+}	20	0.15 (NaCl)	K_1, 4.18
Ba^{2+}	20		K_{SP}, -22.47
Sc^{3+}	25		ev ani cpx
La^{3+}	25	0.5 ($NaClO_4$)	K_{SP}, -22.43
Ce^{3+}	25	0 (corr)	K_1, 18.53
Ce^{4+}	20	dil	ev cpx
Ti^{3+}	20		ev cpx
Zr^{4+}	25		ev ani cpx
Th^{4+}	25		ev ani cpx

Table A-13 (cont.)

Metal	Temp., °C	Medium	Log of Equilibrium Constant
		Phosphate, PO_4^{3-} (continued)	
VO^{2+}	25	$\to 0$	K_{SP}, -24.1
Nb(V)	25		ev ani cpx
Cr^{3+}	20		K_{SP}, -22.62
Cr(VI)	25	0.25 (NaClO$_4$)	ev cpx
Mo(IV)	25	1 (HClO$_4$)	ev cpx
Mo(VI)	25		ev ani cpx
U^{4+}	25		ev cpx
UO_2^{2+}	25		ev ani cpx
Np^{4+}	25		ev ani cpx
NpO_2^+			ev cpx
Pu^{4+}			ev cpx
Mn^{2+}	25	0.2 (Pr$_4$NCl)	K_1, 2.58 (?)
Fe^{3+}			ev cpx
Co^{2+}			no ev cpx (?)
Ni^{2+}	20	dil	K_{SP}, -30.3 (?)
Ir^{4+}			ev ani cpx
Cu^{2+}	20	dil	K_{SP}, -36.9 (?)
Ag^+	20		K_{SP}, -19.89
Zn^{2+}	20		K_{SP}, -32.04 (?)
Cd^{2+}	20		K_{SP}, -32.6 (?)
Hg_2^{2+}	25	0 (corr)	ev cpx
Al^{3+}	25		K_{SP}, -10.41
In^{3+}			ev ani cpx
Pb^{2+}	25	0 (corr)	K_{SP}, -42.10
Bi^{3+}	20		K_{SP}, -22.89
		Nitrite, NO_2^-	
H^+	25		K_1, 3.35
Co^{3+}			ev cpx
Ni^{2+}			no ev cpx
Rh^{3+}			ev cpx
Pd^{2+}			ev cpx
Ir^{3+}			ev cpx
Pt(IV)			ev cpx
Cu^{2+}	20	1 (NaClO$_4$)	K_1, 1.23; K_2, 0.25
Ag^+	25		K_{SP}, -3.8
Zn^{2+}	25	2.5 (NaClO$_4$)	no ev cpx
Cd^{2+}	25	3 (NaClO$_4$)	K_1, 1.80; K_2, 1.21; K_3, 0.80
Hg^{2+}	25		ev cpx
Tl^+	25	0 (corr)	K_1, 0.85
Pb^{2+}	30	2 (NaClO$_4$)	K_1, 2.85

Table A-13 (cont.)

Metal	Temp., °C	Medium	Log of Equilibrium Constant
		Nitrate, NO_3^-	
H^+	0	0 (corr)	K_1, -1.65
	25	0 (corr)	K_1, -1.37
	50	0 (corr)	K_1, -1.24
Li^+	30	0 (corr)	K_1, -1.45
Na^+	18	0 (corr)	K_1, -0.59
K^+	25	$\rightarrow 0$	K_1, -0.24
Cs^+	25	0 (corr)	K_1, 0.04
Be^{2+}	18	0.5 ($NaClO_4$)	K_1, -0.60
Mg^{2+}	18	0 (corr)	no ev cpx
Ca^{2+}	18	0 (corr)	K_1, 0.28 (?)
Sr^{2+}	18	0 (corr)	K_1, 0.82
Ba^{2+}	18	0 (corr)	K_1, 0.92
$Sc^{3+}Dy^{3+}$			ev ani cpx
La^{3+}	20	1 ($NaClO_4$)	K_1, -0.32
	25	1 ($NaClO_4$)	K_1, -0.26
	40	1 ($NaClO_4$)	K_1, -0.23
Ce^{3+}	25	2 ($HClO_4$)	K_1, 1.04; K_2, 0.47
Ce^{4+}			ev cpx
Pr^{3+}		conc HNO_3	ev ani cpx
Nd^{3+}	25	0.35 ($NaClO_4$)	K_1, 0.18
Yb^{3+}			K_1, 0.45; K_2, 0.40
Zr^{4+}	20	4 ($HClO_4$)	K_1, 0.34; K_2, -0.23; K_3, -0.37; K_4, -0.56
ZrO^{2+}	25		ev cpx
Hf^{4+}	25	2 ($HClO_4$)	K_1, 0.34; K_2, -0.34; K_3, -0.72; K_4, -0.80 (?)
Th^{4+}	25	6 ($NaClO_4$)	K_1, 0.45; K_2, -0.30
$Nb(V)$			ev cpx
$Pa(V)$	20		ev cpx
U^{4+}		2 ($HClO_4$)	K_1, 0.20
UO_2^{2+}	10	2 ($NaClO_4$)	K_1, -0.52
	25	2 ($NaClO_4$)	K_1, -0.62
	40	2 ($NaClO_4$)	K_1, -0.77
Np^{4+}	25	1 ($HClO_4$)	K_1, 0.38
Pu^{3+}	25		K_1, 0.67; K_2, -0.02; K_3, -0.48
PuO_2^{2+}			ev cpx
Am^{3+}	22	1 ($HClO_4$)	K_1, 0.26 (?)
Cm^{3+}	20		K_1, 0.57
Fe^{3+}	20	0.6 ($NaClO_4$)	K_1, -0.22 (?)
CO^{2+}	20		ev ion pair
$Ru(IV)$		conc HNO_3	ev cpx
Pd^{2+}			$Pd(NO_3)_4^{2-}$

Table A-13 (cont.)

Metal	Temp., °C	Medium	Log of Equilibrium Constant
		Nitrate NO_3^-	
Pt(IV)			$Pt(NO_3)_4(OH)_2^{2-}$
Cu^{2+}			no ev cpx (?)
Ag^+	18	0 (corr)	K_1, -0.27
Ag^{2+}	25		ev cpx
Zn^{2+}			no ev cpx (?)
Cd^{2+}	25	3 ($NaClO_4$)	K_1, -0.13 (?)
	35	3 ($NaClO_4$)	K_1, -0.21 (?)
	45	3 ($NaClO_4$)	K_1, -0.21 (?)
Hg_2^{2+}	25	3 ($NaClO_4$)	K_1, 0.02; K_2, -0.32
Hg^{2+}	25	3 ($NaClO_4$)	K_1, 0.11; K_2, -0.01
Tl^+	0	0 (corr)	K_1, 0.38
	25	0 (corr)	K_1, 0.33
	40	0 (corr)	K_1, 0.31
Tl^{3+}	10		K_1, 0.41
	18		K_1, 0.18
Pb^{2+}	25	0 (corr)	K_1, 1.08 (?)
		Borate, $B(OH)_4^-$	
H^+	5	3 (NaCl)	K_1, 9.005
	15	3 (NaCl)	K_1, 8.908
	25	3 (NaCl)	K_1, 8.820
Na^+	20		ev cpx
Fe^{3+}			K_1, 8.58
Cd^{2+}	22	0 (corr)	K_1, -8.64
Al^{3+}			K_1, 7.62 (?)
Pb^{2+}	22	0 (corr)	K_{SP}, -10.78
		Sulfide, 2^{2-}	
H^+	20	0 (corr)	K_1, 12.94
	30	0 (corr)	K_1, 12.76
	40	0 (corr)	K_1, 12.61
La^{3+}	25	0 (corr)	K_{SP}, -12.70
Ce^{3+}	25	0 (corr)	K_{SP}, -10.22
V(V)			ev cpx
No(VI)			ev cpx
W(VI)			ev cpx
Mn^{2+}	25	0 (corr)	K_{SP}, -14.96

Table A-13 (cont.)

Metal	Temp., °C	Medium	Log of Equilibrium Constant
		Sulfide, 2^{2-} (continued)	
Re(VII)			ev cpx
Fe^{2+}	25	0 (corr)	K_{SP}, -17.3
CO^{2+}	25	0 (corr)	K_{SP}, -22.51
Ni^{+2}	25	0 (corr)	K_{SP}, -20.7
Cu^+	25	0 (corr)	K_{SP}, -49.44
Cu^{2+}	25	0 (corr)	K_{SP}, -35.10
Ag^+	25	0 (corr)	K_{SP}, -51.21
Au(I)			ev cpx
Zn^{2+}	25	0 (corr)	K_{SP}, -24.10
Cd^{2+}	25	0 (corr)	K_{SP}, -28
Hg_2^{2+}	18		K_{SP}, -47
Hg^{2+}	25	0 (corr)	K_{SP}, -51.05
In^{3+}	25	0	K_{SP}, -73.24
Tl^+	18		K_{SP}, -22.16
Sn^{2+}	25	0 (corr)	K_{SP}, -26
Sn^{4+}			ev cpx
Pb^{2+}	25	0 (corr)	K_{SP}, -29
As(III)			ev cpx
Sb(III)	25	0 (corr)	K_{SP}, -92.77
Sb(V)			SbS_4^{3-}
Bi^{3+}	25	0 (corr)	K_{SP}, -96
Po^{2+}	25	var HCl	K_{SP}, -28

Metal	Temp., °C	Medium	Log of Equilibrium Constant
		Hydroxide, OH^-	
H^+	0	$\rightarrow 0$	K_1, 14.939
	10	$\rightarrow 0$	K_1, 14.533
	20	$\rightarrow 0$	K_1, 14.167
D^+	15	$\rightarrow 0$	K_1, 15.08
	25	$\rightarrow 0$	K_1, 14.71
	35	$\rightarrow 0$	K_1, 14.37
Li^+	5	$\rightarrow 0$	K_1, 0.26 (?)
	15	$\rightarrow 0$	K_1, 0.20 (?)
	25	$\rightarrow 0$	K_1, 0.17 (?)
Na^+	5	$\rightarrow 0$	K_1, -0.81
	15	$\rightarrow 0$	K_1, -0.81
	25	$\rightarrow 0$	K_1, -0.77
K^+	$0 \rightarrow 60$	$\rightarrow 0$	no ev cpx
Rb^+	25		no ev cpx
Cs^+	$0 \rightarrow 60$		no ev cpx
Be^{2+}		0 (corr)	K_{SP}, -17.7

Table A-13 (cont.)

Metal	Temp., °C	Medium	Log of Equilibrium Constant
		Hydroxide, OH$^-$ (continued)	
Mg^{2+}	25	0 (corr)	K_1, 2.58; K_{SP}, -10.6
Ca^{2+}	0	0 (corr)	K_1, 1.37
	25	0 (corr)	K_1, 1.40
	40	0 (corr)	K_1, 1.48
Sr^{2+}	5	0 (corr)	K_1, 0.78
	15	0 (corr)	K_1, 0.80
	25	0 (corr)	K_1, 0.82
Ba^{2+}	5	$\rightarrow 0$	K_1, 0.62
	15	$\rightarrow 0$	K_1, 0.60
	25	$\rightarrow 0$	K_1, 0.64
Sc^{2+}	25	0 (corr)	K_{SP}, -27
Y^{3+}	25	var ClO_4^-	K_{SP}, -22.8
La^{3+}	25		K_{SP}, -19.0
Ce^{3+}	25		K_{SP}, -23
Pr^{3+}	18		K_1, -23.2
Nd^{3+}	25		K_1, -21.5
Pm^{3+}			K_{SP}, -34
Sm^{3+}	25		K_{SP}, -22.1
Eu^{3+}	25		K_{SP}, -23.1
Gd^{3+}	25		K_{SP}, -22.7
Tb^{3+}, Dy^{3+}, Ho^{3+}, Eu^{3+}, Tm^{3+}, Yb^{3+}, Lu^{3+}	25		K_{SP}, -23 to -26
Ti^{3+}	15	0	K_1, 12.94
	25	0	K_1, 12.71
	35	0	K_1, 12.76
Ti(IV)			$TiOH^{3+}$
Zr^{4+}			$ZrOH^{3+}$; K_{SP}, -52
Hf^{4+}	20	0 (corr)	K_{SP}, -53.4
Th^{4+}	22	$\rightarrow 0$	K_{SP}, -44.7
V^{2+}	25	0	K_{SP}, -15.4
V^{3+}	25	0	K_{SP}, -34.4
VO^{2+}	25	0	K_{SP}, -23.5
Pa(V)			ev cpx
Cr^{2+}	18		K_{SP}, -19.7
Cr^{3+}	22	$\rightarrow 0$	K_{SP}, -30.2
U^{4+}			K_{SP}, -51.96
UO_2^{2+}	20	0 (corr)	K_{SP}, -17.22
Np^{4+}	25	2 ($NaClO_4$)	ev cpx
NpO_2^+	25		ev cpx
NpO_2^{2+}	25		K_{SP}, -21.6 (?)
Pu^{3+}			K_{SP}, -19.7

Table A-13 (cont.)

Metal	Temp., °C	Medium	Log of Equilibrium Constant
	Hydroxide, OH^- (continued)		
Pu^{4+}	25		K_{SP}, -55.15
PuO_2^+			ev cpx
PuO_2^{2+}			K_{SP}, -22.7
Mn^{2+}	25	dil	K_{SP}, -12.9
Re^{4+}			$Re(OH)_3^+$
Fe^{2+}	25	0 (corr)	K_{SP}, -14.8
Fe^{3+}	18	0 (corr)	K_{SP}, -37.5
CO^{2+}	25	0 (corr)	K_{SP}, -15.2
CO^{3+}	25	0 (corr)	K_{SP}, -43
Ni^{2+}	25	dil	K_{SP}, -14.5
Ru^{3+}		dil	K_{SP}, -36
Ru^{4+}			K_{SP}, -43.7
Rh^{3+}			ev cpx
Pd^{2+}	25	0 (corr)	K_{SP}, -31
Ir^{4+}			ev cpx
Pt^{2+}	25		K_{SP}, -35
Cu^+	25	0	K_{SP}, -14.7
Cu^{2+}	25	$\rightarrow 0$	K_{SP}, -19.7
Ag^+	18	dil	K_{SP}, -7.89
	20	dil	K_{SP}, -7.84
	25	dil	K_{SP}, -7.71
Au^{3+}	22	0.5 (HNO_3)	K_{SP}, -45.26
Zn^{2+}	25		K_1, 4.4
Cd^{2+}	25	3 ($NaClO_4$)	K_1, 4.3; K_2, 3.4; K_3, 2.6; K_4, 1.7
Hg_2^{2+}	25	0 (corr)	K_{SP}, -23.7 (?)
Hg^{2+}	18		K_1, 11.9; K_2, 10.3
Al^{3+}	25	dil	K_{SP}, -31.7
Ga^{3+}	10		K_{SP}, -36.0
	25		K_{SP}, -35.2
In^{3+}	10	dil	K_{SP}, -34.4
	25	dil	K_{SP}, -33.2
Tl^+	0	0 (corr)	K_1, 0.81
	25	0 (corr)	K_1, 0.82
Tl^{3+}	25	dil	K_{SP}, -34
Ge^{2+}			ev cpx
Sn^{2+}	22	0 (corr)	K_{SP}, -28.1
Pb^{2+}	25	0 (corr)	K_1, 7.82; K_2, 3.06; K_3, 3.06
Sb^{3+}			K_{SP}, -41.4
Bi^{3+}			K_{SP}, -30.4
Po^{4+}			K_{SP}, -37

Table A-14   Oxidation-Reduction Potentials
(From W. M. Latimer, *The Oxidation States of the Elements and Their Potentials in Aqueous Solutions*, Prentice-Hall, Englewood Cliffs, N.J., 2nd ed. 1952, with permission of the publisher)

Oxidation-Reduction Couples in Acid Solution	
Couple	$E°$
$HN_3 = \frac{3}{2} N_2 + H^+ + e^-$	3.09
$Li = Li^+ + e^-$	3.045
$K = K^+ + e^-$	2.925
$Rb = Rb^+ + e^-$	2.925
$As = As^+ + e^-$	2.923
$Ra = Ra^{2+} + 2 e^-$	2.92
$Ba = Ba^{2+} + 2 e^-$	2.90
$Sr = Sr^{2+} + 2 e^-$	2.89
$Ca = Ca^{2+} + 2 e^-$	2.87
$Na = Na^+ + e^-$	2.714
$La = La^{3+} + 3 e^-$	2.52
$Ce = Ce^{3+} + 3 e^-$	2.48
$Nd = Nd^{3+} + 3 e^-$	2.44
$Sm = Sm^{3+} + 3 e^-$	2.41
$Gd = Gd^{3+} + 3 e^-$	2.40
$Mg = Mg^{2+} + 2 e^-$	2.37
$Y = Y^{3+} + 3 e^-$	2.37
$Am = Am^{3+} + 3 e^-$	2.32
$Lu = Lu^{3+} + 3 e^-$	2.25
$H^- = \frac{1}{2} H_2 + e^-$	2.25
$H(g) = H^+ + e^-$	2.10
$Sc = Sc^{3+} + 3 e^-$	2.08
$Pu = Pu^{3+} + 3 e^-$	2.07
$Al + 6 F^- = AlF_6^{3-} + 3 e^-$	2.07
$Th = Th^{4+} + 4 e^-$	1.90
$Np = Np^{3+} + 3 e^-$	1.86
$Be = Be^{2+} + 2 e^-$	1.85
$U = U^{3+} + 3 e^-$	1.80
$Hf = Hf^{4+} + 4 e^-$	1.70
$Al = Al^{3+} + 3 e^-$	1.66
$Ti = Ti^{2+} + 2 e^-$	1.63
$Zr = Zr^{4+} + 4 e^-$	1.53
$Si + 6 F^- = SiF_6^{2-} + 4 e^-$	1.2
$Ti + 6 F^- = TiF_6^{2-} + 4 e^-$	1.19
$Mn = Mn^{2+} + 2 e^-$	1.18
$V = V^{2+} + 2 e^-$	ca. 1.18
$Nb = Nb^{3+} + 3 e^-$	ca. 1.1
$Ti + H_2O = TiO^{2+} + 2 H^+ + 4 e^-$	0.89
$B + 2 H_2O = H_3BO_3 + 3 H^+ + 3 e^-$	0.87
$Si + 2 H_2O = SiO_2 + 4 H^+ + 4 e^-$	0.86

Table A-14 (cont.)

Oxidation-Reduction Couples in Acid Solution

Couple	$E°$
$2 Ta + 5 H_2O = Ta_2O_5 + 10 H^+ + 10 e^-$	0.81
$Zn = Zn^{2+} + 2 e^-$	0.763
$Tl + I^- = TlI + e^-$	0.753
$Cr = Cr^{3+} + 3 e^-$	0.74
$H_2Te = Te + 2 H^+ + 2 e^-$	0.72
$Tl + Br^- = TlBr + e^-$	0.658
$2 Nb + 5 H_2O = Nb_2O_5 + 10 H^+ + 10 e^-$	0.65
$U^{3+} = U^{4+} + e^-$	0.61
$AsH_3 = As + 3 H^+ + 3 e^-$	0.60
$Tl + Cl^- = TlCl + e^-$	0.557
$Ga = Ga^{3+} + 3 e^-$	0.53
$SbH_3 (g) = Sb + 3 H^+ + 3 e^-$	0.51
$P + 2 H_2O = H_3PO_2 + H^+ + e^-$	0.51
$H_3PO_2 + H_2O = H_3PO_3 + 2 H^+ + 2 e^-$	0.50
$Fe = Fe^{2+} + 2 e^-$	0.440
$Eu^{2+} = Eu^{3+} + e^-$	0.43
$Cr^{2+} = Cr^{3+} + e^-$	0.41
$Cd = Cd^{2+} + 2 e^-$	0.403
$H_2Se = Se + 2 H^+ + 2 e^-$	0.40
$Ti^{2+} = Ti^{3+} + e^-$	ca. 0.37
$Pb + 2 I^- = PbI_2 + 2 e^-$	0.365
$Pb + SO_4^{2-} = PbSO_4 + 2 e^-$	0.356
$In = In^{3+} + 3 e^-$	0.342
$Tl = Tl^+ + e^-$	0.3363
$Pt + H_2S = PtS + 2 H^+ + 2 e^-$	0.30
$Pb + 2 Br^- = PbBr_2 + 2 e^-$	0.280
$Co = Co^{2+} + 2 e^-$	0.277
$H_3PO_3 = H_3PO_4 + 2 H^+ + 2 e^-$	0.276
$Pb + 2 Cl^- = PbCl_2 + 2 e^-$	0.268
$V^{2+} = V^{3+} + e^-$	0.255
$V + 4 H_2O = V(OH)_4^+ + 4 H^+ + 5 e^-$	0.253
$Sn + 6 F^- = SnF_6^{2-} + 4 e^-$	0.25
$Ni = Ni^{2+} + 2 e^-$	0.250
$N_2H_5^+ = N_2 + 5 H^+ + 4 e^-$	0.23
$S_2O_6^{2-} = 2 SO_4^{2-} + 4 H^+ + 2 e^-$	0.22
$Mo = Mo^{3+} + 3 e^-$	ca. 0.2
$HCOOH (aq) = CO_2 + 2 H^+ + 2 e^-$	0.196
$Cu + I^- = CuI + e^-$	0.185
$Ag + I^- = AgI + e^-$	0.151
$Sn = Sn^{2+} + 2 e^-$	0.136
$HO_2 = O_2 + H^+ + e^-$	0.13
$Pb = Pb^{2+} + 2 e^-$	0.126
$Ge + 2 H_2O = GeO_2 + 4 H^+ + 4 e^-$	0.15
$W + 3 H_2O = WO_3 (c) + 6 H^+ + 6 e^-$	0.09

Table A-14 (cont.)

Oxidation-Reduction Couples in Acid Solutions	
Couple	$E°$
$HS_2O_4^- + 2\,H_2O = 2\,H_2SO_3 + H^+ + 2\,e^-$	0.08
$Hg + 4\,I^- = HgI_4^{2-} + 2\,e^-$	0.04
$H_2 = 2\,H^+ + 2\,e^-$	0.00
$Ag + 2\,S_2O_3^{2-} = Ag(S_2O_3)_2^{3-} + e^-$	−0.01
$Cu + Br^- = CuBr + e^-$	−0.033
$UO_2^+ = UO_2^{2+} + e^-$	−0.05
$HCHO\ (aq) + H_2O = HCOOH\ (aq) + 2\,H^+ + 2\,e^-$	−0.056
$PH_3\ (g) = P + 3\,H^+ + 3\,e^-$	−0.06
$Ag + Br^- = AgBr + e^-$	−0.095
$Ti^{2+} + H_2O = TiO^{2+} + 2\,H^+ + e^-$	−0.1
$SiH_4 = Si + 4\,H^+ + 4\,e^-$	−0.102
$CH_4 = C + 4\,H^+ + 4\,e^-$	−0.13
$Cu + Cl^- = CuCl + e^-$	−0.137
$H_2S = S + 2\,H^+ + 2\,e^-$	−0.141
$Np^{3+} = Np^{4+} + e^-$	−0.147
$Sn^{2+} = Sn^{4+} + 2\,e^-$	−0.15
$2\,Sb + 3\,H_2O = Sb_2O_3 + 6\,H^+ + 6\,e^-$	−0.152
$Cu^+ = Cu^{2+} + e^-$	−0.153
$Bi + H_2O + Cl^- = BiOCl + 2\,H^+ + 3\,e^-$	−0.16
$H_2SO_3 + H_2O = SO_4^{2-} + 4\,H^+ + 2\,e^-$	−0.17
$CH_3OH\ (aq) = HCHO\ (aq) + 2\,H^+ + 2\,e^-$	−0.19
$Hg + 4\,Br^- = HgBr_4^{2-} + 2\,e^-$	−0.21
$Ag + Cl^- = AgCl + e^-$	−0.222
$(CH_3)_2SO + H_2O = CH_3SO_2 + 2\,H^+ + 2\,e^-$	−0.23
$As + 2\,H_2O = HAsO_2\ (aq) + 3\,H^+ + 3\,e^-$	−0.247
$Re + 2\,H_2O = ReO_2 + 4\,H^+ + 4\,e^-$	−0.252
$Bi + H_2O = BiO^+ + 2\,H^+ + 3\,e^-$	−0.32
$\frac{1}{2}\,C_2N_2 + H_2O = HCNO + H^+ + e^-$	−0.33
$U^{4+} + 2\,H_2O = UO_2^{2+} + 4\,H^+ + 2\,e^-$	−0.334
$Cu = Cu^{2+} + 2\,e^-$	−0.337
$Ag + IO_3^- = AgIO_3 + e^-$	−0.35
$Fe(CN)_6^{4-} = Fe(CN)_6^{3-} + e^-$	−0.36
$V^{3+} + H_2O = VO^{2+} + 2\,H^+ + e^-$	−0.361
$Re + 4\,H_2O = ReO_4^- + 8\,H^+ + 7\,e^-$	−0.363
$HCN(\ aq) = \frac{1}{2}\,C_2N_2 + H^+ + e^-$	−0.37
$S_2O_3^{2-} + 3\,H_2O = 2\,H_2SO_3 + 2\,H^+ + 4\,e^-$	−0.40
$Rh + 6\,Cl^- = RhCl_6^{3-} + 3\,e^-$	−0.44
$2\,Ag + CrO_4^{2-} = Ag_2CrO_4 + 2\,e^-$	−0.446
$S + 3\,H_2O = H_2SO_3 + 4\,H^+ + 4\,e^-$	−0.45
$Sb_2O_4 + H_2O = Sb_2O_5 + 2\,H^+ + 2\,e^-$	−0.48
$2\,Ag + MoO_4^{2-} = Ag_2MoO_4 + 2\,e^-$	−0.49
$2\,NH_3OH^+ = H_2N_2O_2 + 6\,H^+ + 4\,e^-$	−0.496
$ReO_2 + 2\,H_2O = ReO_4^- + 4\,H^+ + 3\,e^-$	−0.51
$S_4O_6^{2-} + 6\,H_2O = 4\,H_2SO_3 + 4\,H^+ + 6e^-$	−0.51

Table A-14 (cont.)

Oxidation-Reduction Couples in Acid Solution	
Couple	E°
$C_2H_6 = C_2H_4 + 2\,H^+ + 2\,e^-$	-0.52
$Cu = Cu^+ + e^-$	-0.521
$Te + 2\,H_2O = TeO_2\,(c) + 4\,H^+ + 4\,e^-$	-0.529
$2\,I^- = I_2 + 2\,e^-$	-0.5355
$3\,I^- = I_3^- + 2\,e^-$	-0.536
$CuCl = Cu^{2+} + Cl^- + e^-$	-0.538
$Ag + BrO_3^- = AgBrO_3 + e^-$	-0.55
$Te + 2\,H_2O = TeOOH^+ + 3\,H^+ + 4\,e^-$	-0.559
$HAsO_2 + 2\,H_2O = H_3AsO_4 + 2\,H^+ + 2\,e^-$	-0.559
$Ag + NO_2^- = AgNO_2 + e^-$	-0.564
$MnO_4^{2-} = MnO_4^- + e^-$	-0.564
$2\,H_2SO_3 = S_2O_6^{2-} + 4\,H^+ + 2\,e^-$	-0.57
$Pt + 4\,Br^- = PtBr_4^{2-} + 2\,e^-$	-0.58
$2\,SbO^+ + 3\,H_2O = Sb_2O_5 + 6\,H^+ + 4\,e^-$	-0.581
$CH_4 + H_2O = CH_3OH\,(aq) + 2\,H^+ + 2\,e^-$	-0.586
$Pd + 4\,Br^- = PdBr_4^{2-} + 2\,e^-$	-0.6
$Ru + 5\,Cl^- = RuCl_5^{2-} + 3\,e^-$	-0.60
$U^{4+} + 2\,H_2O = UO_2^{2+} + 4\,H^+ + 2\,e^-$	-0.62
$Pd + 4Cl^- = PdCl_4^{2-} + 2\,e^-$	-0.62
$CuBr = Cu^{2+} + Br^- + e^-$	-0.640
$Ag + C_2H_3O_2^- = AgC_2H_3O_2 + e^-$	-0.643
$2\,Ag + SO_4^{2-} = Ag_2SO_4 + 2\,e^-$	-0.653
$Au + 4\,CNS^- = Au(CNS)_4^- + 3\,e^-$	-0.66
$PtCl_4^{2+} + 2\,Cl^- = PtCl_6^{2-} + 2\,e^-$	-0.68
$H_2O_2 = O_2 + 2\,H^+ + 2\,e^-$	-0.682
$3\,NH_4^+ = HN_3 + 11\,H^+ + 8\,e^-$	-0.69
$H_2Te = Te + 2\,H^+ + 2\,e^-$	-0.70
$H_2N_2O_2 = 2\,NO + 2\,H^+ + 2\,e^-$	-0.71
$OH + H_2O = H_2O_2 + H^+ + e^-$	-0.72
$Pt + 4\,Cl^- = PtCl_4^{2-} + 2\,e^-$	-0.73
$C_2H_4 = C_2H_2 + 2\,H^+ + 2\,e^-$	-0.73
$Se + 3\,H_2O = H_2SeO_3 + 4\,H^+ + 4\,e^-$	-0.74
$Np^{4+} + 2\,H_2O = NpO_2^+ + 4\,H^+ + e^-$	-0.75
$2\,CNS^- = (CNS)_2 + 2\,e^-$	-0.77
$Ir + 6\,Cl^- = IrCl_6^{3-} + 3\,e^-$	-0.77
$Fe^{2+} = Fe^{3+} + e^-$	-0.771
$2\,Hg = Hg_2^{2+} + 2\,e^-$	-0.789
$Ag = Ag^+ + e^-$	-0.7991
$N_2O_4 + 2\,H_2O = 2\,NO_3^- + 4\,H^+ + 2\,e^-$	-0.80
$Rh = Rh^{3+} + 3\,e^-$	ca. -0.8
$Os + 4\,H_2O = OsO_4\,(c) + 8\,H^+ + 8\,e^-$	-0.85
$H_2N_2O_2 + 2\,H_2O = 2\,HNO_2 + 4\,H^+ + 4\,e^-$	-0.86
$CuI = Cu^{2+} + I^- + e^-$	-0.86
$Au + 4\,Br^- = AuBr_4^- + 3\,e^-$	-0.87

Table A-14 (cont.)

Oxidation-Reduction Couples in Acid Solution	
Couple	$E°$
$Hg_2^{2+} = 2 Hg^{2+} + 2 e^-$	-0.920
$HNO_2 + H_2O = NO_3^- + 3 H^+ + 2 e^-$	-0.94
$PuO_2^+ = PuO_2^{2+} + e^-$	-0.93
$NO + 2 H_2O = NO_3^- + 4 H^+ + 4 e^-$	-0.96
$Au + 2 Br^- = AuBr_2^- + e^-$	-0.96
$Pu^{3+} = Pu^{4+} + e^-$	-0.97
$Pt + 2 H_2O = Pt(OH)_2 + 2 H^+ + 2 e^-$	-0.98
$Pd = Pd^{2+} + 2 e^-$	-0.987
$IrBr_6^{4-} = IrBr_6^{3-} + e^-$	-0.99
$NO + H_2O = HNO_2 + H^+ + e^-$	-1.00
$Au + 4 Cl^- = AuCl_4^- + 3 e^-$	-1.00
$VO^{2+} + 3 H_2O = V(OH)_4^+ + 2 H^+ + e^-$	-1.00
$IrCl_6^{3-} = IrCl_6^{2-} + e^-$	-1.017
$TeO_2 + 4 H_2O = H_6TeO_6 (c) + 2 H^+ + 2 e^-$	-1.02
$NO + 2 H_2O = N_2O_4 + 4 H^+ + 4 e^-$	-1.03
$Pu^{4+} + 2 H_2O = PuO_2^{2+} + 4 H^+ + 2 e^-$	-1.04
$2 Cl^- + \frac{1}{2} I_2 = ICl_2^- + e^-$	-1.06
$2 Br^- = Br_2 (l) + 2 e^-$	-1.0652
$2 HNO_2 = N_2O_4 + 2 H^+ + 2 e^-$	-1.07
$Cu(CN)_2^- = Cu^{2+} + 2 CN^- + e^-$	-1.12
$Pu^{4+} + 2 H_2O = PuO_2^+ + 4 H^+ + e^-$	-1.15
$H_2SeO_3 + H_2O = SeO_4^{2-} + 4 H^+ + 2 e^-$	-1.15
$NpO_2^+ = NpO_2^{2+} + e^-$	-1.15
$4 Cl^- + C + 4 H^+ = CCl_4 + 4 H^+ + 4 e^-$	-1.18
$ClO_3^- + H_2O = ClO_4^- + 2 H^+ + 2 e^-$	-1.19
$\frac{1}{2} I_2 + 3 H_2O = IO_3^- + 6 H^+ + 5 e^-$	-1.195
$HClO + H_2O = ClO_3^- + 3 H^+ + 2 e^-$	-1.21
$2 H_2O = O_2 + 4 H^+ + 4 e^-$	-1.229
$2 S + 2 Cl^- = S_2Cl_2 + 2e^-$	-1.23
$Mn^{2+} + 2 H_2O = MnO_2 + 4 H^+ + 2 e^-$	-1.23
$Tl^+ = Tl^{3+} + 2 e^-$	-1.25
$Am^{4+} + 2 H_2O = AmO_2^+ + 4 H^+ + e^-$	-1.26
$2 NH_4^+ = N_2H_5^+ + 3 H^+ + 2 e^-$	-1.275
$HClO_2 = ClO_2 + H^+ + e^-$	-1.275
$PdCl_4^{2-} + 2 Cl^- = PdCl_6^{2-} + 2 e^-$	-1.288
$N_2O + 3 H_2O = 2 HNO_2 + 4 H^+ + 4 e^-$	-1.29
$2 Cr^{3+} + 7 H_2O = Cr_2O_7^{2-} + 14 H^+ + 6 e^-$	-1.33
$NH_4^+ + H_2O = NH_3OH^+ + 2 H^+ + 2 e^-$	-1.35
$2 Cl^- = Cl_2 + 2 e^-$	-1.3595
$N_2H_5^+ + 2 H_2O = 2 NH_3OH^+ + H^+ + 2 e^-$	-1.42
$Au + 3 H_2O = Au(OH)_3 + 3 H^+ + 3 e^-$	-1.45
$\frac{1}{2} I_2 + H_2O = HIO + H^+ + e^-$	-1.45
$Pb^{2+} + 2 H_2O = PbO_2 + 4 H^+ + 2 e^-$	-1.455
$Au = Au_3^+ + 3 e^-$	-1.50

Table A-14 (cont.)

Oxidation-Reduction Couples in Acid Solution	
Couple	$E°$
$H_2O_2 = HO_2 + H^+ + e^-$	-1.5
$Mn^{2+} = Mn^{3+} + e^-$	-1.51
$Mn^{2+} + 4 H_2O = MnO_4^- + 8 H^+ + 5 e^-$	-1.51
$\frac{1}{2} Br_2 + 3 H_2O = BrO_3^- + 6 H^+ + 5 e^-$	-1.52
$Br_2 + H_2O = HBrO + H^+ + e^-$	-1.59
$2 BiO^+ = Bi_2O_4 + 2 H_2O + 4 H^+ + 2 e^-$	-1.59
$IO_3^- + 3 H_2O = H_5IO_6 + H^+ + 2 e^-$	-1.6
$Bk^{3+} = Bk^{4+} + e^-$	-1.6
$Ce^{3+} = Ce^{4+} + e^-$	-1.61
$\frac{1}{2} Cl_2 + H_2O = HClO + H^+ + e^-$	-1.63
$AmO_2^+ = AmO_2^{2+} + e^-$	-1.64
$HClO + H_2O = HClO_2 + 2 H^+ + 2 e^-$	-1.64
$Au = Au^+ + e^-$	ca. -1.68
$Ni^{2+} + 2 H_2O = NiO_2 + 4 H^+ + 2 e^-$	-1.68
$PbSO_4 + 2 H_2O = PbO_2 + SO_4^{2-} + 4 H^+ + 2 e^-$	-1.685
$Am^{3+} + 2 H_2O = AmO_2^{2+} + 4 H^+ + 3 e^-$	-1.69
$MnO_2 + 2 H_2O = MnO_4^- + 4 H^+ + 3 e^-$	-1.695
$Am^{3+} + 2 H_2O = AmO_2^+ + 4 H^+ + 2 e^-$	-1.725
$2 H_2O = H_2O_2 + 2 H^+ + 2 e^-$	-1.77
$Co^{2+} = Co^{3+} + e^-$	-1.82
$Fe^{3+} + 4 H_2O = FeO_4^{2-} + 8 H^+ + 3 e^-$	-1.9
$NH_4^+ + N_2 = HN_3 + 3 H^+ + 2 e^-$	-1.96
$Ag^+ = Ag^{2+} + e^-$	-1.98
$2 SO_4^{2-} = S_2O_8^{2-} + 2 e^-$	-2.01
$C_2 + H_2O = O_3 + 2 H^+ + 2 e^-$	-2.07
$H_2O + 2 F^- = F_2O + 2 H^+ + 4 e^-$	-2.1
$Am^{3+} = Am^{4+} + e^-$	-2.18
$H_2O = O (g) + 2 H^+ + 2 e^-$	-2.42
$2 F^- = F_2 + 2 e^-$	-2.87
$H_2O = OH + H^+ + e^-$	-2.8
$N_2 + 2 H_2O = H_2N_2O_2 + 2 H^+ + 2 e^-$	-2.85
$2 HF (aq) = F_2 + 2 H^+ + 2 e^-$	-3.06

Oxidation-Reduction Couples in Basic Solution	
Couple	$E°$
$Ca + 2 OH^- = Ca(OH)_2 + 2 e^-$	3.03
$Sr + 2 OH^- + 8 H_2O = Sr(OH)_2 \cdot 8 H_2O + 2 e^-$	2.99
$Ba + 8 H_2O + 2 OH^- = Ba(OH) \cdot 8 H_2O + 2e^-$	2.97
$H (g) + OH^- = H_2O + e^-$	2.93
$La + 3 OH^- = La(OH)_3 + 3 e^-$	2.90
$Lu + 3 OH^- = Lu(OH)_3 + 3 e^-$	2.72
$Mg + 2 OH^- = Mg(OH)_2 + 2 e^-$	2.69
$2 Be + 6 OH^- = Be_2O_3^{2-} + 3 H_2O + 4 e^-$	2.62
$Sc + 3 OH^- = Sc(OH)_3 + 3 e^-$	ca. 2.6

Table A-14 (cont.)

Oxidation-Reduction Couples in Basic Solution	
Couple	$E°$
$Hf + 4\,OH^- = HfO(OH)_2 + H_2O + 4\,e^-$	2.50
$Th + 4\,OH^- = TH(OH)_4 + 4\,e^-$	2.48
$Pu + 3\,OH^- = Pu(OH)_3 + 3\,e^-$	2.42
$U + 4\,OH^- = UO_2 + 2\,H_2O + 4\,e^-$	2.39
$Al + 4\,OH^- = H_2AlO_3 + H_2O + 3\,e^-$	2.35
$Zr + 4\,OH^- = H_2ZrO_3 + H_2O + 4\,e^-$	2.36
$U(OH)_3 + OH^- = U(OH)_4 + e^-$	2.2
$U + 3\,OH^- = U(OH)_3 + 3\,e^-$	2.17
$P + 2\,OH^- = H_2PO_2^- + e^-$	2.05
$B + 4\,OH^- = H_2BO_3^- + 3\,e^-$	1.79
$Si + 6\,OH^- = SiO_3^{2-} + 3\,H_2O + 4\,e^-$	1.70
$U(OH)_4 + 2\,Na^+ + 4\,OH^- = Na_2UO_4 + 4\,H_2O + 2\,e^-$	1.61
$H_2PO_2^- + 3\,OH^- = HPO_3^{2-} + 2\,H_2O + 2\,e^-$	1.57
$Mn + 2\,OH^{1-} = Mn(OH)_2 + 2\,e^-$	1.55
$Mn + CO_3^{2-} = MnCO_3 + 2\,e^-$	1.48
$Zn + S^{2-} = ZnS + 2\,e^-$	1.44
$Cr + 3\,OH^- = Cr(OH)_3 + 3\,e^-$	1.3
$Zn + 4\,CN^- = Zn(CN)_4^{2-} + 2\,e^-$	1.26
$Zn + 2\,OH^- = Zn(OH)_2 + 2\,e^-$	1.245
$Ga + 4\,OH^- = H_2GaO_3^- + H_2O + 3\,e^-$	1.22
$Zn + 4\,OH^- = ZnO_2^{2-} + 2\,H_2O + 2\,e^-$	1.216
$Cr + 4\,OH^- = CrO_2^- + H_2O + 3\,e^-$	1.2
$Cd + S^{2-} = CdS + 2\,e^-$	1.21
$6\,V + 33\,OH^- = 16\,H_2O + HV_6O_{17}^{3-} + 30\,e^-$	1.15
$Te^{2-} = Te + 2\,e^-$	1.14
$HPO_3^{2-} + 3\,OH^- = PO_4^{3-} + 2\,H_2O + 2\,e^-$	1.12
$S_2O_4^{2-} + 4\,OH^- = 2\,SO_3^{2-} + 2\,H_2O + 2\,e^-$	1.12
$Zn + CO_3^{2-} = ZnCO_3 + 2\,e^-$	1.06
$W + 8\,OH^- = WO_4^{2-} + 4\,H_2O + 6\,e^-$	1.05
$Mo + 8\,OH^- = MoO_4^{2-} + 4\,H_2O + 6\,e^-$	1.05
$Cd + 4\,CN^- = Cd(CN)_4^{2-} + 2\,e^-$	1.03
$Zn + 4\,NH_3 = Zn(NH_3)_4^{2+} + 2\,e^-$	1.03
$Fe + S^{2-} = FeS_{(\alpha)} + 2\,e^-$	1.01
$In + 3\,OH^- = In(OH)_3 + 3\,e^-$	1.0
$Pb + S^{2-} = PbS + 2\,e^-$	0.95
$CN^- + 2\,OH^- = CNO^- + H_2O + 2\,e^-$	0.97
$Tl + S^{2-} = Tl_2S + 2\,e^-$	0.96
$Pu(OH)_3 + OH^- = Pu(OH)_4 + e^-$	0.95
$Sn + S^{2-} = SnS + 2\,e^-$	0.94
$SO_3^{2-} + 2\,OH^- = SO_4^{2-} + H_2O + 2\,e^-$	0.93
$Se^{2-} = Se + 2\,e^-$	0.92
$Sn + 3\,OH^- = HSnO_3^- + H_2O + 2\,e^-$	0.91
$Ge + 5\,OH^- = HGeO_3^- + 2\,H_2O + 4\,e^-$	0.9
$HSnO_2^- + H_2O + 3\,OH^- = Sn(OH)_6^- + 2\,e^-$	0.90

Table A-14 (cont.)

Oxidation-Reduction Couples in Basic Solution	
Couple	$E°$
$PH_3 + 3\,OH^- = P + 3\,H_2O + 3\,e^-$	0.89
$Fe + 2\,OH^- = Fe(OH)_2 + 2\,e^-$	0.877
$Ni + S^{2-} = NiS_{(\alpha)} + 2\,e^-$	0.83
$H_2 + 2\,OH^- = 2\,H_2O + 2\,e^-$	0.828
$Cd + 2\,OH^- = Cd(OH)_2 + 2\,e^-$	0.809
$Fe + CO_3^{2-} = FeCO_3 + 2\,e^-$	0.756
$Cd + CO_3^{2-} = CdCO_3 + 2\,e^-$	0.74
$Co + 2\,OH^- = Co(OH)_2 + 2\,e^-$	0.73
$Hg + S^{2-} = HgS + 2\,e^-$	0.72
$Ni + 2\,OH^- = Ni(OH)_2 + 2\,e^-$	0.72
$2\,Ag + S^{2-} = Ag_2S + 2\,e^-$	0.69
$As + 4\,OH^- = AsO_2^- + 2\,H_2O + 3\,e^-$	0.68
$AsO_2^- + 4\,OH^- = AsO_4^{3-} + 2\,H_2O + 2\,e^-$	0.67
$2\,FeS + S^{2-} = Fe_2S_3 + 2\,e^-$	0.67
$Sb + 4\,OH^- = SbO_2^- + 2\,H_2O + 3\,e^-$	0.66
$Co + CO_3^{2-} = CoCO_3 + 2\,e^-$	0.64
$Cd + 4\,NH_3 = Cd(NH_3)_4^{2+} + 2\,e^-$	0.597
$ReO_2 + 4\,OH^- = ReO_4^- + 2\,H_2O + 3\,e^-$	0.594
$Re + 8\,OH^- = ReO_4^- + 4\,H_2O + 7\,e^-$	0.584
$S_2O_3^{2-} + 6\,OH^- = 2\,SO_3^{2-} + 3\,H_2O + 4\,e^-$	0.58
$Re + 4\,OH^- = ReO_2 + H_2O + 4\,e^-$	0.576
$Te + 6\,OH^- = TeO_3^{2-} + 3\,H_2O + 4\,e^-$	0.57
$Fe(OH)_2 + OH^- = Fe(OH)_3 + e^-$	0.56
$O_2^- = O_2 + e^-$	0.56
$2\,Cu + S^{2-} = Cu_2S + 2\,e^-$	0.54
$Pb + 3\,OH^- = HPbO_2^- + H_2O + 2\,e^-$	0.54
$Pb + CO_3^{2-} = PbCO_3 + 2\,e^-$	0.506
$S^{2-} = S + 2\,e^-$	0.48
$Ni + 6\,NH_3\,(aq) = Ni(NH_3)_6^{2+} + 2\,e^-$	0.47
$Ni + CO_3^{2-} = NiCO_3^{2-} + 2\,e^-$	0.45
$2\,Bi + 6\,OH^- = Bi_2O_3 + 3\,H_2O + 6\,e^-$	0.44
$Cu + 2\,CN^- = Cu(CN)_2^- + e^-$	0.43
$Hg + 4\,CN^- = Hg(CN)_4^{2-} + 2\,e^-$	0.37
$Se + 6\,OH^- = SeO_3^{2-} + 3\,H_2O + 4\,e^-$	0.366
$2\,Cu + 2\,OH^- = Cu_2O + H_2O + 2\,e^-$	0.358
$Tl + OH^- = Tl(OH) + e^-$	0.3445
$Ag + 2\,CN^- = Ag(CN)_2^- + e^-$	0.31
$Cu + CNS^- = Cu(CNS) + e^-$	0.27
$OH + 2\,OH^- = HO_2^- + H_2O + e^-$	0.24
$Cr(OH)_3 + 5\,OH^- = CrO_4^{2-} + 4\,H_2O + 3\,e^-$	0.13
$Cu + 2\,NH_3 = Cu(NH_3)_2^+ + e^-$	0.12
$Cu_2O + 2\,OH^- + H_2O = 2\,Cu(OH)_2 + 2\,e^-$	0.080
$HO_2^- + OH^- = O_2 + H_2O + 2\,e^-$	0.076
$TlOH + 2\,OH^- = Tl(OH)_3 + 2\,e^-$	0.05

Table A-14 (cont.)

Oxidation-Reduction Couples in Basic Solution	
Couple	$E°$
$Ag + CN^- = AgCN + e^-$	0.017
$Mn(OH)_2 + 2\,OH^- = MnO_2 + H_2O + 2\,e^-$	0.05
$NO_2^- + 2\,OH^- = NO_3^- + H_2O + 2\,e^-$	-0.01
$Os + 9\,OH^- = HOsO_5^- + 4\,H_2O + 8\,e^-$	-0.02
$2\,Rh + 6\,OH^- = Rh_2O_3 + 3\,H_2O + 6\,e^-$	-0.04
$SiO_3^{2-} + 2\,OH^- = SeO_4^{2-} + H_2O + 2\,e^-$	-0.05
$Pd + 2\,OH^- = Pd(OH)_2 + 2\,e^-$	-0.07
$2\,S_2O_3^{2-} = S_4O_6^{2-} + 2\,e^-$	-0.08
$Hg + 2\,OH^- = HgO\,(r) + H_2O + 2\,e^-$	-0.098
$2\,NH_4OH^- + 2\,OH^- = N_2H_4 + 4\,H_2O + 2\,e^-$	-0.1
$Ir + 6\,OH^- = Ir_2O_3 + 3\,H_2O + 6e^-$	-0.1
$Co(NH_3)_6^{2+} = Co(NH_3)_6^{3+} + e^-$	-0.1
$Mn(OH)_2 = Mn(OH)_3 + e^-$	-0.1
$Pt + 2\,OH^- = Pt(OH)_2 + 2\,e^-$	-0.15
$Co(OH)_2 + OH^- = Co(OH)_3 + e^-$	-0.17
$PbO\,(r) + 2\,OH^- = PbO_2 + H_2O + 2\,e^-$	-0.248
$I^- + 6\,OH^- = IO_3^- + 3\,H_2O + 6\,e^-$	-0.26
$PuO_2OH + OH^- = PuO_2(OH)_2 + e^-$	-0.26
$Ag + 2\,SO_3^{2-} = Ag(SO_3)_2^{3-} + e^-$	-0.30
$ClO_2^- + 2\,OH^- = ClO_3^- + H_2O + 2\,e^-$	-0.33
$2\,Ag + 2\,OH^- = Ag_2O + H_2O + 2\,e^-$	-0.344
$ClO_3^- + 2\,OH^- = ClO_4^- + H_2O + 2\,e^-$	-0.36
$Ag + 2\,NH_3 = Ag(NH_3)_2^+ + e^-$	-0.373
$TeO_3^{2-} + 2\,OH^- = TeO_4^{2-} + H_2O + 2\,e^-$	-0.4
$OH^- + HO_2^- = O_2^- + H_2O + e^-$	-0.4
$4\,OH^- = O_2 + 2\,H_2O + 4\,e^-$	-0.401
$2\,Ag + CO_3^{2-} = Ag_2CO_3 + 2\,e^-$	-0.47
$Ni(OH)_2 + 2\,OH^- = NiO_2 + 2\,H_2O + 2\,e^-$	-0.49
$I^- + 2\,OH^- = IO^- + H_2O + 2\,e^-$	-0.49
$Ag_2O + 2\,OH^- = 2\,AgO + H_2O + 2\,e^-$	-0.57
$MnO_2 + 4\,OH^- = MnO_4^{2-} + 2\,H_2O + 2\,e^-$	-0.60
$RuO_4^{2-} = RuO_4^- + e^-$	-0.60
$Br^- + 6\,OH^- = BrO_3^- + 3\,H_2O + 6\,e^-$	-0.61
$ClO^- + 2\,OH^- = ClO_2^- + H_2O + 2\,e^-$	-0.66
$IO_3^- + 3\,OH^- = H_3IO_6^{2-} + 2\,e^-$	-0.7
$N_2H_4 + 2\,OH^- = 2\,NH_2OH + 2\,e^-$	-0.73
$2\,AgO + 2\,OH^- = Ag_2O_3 + H_2O + 2\,e^-$	-0.74
$Br^- + 2\,OH^- = BrO^- + H_2O + 2\,e^-$	-0.76
$3\,OH^- = HO_2^- + H_2O + 2\,e^-$	-0.88
$Cl^- + 2\,OH^- = ClO^- + H_2O + 2\,e^-$	-0.89
$FeO_2^- + 4\,OH^- = FeO_4^{2-} + 2\,H_2O + 3\,e^-$	-0.9
$ClO_2^- = ClO_2 + e^-$	-1.16
$O_2 + 2\,OH^- = O_3 + H_2O + 2\,e^-$	-1.24
$OH^- = OH + e^-$	-2.0

Table A-15   Activities and Activity Coefficients in Seawater

Species	Temp., °C	Pressure, kg/cm²	Salinity, ‰	Comment	γ	a	Table Ref.
Na₂SO₄	25	1	34.8		0.378		1
Na₂SO₄	25	1	25		0.405		1
Na₂SO₄	25	1	15	Artificial	0.435		1
Na₂SO₄	25	1	5	seawater	0.620		1
Na₂SO₄	20	1	34.8		0.385		1
Na₂SO₄	15	1	34.8		0.440		1
NaCl	25	1	34.8		0.672		2
NaCl	25	1	25		0.690		2
NaCl	25	1	15	Artificial	0.730		2
NaCl	25	1	5	seawater	0.795		2
NaCl	35	1	34.8		0.685		2
NaCl	15	1	34.8		0.679		2
NaCl	0	1	34.8		0.650		2
NaHCO₃	25	1	34.3		1.13		3
MgCO₃	25	1	34.3		1.13		3
CaCO₃	25	1	34.3		1.13		3
MgSO₄	25	1	34.3		1.13		3
CaSO₄	25	1	34.3		1.13		3
Na⁺	25	1	34.3		0.76		3
HCO₃⁻	25	1	34.3		0.68		3
NaCO₃⁻	25	1	34.3		0.68		3
NaSO₄⁻	25	1	34.3		0.68		3
KSO₄⁻	25	1	34.3		0.68		3
MgHCO₃⁺	25	1	34.3		0.68		3
CaHCO₃⁺	25	1	34.3		0.68		3
K⁺	25	1	34.3		0.64		3
Mg²⁺	25	1	34.3		0.36		3
Ca²⁺	25	1	34.3		0.28		3
SO₄²⁻	25	1	34.3		0.12		3
CO₃²⁻	25	1	34.3		0.20		3
HCO₃⁻	25	1	34.3	Standard	0.550		4
CO₃²⁻	25	1	34.3	seawater	0.021		4
Ca²⁺	25	1	34.3		0.203		4
HCO₃⁻	25	1	31	Atlantic	0.561		4
CO₃²⁻	25	1	31	(Woods	0.024		4
Ca²⁺	25	1	31	Hole)	0.223		4
CaCO₃	10	100			1.04 γ 1 atm		4
CaCO₃	10	200	Cal. for same		1.075 γ 1 atm		4
CaCO₃	10	300	ionic strength		1.12 γ 1 atm		4
CaCO₃	10	400	as seawater		1.16 γ 1 atm		4
CaCO₃	10	500			1.20 γ 1 atm		4
Mg²⁺	25	1	35		0.063	0.00346	5
Mg²⁺	25	1	25	Artificial	0.067	0.00263	5
Mg²⁺	25	1	15	seawater	0.087	0.00182	5
Mg²⁺	25	1	5		0.17	0.00132	5

Table A-15 (cont.)

Species	Temp., °C	Pressure, hgm/cm²	Salinity, ‰	Comment	γ	a	Table Ref.
OH^-	23	1	32.5⎤	Pacific		0.000029	6
Mg^{2+}	23	1	32.5⎦	(Oregon)	0.17	0.0084	6
Mg^{2+}	25	1	35.7	Standard seawater	0.36	0.07	6

1. R. F. Platford, and T. Dafoe, *J. Marine Res.* (*Sears Found. Marine Res.*), **23**, 63 (1965).
2. R. F. Platford, *J. Marine Res.* (*Sears Found. Marine Res.*), **23**, 55 (1965).
3. R. M. Garrels and M. E. Thompson, *Am. J. Sci.*, **260**, 57 (1962).
4. R. A. Berner, *Geochim. Cosmochim. Acta*, **29**, 947 (1965).
5. R. F. Platford, *J. Fisheries Res. Board Can.*, **22**, 113 (1965).
6. R. M. Pytkowicz, I. W. Duedall, and D. N. Connors, *Science*, **152**, 640 (1966).

Table A-16   Crystal Radii of Some of the Principal Ionic Constituents of Seawater

Na^+	0.95 Å	Cl^-	1.81 Å
Mg^{2+}	0.65	Br^-	1.95
Ca^{2+}	0.99	F^-	1.36
K^+	1.33	I^-	2.16
Sr^{2+}	1.13		
Al^{3+}	0.50		
Rb^+	1.48		
Li^+	0.60		
Ba^{2+}	1.35		

Table A-17   Relative Frequency Distribution of Elements[a]
(From G. Dietrich, *General Oceanography*, Interscience, New York, 1963, with permission of John Wiley & Sons)

Atomic Number	Element	Universe	Earth's Crust	Ocean	Earth's Crust —Ocean	Atomic Number	Element	Universe	Earth's crust	Ocean	Earth's crust —Ocean
		$\log N\mu$						$\log N\mu$			
1	H	14.20	9.94	9.04	0.90	43	Tc				
2	He	14.06	3.90			44	Ru	6.72	4.60		
3	Li	6.84	7.60	2.85	4.75	45	Rh	6.15	4.00		
4	Be	6.25	6.70			46	Pd	6.78	4.70		
5	B	6.41	7.00	4.67	2.33	47	Ag	6.40	4.78	0.48	4.30
6	C	11.68	8.90	5.87	3.03	48	Cd	6.99	5.78		

Table A-17 (cont.)

Atomic Number	Element	Universe	Earth's crust	Ocean	Earth's crust —Ocean	Atomic Number	Element	Universe	Earth's crust	Ocean	Earth's crust —Ocean
			log $N\mu$						log $N\mu$		
	M	8.76				49	In	6.17	5.00		
7	N	12.08	8.48	4.00	4.48	50	Sn	7.27	6.78	1.48	5.30
	M	6.83				51	Sb	5.95	5.48		
8	O	12.39	11.69	9.95	1.74	52	Te	7.25	4.00		
	M	11.60				53	I	6.23	4.85	2.70	2.15
9	F	9.23	8.43	4.11	4.32	54	X	6.92	1.48		
10	Ne	12.56	3.70			55	Cs	5.67	5.85	1.30	4.55
11	Na	9.92	10.42	8.03	2.39	56	Ba	7.06	8.60	2.70	5.90
12	Mg	11.38	10.28	7.11	3.17	57	La	6.42	6.65	0.48	6.17
13	Al	10.21	10.88	3.08	7.80	58	Ce	6.87	7.30	0.60	6.70
14	Si	11.40	11.40	4.00	7.40	59	Pr	6.13	6.54		
15	P	9.60	9.08	2.78	6.30	60	Nd	6.68	7.08		
16	S	10.88	8.78	6.83	1.95	61	Pm				
17	Cl	9.78	9.28	8.28	1.00	62	Sm	6.24	6.65		
18	Ar	9.70	6.54			63	Eu	5.53	5.18		
19	K	9.43	10.38	6.59	3.79	64	Gd	6.42	6.65		
20	Ca	10.28	10.53	6.62	3.91	65	Tb	5.77	5.85		
21	Sc	6.91	6.78	1.60	7.18	66	Dy	6.52	6.65		
22	Ti	9.10	9.76			67	Ho	5.92	5.85		
23	V	8.11	8.26	0.48	7.78	68	Er	6.43	6.60		
24	Cr	9.70	8.52			69	Tm	5.69	5.85		
25	Mn	9.63	8.95	1.70	7.25	70	Yb	6.48	6.54		
26	Fe	11.92	10.67	2.70	7.97	71	Lu	5.92	5.95		
27	Co	9.77	7.08	0.00	7.08	72	Hf	6.63	7.40		
28	Ni	10.90	8.26	0.00	8.26	73	Ta	5.86	5.30		
29	Cu	8.16	8.00	1.70	6.30	74	W	7.26	7.70		
30	Zn	9.00	8.23	1.70	6.53	75	Re	5.62	3.00		
31	Ga	7.63	5.30	0.70	4.60	76	Os	6.52	4.60		
32	Ge	8.13	6.00			77	Ir	6.29	4.30		
33	As	7.40	6.65	2.18	4.47	78	P	6.75	4.70		
34	Se	8.31	5.90	1.60	4.30	79	Au	6.05	3.78	2.60	5.18
35	Br	7.53	6.78	5.82	0.96	80	Hg	6.18	4.48	1.48	5.00
36	Kr	8.10	2.30			81	Tl	6.14	5.00		
37	Rb	7.16	7.54	3.30	4.24	82	Pb	7.75	7.30	1.70	5.60
38	Sr	7.60	8.30	5.11	3.19	83	Bi	5.64	4.48	0.30	4.18
39	Y	6.94	7.85	0.48	7.37						
40	Zr	7.76	8.36			90	Th	6.45	7.08	0.60	6.48
41	Nb	6.72	5.78								
42	Mo	7.40	6.88	0.85	6.03	92	U	5.80	6.62	1.30	5.32

[a] μ = atomic weight; N = number of molecules; M = occurrence in meteorites.

Table A-18   Henry's Law Constants for the Dissolution of CO_2 in Pure Water and Aqueous NaCl Solutions
(From H. S. Harned and R. Davis, Jr., *J. Am. Chem. Soc.*, **65**, 2030 (1943), with permission of American Chemical Society)

	$-\log C$ (defined by Eq. 7.1)			
Temp., °C	Pure Water	0.2 M, NaCl	0.5 M NaCl	1 M NaCl
0	1.1144	1.1383	1.1738	1.2313
5	1.1938	1.2167	1.2510	1.3065
10	1.2695	1.2920	1.3259	1.3800
15	1.3412	1.3638	1.3970	1.4482
20	1.4063	1.4285	1.4605	1.5110
25	1.4635	1.4842	1.5152	1.5645
30	1.5209	1.5410	1.5712	1.6191

Table A-19   First and Second Dissociation Constants of Boric Acid in Seawater
(From J. Lyman, Ph.D. Thesis, Univ. of Calif., 1956)

	$-\log \dfrac{a_{H^+}(H_2BO_3{}^-)}{(H_3BO_3)}$							
				t, °C				
Cl, ‰	0	5	10	15	20	25	30	35
0	9.50	9.44	9.38	9.33	9.28	9.24	9.20	9.16
1	9.40	9.34	9.28	9.23	9.18	9.14	9.10	9.06
4	9.28	9.22	9.16	9.11	9.06	9.02	8.98	8.94
9	9.14	9.08	9.03	8.98	8.93	8.88	8.85	8.82
16	9.00	8.95	8.89	8.84	8.80	8.76	8.72	8.69
17	8.98	8.93	8.88	8.83	8.78	8.74	8.70	8.67
18	8.96	8.91	8.86	8.81	8.76	8.72	8.69	8.66
19	8.95	8.90	8.85	8.80	8.75	8.71	8.67	8.64
20	8.94	8.88	8.83	8.78	8.74	8.69	8.65	8.63
21	8.92	8.87	8.82	8.77	8.72	8.68	8.64	8.61
25	8.85	8.80	8.75	8.70	8.66	8.62	8.59	8.56
36	8.71	8.66	8.61	8.57	8.53	8.49	8.46	8.43
49	8.56	8.52	8.47	8.43	8.39^5	8.36	8.33	8.30
64	8.41	8.37	8.33	8.30	8.26^5	8.23	8.20	8.17

Table A-19 (cont.)

$$-\log \frac{a_{H^+}(HBO_3^{2-})}{(H_2BO_3^-)}$$

Cl, ‰	t, °C							
	0	5	10	15	20	25	30	35
16	11.17	10.91	10.65	10.40	10.16	9.97	9.80	9.64
17	11.14	10.88	10.63	10.38	10.13	9.94	9.78	9.62
18	11.11	10.85	10.60	10.35	10.10	9.92	9.75	9.60
19	11.08	10.82	10.57	10.32	10.08	9.89	9.73	9.58
20	11.05	10.79	10.54	10.30	10.05	9.87	9.71	9.57
21	11.01	10.76	10.52	10.27	10.03	9.85	9.69	9.55
25	10.92	10.68	10.43	10.18	9.96	9.78	9.63	9.49
36	10.71	10.46	10.23	10.01	9.80	9.62	9.48	9.35
49	10.52	10.28	10.04	9.84	9.65	9.48	9.34	9.23
64	10.35	10.10	9.88	9.69	9.52	9.36	9.23	9.11

Table A-20 Solubility of Carbon Monoxide in Seawater
(From E. Douglas, *J. Phys. Chem.*, **71**, 1931 (1967), with permission of
American Chemical Society)

Temp., °C	Chlorinity						
	15	16	17	18	19	20	21
	α, Carbon Monoxide						
−2	0.03162	0.03124	0.03084	0.03044	0.03004	0.02966	0.02926
−1	0.03090	0.03052	0.03014	0.02976	0.02938	0.02900	0.02862
0	0.03024	0.02986	0.02948	0.02910	0.02872	0.02835	0.02797
1	0.02949	0.02913	0.02878	0.02842	0.02807	0.02772	0.02736
2	0.02880	0.02846	0.02811	0.02776	0.02743	0.02709	0.02675
3	0.02812	0.02779	0.02746	0.02713	0.02680	0.02648	0.02614
4	0.02750	0.02717	0.02684	0.02652	0.02620	0.02588	0.02556
5	0.02686	0.02656	0.02625	0.02594	0.02564	0.02532	0.02501
6	0.02632	0.02602	0.02572	0.02541	0.02510	0.02480	0.02450
7	0.02578	0.02548	0.02519	0.02490	0.02460	0.02432	0.02402
8	0.02524	0.02496	0.02468	0.02440	0.02412	0.02384	0.02356
9	0.02475	0.02448	0.02421	0.02394	0.02366	0.02339	0.02312
10	0.02428	0.02402	0.02376	0.02350	0.02322	0.02296	0.02270
11	0.02385	0.02359	0.02334	0.02308	0.02282	0.02256	0.02230
12	0.02343	0.02318	0.02292	0.02267	0.02242	0.02216	0.02192
13	0.02302	0.02278	0.02252	0.02228	0.02202	0.02178	0.02153
14	0.02262	0.02238	0.02213	0.02188	0.02164	0.02140	0.02116
15	0.02224	0.02200	0.02176	0.02152	0.02128	0.02104	0.02080
16	0.02187	0.02164	0.02140	0.02116	0.02092	0.02070	0.02046
17	0.02152	0.02129	0.02106	0.02082	0.02059	0.02036	0.02014
18	0.02118	0.02094	0.02072	0.02048	0.02026	0.02002	0.01980

Table A-20 (cont.)

Temp., °C	Chlorinity						
	15	16	17	18	19	20	21
			α, Carbon Monoxide				
19	0.02084	0.02061	0.02038	0.02016	0.01994	0.01971	0.01949
20	0.02052	0.02030	0.02008	0.01985	0.01962	0.01940	0.01917
21	0.02022	0.02000	0.01978	0.01955	0.01932	0.01910	0.01888
22	0.01993	0.01971	0.01948	0.01926	0.01904	0.01882	0.01860
23	0.01965	0.01942	0.01920	0.01898	0.01876	0.01854	0.01832
24	0.01936	0.01915	0.01893	0.01872	0.01850	0.01828	0.01806
25	0.01912	0.01890	0.01868	0.91847	0.01825	0.01804	0.01782
26	0.01886	0.01864	0.01844	0.01822	0.01802	0.01780	0.01760
27	0.01862	0.01842	0.01822	0.01880	0.01780	0.01760	0.01740
28	0.01839	0.01820	0.01800	0.01780	0.01760	0.01740	0.01720
29	0.01819	0.01800	0.01780	0.01760	0.01741	0.01722	0.01702
30	0.01796	0.01778	0.01760	0.01742	0.01724	0.01706	0.01688

[a] Note that chlorinity is expressed in grams of chlorine per kilogram of seawater, while α is given as volume of gas (STPD) absorbed by a unit volume of water when the pressure of the gas equals 760 mm.

Table A-21   Some Optical Properties of Seawater
A.   Light Absorption of Typical Seawaters—Extinction for 10 cm Path Length
(From G. L. Clarke and H. R. James, *J. Opt. Soc. Am.* **29**, 43 (1939).

Sample	Wavelength, Å							
	3600	4000	5000	5200	6000	7000	7500	8000
Pure water	0.001	0.001	0.002	0.002	0.010	0.025	0.115	0.086
Artificial seawater	0.011	0.003	0.005	0.007	0.010	0.025	0.115	0.086
Ocean water, unfiltered	0.012	0.009	0.007	0.008	0.011	0.025	0.115	0.086
Continental slope waters, unfiltered	0.052	0.030	0.011	0.010	0.012	0.035	0.130	0.088
Continental slope waters, filtered	0.016	0.010	0.005	0.005	0.012	0.030	0.115	0.086
Inshore water, unfiltered	0.055	0.042	0.028	0.026	0.035	0.052	0.140	0.100
Inshore water, filtered	0.015	0.010	0.005	0.005	0.010	0.025	0.110	0.086

B. Differences between the Extinctions of Seawaters and Pure Water
(From G. L. Clarke and H. R. James, *J. Opt. Soc. Am.* **29**, 43 (1939).

Sample	Wavelength, Å							
	3600	4000	5000	5200	6000	7000	75000	8000
Artificial seawater	0.010	0.002	0.003	0.005	nil	nil	nil	nil
Ocean water, unfiltered	0.011	0.008	0.005	0.006	0.001	nil	nil	nil
Continental slope water, unfiltered	0.051	0.029	0.009	0.008	0.002	0.010	0.015	0.002
Continental slope water, filtered	0.015	0.009	0.003	0.003	0.002	0.005	nil	nil
Inshore water, unfiltered	0.054	0.041	0.026	0.024	0.025	0.027	0.025	0.015
Inshore water, filtered	0.014	0.009	0.003	0.003	nil	nil	nil	nil

[a] E_{10} cm (seawater) $- E_{10}$ cm (pure water) as a function of wavelength.

C. Refractive Index of Seawater[a]
(From C. L. Utterback, T. G. Thompson, and B. D. Thomas, *J. Conseil Exploration Mer*, **9**, 35 (1934) with permission of Conseil International pour l'Exploration.)

$S, \%_{00}$	Temperature, °C											
	0		5		10		15		20		25	
0	1.3	3395	1.3	3385	1.3	3370	1.3	3340	1.3	3300	1.3	3250
5		3500		3485		3465		3435		3395		3345
10		3600		3585		3565		3530		3485		3435
15		3700		3685		3660		3625		3580		3525
20		3795		3780		3750		3715		3670		3620
25		3895		3875		3845		3805		3760		3710
30		3991		3966		3935		3898		3851		3798
31		4011		3985		3954		3916		3869		3816
32		4030		4004		3973		3934		3886		3834
33		4049		4023		3992		3953		3904		3851
34		4068		4042		4011		3971		3922		3868
35		4088		4061		4030		3990		3940		3886
36		4107		4080		4049		4008		3958		3904
37		4127		4099		4068		4026		3976		3922
38		4146		4118		4086		4044		3994		3940
39		4166		4139		4105		4062		4012		3958
40		(4185)		(4157)		(4124)		(4080)		(4031)		(3976)
41		(4204)		(4716)		(4143)		(4098)		(4049)		(3994)

[a] For sodium D light.

Table A-22   The Area, Volume, and Mean Depth of the Oceans and Seas
(From H. U. Sverdrup, M. W. Johnson, and R. H. Fleming, *The Oceans*, Prentice-Hall, Inc., Englewood Cliffs, N.J., 1943, with permission of the publisher)

Body	Area, 10^6 km^2	Volume, 10^6 km^3	Mean Depth, m
Atlantic Ocean ⎫	82.441	323.613	3926
Pacific Ocean ⎬ excluding adjacent seas	165.246	707.555	4282
Indian Ocean ⎭	73.443	291.030	3963
All oceans (excluding adjacent seas)	321.130	1322.198	4117
Arctic Mediterranean	14.090	16.980	1205
American Mediterranean	4.319	9.573	2216
Mediterranean Sea and Black Sea	2.966	4.238	1429
Asiatic Mediterranean	8.143	9.873	1212
Large Mediterranean Seas	29.518	40.664	1378
Baltic Sea	0.422	0.023	55
Hudson Bay	1.232	0.158	128
Red Sea	0.438	0.215	491
Persian Gulf	0.239	0.006	25
Small Mediterranean seas	2.331	0.402	172
All Mediterranean seas	31.849	41.066	1289
North Sea	0.575	0.054	94
English Channel	0.075	0.004	54
Irish Sea	0.103	0.006	60
Gulf of St. Lawrence	0.238	0.030	127
Andaman Sea	0.798	0.694	870
Bering Sea	2.268	3.259	1437
Okhotsk Sea	1.528	1.279	838
Japan Sea	1.008	1.361	1350
East China Sea	1.249	0.235	188
Gulf of California	0.162	0.132	813
Bass Strait	0.075	0.005	70
Marginal seas	8.079	7.059	874
All adjacent seas	39.928	48.125	1205
Atlantic Ocean ⎫	106.463	354.679	3332
Pacific Ocean ⎬ including adjacent seas	179.679	723.699	4028
Indian Ocean ⎭	74.917	291.945	3897
All oceans (including adjacent seas)	361.059	1370.323	3795

Table A-23 Chemical Analyses of Some Natural Waters, concentration in ppm

Description	Na	Mg	Ca	K	Cl	SO₄	CO₃
Seawater[a]	10,770	1,300	409		19,070	2,710	139
Natural brine, Deep-Springs, Calif.[b]	101,000			15,700	58,100	40,500	75,800
Webster, S. Dakota[a]	81	79	216		1	718	355
Dalpra Farm, Denver, Colo.[a]	886	72	118	16	131	1,943	473
Coastal Texas[a]	2,130	53	87		3,400	0	330
Saline Red Sea waters[c]	140,000	1,300	4,500	1,700	150,000		
Oil field brine, Bradford, Penn.[d]	32,600*	1,940	13,260		77,340	730	
Oil field brine, Oklahoma City[d]	91,603	3,468	18,753		184,387	268	18
Oil field brine, Burgan, Kuwait[d]	46,191	2,206	10,158		95,275	198	360

[a] Mason-Rust, An Engineering Evaluation of the Electrodialysis Process Adapted to Computer Methods for Water Desalination Plants, Rept. for OSW (Jan. 1965).
[b] R. M. Garrels and D. L. Christ, *Solutions, Minerals, and Equilibria*, Harper & Row, New York, 1965.
[c] A. C. Neumann, and K. E. Chave, *Nature*, **206**, 1346 (1965).
[d] A. I. Levorsen, *Geology of Petroleum*, W. H. Freeman & Co., San Francisco, Calif., 1956.
[e] (Na + K).

Table A-24 Densities of Solutions of the Major Seawater Constituents (From B. M. Fabuss, A. Korosi, and A. K. M. Huq, *J. Chem. Eng. Data*, **11**, 325 (1966), with permission of American Chemical Society)

Densities of Solutions Containing NaCl and NaSO₄, g/cm³

Temp., °C	NaCl Molality								
	0.1			1.0			2.5		
	Na₂SO₄ Molality								
	0.01	0.05	0.15	0.01	0.05	0.15	0.01	0.05	0.15
25	1.0024	1.0072	1.0196	1.0372	1.0419	1.0532	1.0894	1.0942	1.1046
45	0.9954	1.0000	1.0120	1.0292	1.0337	1.0449	1.0801	1.0849	1.0950
65	0.9856	0.9904	1.0024	1.0193	1.0236	1.0347	1.0695	1.0742	1.0843
75	0.9802	0.9848	0.9968	1.0136	1.0180	1.0292	1.0638	1.0684	1.0786
100	0.9641	0.9686	0.9803	0.9977	1.0021	1.0134	1.0478	1.0524	1.0627
125	0.9450	0.9496	0.9613	0.9791	0.9836	0.9949	1.0300	1.0344	1.0448
150	0.9237	0.9284	0.9402	0.9588	0.9633	0.9746	1.0107	1.0153	1.0257
175	0.8995	0.9045	0.9167	0.9363	0.9408	0.9522	0.9898	0.9944	1.0049

Table A-24 (cont.)

Densities of Solutions Containing NaCl and $MgSO_4$, g/cm^3

	NaCl Molality								
	0.1			1.0			2.5		
	$MgSO_4$ Molality								
Temp., °C	0.00972	0.09722	0.29165	0.00972	0.09722	0.29165	0.00972	0.09722	0.29165
25	1.0022	1.0129	1.0360	1.0372	1.0474	1.0690	1.0901	1.0993	1.1195
45	0.9951	1.0058	1.0286	1.0290	1.0392	1.0608	1.0809	1.0900	1.1102
65	0.9855	0.9962	1.0189	1.0192	1.0291	1.0506	1.0702	1.0794	1.0995
75	0.9801	0.9906	1.0133	1.0136	1.0235	1.0450	1.0645	1.0737	1.0938
100	0.9639	0.9745	0.9975	0.9978	1.0078	1.0293	1.0488	1.0583	1.0778
125	0.9447	0.9554	0.9785	0.9792	0.9893	1.0110	1.0309	1.0405	1.0602
150	0.9233	0.9340	0.9572	0.9588	0.9690	0.9908	1.0116	1.0214	1.0412
175	0.8992	0.9101	0.9334	0.9363	0.9467	0.9686	0.9908	1.0007	1.0207

Densities of Solutions Containing Na_2SO_4 and $MgSO_4$ g/cm^3

	Na_2SO_4 Molality								
	0.01			0.05			0.15		
	$MgSO_4$ Molality								
Temp., °C	0.01	0.10	0.30	0.01	0.10	0.30	0.01	0.10	0.30
25	0.9994	1.0101	1.0336	1.0046	1.0150	1.0382	1.0170	1.0273	1.0500
45	0.9924	1.0031	1.0262	0.9975	1.0079	1.0307	1.0096	1.0199	1.0423
65	0.9829	0.9935	1.0164	0.9878	0.9982	1.0210	1.0000	1.0101	1.0322
75	0.9774	0.9880	1.0110	0.9821	0.9926	1.0154	0.9942	1.0043	1.0268
100	0.9611	0.9720	0.9949	0.9660	0.9767	0.9994	0.9782	0.9883	1.0106
125	0.9419	0.9528	0.9759	0.9467	0.9574	0.9802	0.9592	0.9693	0.9915
150	0.9203	0.9312	0.9544	0.9253	0.9360	0.9588	0.9379	0.9478	0.9703
175	0.8963	0.9071	0.9304	0.9011	0.9119	0.9348	0.9140	0.9241	0.9468

Table A-25 Partial Equivalent Conductances of Seawater Constituent Salts
(From D. N. Connors, Ph.D. Thesis, Oregon State Univ., 1967, with permission of the author)

Salt	Salinity, ‰	Temp., °C	Partial Equiv. Cond., cm²/(ohm-equiv)
NaCl	34.325	24.958	80.13 ± 0.04
	30.126	24.958	82.38 ± 0.03
	34.325	14.986	64.96 ± 0.04
	34.325	0.00	44.54 ± 0.12
	30.126	0.00	45.36 ± 0.32
KCl	34.325	24.958	102.64 ± 1.42
	30.126	24.958	104.74 ± 0.02
	34.325	14.986	84.36 ± 0.10
	34.325	0.00	59.40 ± 0.12
	30.126	0.00	59.86 ± 0.26
Na₂SO₄	34.325	24.958	46.93 ± 0.33
	30.126	24.958	49.33 ± 0.08
	34.325	14.986	37.48 ± 0.02
	34.325	0.00	25.04 ± 0.04
	30.126	0.00	26.09 ± 0.25
K₂SO₄	34.325	24.958	69.14 ± 0.40
	30.126	24.958	71.62 ± 0.02
	34.325	14.986	56.98 ± 0.02
	34.325	0.00	39.86 ± 0.06
	30.126	0.00	40.28
KHCO₃	35.567	24.920	64.25 ± 0.33
	30.180	24.920	67.19
	33.953	14.950	53.96 ± 0.14
	35.567	0.00	36.68
	30.180	0.00	37.92 ± 0.14
Ca(NO₃)₂	35.567	24.920	61.34 ± 0.42
	30.180	24.920	63.86 ± 0.32
	33.953	14.950	51.42 ± 0.09
	35.567	0.00	35.18
	30.180	0.00	36.70 ± 0.06
MgSO₄	35.567	24.920	29.79 ± 0.10
	30.180	24.920	32.82 ± 0.18
	33.953	14.950	24.46 ± 0.06
	35.567	0.00	15.24
	30.180	0.00	17.20 ± 0.07
NaNO₃	35.567	24.920	71.64 ± 0.38
	30.180	24.920	74.72 ± 0.02
	33.953	14.950	59.47 ± 0.17
	35.567	0.00	40.28
	30.180	0.00	42.02 ± 0.15

Table A-26   Vapor Pressure of Pure Water as a Function of Temperature[a] (From N. E. Dorsey, *Properties of Ordinary Water-Substance*, Reinhold Pub. Corp., New York, 1940, with permission of the publisher)

t_1	0	1	2	3	4
$t_2\rightarrow$			P		
−10					
0	0.0060273	0.006479	0.006960	0.007473	0.008019
+10	0.012102	0.012936	0.013821	0.014759	0.015752
20	0.023042	0.024508	0.026056	0.027688	0.029409
30	0.041831	0.044293	0.046881	0.049599	0.052452
40	0.072748	0.076718	0.080873	0.085222	0.089770
50	0.121698	0.12787	0.13431	0.14102	0.14802
60	0.196560	0.20584	0.21549	0.22553	0.23596
70	0.307520	0.32107	0.33512	0.34969	0.36479
80	0.467396	0.48665	0.50657	0.52717	0.54846
90	0.691923	0.71863	0.74619	0.77463	0.80396
100	1.00000	1.0362	1.0735	1.1120	1.1515

t_1	5	6	7	8	9
$t_2\rightarrow$			P		
−10	0.004162	0.004487	0.004835	0.005207	0.005604
0	0.008600	0.009218	0.009875	0.010574	0.011315
+10	0.016804	0.017917	0.019094	0.020338	0.021653
20	0.031222	0.033133	0.035144	0.037261	0.39489
30	0.055446	0.58588	0.061883	0.065337	0.068956
40	0.094526	0.099497	0.10469	0.11012	0.11578
50	0.15531	0.16291	0.17082	0.17906	0.18764
60	0.24679	0.25805	0.26973	0.28186	0.29446
70	0.38043	0.39664	0.41342	0.43080	0.44878
80	0.57047	0.59322	0.61672	0.64099	0.66605
90	0.83421	0.86540	0.89755	0.93068	0.96482
100	1.1922	1.2341	1.2772	1.3215	1.3670

[a] Unit of $P = 1$ atm $= 1.01325$ bars $= 1.03323$ kg*/cm². Temp. $= (t_1 + t_2)$°C, int. centigrade scale.

Table A-27   Gas Molecular Diffusion
Coefficients in Pure Water at 1 atm

Gas	T	D
CO_2^a	25°C	1.92×10^{-5} cm²/sec
N_2^a	10	1.29
	25	2.01
	40	2.83
	55	3.80
O_2^a	10	1.54
	25	2.20
	40	3.33
	55	4.50
O_2^b	30	3.49
N_2^b	30	3.47
H_2^b	30	5.42

[a] From R. T. Ferrell and D. M. Himmelbau, *J. Chem. Eng. Data*, **12**, 111 (1967).
[b] From I. M. Kreiger, G. W. Mulholland, and C. S. Dickey, *J. Phys. Chem.*, **71**, 1123 (1967).
[c] These coefficients were determined from measurements in the rate of decrease in size of small gas bubbles.

Table A-28 Velocity of Sound in Seawater
(From W. D. Wilson, *J. Acoust. Soc. Am.* **32**, 641 (1960b), with permission of American Institute of Physics)

For the equation $V = 1449.22 + \Delta V_t + \Delta V_p + \Delta V_s + \Delta V_{stp}$ (V and ΔV in m/sec; cf. Eq. 13.2)

T	0.0	0.1	0.2	0.3	0.4	0.5	0.6	0.7	0.8	0.9
					ΔV_t					
−3	−14.37	−14.87	−15.36	−15.86	−16.36	−16.86	−17.36	−17.87	−18.37	−18.88
−2	−9.47	−9.95	−10.44	−10.93	−11.41	−11.90	−12.39	−12.89	−13.38	−13.87
−1	−4.68	−5.15	−5.63	−6.10	−6.58	−7.06	−7.54	−8.02	−8.50	−8.98
0	0.00	−0.46	−0.93	−1.39	−1.86	−2.33	−2.79	−3.36	−3.73	−4.21
0	0.00	0.46	0.92	1.38	1.84	2.30	2.75	3.21	3.66	4.12
1	4.57	5.02	5.47	5.92	6.37	6.81	7.26	7.70	8.15	8.59
2	9.03	9.47	9.91	10.35	10.79	11.22	11.66	12.09	12.52	12.96
3	13.39	13.82	14.24	14.67	15.10	15.53	15.95	16.37	16.80	17.22
4	17.64	18.06	18.48	18.89	19.31	19.73	20.14	20.55	20.97	21.38
5	21.79	22.20	22.60	23.01	23.42	23.82	24.23	24.63	25.03	25.43
6	25.84	26.23	26.63	27.03	27.43	27.82	28.22	28.61	29.00	29.39
7	29.78	30.17	30.56	30.95	31.34	31.72	32.11	32.49	32.87	33.25
8	33.64	34.02	34.39	34.77	35.15	35.53	35.90	36.27	36.65	37.02
9	37.39	37.76	38.13	38.50	38.87	39.23	39.60	39.96	40.33	40.69
10	41.05	41.41	41.77	42.13	42.49	42.85	43.20	43.56	43.91	44.27
11	44.62	44.97	45.32	45.67	46.02	46.37	46.72	47.06	47.41	47.75
12	48.10	48.44	48.78	49.12	49.46	49.80	50.14	50.48	50.81	51.15
13	51.48	51.82	52.15	52.48	52.81	53.14	53.47	53.80	54.13	54.46
14	54.78	55.11	55.43	55.76	56.08	56.40	56.72	57.04	57.36	57.68
15	58.00	58.31	58.63	58.94	59.26	59.57	59.88	60.19	60.50	60.81
16	61.12	61.43	61.74	62.04	62.35	62.65	62.96	63.26	63.56	63.87
17	64.17	64.47	64.76	65.06	65.36	65.66	65.95	66.25	66.54	66.83
18	67.13	67.42	67.71	68.00	68.29	68.58	68.87	69.15	69.44	69.72
19	70.01	70.29	70.57	70.86	71.14	71.42	71.70	71.98	72.26	72.53
20	72.81	73.09	73.36	73.63	73.91	74.18	74.45	74.72	74.99	75.26
21	75.53	75.80	76.07	76.34	76.60	76.87	77.13	77.39	77.66	77.92

ΔV_p

P^a	0	10	20	30	40	50	60	70	80	90
22	78.18	78.44	78.70	78.96	79.22	79.48	79.73	79.99	80.24	80.50
23	80.75	81.01	81.26	81.51	81.76	82.01	82.26	82.51	82.76	83.01
24	83.25	83.50	83.74	83.99	84.23	84.47	84.72	84.96	85.20	85.44
25	85.68	85.92	86.16	86.39	86.63	86.87	87.10	87.34	87.57	87.80
26	88.04	88.27	88.50	88.73	88.96	89.19	89.42	89.64	89.87	90.10
27	90.32	90.55	90.77	91.00	91.22	91.44	91.66	91.88	92.10	92.32
28	92.54	92.76	92.98	93.19	93.41	93.63	93.84	94.06	94.27	94.48
29	94.69	94.91	95.12	95.33	95.54	95.75	95.96	96.16	96.37	96.58
30	96.78	96.99	97.19	97.40	97.60	97.80	98.00	98.21	98.41	98.61
0	—	1.61	3.21	4.82	6.44	8.05	9.67	11.29	12.91	14.53
100	16.16	17.79	19.42	21.05	22.68	24.32	25.96	27.60	29.24	30.89
200	32.54	34.19	35.84	37.50	39.15	40.81	42.47	44.14	45.81	47.47
300	49.15	50.82	52.49	54.17	55.85	57.54	59.22	60.91	62.60	64.29
400	65.98	67.68	69.38	71.08	72.78	74.49	76.19	77.90	79.61	81.33
500	83.04	84.76	86.48	88.20	89.92	91.65	93.38	95.10	96.84	98.57
600	100.30	102.04	103.78	105.52	107.26	109.00	110.75	112.49	114.24	115.99
700	117.74	119.49	121.25	123.00	124.76	125.52	128.28	130.04	131.80	133.56
800	135.33	137.09	138.86	140.62	142.39	144.16	145.93	147.70	149.47	151.24
900	153.01	154.78	156.56	158.33	160.10	161.88	163.65	165.42	167.20	168.97

[a] At atmospheric pressure, $P = 1.0332$ kg/cm², $\Delta V_P = 0.16$ m/sec.

ΔV_S

S	0.0	0.1	0.2	0.3	0.4	0.5	0.6	0.7	0.8	0.9
33	-3.09	-2.93	-2.76	-2.59	-2.43	-2.26	-2.10	-1.94	-1.78	-1.62
34	-1.47	-1.32	-1.16	-1.01	-0.86	-0.72	-0.57	-0.42	-0.28	-0.14
35	0.00	0.14	0.28	0.41	0.54	0.68	0.81	0.94	1.06	1.19
36	1.31	1.44	1.56	1.68	1.79	1.91	2.03	2.14	2.25	2.36
37	2.47	2.58	2.68	2.79	2.89	2.99	3.09	3.19	3.28	3.38

Table A-28 (cont.)

		ΔV_{stp}							
					T				
P	S	-4	-2	0	2	4	6	8	10
1.0332	33	-0.09	-0.05	0.00	0.05	0.09	0.14	0.19	0.24
	34	-0.05	-0.02	0.00	0.02	0.05	0.07	0.09	0.12
	35	0.00	0.00	0.00	0.00	0.00	0.00	0.00	0.00
	36	0.05	0.02	0.00	-0.02	-0.05	-0.07	-0.10	-0.12
	37	0.10	0.05	0.00	-0.05	-0.10	-0.14	-0.19	-0.24
100	33	0.00	-0.03	-0.05	-0.06	-0.05	-0.04	-0.02	0.02
	34	0.07	0.02	-0.02	-0.06	-0.08	-0.09	-0.09	-0.08
	35	0.14	0.07	0.00	-0.06	-0.10	-0.14	-0.16	-0.18
	36	0.22	0.11	0.02	-0.06	-0.13	-0.19	-0.24	-0.28
	37	0.29	0.16	0.05	-0.06	-0.15	-0.23	-0.31	-0.37
200	33	0.12	0.01	-0.09	-0.16	-0.21	-0.25	-0.26	-0.25
	34	0.22	0.08	-0.04	-0.14	-0.22	-0.28	-0.31	-0.33
	35	0.31	0.14	0.00	-0.12	-0.23	-0.31	-0.37	-0.41
	36	0.41	0.21	0.04	-0.10	-0.23	-0.34	-0.42	-0.49
	37	0.50	0.28	0.09	-0.08	-0.24	-0.37	-0.48	-0.57
300	33	0.29	0.07	-0.12	-0.27	-0.40	-0.49	-0.55	-0.58
	34	0.40	0.15	-0.06	-0.24	-0.39	-0.50	-0.59	-0.65
	35	0.51	0.24	0.00	-0.20	-0.38	-0.52	-0.63	-0.71
	36	0.62	0.32	0.06	-0.17	-0.37	-0.53	-0.67	-0.78
	37	0.73	0.41	0.12	-0.13	-0.36	-0.55	-0.71	-0.84
400	33	0.49	0.15	-0.15	-0.40	-0.60	-0.77	-0.89	-0.98
	34	0.61	0.25	-0.97	-0.35	-0.58	-0.77	-0.92	-1.04
	35	0.73	0.34	0.00	-0.30	-0.56	-0.78	-0.95	-1.09
	36	0.86	0.44	0.07	-0.25	-0.54	-0.78	-0.98	-1.15
	37	0.98	0.54	0.15	-0.21	-0.52	-0.78	-1.01	-1.20
500	33	0.73	0.26	-0.16	-0.53	-0.84	-1.09	-1.30	-1.46
	34	0.86	0.36	-0.08	-0.47	-0.81	-1.09	-1.32	-1.51
	35	1.00	0.47	0.00	-0.42	-0.78	-1.09	-1.35	-1.56
	36	1.13	0.58	0.08	-0.36	-0.75	-1.08	-1.37	-1.61
	37	1.26	0.68	0.16	-0.31	-0.72	-1.08	-1.39	-1.66
600	33	1.02	0.39	-0.17	-0.67	-1.10	-1.47	-1.78	-2.03
	34	1.16	0.50	-0.09	-0.61	-1.07	-1.47	-1.80	-2.08
	35	1.30	0.62	0.00	-0.55	-1.04	-1.46	-1.82	-2.12
	36	1.44	0.73	0.09	-0.49	-1.00	-1.45	-1.84	-2.17
	37	1.58	0.84	0.17	-0.43	-0.97	-1.45	-1.86	-2.22
700	33	1.37	0.56	-0.17	-0.83	-1.40	-1.91	-2.34	-2.70
	34	1.51	0.67	-0.09	-0.77	-1.37	-1.90	-2.36	-2.75
	35	1.65	0.79	0.00	-0.71	-1.34	-1.90	-2.38	-2.80
	36	1.79	0.90	0.09	-0.65	-1.31	-1.89	-2.40	-2.85
	37	1.93	1.01	0.17	-0.59	-1.27	-1.88	-2.42	-2.89
800	33	1.77	0.76	-0.17	-1.00	-1.74	-2.41	-2.98	-3.48
	34	1.91	0.87	-0.08	-0.94	-1.72	-2.40	-3.01	-3.54

Table A-28 (cont.)

$$\Delta V_{stp}$$

P	S	T							
		−4	−2	0	2	4	6	8	10
	35	2.05	0.98	0.00	−0.89	−1.69	−2.40	−3.04	−3.59
	36	2.19	1.09	0.08	−0.83	−1.66	−2.40	−3.06	−3.64
	37	2.32	1.20	0.17	−0.78	−1.63	−2.40	−3.09	−3.70
900	33	2.24	0.99	−0.15	−1.19	−2.13	−2.98	−3.73	−4.39
	34	2.37	1.09	−0.08	−1.14	−2.11	−2.98	−3.76	−4.45
	35	2.50	1.20	0.00	−1.09	−2.09	−2.99	−3.80	−4.51
	36	2.63	1.30	0.08	−1.05	−2.07	−3.00	−3.83	−4.58
	37	2.76	1.40	0.15	−1.00	−2.05	−3.01	−3.87	−4.64
1000	33	2.76	1.26	−0.13	−1.40	−2.57	−3.62	−4.57	−5.42
	34	2.88	1.35	−0.06	−1.37	−2.56	−3.64	−4.62	−5.50
	35	3.01	1.44	0.00	−1.33	−2.55	−3.66	−4.67	−5.58
	36	3.13	1.54	0.06	−1.29	−2.54	−3.68	−4.72	−5.65
	37	3.25	1.63	0.13	−1.26	−2.53	−3.70	−4.77	−5.73

12	14	16	18	20	22	24	26	28	30
0.29	0.33	0.38	0.43	0.48	0.53	0.57	0.62	0.67	0.72
0.14	0.17	0.19	0.21	0.24	0.26	0.29	0.31	0.34	0.36
0.00	0.00	0.00	0.00	0.00	0.00	0.00	0.00	0.00	0.00
−0.15	−0.17	−0.19	−0.22	−0.22	−0.26	−0.26	−0.31	−0.33	−0.36
−0.29	−0.34	−0.38	−0.43	−0.48	−0.53	−0.57	−0.62	−0.67	−0.72
0.06	0.11	0.17	0.24	0.31	0.39	0.48	0.58	0.68	0.79
−0.06	−0.04	−0.00	−0.04	0.09	0.15	0.22	0.29	0.37	0.45
−0.18	−0.18	−0.17	−0.15	−0.13	−0.09	−0.05	−0.01	0.05	0.11
−0.31	−0.33	−0.34	−0.35	−0.35	−0.34	−0.32	−0.30	−0.27	−0.23
−0.43	−0.48	−0.51	−0.54	−0.57	−0.58	−0.59	−0.59	−0.58	−0.57
−0.23	−0.19	−0.13	−0.05	0.04	0.15	0.27	0.41	0.56	0.72
−0.33	−0.32	−0.28	−0.23	−0.16	−0.08	0.02	0.13	0.25	0.39
−0.44	−0.44	−0.43	−0.41	−0.37	−0.31	−0.24	−0.15	−0.05	0.06
−0.54	−0.57	−0.59	−0.59	−0.57	−0.54	−0.49	−0.43	−0.35	−0.26
−0.65	−0.70	−0.74	−0.76	−0.77	−0.76	−0.74	−0.70	−0.65	−0.59
−0.59	−0.57	−0.52	−0.45	−0.35	−0.23	−0.09	0.07	0.26	0.46
−0.68	−0.68	−0.66	−0.61	−0.54	−0.45	−0.33	−0.19	−0.03	0.14
−0.77	−0.80	−0.80	−0.78	−0.73	−0.66	−0.57	−0.46	−0.33	−0.18
−0.86	−0.91	−0.94	−0.94	−0.92	−0.88	−0.81	−0.73	−0.62	−0.49
−0.95	−1.03	−1.08	−1.11	−1.12	−1.10	−1.06	−0.99	−0.91	−0.81
−1.03	−1.05	−1.03	−0.97	−0.88	−0.77	−0.62	−0.44	−0.24	−0.01
−1.11	−1.15	−1.16	−1.13	−1.07	−0.98	−0.85	−0.70	−0.53	−0.32
−1.19	−1.26	−1.29	−1.29	−1.25	−1.18	−1.09	−0.96	−0.81	−0.63
−1.27	−1.37	−1.42	−1.44	−1.43	−1.39	−1.32	−1.22	−1.10	−0.95
−1.36	−1.47	−1.55	−1.60	−1.62	−1.60	−1.56	−1.48	−1.38	−1.26
−1.57	−1.64	−1.66	−1.64	−1.58	−1.48	−1.35	−1.17	−0.97	−0.73

Table A-28 (cont.)

12	14	16	18	20	22	24	26	28	30
−1.65	−1.74	−1.79	−1.80	−1.76	−1.69	−1.58	−1.43	−1.25	−1.04
−1.72	−1.84	−1.92	−1.95	−1.94	−1.89	−1.81	−1.69	−1.53	−1.34
−1.80	−1.94	−2.04	−2.10	−2.12	−2.10	−2.04	−1.94	−1.82	−1.65
−1.87	−2.04	−2.17	−2.26	−2.30	−2.30	−2.27	−2.20	−2.10	−1.96
−2.22	−2.36	−2.45	−2.48	−2.47	−2.40	−2.30	−2.14	−1.95	−1.72
−2.30	−2.46	−2.57	−2.63	−2.65	−2.61	−2.53	−2.40	−2.24	−2.03
−2.37	−2.56	−2.70	−2.79	−2.82	−2.81	−2.76	−2.66	−2.52	−2.34
−2.44	−2.66	−2.83	−2.94	−3.00	−3.02	−2.99	−2.92	−2.80	−2.65
−2.52	−2.76	−2.95	−3.09	−3.18	−3.22	−3.22	−3.18	−3.09	−2.96
−3.00	−3.23	−3.40	−3.51	−3.56	−3.55	−3.49	−3.38	−3.22	−3.02
−3.07	−3.33	−3.53	−3.66	−3.74	−3.76	−3.73	−3.64	−3.51	−3.33
−3.15	−3.43	−3.65	−3.82	−3.92	−3.97	−3.96	−3.91	−3.80	−3.65
−3.22	−3.53	−3.78	−3.97	−4.10	−4.18	−4.20	−4.17	−4.09	−3.97
−3.30	−3.64	−3.91	−4.13	−4.29	−4.39	−4.44	−4.43	−4.38	−4.28
−3.91	−4.25	−4.53	−4.73	−4.87	−4.95	−4.96	−4.91	−4.81	−4.65
−3.99	−4.36	−4.66	−4.90	−5.06	−5.17	−5.21	−5.19	−5.11	−4.98
−4.07	−4.47	−4.80	−5.06	−5.25	−5.38	−5.45	−5.46	−5.41	−5.31
−4.15	−4.58	−4.93	−5.22	−5.44	−5.60	−5.69	−5.73	−5.71	−5.63
−4.23	−4.69	−5.07	−5.39	−5.63	−5.82	−5.94	−6.00	−6.01	−5.96
−4.96	−5.45	−5.86	−6.19	−6.44	−6.62	−6.73	−6.77	−6.75	−6.66
−5.05	−5.57	−6.00	−6.36	−6.64	−6.85	−6.98	−7.05	−6.06	−7.00
−5.14	−5.69	−6.15	−6.53	−6.84	−7.08	−7.24	−7.34	−7.37	−7.34
−5.23	−5.81	−6.30	−6.71	−7.04	−7.31	−7.50	−7.62	−7.68	−7.68
−5.32	−5.92	−6.44	−6.88	−7.24	−7.53	−7.75	−7.91	−7.99	−8.02
−6.18	−6.83	−7.40	−7.88	−8.27	−8.58	−8.81	−8.97	−9.06	−9.07
−6.28	−6.97	−7.56	− 8.07	−8.49	−8.83	−9.09	−9.27	−9.39	−9.43
−6.39	−7.10	−7.72	−8.26	−8.71	−9.07	−9.36	−9.57	−9.71	−9.79
−6.49	−7.23	−7.88	−8.45	−8.92	−9.32	−9.63	−9.88	−10.04	−10.14
−6.59	−7.36	−8.04	−8.63	−9.14	−9.56	−9.91	−10.18	−10.37	−10.50

Table A-29   Properties of Sea Ice[a]
A.   Density of Sea Ice
(From G. Dietrich, *General Oceanography*, Interscience,
New York, 1963, with permission of John Wiley & Sons)

Air Content, % of vol	S, ‰			
	0	10	20	30
0	0.918	0.925	0.934	0.942
3	0.890	0.898	0.906	0.914
6	0.863	0.871	0.879	0.887
9	0.835	0.843	0.851	0.859

B. Specific Heat of Sea Ice
(From H. U. Sverdrup, M. W. Johnson, and R. H. Fleming, *The Oceans*, Prentice-Hall, Inc., Englewood Cliffs, N.J., 1942, with permission of the publisher)

Salinity, ‰	Temperature, °C										
	−2	−4	−6	−8	−10	−12	−14	−16	−18	−20	−22
0	0.48	0.48	0.48	0.48	0.48	0.47	0.47	0.47	0.47	0.47	0.46
2	2.47	1.00	0.73	0.63	0.57	0.55	0.54	0.53	0.53	0.52	0.52
4	4.63	1.50	0.96	0.76	0.64	0.59	0.57	0.57	0.56	0.55	0.54
6	6.70	1.99	1.20	0.88	0.71	0.64	0.61	0.60	0.58	0.57	0.56
8	8.76	2.49	1.43	1.01	0.78	0.68	0.64	0.64	0.61	0.60	0.58
10	10.83	2.99	1.66	1.14	0.85	0.73	0.68	0.67	0.64	0.62	0.60
15	16.01	4.24	2.24	1.46	1.02	0.85	0.77	0.76	0.71	0.68	0.65

C. Latent Heat of Melting of Sea Ice
(Source: Same as for Part B)

Temperature, °C	Salinity, ‰						
	0	2	4	6	8	10	15
−1	80	72	63	55	46	37	16
−2	81	77	72	68	63	59	48

D. Coefficient of Thermal Expansion for Sea Ice ($\times 10^4$)
(Source: Same as for Part B)

S, ‰	Temperature, °C										
	−2	−4	−6	−8	−10	−12	−14	−16	−18	−20	−22
2	−22.10	−4.12	−1.06	0.16	0.83	1.13	1.23	1.27	1.33	1.38	1.44
4	−45.89	−9.92	−3.81	−1.37	−0.02	0.57	0.78	0.85	0.96	1.07	1.88
6	−69.67	−15.73	−6.55	−2.90	−0.88	0.00	0.33	0.43	0.60	0.76	0.93
8	−93.46	−21.53	−9.30	−4.42	−1.73	−0.57	−0.13	0.02	0.23	0.54	0.67
10	−117.25	−27.34	−12.05	−5.95	−2.59	−1.13	−0.59	−0.40	−0.13	0.14	0.42
15	−176.72	−41.85	−18.92	−9.78	−4.73	−2.54	−1.72	−1.45	−1.04	−0.63	−0.22

Table A-29 (cont.)

> E.   Thermal Conductivity of Sea Ice at Different Depths
> (From A. Defant, *Physical Oceanography*, Pergamon
> Press, New York, 1961, Vol. 1, with permission of
> the publisher)

			Depth, m	
	0	25	60 and 75, resp.	125
Winter				
1922–1923	2.4	3.6	4.0	4.2×10^{-3}
1923–1924	1.7	3.3	4.5	5.0×10^{-3}

[a] The primary source of the data in these tables (except Part A) is
the *Maud* expedition as reported by Malmgren (1927).

Table A-30   Adiabatic Cooling
(H. U. Sverdrup, M. W. Johnson, and R. H. Fleming, *The Oceans*,
Prentice-Hall, Inc. Englewood Cliffs, N.J., 1942, with permission of the
publisher)

A.   Adiabatic Temperature Gradient in the Sea, °C/1000 m at Salinity
34.85‰

Depth, m	Temperature, °C								
	−2	0	2	4	6	8	10	15	20
0	0.016	0.035	0.053	0.078	0.087	0.103	0.118	0.155	0.190
1,000	0.036	0.054	0.071	0.087	0.103	0.118	0.132	0.166	0.199
2,000	0.056	0.073	0.089	0.104	0.118	0.132	0.146	0.177	0.207
3,000	0.075	0.091	0.106	0.120	0.133	0.146	0.159	0.188	
4,000	0.093	0.108	0.122	0.135	0.147	0.159	0.170	0.197	
5,000	0.110	0.124	0.137	0.149					
6,000	0.120	0.140	0.152	0.163					
7,000		0.155	0.165	0.175					
8,000		0.169	0.178	0.187					
9,000		0.182	0.191	0.198					
10,000		0.194	0.202	0.209					

B. Adiabatic Cooling (in 0.01°C) when Seawater ($S = 34.85\permil$) which has a Temperature at the Depth of m meters is Raised from that Depth to the Surface

Depth,	Temperature, °C												
	−2	−1	0	1	2	3	4	5	6	7	8	9	10
1,000	2.6	3.5	4.4	5.3	6.2	7.0	7.8	8.6	9.5	10.2	11.0	11.7	12.4
2,000	7.2	8.9	10.7	12.4	14.1	15.7	17.2	18.8	20.4	21.9	23.3	24.8	26.2
3,000	13.6	16.1	18.7	21.2	23.6	25.9	28.2	30.5	32.7	34.9	37.1	39.2	41.2
4,000	21.7	25.0	28.4	31.6	34.7	37.7	40.6	43.5	46.3	49.1	51.9	54.6	57.2
5,000	31.5	35.5	39.6	43.4	47.2	50.9	54.4						
6,000	42.8	47.5	52.2	56.7	61.1	65.3	69.4						
7,000			66.2	71.3	76.2	80.9	85.5						
8,000			81.5	87.1	92.5	97.7	102.7						
9,000			98.1	104.1	109.9	115.6	121.0						
10,000			115.7	122.1	128.3	134.4	140.2						

C. Adiabatic Cooling (in 0.01°C) when Water of the Indicated Temperature and Salinity is Raised from 1000 m to the Surface

Salinity,	Temperature, °C											
$\permil$	0	2	4	6	8	10	12	14	16	18	20	22
30.0	3.5	5.3	7.0	8.7	10.3	11.8	13.2	14.7	16.1	17.6	18.9	20.3
32.0	3.9	5.7	7.3	9.0	10.6	12.1	13.5	15.0	16.4	17.8	19.1	20.5
34.0	4.3	6.0	7.7	9.4	10.9	12.4	13.8	15.3	16.6	18.0	19.3	20.7
36.0	4.7	6.4	8.1	9.7	11.2	12.7	14.1	15.5	16.9	18.3	19.6	20.9
38.0	5.1	6.8	8.4	10.0	11.6	13.0	14.4	15.8	17.2	18.5	19.8	21.1

Table A-31   Some Colligative Properties of Seawater[a]
(From G. Dietrich, *General Oceanography*, Interscience, New York, 1963, with permission of John Wiley & Sons)

$S, \permil$	Φ_0, atm	Φ_{20}, atm	ϑ, °C	ΔT_s, °C	b, mm Hg
4	2.59	2.78	−0.214	0.06	1.69
8	5.16	5.54	−0.427	0.12	3.39
12	7.73	8.30	−0.640	0.19	5.08
16	10.34	11.10	−0.856	0.25	6.82
20	12.97	13.92	−0.074	0.31	8.55
24	15.63	16.78	−1.294	0.38	10.30
28	18.31	19.65	−1.516	0.44	12.13
32	21.02	22.56	−1.740	0.51	13.97
36	23.76	25.50	−1.967	0.57	15.79
40	26.53	28.48	−2.196	0.64	17.73

[a] Osmotic pressure Φ, freezing point ϑ, boiling point elevation ΔT_s, and vapor pressure lowering b as functions of salinity.

Table A-32   Boiling Point Elevation for Seawater
A.   (From B. M. Fabuss and A. Korosi, *J. Chem. Eng. Data*, **11**, 606 (1966), with permission of American Chemical Society)

		Temperature								
Cl, ‰	*S*, ‰	°C, 20 / °F, 68	40 / 104	60 / 140	80 / 176	100 / 212	120 / 248	140 / 284	160 / 320	180 / 356
19.00	34.46	0.30	0.34	0.40	0.47	0.53	0.59	0.66	0.71	0.77
22.80	41.35	0.36	0.42	0.49	0.57	0.67	0.72	0.80	0.87	0.93
26.61	48.24	0.43	0.50	0.58	0.68	0.77	0.86	0.95	1.03	1.10
30.41	55.13	0.51	0.58	0.68	0.79	0.89	1.00	1.10	1.19	1.28
34.21	62.03	0.57	0.67	0.79	0.90	1.01	1.14	1.26	1.36	1.46
38.01	68.92	0.65	0.75	0.88	1.02	1.16	1.29	1.42	1.54	1.65
41.81	75.81	0.72	0.85	1.98	1.13	1.27	1.44	1.58	1.72	1.84
45.61	82.70	0.81	0.94	1.07	1.26	1.43	1.59	1.75	1.90	2.04
49.41	89.59	0.89	1.04	1.21	1.39	1.59	1.75	1.92	2.09	2.24
53.21	96.48	0.98	1.13	1.32	1.52	1.70	1.91	2.10	2.28	2.45
57.01	103.38	1.06	1.23	1.44	1.66	1.86	2.08	2.28	2.48	2.66

B.   [From R. W. Stoughton and M. H. Lietzke, *J. Chem. Eng. Data*, **12**, 101 (1967), with permission of American Chemical Society]

		Wt % Sea Salt									
T, °C	p_0, atm	2.0	3.45[a]	4.0	6.0	8.0	12.0	16.0	20.0	25.0	28.0
25	0.031	0.177	0.312	0.366	0.570	0.795	1.324	1.991	2.858	4.36	5.58
30	0.042	0.184	0.325	0.380	0.594	0.829	1.381	2.077	2.977	4.52	5.77
40	0.073	0.198	0.350	0.410	0.642	0.898	1.497	2.250	3.216	4.86	6.16
50	0.122	0.214	0.377	0.442	0.692	0.969	1.616	2.426	3.458	5.19	6.56
60	0.197	0.229	0.405	0.474	0.744	1.041	1.737	2.604	3.704	5.53	6.96
70	0.309	0.245	0.433	0.508	0.797	1.115	1.860	2.786	3.952	5.88	7.36
80	0.469	0.262	0.463	0.542	0.851	1.191	1.986	2.969	4.202	6.22	7.77
90	0.694	0.279	0.493	0.578	0.907	1.269	2.113	3.155	4.454	6.57	8.18
100	1.003	0.296	0.524	0.615	0.964	1.348	2.242	3.342	4.708	6.92	8.60
110	1.418	0.315	0.556	0.652	1.022	1.429	2.374	3.532	4.963	7.27	9.01
120	1.965	0.334	0.590	0.691	1.082	1.512	2.508	3.723	5.22	7.62	9.43
130	2.673	0.354	0.624	0.731	1.144	1.597	2.643	3.917	5.48	7.98	9.86
140	3.577	0.375	0.660	0.773	1.208	1.684	2.782	4.113	5.74	8.34	10.28
150	4.711	0.396	0.697	0.816	1.274	1.774	2.923	4.311	6.00	8.70	10.71
160	6.119	0.418	0.735	0.861	1.341	1.866	3.066	4.511	6.27	9.06	11.15
180	9.931	0.466	0.817	0.955	1.484	2.057	3.361	4.920	6.81	9.79	12.03
200	15.407	0.519	0.906	1.058	1.637	2.261	3.670	5.342	7.36	10.54	12.94
220	22.993	0.577	1.003	1.170	1.802	2.480	3.995	5.777	7.92	11.30	13.87
240	33.184	0.642	1.111	1.293	1.983	2.715	4.338	6.231	8.49	12.08	14.84
260	46.520	0.716	1.232	1.431	2.181	2.971	4.703	6.700	9.08	12.88	15.83

[a] Standard seawater.

Table A-33   Vapor Pressure and Osmotic Equivalent of Seawater at 25°C
(From R. A. Robinson *J. Marine Biol. Assoc. U.K.*, **33**, 449 (1954), with
permission of Marine Biological Association of the United Kingdom)[a]

Cl, ‰	R	NaCl	KCl	CaCl$_2$	MgCl$_2$	MgSO$_4$	Na$_2$SO$_4$	Sucrose	Urea	V.P. Lowering[b]	Osmotic Pressure, atm
10	0.02861	0.2861	0.2908	0.2039	0.2005	0.5056	0.2374	0.5065	0.5400	0.00946	12.87
11	0.02869	0.3156	0.3211	0.2240	0.2199	0.5597	0.2643	0.5560	0.5965	0.01042	14.19
12	0.02877	0.3452	0.3516	0.2441	0.2393	0.6138	0.2918	0.6053	0.6534	0.01139	15.51
13	0.02885	0.3751	0.3825	0.2642	0.2588	0.6675	0.3196	0.6546	0.7112	0.01237	16.85
14	0.02893	0.4050	0.4134	0.2481	0.2780	0.7206	0.3477	0.7040	0.7695	0.01334	18.19
15	0.02901	0.4352	0.4447	0.3043	0.2975	0.7738	0.3762	0.7534	0.8285	0.01433	19.55
16	0.02908	0.4653	0.4760	0.3243	0.3165	0.8264	0.4051	0.8025	0.8880	0.01532	20.91
17	0.02916	0.4957	0.5077	0.3445	0.3356	0.8786	0.4347	0.8516	0.9482	0.01631	22.28
18	0.02924	0.5263	0.5397	0.3645	0.3546	0.9300	0.4648	0.9008	1.010	0.01732	23.66
19	0.02932	0.5571	0.5719	0.3845	0.3738	0.9803	0.4954	0.9497	1.071	0.01833	25.06
20	0.02940	0.5880	0.6043	0.4044	0.3929	1.028	0.5264	1.133		0.01936	26.47
21	0.02948	0.6191	0.6370	0.4243	0.4122	1.076	0.5578	1.047	1.197	0.02039	27.89
22	0.02956	0.6503	0.6698	0.4440	0.4313	1.123	0.5896	1.095	1.260	0.02142	29.33

[a] This column gives the relative pressure lowering $\Delta p/p° = (p° - p)/p°$, where p is the vapor pressure of
seawater and $p°$ is the vapor pressure of pure water. $p° = 23.756$ mm at 25°C.
[b] A. B. Arons and C. F. Kientzler, *Trans. Am. Geophys. Union*, 35, 722 (1954), also give vapor pressure information for sea salt solutions.

Table A-34   Osmotic Pressure (in atm) of Sea Salt Solutions
(From R. W. Stoughton and M. H. Lietzke, *J. Chem. Eng. Data*, **12**, 101
(1967), with permission of American Chemical Society)

	Wt % Salts								
T, °C	1.00	2.00	3.45[a]	5.00	7.50	10.00	15.00	20.00	25.00
25	7.11	14.29	25.15	37.49	59.30	83.97	144.93	228.59	348
40	7.42	14.93	26.34	39.35	62.41	88.53	152.81	239.86	362
60	7.80	15.70	27.74	41.51	65.98	94.69	161.53	252.06	376
80	8.13	16.37	28.94	43.32	68.69	97.82	168.27	261.17	386
100	8.42	16.94	29.92	44.79	71.17	100.97	173.16	267.47	393

Author Index

Subject Index

A indicates material in an appendix, n in a note. Unless otherwise specified, a listed property is for either pure water or seawater.